Anti-inflammatory effect of fullerene C_{60} in a mice model of atopic dermatitis

Nadezda Shershakova*, Elena Baraboshkina, Sergey Andreev*, Daria Purgina, Irina Struchkova, Oleg Kamyshnikov, Alexandra Nikonova and Musa Khaitov

Abstract

Background: Water-soluble form of fullerene C_{60} is a promising tool for the control of ROS-dependent inflammation including allergic diseases. Anti-inflammatory effects of C_{60} (nC_{60}) aqueous dispersion were evaluated in the mouse models of atopic dermatitis using subcutaneous (SC) and epicutaneous (EC) applications during 50 days period. A highly stable nC_{60} was prepared by exhaustive dialysis of water-organic C_{60} solution against water, where the size and ζ-potential of fullerene nanoparticles are about 100 nm and -30 mV, respectively.

Results: To induce skin inflammation, female BALB/c mice were EC sensitized with ovalbumin three times during one-weekly exposures. The nC_{60} solution was administrated in mice subcutaneously (SC) (0.1 mg/kg) and epicutaneously (EC) (1 mg/kg). Significant suppression of IgE and Th2 cytokines production and a concomitant rise in concentrations of Th1 cytokines were observed in nC_{60}-treated groups. In addition, a significant increase in the levels of Foxp3^{+} and filaggrin mRNA expression was observed at EC application. Histological examination of skin samples indicated that therapeutic effect was achieved by both EC and SC treatment, but it was more effective with EC. Pronounced reduction of the eosinophil and leukocyte infiltration in treated skin samples was observed.

Conclusions: We suppose that nC60 treatment shifts immune response from Th2 to Th1 and restores to some extent the function of the skin barrier. This approach can be a good alternative to the treatment of allergic and other inflammatory diseases.

Keywords: Fullerene C_{60}, Atopic dermatitis, Mouse model, Cytokine

Background

Atopic dermatitis (AD) is a chronic inflammatory skin disease that predominantly affects children and is characterized by skin lesions, persistent erythema, scaling, excoriations, and pruritus. In addition, the disease is commonly associated with allergic rhinitis and asthma. The number of AD patients increased by 10–30 % in children and 2–10 % in adults in the last 30 years worldwide [1–3]. AD is the result of complex interactions among several genetic factors, deficiencies in skin barrier function, exposure to various allergens and infectious agents and features of the immune response [4–6]. In 60 % of patients, AD starts before the age of 6 years, in 18 % the onset is after the age of 20 [7]. Its pathogenesis involves impairments in the skin barrier, allowing abnormally enhanced dermal presentation of antigens/allergens to the immune system [8, 9]. In many AD patients Th2/IgE-mediated allergic reactions play leading role, however, Th2 cells predominate in the initial stage lesions with a switch to Th1-cells in a chronic phase. As a rule, diseases arising from a dysfunction of immune cells and/or their products often manifest with skin symptoms [10]. Many observations suggest that allergic and inflammatory skin diseases like AD are mediated by oxidative stress [11–13]. Mast cells generate mainly intracellular reactive oxygen species (ROS) following the aggregation of FceRI; these ROS may act as secondary messengers in the induction of several biological responses [14]. A generation of ROS induces oxidative protein damage in the stratum cornea, which leads to the disruption of barrier function and the exacerbation of AD [15, 16].

*Correspondence: nn.shershakova@nrcii.ru; sm.andreev@nrcii.ru
NRC Institute of Immunology FMBA of Russia, Moscow, Russia

Fullerene is a molecular carbon form with cage spheroid structure, and it possesses strong antioxidant activity. At present, the fullerene C_{60} is produced industrially in large quantities and is commercially available. Its unique properties include strong electron-acceptor activity, high polarizability and hydrophobicity; a large surface of C_{60} molecule allows attachment of various hydrophilic addends, providing prospects for the design of new biomedical products. Fullerene and some of its derivatives as potent antioxidants (in vitro and in vivo) are capable of effective ROS inactivation [17]. An important feature of fullerene C_{60} is its low toxicity [18, 19], a permeability through biological membranes [20, 21], and lack of immunogenicity [22]. Thus, fullerene presents a promising therapeutic agent for the treatment of allergies and other inflammatory diseases such as Parkinson's, Alzheimer's and AD.

Anti-inflammatory activity of fullerene derivatives was recorded by several investigators, and, in particular, researchers from the Luna group (led by Dr. C. L. Kepley) were the pioneers in this field [23–26]. Some C60-drived compounds were able to inhibit an IgE-dependent allergic response. For instance, fullerenol ($C_{60}(OH)_n$) and amino-fullerene ($C_{60}(NHCH_2CH_3)_n$) inhibited in vitro IgE-dependent degranulation of mast cells and secretion of cytokines and prostaglandins in response to allergen and authors suggested that this effect is associated with lower levels of ROS in the stimulated cells [23]. On the other hand, the C_{60} aqueous dispersion obtained by Andrievsky's method [27] induced gene expression of proinflammatory cytokines (IL-1, TNF-α, IL-6) and Th1 cytokines (IL-12, IFN-γ) in mice via aerosol administration to the airways [28]. Aqueous dispersion of C_{60} (nC_{60}) provided inhibitory effect on IgE-mediated histamine release from peripheral blood basophils in vitro and suppressed anaphylaxis in mice caused by administration of ovalbumin [29]. Similar C_{60} dispersion exhibited significant modulator activity on the DTH reaction suppressing Th1 cytokines release [30, 31].

Fullerene C_{60} in aqueous media always exists in form of nanoparticles (clusters), and hence, the degree of their possible undesirable effect is still being discussed [32, 33]. In several studies, nanoparticles are perceived as toxic agents, in others—the damage is not detected. As was shown, the polystyrene nanoparticles aggravated AD-like skin lesions [34] and metal oxide nanoparticles such as ZnO induced systemic production of IgE antibodies [35]. Currently, nanoparticles were used for transcutaneous drugs delivery, but not for the AD therapy [36, 37]. It should be noted, that the use of fullerene at the AD has not been described.

The purpose of this study was to evaluate the therapeutic properties of the nC_{60} obtained by novel biocompatible method for the treatment of AD induced in a mouse model [38].

Results

Aqueous fullerene solution, nC_{60}

Recently, we have proposed a new simple method for preparation of C_{60} aqueous solution using a "dialysis principle" [39]. Shortly, the protocol design includes the dissolution of crystalline fullerene in N-methylpyrrolidone (NMP), dilution of this solution with distilled water or aqueous solution of an L-amino acid used as a stabilizing agent with subsequent exhaustive dialysis against deionized water. Thus, the protocol excludes the use of toxic organic solvents (toluene, tetrahydrofuran) as well as sonication, heating and durable mixing frequently used in the known methods [27, 40, 41]. This approach provides a high conversation of the fullerene C_{60} from the crystalline state to the solution with concentration of C_{60} up to 1 g/l (at additional vacuum concentration) and hydrodynamic size of C_{60} particles about 80–100 nm (Table 1). Absorption spectrum of nC_{60} is characterized by three intense maxima at 219, 265 and 344 nm and weak broad band between 400 and 500 nm, and it practically did not differ from those of nC60 obtained through other methods. FTIR spectra of dried nC_{60} shows vibration bands of free C_{60} molecules at 1182 and 1428 cm^{-1} and additional ones at 3500–3200, 1650–1660, and 1000–1100 cm^{-1} due to obvious presence of residual water molecules and NMP (as a donor–acceptor complex). Calculations on the molar ratio of NMP to fullerene displayed the value 0.7–1.1, while a calculation of the ratio H_2O to C_{60} based on the hydrogen content gave values from 5 to 10. The NMP is known to have low toxicity via oral, dermal and inhalation routes of delivery. In medicine, it has a long track recorded as a constituent in medical devices approved by the European Commission and FDA and thus can be considered as safe [42]. In addition, good heat resistance

Table 1 Physical parameters of the nC_{60}

Parameter	Value
Feasible concentration[a]	~1 mg/ml
Average size (nm)	80–100 nm
PDI (polydispersity index)	0.175 ± 5
Zeta potential (mV)	−27/30
UV–Vis major absorption peaks (nm)	219 (s), 265 (s), 344 (s), 450 (bw)
FTIR spectra (cm^{-1})	525, 1000–1100, 1182, 1428, 1650–1660, 3200–3500
Mol. ratio C_{60}:NMP	~1
Mol. ratio C_{60}:H_2O	5–10

s strong, bw broad weak

[a] With optional vacuum evaporation

enables to sterilize the C_{60} solution prior to its administration into the body. In this study, concentration of C_{60} in the stock nC_{60} was 120 µg/ml.

nC_{60} modulates allergen-specific antibody production

To induce a mouse AD [38, 43, 44], female BALB/c mice were EC sensitized by skin application with 100 µg doses of OVA in PBS (1 mg/ml) during three one-week exposures with two-week intervals as described in "Materials and methods" and showed in Fig. 1 ("AD" group). Mice received nC_{60} by both EC ("nC_{60} EC") and SC ("nC_{60} SC") administration routes in doses 1 mg/kg and 0.1 mg/kg of nC_{60}, respectively, between OVA-applications. Control group received PBS-application only ("placebo"). Figure 2 shows that OVA-specific IgE and IgG1 levels in sera were elevated in OVA-sensitized mice ("AD" group). The

main question was whether nC_{60} treatment could change OVA-specific antibody level in OVA-sensitized mice.

To assess the effect of nC_{60}-treatment on specific IgE and IgG production sera were collected after 2-nd and 3-rd sensitizations. As can be seen from Fig. 2a, specific IgE concentrations in OVA-sensitized mice were reduced after nC_{60} treatment. SC administration of nC_{60} had stronger inhibition effect on IgE production compared with the EC one. This tendency was observed both after the 2-nd and 3-rd OVA-sensitizations (2 and 3 sens, Fig. 2). After the last sensitization the reduction of IgE concentration was about 50 % (statistically significant). OVA-specific IgG1 and IgG2a levels and IgG1/IgG2a ratio as markers for Th2 and Th1 responses [45] were measured for each group, and results (Fig. 2b) suggest that the nC_{60} treatment had no statistically significant

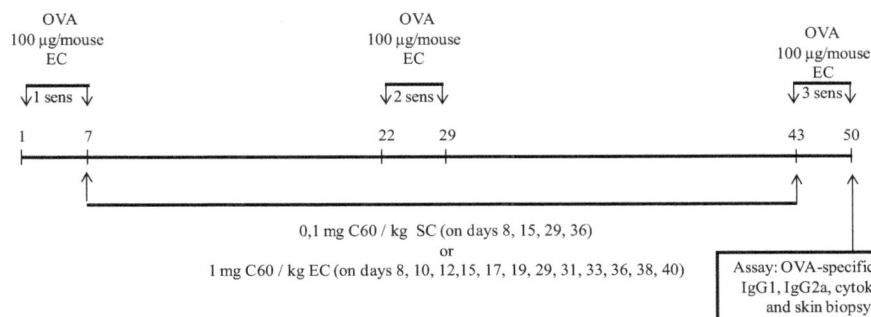

Fig. 1 Sensitization protocol and nC_{60} treatment. Mice were sensitized with OVA (100 µg) or saline applied at 100 µl to a sterile patch. The patch was applied for 1 week and then removed. nC_{60} treatment was done between the applications (sens): SC (0,1 mg nC_{60}/kg) on days 8, 15, 29, 36; EC (1 mg nC_{60}/kg) on days 8, 10, 12, 15, 17, 19, 29, 31, 33, 36, 38, 40. The control group ("placebo") had no any treating

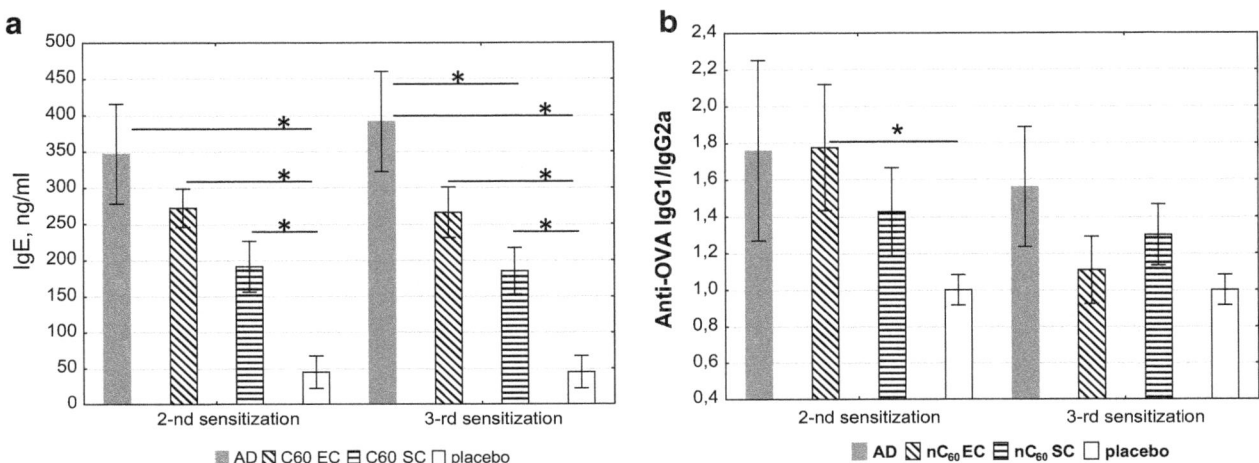

Fig. 2 Changes in concentrations of OVA-specific antibodies in sera obtained after treatment with nC_{60} detected by ELISA. The results are presented as mean value (mean ± SE, n = 8 for each). AD, OVA-sensitized mice (AD model); nC_{60} EC, OVA-sensitized mice treated with nC_{60} by EC; nC_{60} SC, OVA-sensitized mice treated with nC_{60} by SC; placebo, PBS-sensitized mice. **a** OVA-specific IgE response, **b** OVA-specific IgG1/IgG2a ratio ($*p < 0.05$)

effect on the specific antibody production (although a decreasing trend, especially after 3-rd sensitization was observed).

Fullerene-induced cytokine profile alterations

Analysis of cytokine expression was carried out after the 3-rd sensitization in supernatants of OVA-stimulated mouse splenocytes (3 sensitization, Fig. 1). It was shown that the nC_{60} exerts a strong effect as the IL-4 concentration was decreased by 2.5 times compared with "AD" group regardless of the route of administration (Fig. 3a). The same effect we also observed for IL-5. In this case, a slightly stronger effect was observed via EC application (Fig. 3b).

Figure 3c shows that the IL-12 concentration was significantly higher in group "nC_{60} EC" and "placebo" compared with "AD" and "nC60 SC" group. IFN-γ expression was significantly increased in nC_{60} group compared with

"AD" and normal-control ("placebo") Fig. 3d. Moreover, the strongest effect was observed after EC/nC_{60} treatment.

Foxp3 + regulatory cells (Foxp3 + Tregs) induction

In this study, we have measured the Foxp3 expression in OVA-stimulated mouse splenocytes to evaluate the level of Foxp3 + Tregs after the nC_{60} treatment. As we can see a significant increase in Foxp3 expression was observed in the "nC60 EC" group compared with "AD" and "placebo" (Fig. 4). The Foxp3 level was only slightly higher in group "nC_{60} SC" than in "AD".

Analysis of filaggrin expression

We used a quantitative real-time PCR to evaluate the effect of nC_{60} on expression the FLG in AD-induced mice (Fig. 5). It was shown that the nC_{60} potently promoted the FLG expression. The most significant increase of FLG

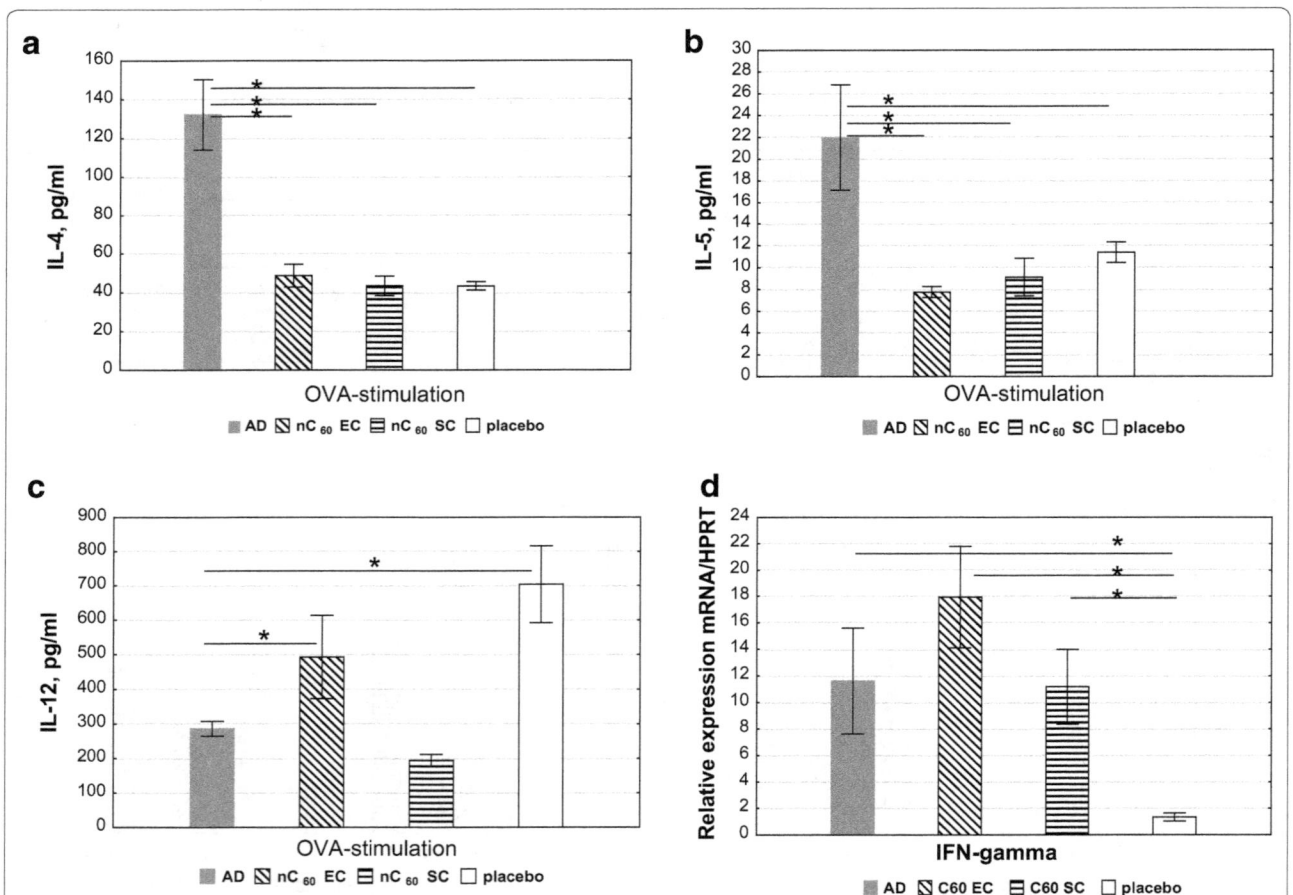

Fig. 3 The levels of cytokines IL-4 (**a**), IL-5 (**b**) and IL-12 (**c**) in supernatants of OVA-stimulated mouse splenocytes taken after 3-rd sensitization and incubated with OVA for 72 h (quantified by ELISA). Relative expression of IFN-γ mRNA (**d**) in mouse splenocytes taken after 3-rd sensitization and incubated with OVA for 72 h (quantified by real-time PCR). The results are presented as a mean concentration (mean ± SE, n = 8 for each). Designations of *columns* in the diagram are the same as those in Fig. 2; *p < 0.05

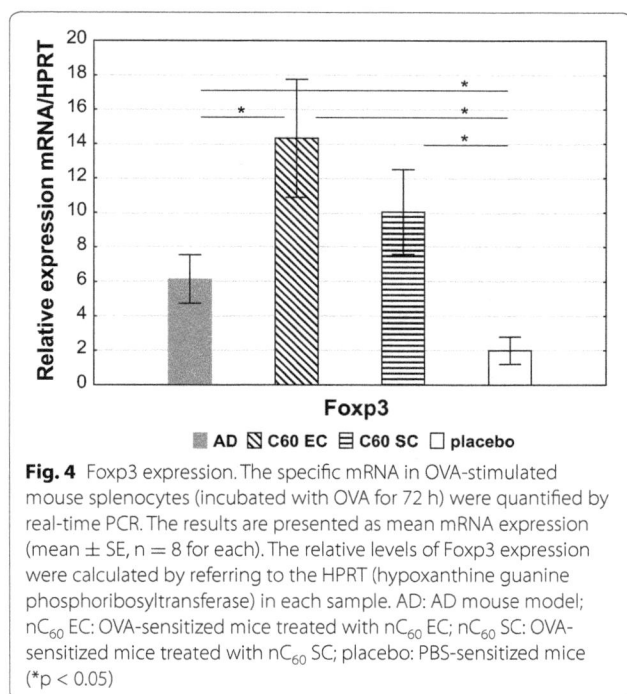

Fig. 4 Foxp3 expression. The specific mRNA in OVA-stimulated mouse splenocytes (incubated with OVA for 72 h) were quantified by real-time PCR. The results are presented as mean mRNA expression (mean ± SE, n = 8 for each). The relative levels of Foxp3 expression were calculated by referring to the HPRT (hypoxanthine guanine phosphoribosyltransferase) in each sample. AD: AD mouse model; nC_{60} EC: OVA-sensitized mice treated with nC_{60} EC; nC_{60} SC: OVA-sensitized mice treated with nC_{60} SC; placebo: PBS-sensitized mice (*p < 0.05)

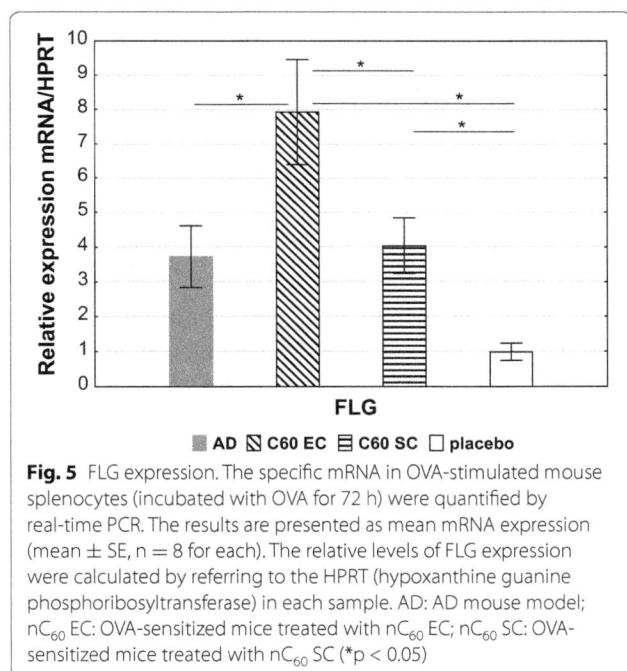

Fig. 5 FLG expression. The specific mRNA in OVA-stimulated mouse splenocytes (incubated with OVA for 72 h) were quantified by real-time PCR. The results are presented as mean mRNA expression (mean ± SE, n = 8 for each). The relative levels of FLG expression were calculated by referring to the HPRT (hypoxanthine guanine phosphoribosyltransferase) in each sample. AD: AD mouse model; nC_{60} EC: OVA-sensitized mice treated with nC_{60} EC; nC_{60} SC: OVA-sensitized mice treated with nC_{60} SC (*p < 0.05)

was observed at EC/nC_{60} application, in this case, FLG expression was increased about 3 times (p < 0.05).

Histological assay

Skin biopsies from sensitized mice were taken after the final 3-rd sensitization. Skin sections were stained with H&E and examined at 100-fold magnification. The skin samples from the "AD" group had dermal and epidermis necrosis from severe to medium degree. There were visible marked epidermis hyperplasia, pronounced diffuse mixed leukocyte infiltrates with eosinophils prevalence both in dermis and epidermis (Fig. 6a). The cellular infiltrate consisted of neutrophils, eosinophils, and lymphocytes, while the group "placebo" did not present any pathologic changes showing almost normal appearance (Fig. 6d). The infiltration degree of leukocytes in dermis and epidermis were decreased in SC/nC_{60}-treated mice ("nC_{60} SC") compared with "AD", however, the eosinophils amount was the same as in the "AD" (Fig. 6c). In this case, we observed decrease in the epidermis necrosis and destructive hemorrhage in derma. There was a moderate epidermal hyperkeratosis obviously as a compensatory protection mechanism against damage. Figure 6b shows that the most pronounced nC_{60} therapeutic effect was in the case of EC nC_{60}-treated mice ("nC_{60} EC"). The leukocytes infiltration degree and the eosinophils number were significantly reduced in "nC_{60} EC" group compared with "AD" and "nC_{60} SC". It was shown, that the epidermal necrosis, destructive hemorrhage in dermis and hyperkeratosis either were absent or were mild.

We evaluated the histologic pictures in semi-quantitative histological index (score) comparing nC_{60}-treated and non-treated groups with each other based on the microscopic features. Epidermal thickening, epidermal necrosis, epidermal hyperkeratosis, dermal and subcutaneous fat necrosis, swelling, hemorrhage, and cell infiltration of dermis and subcutaneous fat were the main evaluation criteria for the histologic skin lesions (Table 2). Each parameter had a degree of manifestation, estimated notional value and appropriate score.

Figure 6e shows that the maximum inflammatory index was observed in the "AD" (score = 24) while the "placebo" shows the value of 4. The inflammatory indexes for groups "nC_{60} EC" and "nC_{60} SC" were 14 and 18, respectively. Thus, the skin samples histological analysis showed that the nC_{60} therapy improved histological picture reducing an allergic inflammation by approximately 42 and 25 %, respectively for EC and SC applications compared with the untreated mice.

Discussion

The investigation of fullerene C_{60} biological effects has attracted increasing attention in recent years. An important issue is that a crystalline fullerene C_{60} is practically insoluble in an aqueous medium without special processing. In this study, an aqueous solution of the fullerene was prepared by novel dialysis method [46]. Based on the spectral and elemental analyzes, we can speculate that process underlying the C_{60} solubilization can involve a

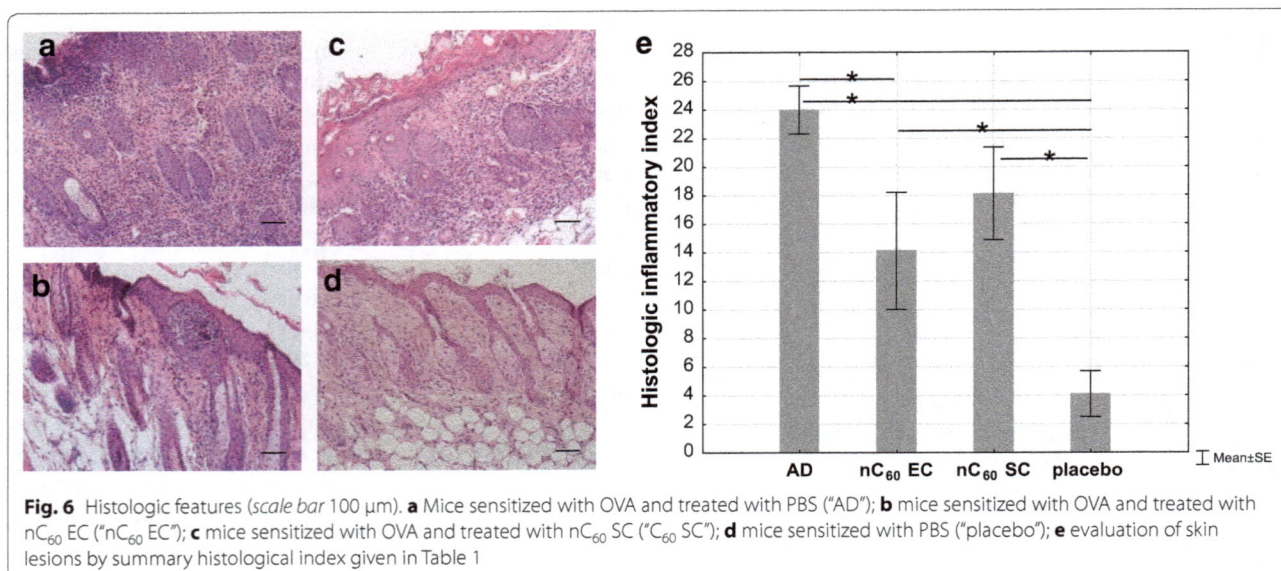

Fig. 6 Histologic features (*scale bar* 100 μm). **a** Mice sensitized with OVA and treated with PBS ("AD"); **b** mice sensitized with OVA and treated with nC_{60} EC ("nC_{60} EC"); **c** mice sensitized with OVA and treated with nC_{60} SC ("C_{60} SC"); **d** mice sensitized with PBS ("placebo"); **e** evaluation of skin lesions by summary histological index given in Table 1

formation of the C_{60} molecule complexes or their clusters with NMP followed by partial hydroxylation of nanoparticle surface that is capable of stabilizing the nC_{60} aggregates. Apparently, these aggregates are surrounded by firmly bound water envelope; the water is not removed on drying under high vacuum. We suppose, that the bond between C_{60} and NMP has definitely non-covalent character (donor–acceptor bond) as evidenced by experimental results and theoretical calculations [47]. It should be noted that in course of time (>6 months), we have observed a decrease in the heterogeneity of nanoparticles size, they have become more uniform in size (100 nm). Perhaps, it reflects the establishment of a dynamic equilibrium in a system where a process of fullerene molecules distribution from cluster to cluster takes place.

We suggest that a new method for the nC_{60} preparing as a biocompatible process is potentially very suitable for medical use. The relationship between basic physicochemical characteristics and toxic effects (in vivo) of nC_{60} practically have not been covered in scientific literature. The relationship between zeta potentials and size of functionalized C_{60} aggregates and their influence on the toxicity against bacterial cells was recently described. It was shown, that an increase of surface charge (indifferently, + or −) always leads to a decrease in the size of nanoparticles, but the toxicity is always associated with positively charged C_{60} aggregates [48]. Nanoparticles in the nC_{60} always bear a negative charge. In other studies, dimensional effects have been shown to affect the cytotoxicity of nC_{60}. Authors speculate that this effect may be associated with cell-contacting surface of nanoparticles. The size of nanoparticles is negatively correlated with their toxicity [49]. The question

of C_{60} water-soluble form toxicity is closely linked to the nanoparticle morphology and its surface chemistry, which is also determined by the method of its preparation. The surface of the nC_{60} particles differs from those described in the literature, in view of the presence of complexed hydrophilic component, NMP, and partial hydroxylation [46]. Our studies on safety of the nC_{60} upon acute intravenous administration to BALB/c mice demonstrated that no lethal outcomes were observed and the body weight was increased in a pattern similar to the control group (unpublished observations). Exact dose for humans will be defined at the stage of clinical trials only.

Some researchers have found that fullerene derivatives, as fullerenol and amino-fullerene, inhibit in vitro the IgE-dependent degranulation of mast cells, secretion of Th2 cytokines and prostaglandins in response to an allergen stimulation that appears to be, in part, through the cellular ROS levels inhibition [23]. Certain fullerene derivatives have been able to prevent the development of inflammation and edema in mice after administration of phorbol-myristate-acetate (PMA) [24]. The inhibition has been shown to depend on the structure of addends attached to carbon skeleton. For example, fullerene C_{70}, containing 4 glycolic acid molecules, noticeably inhibited artificially induced anaphylaxis, and at the same time, fullerene with 4 attached inositol molecules showed no activity. It was shown that the fullerene material in contact with serum medium or in the cells is capable of binding to albumin and other proteins, including some enzymes. Thus, perhaps it modulates both enzymatic and signaling redox processes in the cell, but available data are very limited [25].

Table 2 The main assessment criteria for histological skin lesions

No.	Criterion	The degree of manifestation	Score
1	Epidermal thickening	Absent	0
		Mild	1
		Moderate	2
		Pronounced	3
2	Epidermal necrosis	Absent	0
		Present	1
3	Epidermal hyperkeratosis	Absent	0
		Present	1
4	Connective tissue like dermal proliferation	Absent	0
		Mild	1
		Moderate	2
		Pronounced	3
5	Dermal necrosis	Absent	0
		Present	1
6	Dermal swelling	Absent	0
		Present	1
7	Dermal hemorrhage	Absent	0
		Present	1
8	Connective tissue like subcutaneous fat proliferation	Absent	0
		Mild	1
		Moderate	2
		Pronounced	3
9	Subcutaneous fat necrosis	Absent	0
		Present	1
10	Subcutaneous fat swelling	Absent	0
		Present	1
11	Subcutaneous fat hemorrhage	Absent	0
		Present	1
12	Cell infiltration of dermis	Absent	0
		Mild	1
		Moderate	2
		Pronounced	3
13	Cell infiltration of dermis (eosinophils)	Absent	0
		Mild	1
		Moderate	2
		Pronounced	3
14	Cell infiltration of dermis (polynuclear leukocytes)	Absent	0
		Mild	1
		Moderate	2
		Pronounced	3
15	Cell infiltration of dermis (lymphocytes)	Absent	0
		Mild	1
		Moderate	2
		Pronounced	3

Table 2 continued

No.	Criterion	The degree of manifestation	Score
16	Cell infiltration of subcutaneous fat	Absent	0
		Mild	1
		Moderate	2
		Pronounced	3
17	Cell infiltration of subcutaneous fat (eosinophils)	Absent	0
		Mild	1
		Moderate	2
		Pronounced	3
18	Cell (polynuclear leukocytes) infiltration of subcutaneous fat	Absent	0
		Mild	1
		Moderate	2
		Pronounced	3
19	Cell (lymphocytes) infiltration of subcutaneous fat	Absent	0
		Mild	1
		Moderate	2
		Pronounced	3

Earlier, based on a DTH mouse model, it was shown that the fullerene treatment significantly attenuated the footpad inflammation compared with DTH-control and switched the cytokine balance towards Th1-dominance [50]. Later we have also shown that Th1 cytokines production was significantly suppressed, however it was intriguing that Th2 cytokines IL-4 and IL-5 were also significantly suppressed by a fullerene treatment [31].

These studies were undertaken in order to evaluate the OVA-induced Th1 and Th2 cytokine secretion and to determine whether the treatment with nC_{60} can shift Th2 to Th1 response and modulate the proinflammatory cytokine production. Th2 cells produce IL-4 and IL-5 that have a prominent role in immediate-type hypersensitivity and apparently are involved in the initial stages of AD.

The level of Th1 cytokines, including IL-12 and IFN-γ was also examined. IL-12, key factor of T-cell differentiation into Th1 cells, plays a dominant role in many inflammatory diseases like AD and stimulates the IFN-γ production [51]. Thus, we demonstrated that therapeutic treatment of AD mice with nC_{60} shifts immune response from Th2 to Th1 for example changing the IgG1/IgG2a ratio as markers for Th2 and Th1 responses. We do not know the exact mechanism, but one can assume that this phenomenon is associated with the action of the nC_{60} on a redox homeostasis. ROS and other redox active molecules fulfill key functions in immunity and Th1/Th2 shift which appears to be crucially dependent on the activation of redox-sensitive signaling cascades [52].

Immune regulation and tolerance are essential functions of the immune system to prevent and limit harmful immune responses to self- and non-self-antigens. CD4/CD25 regulatory T cells (Tregs) [53] represent a unique lineage of immunoregulatory cells in both humans and animals and play a central role in the maintenance of immunologic self-tolerance and are involved in the release of anti-inflammatory cytokine IL-10 [54]. Lack of Foxp3 + CD4$^+$CD25$^+$ T cells leads to immune dysregulation and affected patients often have AD-like skin lesions, increased IgE levels, and enhanced Th2 responses. However, conflicting results regarding the numbers and functions of Tregs in AD have been reported. Some investigators [55] demonstrated the absence of Foxp3 + Tregs in patient's skin that suggests a disregulation in process of inflammation. In contrast, other authors [56] showed the elevated number of circulating CD4$^+$CD25$^+$ Tregs with a normal suppressive function in patients with AD. However, they used the CD25 molecule expression only (without Foxp3) as a marker for Tregs.

This experiment clearly shows that the EC treatment by fullerene leads to increased expression of Foxp3 and may shed light on the mechanism of nC$_{60}$ therapeutic effect.

The weakening of the skin barrier function in patients with mutations in filaggrin gene (FLG) probably promotes increased penetration of allergens by transdermal route. There is a direct interrelation between the AD and nonsense mutations in a filaggrin encoding gene [3]. Hence, one of possible therapeutic strategy to regulate AD is upregulating the FLG expression [57]. Recently, Otsuka et al. have screened more than 1000 compounds in a bioactive chemical library to find candidates to stimulate FLG mRNA expression and have revealed the compound JTC801 promoted the FLG mRNA and protein expression in vitro and in vivo. Potential utility of such therapy is indicated by the fact that a modest 20 % increase in filaggrin copy number leads to the 40 % reduction in AD susceptibility [25, 58].

Based on histological data and the cytokine secretion profile switch from Th2 to Th1 pattern we can conclude that fullerene nC$_{60}$ has significant anti-allergic and anti-inflammatory activity. We observed a significant suppression of IgE and Th2 cytokines (IL-4 and IL-5) production, and with a concomitant increase of Th1 cytokines production: IL-12 (at EC application only) and IFN-γ. However, there was a difference in nC$_{60}$ effect depended on the administration route. More intense specific IgE and Th2 cytokines suppression were observed at the EC application. In addition, this treatment significantly increased the IL-12 level compared with "AD". Based on this facts, we can hypothesize that nC60 reduces AD

inflammation by activating the IFN-γ production and Th2 response suppression [58]. These results suggest that nC$_{60}$ might be used as an agent to suppress proinflammatory cytokine production.

Since AD is a chronic disease, apparently the barrier dysfunction is a leading primary cause of AD. Surprisingly, it turned out that the nC$_{60}$ quite markedly increases the filaggrin expression in vivo (Fig. 5). We could not find any scientific publications on the effect of fullerene on filaggrin gene activation except the data from presentation of "Vitamin C60 BioResearch Corporation", where C$_{60}$/PVP complex ("Radical Sponge") increased FLG expression in RS-treated cells control cells about a four times (http://www.novac60.com/wp-content/uploads/2013/10/VC60-Fullerene-skin-barrier-effect.pdf). However, these data once again point out the potential of fullerene C$_{60}$ as a stimulator of the filaggrin production.

Histological analysis revealed that positive therapeutic effect was achieved both at EC and SC nC$_{60}$-treatment, but the former was more effective. The main dermal inflammatory response component was an eosinophilic infiltration and a pronounced reduction of eosinophils number was observed in "nC$_{60}$ EC" group. It should be noted that this result correlated with the decrease in the IL-5 concentration in the same group combined with the regenerative processes in the skin as opposed to "AD".

Thus, in this study we have demonstrated that nC$_{60}$ application inhibits the inflammatory process and may represent a perspective therapeutic approach to control allergic inflammation. However further studies are needed to understand the mechanism of fullerene activity.

Conclusions

We have found that the nC$_{60}$ inhibits significantly specific IgE production in mouse AD model. In addition, it was shown that IL-4, IL-5 levels were significantly decreased after EC and SC C60-treatment. It was observed that EC C$_{60}$-treatment shifts immune response from Th2 to Th1, markedly increasing the production of IL-12 and IFN-γ. We have also revealed that the use of nC$_{60}$ in EC route increases Foxp3 and FLG expression. Thus, simultaneous increase of Foxp3 and filaggrin expression leads to reduction in AD susceptibility. The histological analysis of skin samples showed that the nC$_{60}$ therapy improved histological picture reducing an allergic inflammation via EC as well as SC applications compared with the untreated mice. We do not know exactly why EC application is more effective then SC one. The possible explanation is that EC application with nC$_{60}$ increases the fullerene availability to the immune system due to the presence of a large amounts of immune cells in the skin with allergic inflammation (AD).

Thus, the use of nC_{60} in form of EC application is a promising alternative therapeutic approach to control allergy inflammation. Exact dose for humans will be defined at the stage of clinical trials only.

Methods

Reagents

Ovalbumin (OVA) (Grade V, 99 %) and L-alanine were purchased from Sigma-Aldrich (USA). N-methylpyrrolidone (NMP, 99 %) was from Panreac (Spain). Crystalline fullerene C_{60} was purchased from SES Research (99.9 %, catalog 600–9969, USA).

nC_{60} preparation

Aqueous fullerene dispersion, nC_{60}, was obtained by method described earlier [39]. Briefly, 20 mg of C_{60} were dissolved in 25 ml of N-methylpyrrolidone (magnetic stirrer) and resulting dark brown-purple solution was mixed with solution of 40 mg L-alanine in 100 ml of deionized water. The obtained dark-red transparent solution was stirred for 1 h and then subjected to exhaustive dialysis (cut off 10/50 kDa) against deionized water. Final dialysis solution was filtered through 0.45 mm nitrocellulose membrane resulting in a clear transparent solution with brownish-yellow color with concentration 120 µg/ml.

Mice sensitization (AD model) and nC_{60} treatment

Female BALB/c mice ages 4–6 weeks were purchased from the animal nursery Filial SCBMT "Stolbovaya" (Moscow region, Russia) and kept in a pathogen-free environment with an OVA-free diet. All experimental procedures were carried out according to order no. 708 of the Ministry of Health of the Russian Federation and "Regulations on the ethical attitudes to laboratory animals NRC Institute of Immunology FMBA of Russia (Moscow, Russia)".

EC sensitization of mice to induce a skin inflammation was carried out as described by Spergel et al. (1998). Briefly, mice were shaved with an electric razor. OVA (100 µg) in PBS (100 µl) or PBS as placebo was placed on a 1×1 cm^2 patch of sterile gauze, which was then secured onto the skin with a transparent bioclusive dressing (Systagenix Wound Management Limited, United Kingdom). The patch was applied thrice over a 1-week period. An inspection at the end of each sensitization period confirmed that the patch remained in place. C_{60} treatment was carried out by subcutaneous (SC; 0.1 mg/kg) and epicutaneous nC_{60} (EC; 1 mg/kg) administrations, between OVA-applications as shown in Fig. 1.

Antibody and cytokine assay

The anti-OVA IgE, IgG1, and IgG2a antibodies levels in sera obtained before, during, and after nC_{60}-treatment were detected by ELISA (ELISA kits from BD, USA) according to the manufacturer's protocol. OVA was used for coating the plates. Mouse anti-ovalbumin IgE mAb (AbD Serotec, UK) and biotin rat anti-mouse IgE (BD) were used for detection anti-OVA mouse IgE. These components were used to construct the calibration curve and then to analyze sera.

The spleens were taken after the last allergen application, and levels of IL-4, IL-5, and IL-12 (p40) in supernatants of OVA-stimulated splenocytes were determined by using ELISA [Duo-Set from R&D Systems (UK) and ELISA set from BD (USA)] according to the manufacturer's protocol.

Real-time PCR

The total RNA from OVA-stimulated mice splenocytes was extracted using the RNeasy Mini Kit (Qiagen, Courtaboeuf, France) according to the manufacturer's instructions. The RNA concentration was determined, and cDNA was synthesized through a reverse transcription reaction ("Reverta-L", Interlabservice, Russia). Quantitative real-time PCR analysis of mRNA expression was done by the iQ5 system (Bio-Rad, USA) and the PCR-Mix kit (Sintol, Russia).

The results are presented as mRNA expression (Foxp3, FLG). Calculations to determine the relative level of gene expression were made using the comparative C_t method ($\Delta\Delta C_t$) referring to the mHPRT in each sample; the results are presented as arbitrary units.

Histological analysis

The skin specimens from patch areas were removed for histologic examination immediately after the last EC application with OVA. Skin biopsies were taken from similar body sites, fixed overnight with 10 % paraformaldehyde at 4 °C, and embedded in paraffin. Four-micrometer sections were stained with hematoxylin and eosin (H&E). The histological preparations were analyzed under a light microscope (Leica DM2000, Germany) with $50\times$, $100\times$, and $400\times$ lenses.

Statistical analysis

The data are shown as mean $\pm$ SE. Statistical analysis was done with the program Statistica 8.0 (StatSoft Inc., USA). The significance of the results was determined by using Student's t test. Differences were considered significant at $p < 0.05$. The Quantitative RT-PCR data were calculated by using the comparative C_t method ($\Delta\Delta C_t$).

Authors' contributions
NS, EB, IS, AN carried out the molecular genetic studies and the immunoassays. DP have prepared the C_{60} aqueous solution. OK carried out the histological analysis. NS, SA, MK have made substantial contributions to conception, design, analysis, interpretation of data and given final approval of the version to be published. All authors read and approved the final manuscript.

Acknowledgements

We would like to thank Dr. Svetlana Korobova for critical reading of the manuscript. This work was supported by the Ministry of Science and Education of the Russian Federation, the Federal Target Program "Research and development on priority directions of scientific-technological complex of Russia in 2014–2020" (agreement 14.604.21.0059, the unique identifier RFMEFI60414X0059).

Competing interests

The authors declare that they have no competing interests.

References

1. Bieber T. Atopic dermatitis. N Engl J Med. 2008;358:1483–94.
2. Bieber T, Novak N. Pathogenesis of atopic dermatitis: new developments. Curr Allergy Asthma Rep. 2009;9:291–4.
3. Palmer CN, Irvine AD, Terron-Kwiatkowski A, Zhao Y, Liao H, Lee SP, et al. Common loss-of-function variants of the epidermal barrier protein filaggrin are a major predisposing factor for atopic dermatitis. Nat Genet. 2006;38:441–6.
4. Kezic S, Novak N, Jakasa I, Jungersted JM, Simon M, Brandner JM, et al. Skin barrier in atopic dermatitis. Front Biosci. 2014;19:542–56.
5. Arkwright PD, Motala C, Subramanian H, Spergel J, Schneider LC, Wollenberg A, et al. Management of difficult-to-treat atopic dermatitis. J Allergy Clin Immunol Pract. 2013;1:142–51.
6. Shershakova NN, Babakhin AA, Elisyutina OG, Khaitov MR. Atopic dermatitis: experimental models for study of pathogenesis and development new methods of treatment. RAG. 2011;6:3–11 **(in Russian)**.
7. Garmhausen D, Hagemann T, Bieber T, Dimitriou I, Fimmers R, Diepgen T, et al. Characterization of different courses of atopic dermatitis in adolescent and adult patients. Allergy. 2013;68:498–506.
8. Bieber T. Atopic dermatitis. Ann Dermatol. 2010;22:125–37.
9. Bonness S, Bieber T. Molecular basis of atopic dermatitis. Curr Opin Allergy Clin Immunol. 2007;7:382–6.
10. Schlapbach C, Simon D. Update on skin allergy. Allergy. 2014;69:1571–81.
11. Nakai K, Yoneda K, Maeda R, Munehitro A, Fujita N, Yokoi I, et al. Urinary biomarker of oxidative stress in patients with psoriasis and atopic dermatitis. J Eur Acad Dermatol Venerol. 2009;23:1405–8.
12. Kapun AP, Salobir J, Levart A, Kotnik T, Svete AN. Oxidative stress markers in canine atopic dermatitis. Res Vet Sci. 2012;92(3):469–70.
13. Koren Carmi I, Haj R, Yehuda H, Tamir S, Reznick AZ. The Role of oxidation in FSL-1 induced signaling pathways of an atopic dermatitis model in hHaCaT keratinocytes. Adv Exp Med Biol. 2015;849:1–10.
14. Okayama Y. Oxidative stress in allergic and inflammatory skin diseases. Curr Drug Targets Inflamm Allergy. 2005;4(4):517–9.
15. Niwa Y, Sumi H, Kawahira K, Terashima T, Nakamura T, Akamatsu H. Protein oxidative damage in the stratum corneum: evidence for a link between environmental oxidants and the changing prevalence and nature of atopic dermatitis in Japan. Br J Dermatol. 2003;149:248–54.
16. Bito T, Nishigori C. Impact of reactive oxygen species on keratinocyte signaling pathways. J Dermatol Sci. 2012;68(1):3–8.
17. Gharbi N, Pressac M, Hadchouel M, Szwarc H, Wilson SR, Moussa F. [60] Fullerene is a powerful antioxidant in vivo with no acute or subacute toxicity. Nano Lett. 2005;5:2578–85.
18. Hendrickson OD, Morozova OV, Zherdev AV, Yaropolov AI, Klochkov SG, Bachurin SO, Dzantiev BB. Study of distribution and biological effects of fullerene C60 after single and multiple intragastrical administrations to rats. Fuller Nanotub Carbon Nanostruct. 2015;23(7):658–68.
19. Baati T, Bourasset F, Gharbi N, Njim L, Abderrabba M, Kerkeni A, Szwarc H, Moussa F. The prolongation of the lifespan of rats by repeated oral administration of 60 fullerene. Biomaterials. 2012;33:4936–46.
20. Li W, Chen C, Ye C, Wei T, Zhao Y, Lao F, Chen Z, Meng H, Gao Y, Yuan H, Xing G, Feng Z, Chai Z, Zhang X, Yang F, Han D, Tang X, Zhang Y. The

21. translocation of fullerenic nanoparticles into lysosome via the pathway of clathrin-mediated endocytosis. Nanotechnology. 2008;19(14):145102.
21. Andreev I, Petrukhina A, Garmanova A, Babakhin A, Andreev S, Romanova V, Troshin P, Troshina O, DuBuske L. Penetration of fullerene C60 derivatives through biological membranes. Fuller Nanotub Carbon Nanostruct. 2008;16:89–102.
22. Andreev SM, Babakhin AA, Petrukhina AO, Romanova VS, Parnes ZN, Petrov RV. Immunogenic and allergic properties of fullerene conjugates with amino acid and proteins. Dokl Biochem. 2000;370:4–7.
23. Ryan JJ, Bateman HR, Stover A, Gomez G, Norton SK, Zhao W, Schwartz LB, Lenk R, Kepley CL. Fullerene nanomaterials inhibit the allergic response. J Immunol. 2007;179:665–72.
24. Dellinger A, Zhou Z, Lenk R, MacFarland D, Kepley CL. Fullerene nanomaterials inhibit phorbol myristate acetate-induced inflammation. Exp Dermatol. 2009;18:1079–81.
25. Norton SK, Dellinger A, Zhou Z, Lenk R, MacFarland D, Vonakis B, Conrad D, Kepley CL. A new class of human mast cell and peripheral blood basophil stabilizers that differentially control allergic mediator release. Clin Trans Sci. 2010;3:158–69.
26. Magoulas GE, Garnelis T, Athanassopoulos CM, Papaioannou D, Mattheolabakis G, Avgoustakis K, Hadjipavlou-Litina D. Synthesis and antioxidative/anti-inflammatory activity of novel fullerene–polyamine conjugates. Tetrahedron. 2012;68(35):7041–9.
27. Andrievsky GV, Kosevich MV, Vovk OM, Shelkovsky VS, Vashcenko LA. On the production of an aqueous colloidal solution of fullerenes. J Chem Soc. 1995;12:1281–2.
28. Bunz H, Plankenhorn S, Klein R. Effect of buckminsterfullerenes on cells of the innate and adaptive immune system: an in vitro study with human peripheral blood mononuclear cells. Intern J Nanomed. 2012;7:4571–80.
29. Babakhin AA, Andrievsky G, DuBuske LM. Inhibition of systemic and passive cutaneous anaphylaxis by water-soluble fullerene C60. J Allergy Clin Immunol. 2009;123(2):118.
30. Bashkatova YeN, Andreev SM, Shershakova NN, Babakhin AA, Shilovsky IP, Khaitov MR. Study of modulatory activity of fullerene C60 derivatives on DTH reaction. Phiziologiya i patologiya immunnoy sistemy. 2012; 2:17–27 **(in Russian)**.
31. Bashkatova E, Shershakova N, Shilovski I, Babakhin A, Andreev S, Khaitov M. Fullerene adducts attenuate delayed-type hypersensitivity reactions induced in mice. Allergy. 2013;68:442.
32. Yanagisawa R, Takano H, Inoue KI, Koike E, Sadakane K, Ichinose T. Size effects of polystyrene nanoparticles on atopic dermatitis-like skin lesions in NC/NGA mice. Int J Immunopathol Pharmacol. 2010;23(1):131–41.
33. Ilves M, Palomäki J, Vippola M, Lehto M, Savolainen K, Savinko T, Alenius H. Topically applied ZnO nanoparticles suppress allergen induced skin inflammation but induce vigorous IgE production in the atopic dermatitis mouse model. Part Fibre Toxicol. 2014;11:38. doi:10.1186/s12989-014-0038-4.
34. Wiesenthal A, Hunter L, Wang S, Wickliffe J, Wilkerson M. Nanoparticles: small and mighty. Int J Dermatol. 2011;50(3):247–54. doi:10.1111/j.1365-4632.2010.04815.x.
35. Rancan F, Gao Q, Graf C, Troppens S, Hadam S, Hackbarth S, Kembuan C, Blume-Peytavi U, Rühl E, Lademann J, Vogt A. Skin penetration and cellular uptake of amorphous silica nanoparticles with variable size, surface functionalization, and colloidal stability. ACS Nano. 2012;6(8):6829–42. doi:10.1021/nn301622h.
36. Hussain Z, Katas H, Mohd Amin MC, Kumolosasi E, Sahudin S. Downregulation of immunological mediators in 2,4-dinitrofluorobenzene-induced atopic dermatitis-like skin lesions by hydrocortisone-loaded chitosan nanoparticles. Int J Nanomed. 2014;9:5143–56. doi:10.2147/IJN.S71543.
37. Kang MJ, Eum JY, Jeong MS, Park SH, Moon KY, Kang MH, Kim MS, Choi SE, Lee MW, do Lee I, Bang H, Lee CS, Joo SS, Li K, Lee MK, Seo SJ, Choi YW. Tat peptide-admixed elastic liposomal formulation of hirsutenone for the treatment of atopic dermatitis in NC/Nga mice. Int J Nanomed. 2011;6:2459–67. doi:10.2147/IJN.S24350.
38. Shershakova NN, Babakhin AA, Bashkatova EN, Kamyshnikov OY, Andreev SM, Shilovsky IP, et al. Allergen-specific immunotherapy of experimental atopic dermatitis. Immunologia. 2014;35(3):155–60 **(in Russian)**.
39. Andreev SM, Purgina DD, Bashkatova EN, Garshev AV, Maerle AV, Khaitov MR. Facile preparation of aqueous fullerene C60 nanodispersions. Nanotechnol Russ. 2014;9(7–8):369–79.

40. Deguchi SH, Mukai SA. Top-down preparation of dispersions of C60 nanoparticles in organic solvents. Chem Lett. 2006;35(4):396–3974.

41. Mchedlov-Petrossyan NO. Fullerene C_{60} solutions: colloid aspect. Chem Phys Technol Surf. 2010;1(1):19–37.

42. Opinion on NMP. Scientific Committee on Consumer Safety. 2011. SCCS/1413/11. http://ec.europa.eu/health/scientific_committees/consumer_safety/docs/sccs_o_050.pdf.

43. Spergel JM, Mizoguchi E, Brewer JP, Martin TR, Bhan AK, Geha RS. Epicutaneous sensitization with protein antigen induces localized allergic dermatitis and hyperresponsiveness to methacholine after single exposure to aerosolized antigen in mice. J Clin Invest. 1998;101:1614–22.

44. Shershakova N, Bashkatova E, Babakhin A, Andreev S, Nikonova A, Shilovsky I, Kamyshnikov O, Buzuk A, Elisyutina O, Fedenko E, Khaitov M. Allergen-specific immunotherapy with monomeric allergoid in a mouse model of atopic dermatitis. PLoS ONE. 2015;10(8):e0135070. doi:10.1371/journal.pone.0135070.

45. Mountford AP, Fisher A, Wilson RA. The profile of IgG1 and IgG2a antibody responses in mice exposed to *Schistosoma mansoni*. Parasit Immunol. 1994;16(10):521–7.

46. Andreev S, Purgina D, Bashkatova E, Garshev A, Maerle A, Andreev I, Osipova N, Shershakova N, Khaitov M. Study of fullerene aqueous dispersion prepared by novel dialysis method: simple way to fullerene aqueous solution. Fuller Nanotub Carbon Nanostruct. 2015;23(9):792–800. doi:10.1080/1536383X.2014.998758.

47. Karpenko OB, Trachevskij VV, Filonenko OV, Lobanov VV, Avdeev MV, Tropin TV, Kyzyma OA, Snegir SV. NMR study of non-equilibrium state of fullerene C60 in N-methyl-2-pyrrolidone. Ukr J Phys. 2012;8:860–3.

48. Deryabin DG, Efremova LV, Vasilchenko AS, Saidakova EV, Sizova EA, Troshin PA, Zhilenkov AV, Khakina EE. J Nanobiotechnol. 2015;13:50. doi:10.1186/s12951-015-0112-6.

49. Song M, Yuan S, Yin J, Wang X, Meng Z, Wang H, Jiang G. Size-dependent toxicity of nano-C_{60} aggregates: more sensitive indication by apoptosis-related Bax translocation in cultured human cells. Environ Sci Technol. 2012;46:3457–64.

50. Yamashita K, Sakai M, Takemoto N, Tsukimoto M, Uchida K, Yajima H, et al. Attenuation of delayed-type hypersensitivity by fullerene treatment. Toxicology. 2009;261:19–24.

51. Sun L, He C, Nair L, Yeung J, Egwuagu CE. Interleukin 12 (IL-12) family cytokines: role in immune pathogenesis and treatment of CNS autoimmune disease. Cytokine. 2015. doi:10.1016/j.cyto.2015.01.030.

52. Gostner JM, Becker K, Fuchs D, Sucher R. Redox regulation of the immune response. Redox Rep. 2013;18(3):88–94.

53. Lu LF, Lind EF, Gondek DC, Bennett KA, Gleeson MW, Pino-Lagos K, et al. Mast cells are essential intermediaries in regulatory T-cell tolerance. Nature. 2006;442(7106):997–1002.

54. Moore KW, de WaalMalefyt R, Coffman RL, O'Garra A. Interleukin-10 and the interleukin-10 receptor. Annu Rev Immunol. 2001;19:683–765.

55. Verhagen J, Akdis M, Traidl-Hoffmann C, Schmid-Grendelmeier P, Hijnen D, Knol EF, et al. Absence of T regulatory cell expression and function in atopic dermatitis skin. J Allergy Clin Immunol. 2006;117:176–83.

56. Ou LS, Goleva E, Hall C, Leung DY. T regulatory cells in atopic dermatitis and subversion of their activity by superantigens. J Allergy Clin Immunol. 2004;113:756–63.

57. Otsuka A, Doi H, Egawa G, Maekawa A, Fujita T, Nakamizo S, et al. Possible new therapeutic strategy to regulate atopic dermatitis through upregulating filaggrin expression. J Allergy Clin Immunol. 2014;133(1):139–46.

58. Toda M, Leung DY, Molet S, Boguniewicz M, Taha R, Christodoulopoulos P, et al. Polarized in vivo expression of IL-11 and IL-17 between acute and chronic skin lesions. J Allergy Clin Immunol. 2003;111:875–81.

2

A versatile papaya mosaic virus (PapMV) vaccine platform based on sortase-mediated antigen coupling

Ariane Thérien[1], Mikaël Bédard[1], Damien Carignan[1], Gervais Rioux[1], Louis Gauthier-Landry[1], Marie-Ève Laliberté-Gagné[1], Marilène Bolduc[1], Pierre Savard[2] and Denis Leclerc[1]* ⓘ

Abstract

Background: Flexuous rod-shaped nanoparticles made of the coat protein (CP) of papaya mosaic virus (PapMV) have been shown to trigger innate immunity through engagement of toll-like receptor 7 (TLR7). PapMV nanoparticles can also serve as a vaccine platform as they can increase the immune response to fused peptide antigens. Although this approach shows great potential, fusion of antigens directly to the CP open reading frame (ORF) is challenging because the fused peptides can alter the structure of the CP and its capacity to self assemble into nanoparticles—a property essential for triggering an efficient immune response to the peptide. This represents a serious limitation to the utility of this approach as fusion of small peptides only is tolerated.

Results: We have developed a novel approach in which peptides are fused directly to pre-formed PapMV nanoparticles. This approach is based on the use of a bacterial transpeptidase (sortase A; SrtA) that can attach the peptide directly to the nanoparticle. An engineered PapMV CP harbouring the SrtA recognition motif allows efficient coupling. To refine our engineering, and to predict the efficacy of coupling with SrtA, we modeled the PapMV structure based on the known structure of PapMV CP and on recent reports revealing the structure of two closely related potexviruses: pepino mosaic virus (PepMV) and bamboo mosaic virus (BaMV). We show that SrtA can allow the attachment of long peptides [Influenza M2e peptide (26 amino acids) and the HIV-1 T20 peptide (39 amino acids)] to PapMV nanoparticles. Consistent with our PapMV structural model, we show that around 30% of PapMV CP subunits in each nanoparticle can be fused to the peptide antigen. As predicted, engineered nanoparticles were capable of inducing a strong antibody response to the fused antigen. Finally, in a challenge study with influenza virus, we show that mice vaccinated with PapMV-M2e are protected from infection.

Conclusions: This technology will allow the development of vaccines harbouring long peptides containing several B and/or T cell epitopes that can contribute to a broad and robust protection from infection. The design can be fast, versatile and can be adapted to the development of vaccines for many infectious diseases as well as cancer vaccines.

Keywords: Papaya mosaic virus, Flexuous rod shape nanoparticles, Vaccine platform, Influenza M2e based vaccine, Sortase, Transpeptidase

Background

The use of nanoparticles as vaccine platforms is a very promising approach to improve immune response directed against poorly immunogenic antigens. Nanoparticles can be made of viral structural proteins that self-assemble to mimic the native organisation and conformation of a viral pathogen [1–9]. They can also present peptide antigens on their surface and induce a potent humoral response [2–4, 10]. Being of a size similar to viruses, nanoparticles are efficiently phagocyted and processed by antigen presentation cells (APCs) such as

*Correspondence: Denis.Leclerc@crchudequebec.ulaval.ca
[1] Department of Microbiology, Infectiology and Immunology, Infectious Disease Research Center, Laval University, 2705 Boul. Laurier, Quebec City, PQ G1V 4G2, Canada
Full list of author information is available at the end of the article

dendritic cells (DC) [11, 12]. They can also activate innate immunity through pathogen associated molecular patterns (PAMPs), which are recognized by innate receptors (pathogen recognition receptors; PRRs) [13, 14]. Therefore, the use of nanoparticles to display heterologous epitopes is a promising strategy for the development of novel vaccines.

We have developed a novel rod-shaped viral nanoparticle made of the coat protein (CP) of papaya mosaic virus (PapMV) that is self-assembled around a ssRNA. PapMV nanoparticles trigger innate immunity efficiently [14] through engagement of toll-like receptor 7 (TLR7) [15]. After phagocytosis, nanoparticles reach the endosome of immune cells and liberate the ssRNA, which engages and activates TLR7 [15]. PapMV nanoparticles can be used as an adjuvant for improvement of vaccines [16], an immune enhancer for the treatment of cancer in immunotherapy [17], or as a vaccine platform to trigger an immune response to a specific peptide antigen [18]. Fusion of peptides directly to the open reading frame (ORF) of the PapMV CP leads to the formation of chimeric nanoparticles that can trigger either a humoral [19, 20] or a CTL response against the fused antigen [21–24]. This approach has shown the great potential of PapMV nanoparticles as a vaccine platform, but has also revealed the challenges involved in their engineering. Indeed, the fusion of peptides to the CP ORF can alter its structure and affect its capacity to self-assemble into immunogenic nanoparticles. We showed previously that the fusion of long peptides interferes with self-assembly [20].

To decrease this stress on the structure of the nanoparticles, we propose here a novel method for coupling of peptides to already self-assembled nanoparticles using the bacterial transpeptidase, sortase A (SrtA). SrtA anchors surface proteins, such as virulence factors, to the bacterial cell wall peptidoglycans of Gram-positive bacteria [25–27]. Target proteins become linked to a polyG motif at the N-terminus of the acceptor peptidoglycan through the recognition and cleavage of their LPXTG recognition motif by SrtA. SrtA has been used widely for the site-specific labelling of proteins with a wide range of functional groups, such as fluorescent labels and green fluorescent protein (GFP) [28–31], biotin [32], PEG [29, 31, 33], peptide nucleic acids [34], lipids [35], sugars [36], and other proteins [30, 37, 38]. A soluble form of SrtA lacking the transmembrane domain can be produced efficiently in *E. coli* and used in in vitro transpeptidation reactions.

Based on in silico modelling of the 3D structure of full length PapMV CP and of PapMV, we engineered the PapMV vaccine platform with the receptor motif of SrtA, and used SrtA to attach long peptides to the nanoparticles. This technology allows rapid and efficient coupling

of peptide antigens to PapMV nanoparticles without affecting their structure. PapMV nanoparticles with SrtA-conjugated influenza M2e peptide were shown to be immunogenic, and induced protection against influenza infection.

Results
Structural models of PapMV CP and PapMV
The recent solving of the near-atomic structure of Bamboo mosaic virus (BaMV) [39] and Pepino mosaic virus (PepMV) [40] by cryo-electron microscopy (cryo-EM), has allowed the structural details of *Potexvirus* viruses to be revealed at an atomic level. Interestingly, the CPs of these two *Potexviruses* share high structural conservation with the structure of the truncated CP of PapMV (PDB ID 4DOX) [41] with root mean square deviations (RMSD) of around 1.6 and 3 Å for the backbone heavy atoms of PepMV (PDB ID 5FN1) and BaMV (PDB ID 52AT), respectively (Fig. 1a). Moreover, previous low resolution cryo-EM data reported by our group on PapMV [41] demonstrates that it adopts a capsid symmetry similar to that adopted by BaMV and PapMV, consisting of a left-handed helix of ~130Å in diameter, with ~10 CP subunits per turn and a pitch of ~35 Å. These structural similarities prompted us to elaborate a structural model for PapMV based on BaMV and PapMV to guide the development of our nanoparticles following a rational design approach. Briefly, a homology model of the complete PapMV CP based on the structure of the truncated PapMV CP (PDB ID 4DOX), BaMV CP and PepMV CP was generated using I-TASSER [42]. The complete PapMV CP was then aligned on the subunit CPs of PepMV (PDB 5FN1) to construct the model (Fig. 1b). The viral ssRNA (in orange) was positioned by homology with PepMV (Fig. 1b, c).

As observed for the structures of BaMV and PepMV, the N-terminal ends of PapMV CPs are exposed on the surface of the virus, while the C-terminii are located in the central cavity (Fig. 1c). Therefore, fusions to the N-terminus can modify the surface of PapMV nanoparticles and affect the interface of interaction between nanoparticles and immune cells. Fusions made at the C-terminus could cause steric hindrance in the central cavity, interfering with nanoparticle assembly, as indicated by the cross-section view of the PapMV nanoparticle model displayed in Fig. 1c (right illustration) and as previously reported [20]. Considering that the C-terminal sections of two CPs separated by one capsid turn define a cavity of around 25 Å (Fig. 1c), it is anticipated that the longest C-terminal fusion allowed in the core of the nanoparticles would be ~17 amino acids if arranged as an α-helix. But, fusion of long peptides on each side of the nanoparticles where the C-terminus are accessible

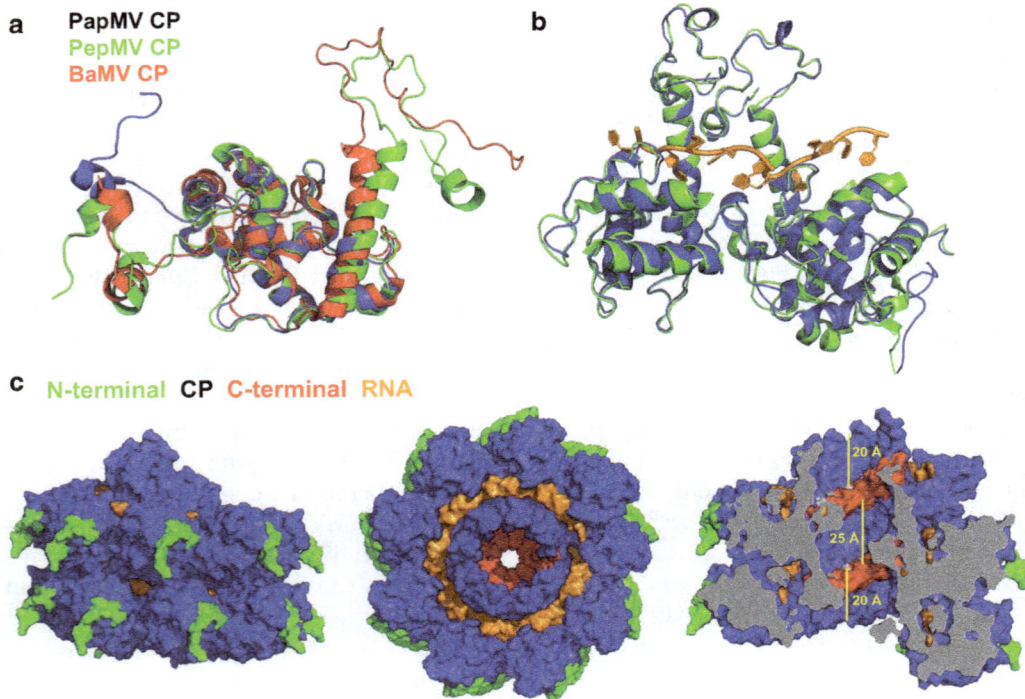

Fig. 1 Modelling of the PapMV CP and PapMV structure. **a** The full-length structure of the PapMV CP was modelled based on the recently published structure of two members of the potexvirus group: BaMV CP and PepMV CP. The core region of PapMV CP (PDB 4DOX, *blue*) superimposes well on the core region of PepMV CP (PDB 5FN1, *green*) and BaMV CP (PDB 5A2T, *red*). **b** Superposition of two subunits of the PapMV model (*blue*) with two CP subunits of the PepMV structure (*green*, PDB 5FN1) demonstrates the concordance between the two structures. **c** To show how each of the PapMV CP interacts with each other, and with the ssRNA in the nanoparticle, we modeled the self assembly of the 18 subunits (~2 turns) that comprise PapMV nanoparticles. The CP N-terminal residues (10 first) are shown in *green*, the CP core in *blue*, the CP C-terminus in *red* (10 last residues), and the RNA is in *orange*. The last C-terminal residue of the CP is displayed in *light grey* in the cutaway view on the *right*, and the *bars* represent 20 Å—the distance separating CP C-terminal residues from the PapMV nanoparticle exterior at both extremities, and 32 Å—the distance between two CP C-terminal residues separated by one capsid turn

is possible without changing the surface of the nanoparticle. Therefore, it suggests that the development of a method allowing peptide coupling to available C-terminus at the extremity of already-assembled nanoparticles is an attractive approach for the coupling of long peptides to our vaccine platform.

Engineering of the PapMV nanoparticles with the SrtA recognition motif

Based on the structural model of the PapMV nanoparticle presented in Fig. 1c, we chose to fuse the sortase A (SrtA) LPETGG recognition motif followed by a 6×H tag to the C-terminus of the PapMV CP. This fusion is preceded by the linker TSTTR, which was introduced to allow the LPETGG motif to reach the exterior of the nanoparticles at both ends (Fig. 1c-right panel), making it available for the transpeptidation reaction. Indeed, assuming a maximal length of 3.5 Å per amino acid, the PapMV structural model suggests that around 5 amino acids are required to span the 20 Å distance separating

CP C-terminal residues from the nanoparticle exterior at both extremities (Fig. 1c-right panel). Fusion of the receptor motif did not alter the capacity of the CP to self-assemble around the ssRNA, leading to generation of nanoparticles (PapMV-SrtA) with an average length of 40 nm vs 54 nm for the WT PapMV nanoparticle, as shown by dynamic light scattering (DLS) (Fig. 2a). The width of the DLS curve in the two types of nanoparticles was comparable, suggesting that nanoparticles are of different lengths ranging from ~20 to ~100 nm. Electron microscopy (EM) confirmed that PapMV-WT and PapMV-SrtA are similar in appearance (Fig. 2b). As a control, we engineered another PapMV CP lacking the TSTTR linker preceding the SrtA recognition motif [PapMV-SrtA (short)] (Additional file 1: Figure S1). This construct is predicted to not permit coupling because the SrtA recognition motif is hidden too far within the core cavity of the PapMV nanoparticle. As with PapMV-SrtA, the fusion did not interfere with self-assembly, and generated nanoparticles comparable to those of WT PapMV,

a

PapMV WT: ...IQFLPPPE
PapMV-SrtA: ...IQFLPPPETSTTRLPETGGHHHHHH

b

c

Fig. 2 Engineering of PapMV coat protein carrying the SrtA recognition motif. **a** Amino acid sequence of the PapMV CP C-terminus of the WT CP as compared to the engineered PapMV CP that harbours a linker of 5 amino acids (TSTTR) followed by the SrtA recongnition motif (LPETGG); a 6×His tag was inserted at the C-terminus of the PapMV coat protein (named PapMV-SrtA). **b** The length of the rod-shaped nanoparticles was assessed by dynamic light scattering (DLS). PapMV-SrtA (40 nm) was showed to be slightly smaller than the WT PapMV (54 nm). **c** Transmission electron microscopy (TEM) of WT PapMV (left) and PapMV-SrtA (right) nanoparticles shows the rod shape of the nanoparticles

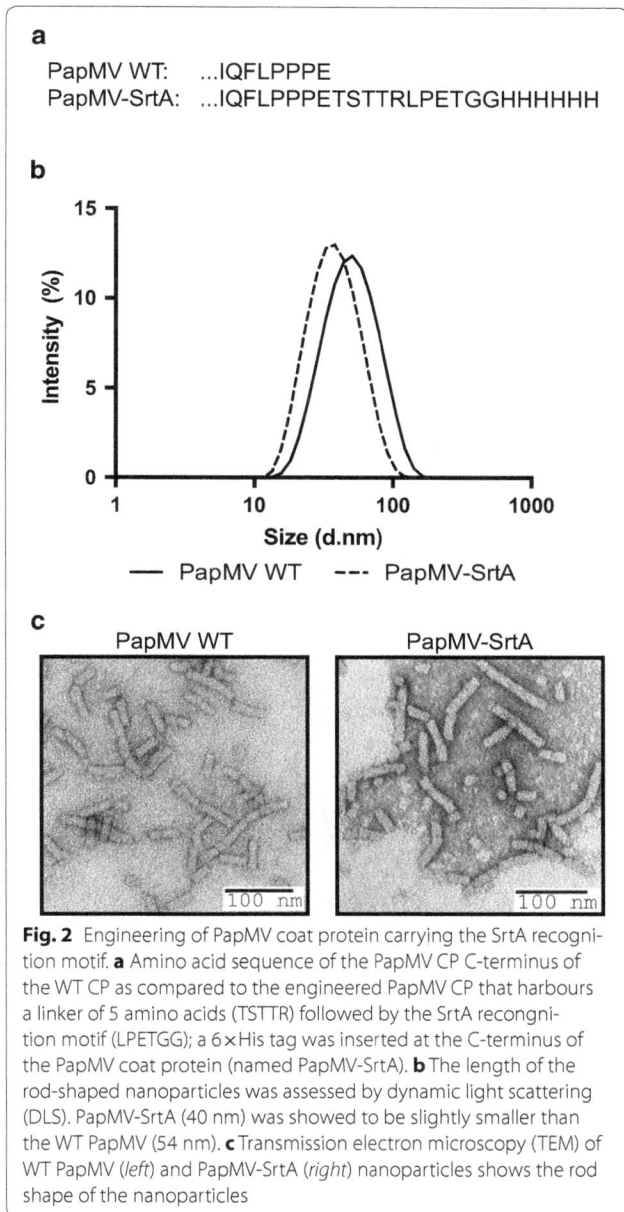

as confirmed by DLS (Additional file 1: Figure S1B) and electron microscopy (Additional file 1: Figure S1C).

SrtA-mediated coupling of peptides onto PapMV nanoparticles

To achieve SrtA-mediated coupling onto PapMV nanoparticles, we chose two long peptides of 26 amino acids (influenza M2e peptide) and 39 amino acids (HIV-1 T20 peptide), respectively. The peptide M2e is derived from the extracellular domain of the matrix protein 2 (M2e) of influenza virus. M2e is highly conserved in most influenza A strains [43] and is a valuable antigen in inducing

protection to influenza infection [19, 44]. The T20 peptide is derived from the surface glycoprotein gp41 of human immunodeficiency virus (HIV-1) [45]. A monoclonal antibody (2F5) directed towards this peptide was shown to neutralise HIV-1 infection in vitro, and has been proposed for use as a vaccine antigen [46].

We performed transpeptidation reactions using 25 μM of PapMV-SrtA nanoparticles (based on the amount of PapMV CP evaluated by BCA) incubated with 50 μM of either the M2e or T20 peptide and 50 μM of SrtA. Following the coupling reaction, the SrtA and the free peptides were removed by filtration, and the samples were loaded on SDS-PAGE to reveal the coupling efficiency (Fig. 3a-top panel). PapMV CP coupled to either the M2e or the T20 peptide migrated as slightly higher molecular weight proteins than the uncoupled PapMV-SrtA CP on SDS-PAGE (Fig. 3a). The identity of the higher band was further confirmed by immunoblotting using specific antibodies directed towards PapMV CP (Fig. 3a-second panel), or the M2e peptide (Fig. 3a-fourth panel). Also, we used a specific antibody directed to the 6×H tag that, as expected, reacted only with the uncoupled PapMV-SrtA CP (Fig. 3a-third panel). To assess the coupling efficacy precisely, we calculated the ratio between the intensity of the coupled CP and that of the total CP (coupled + uncoupled) on SDS-PAGE (Fig. 3a), which revealed a coupling efficacy of 35% with the M2e peptide and 32% with the T20 peptide. We performed six coupling reactions and always obtained a similar coupling efficiency, which ranged from 26 to 38% for M2e and 30 to 34% for the T20 peptide (not shown). To evaluate whether the structure of the nanoparticles was affected by the coupling reaction, we compared the length of the PapMV-SrtA nanoparticles with those conjugated with M2e or T20 peptides (Fig. 3b). The nanoparticles were similar in length compared to PapMV-SrtA, as shown by DLS (Fig. 3b), and in their appearance in EM (Fig. 3c).

As expected from our prediction based on the structure of the PapMV-SrtA CP, only a fraction (about 1/3) of the CP reacted with SrtA. We observed that coupling was saturated with 50 μM of peptide and no additional coupling was observed when 100 or 200 μM was used (Additional file 2: Figure S2). Finally, we confirmed that coupling reactions with the PapMV-SrtA(short) (Additional file 3: Figure S3A) were very inefficient as compared to those with PapMV-SrtA, which shows the importance of the linker (TSTTR) in making the SrtA recognition motif solvent-accessible (Additional file 3: Figure S3B). Coupling on PapMV-SrtA(short) nanoparticles was consistently ineffective (four repeated experiments), unlike coupling on the PapMV-SrtA construct as revealed by immunoblot using an antibody directed against the M2e peptide (Additional file 3: Figure S3B).

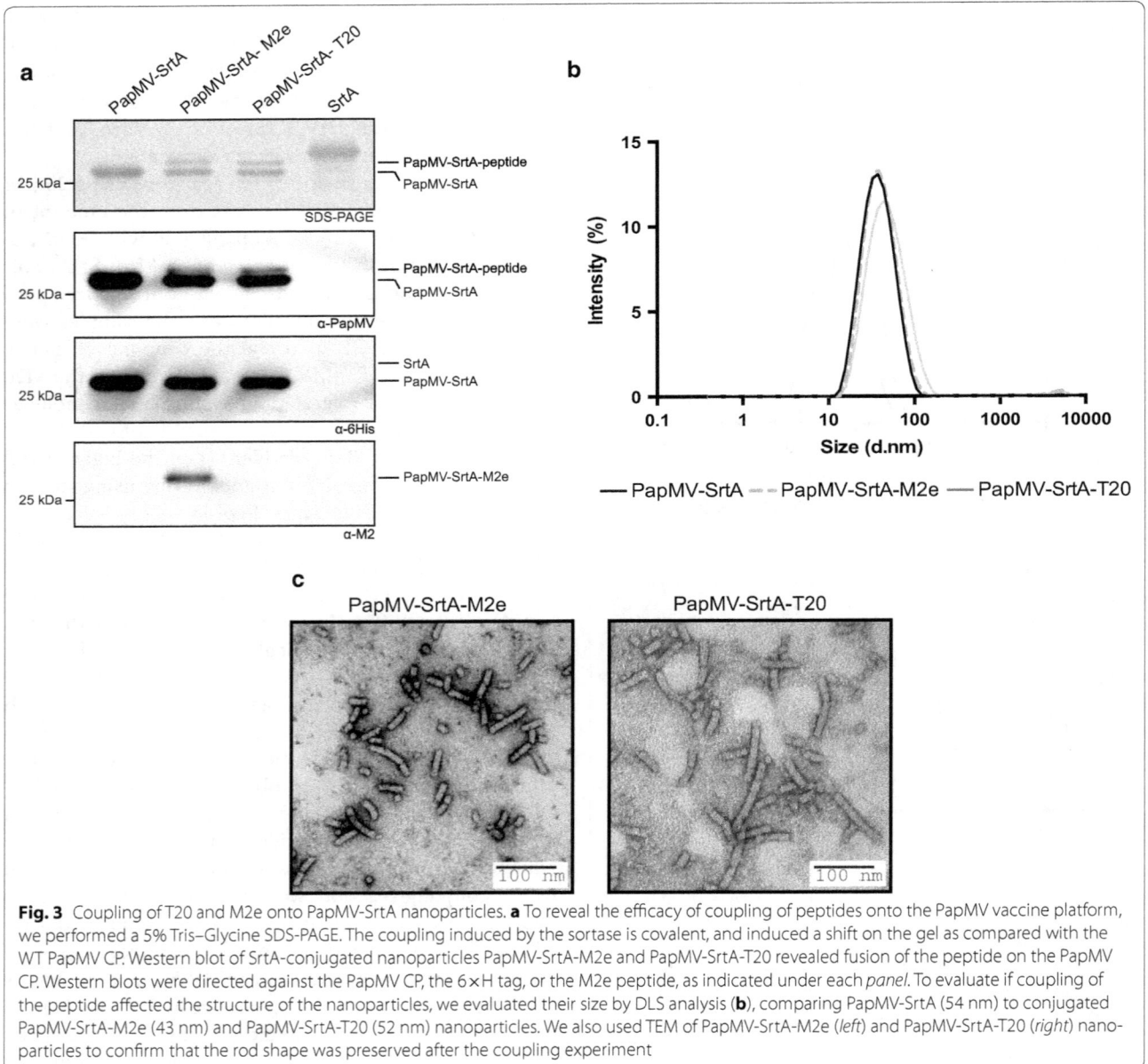

Fig. 3 Coupling of T20 and M2e onto PapMV-SrtA nanoparticles. **a** To reveal the efficacy of coupling of peptides onto the PapMV vaccine platform, we performed a 5% Tris–Glycine SDS-PAGE. The coupling induced by the sortase is covalent, and induced a shift on the gel as compared with the WT PapMV CP. Western blot of SrtA-conjugated nanoparticles PapMV-SrtA-M2e and PapMV-SrtA-T20 revealed fusion of the peptide on the PapMV CP. Western blots were directed against the PapMV CP, the 6×H tag, or the M2e peptide, as indicated under each *panel*. To evaluate if coupling of the peptide affected the structure of the nanoparticles, we evaluated their size by DLS analysis (**b**), comparing PapMV-SrtA (54 nm) to conjugated PapMV-SrtA-M2e (43 nm) and PapMV-SrtA-T20 (52 nm) nanoparticles. We also used TEM of PapMV-SrtA-M2e (*left*) and PapMV-SrtA-T20 (*right*) nanoparticles to confirm that the rod shape was preserved after the coupling experiment

The PapMV-SrtA vaccine platform enhances the humoral response to the T20 peptide

To assess the capacity of the PapMV-SrtA vaccine platform to improve the humoral response to the T20 peptide, we immunized Balb/C mice (5/group) twice, with a 14-day interval, with 90 µg of PapMV-SrtA, 30 or 90 µg of PapMV-SrtA-T20, or 90 µg of PapMV-SrtA with 40 µg of free T20 peptide. Blood levels of total anti-T20 IgG and IgG2a were assessed by ELISA 13 days after each immunization (Fig. 4a–d). Mice vaccinated with PapMV-SrtA or PapMV-SrtA and the free T20 peptide did not develop any detectable T20 immunoglobulins. However, 30 µg of PapMV-SrtA-T20 was sufficient to mount a robust antibody response against the T20 peptide after a single immunization (Fig. 4ab). This result confirms that coupling of the peptide to the PapMV vaccine platform improved the immunogenicity of the peptide significantly. We also observed a rapid class switch towards a TH1 response, since a strong IgG2a response specific to the T20 peptide was recorded after the first immunization (Fig. 4b), which is expected with the PapMV vaccine platform as previously reported [19, 44, 47, 48]. The IgG2a isotype is known to show a higher avidity for the antigen, and is therefore more efficient for neutralization

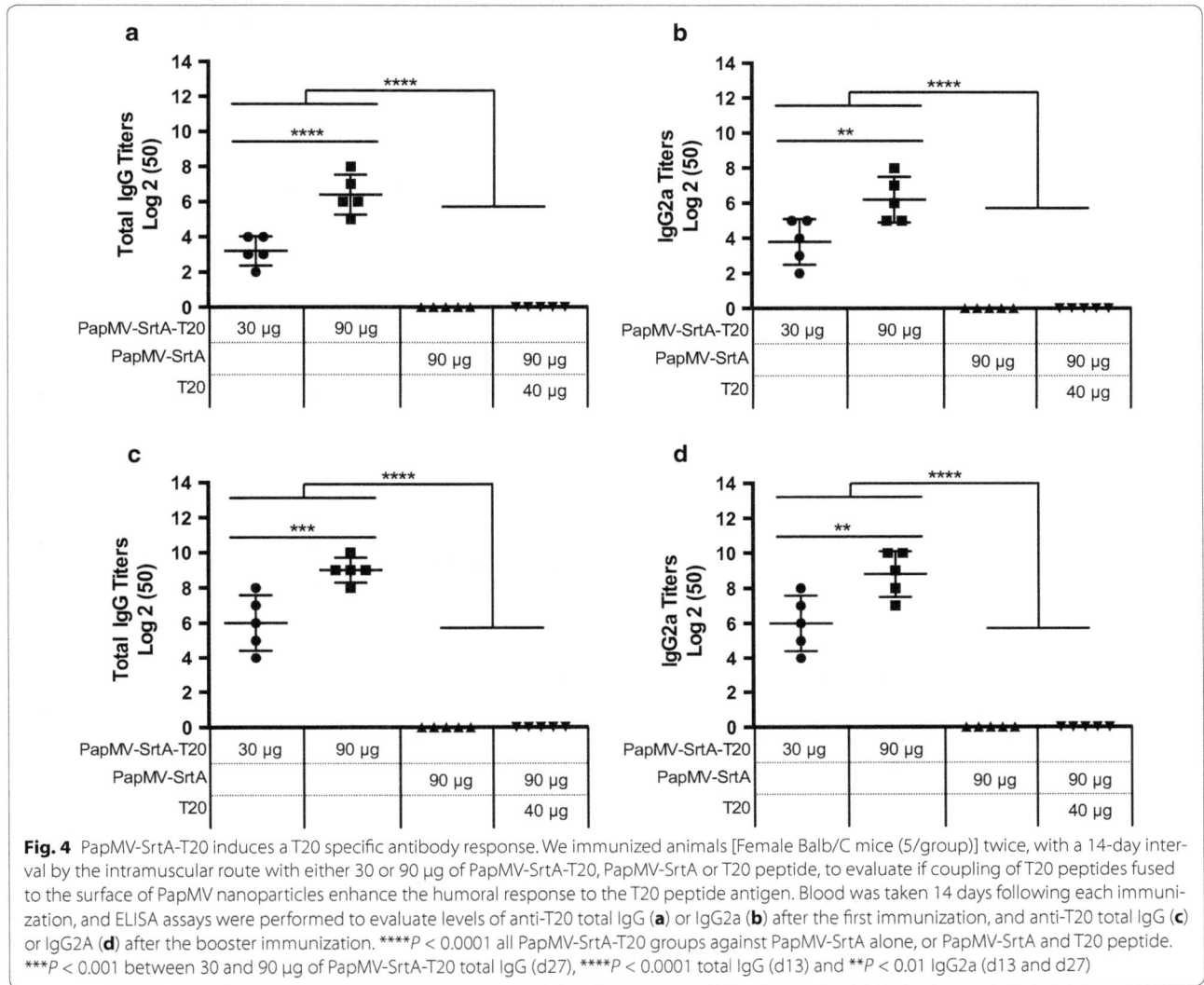

Fig. 4 PapMV-SrtA-T20 induces a T20 specific antibody response. We immunized animals [Female Balb/C mice (5/group)] twice, with a 14-day interval by the intramuscular route with either 30 or 90 μg of PapMV-SrtA-T20, PapMV-SrtA or T20 peptide, to evaluate if coupling of T20 peptides fused to the surface of PapMV nanoparticles enhance the humoral response to the T20 peptide antigen. Blood was taken 14 days following each immunization, and ELISA assays were performed to evaluate levels of anti-T20 total IgG (**a**) or IgG2a (**b**) after the first immunization, and anti-T20 total IgG (**c**) or IgG2A (**d**) after the booster immunization. ****P < 0.0001 all PapMV-SrtA-T20 groups against PapMV-SrtA alone, or PapMV-SrtA and T20 peptide. ***P < 0.001 between 30 and 90 μg of PapMV-SrtA-T20 total IgG (d27), ****P < 0.0001 total IgG (d13) and **P < 0.01 IgG2a (d13 and d27)

of a viral infection. Finally, and as expected, the boost immunization raised the total IgG and IgG2a titers significantly (Fig. 4c, d).

The PapMV vaccine platform enhances the humoral response to the M2e peptide and induces protection to an influenza challenge

To evaluate the efficacy of our approach in an animal infectious model, we immunized mice with PapMV-SrtA-M2e, and evaluated the humoral response directed towards the M2e peptide, and the capacity to provide protection to an influenza challenge. We also included in this study, as a reference, the PapMV-sM2e construct previously generated in our laboratory [19]. PapMV-sM2e harbours a fusion of a 9-amino-acid long peptide (EVETPIRNE) corresponding to the central region of the M2e peptide. In these nanoparticles, every PapMV CP

molecule harbours a fusion of the small M2e peptide. This construct was previously shown to induce production of antibodies to the full length M2e peptide, and to provide protection to an influenza challenge [19]. To compare the efficacy of the two constructs, five mice per group received prime and booster immunizations at intervals of 14 days with either 30 or 90 μg of PapMV-SrtA-M2e, 30 μg PapMVCP-sM2e or formulation buffer. Thirteen days following the first and the second immunization, sera were collected, and levels of total IgG and IgG2a directed against the M2e peptide were measured by ELISA. Interestingly, the levels of antibodies directed against the M2e (total IgG and IgG2a) were similar between all groups following the first immunization (Additional file 4: Figure S4). After two immunizations, total IgG and IgG2a were comparable between groups immunized with 30 μg of PapMV-SrtA-M2e or 30 μg of

PapMV-sM2e (Fig. 5a, b). However, immunization with 90 µg of PapMV-SrtA-M2e triggered a humoral response against the M2e peptide that was significantly higher than that of 30 µg of either PapMV-SrtA-M2e or PapMV-sM2e (Fig. 5a).

To evaluate the capacity of the candidate vaccine to protect against influenza, we challenged immunized mice 14 days after the second immunization with $1 \times$LD80 of influenza A/WSN/33 virus. This experiment was critical because it allowed us to demonstrate that the nanoparticle vaccine platform not only improves the humoral response to a peptide antigen, but also induces production of antibodies that protect the animal against an influenza infection. After the challenge, three different assessments were used to confirm the efficacy of the vaccine formulation: (1) weight loss, which gives a reliable readout because infected mice loose weight rapidly following an influenza challenge. Because weight can be evaluated precisely each day following the challenge, it becomes an important readout to evaluate the efficacy of the protection; (2) assessment of symptoms, which is an empirical measurement that usually correlates with weight loss. Symptoms were assessed by technicians blind to treatment group to decrease the risk of assessment bias; (3) survival, which is the ultimate proof of protection induced by a vaccine formulation. To comply with the animal ethics protocol of our institution, we sacrificed animals that lost more than 20% of their body weight.

In brief, mice immunized with PapMV-SrtA-M2e (30 and 90 µg) and PapMV-sM2e (30 µg) showed minimal weight loss as compared to mice immunized with formulation buffer (Fig. 5c). Clinical signs of infection (ruffled fur, curved back, mobility loss) were significantly reduced in mice immunized with 90 µg of PapMV-SrtA-M2e, although weight loss was not significantly different between mice immunized with PapMV-SrtA-M2e or PapMV-sM2e (Fig. 5c, d). Formulation buffer immunized mice showed severe clinical symptoms and significant weight loss, reaching a mean of 79.3% of their initial weight, at day 8 post-infection. Only groups immunized with 90 µg of PapMV-SrtA-M2e or with PapMV-sM2e showed 100% survival. While 60% of mice immunized with 30 µg of PapMV-SrtA-M2e survived the infection, only 20% of formulation buffer immunized mice survived (Fig. 5e). To confirm this result, we repeated the same experiment with 10 mice per group, evaluated the humoral response after two immunizations (Additional file 5: Figure S5AB), and performed the challenge with 1LD80 of the WSN/33 influenza strain. We monitored weight loss (Additional file 5: Figure S5C), symptoms (Additional file 5: Figure S5D) and survival (Additional file 5: Figure S5E). Consistent with the previous experiment (Fig. 5), the two best performing vaccine formulations were PapMV-sM2e and PapMV-SrtA-M2e, which induced a significant protection to influenza challenge.

Discussion

In this study, we built a 3D model of PapMV CP and PapMV nanoparticles based on the recently published structures of two other members of the potexvirus family: BaMV [39] and PepMV [40]. This model was used to refine engineering of the PapMV vaccine platform. We confirmed experimentally that a fusion of a 17 amino acid peptide harbouring the SrtA recognition motif to the C-terminus of the CP was tolerated, and did not interfere with self-assembly. As anticipated from our PapMV structure prediction, only a fraction of CP subunits in the nanoparticle seem to present the SrtA recognition motif to the sortase, leading to coupling efficiencies that reached a plateau at around 30% for both peptides tested. Coupling probably occurs on the extremities of the nanoparticles, leaving the surface free for interactions with immune cells. In fact, even if the reaction was pushed by increasing the amount of peptide, the amount of coupling did not increase, suggesting that we had reached saturation of all available sites. According to the structural model, the C-termini of the CP units are located in the interior cavity of the nanoparticle, and are thus not available for the coupling reaction. The coupling of peptide antigens to already pre-assembled nanoparticles allowed attachment of longer peptides (26 and 39 amino acids) without affecting the structure of the nanoparticle. Consistent with our model, fusion of a 24-amino-acid peptide to the CP C-terminus led to the formation of unstable nanoparticles that were less immunogenic [20]. With the SrtA approach, we can fuse longer peptides containing multiple epitopes—a major advantage when developing new vaccines.

While SrtA has been widely used for the modification and conjugation of diverse proteins including antibodies, protein–protein, and protein-to-solid-support conjugates, few studies have investigated the modification of nanoparticles by SrtA-mediated conjugation. A recent study showed that SrtA could be used to link FITC or GFP to the N-terminus of cowpea chlorotic mottle virus (CCMV) coat protein expressing an N-terminal glycine residue [28]. Although the authors of the latter study did not use SrtA to functionalize the coat proteins post-assembly, they were able to co-assemble conjugated and unconjugated CCMV into VLPs to encapsulate the cargo. In another study, M13 bacteriophage surface proteins were modified using SrtA from *Staphylococcus aureus* and *Streptococcus pyogenes* [49]. Various proteins and small molecules have been attached to either the pIII,

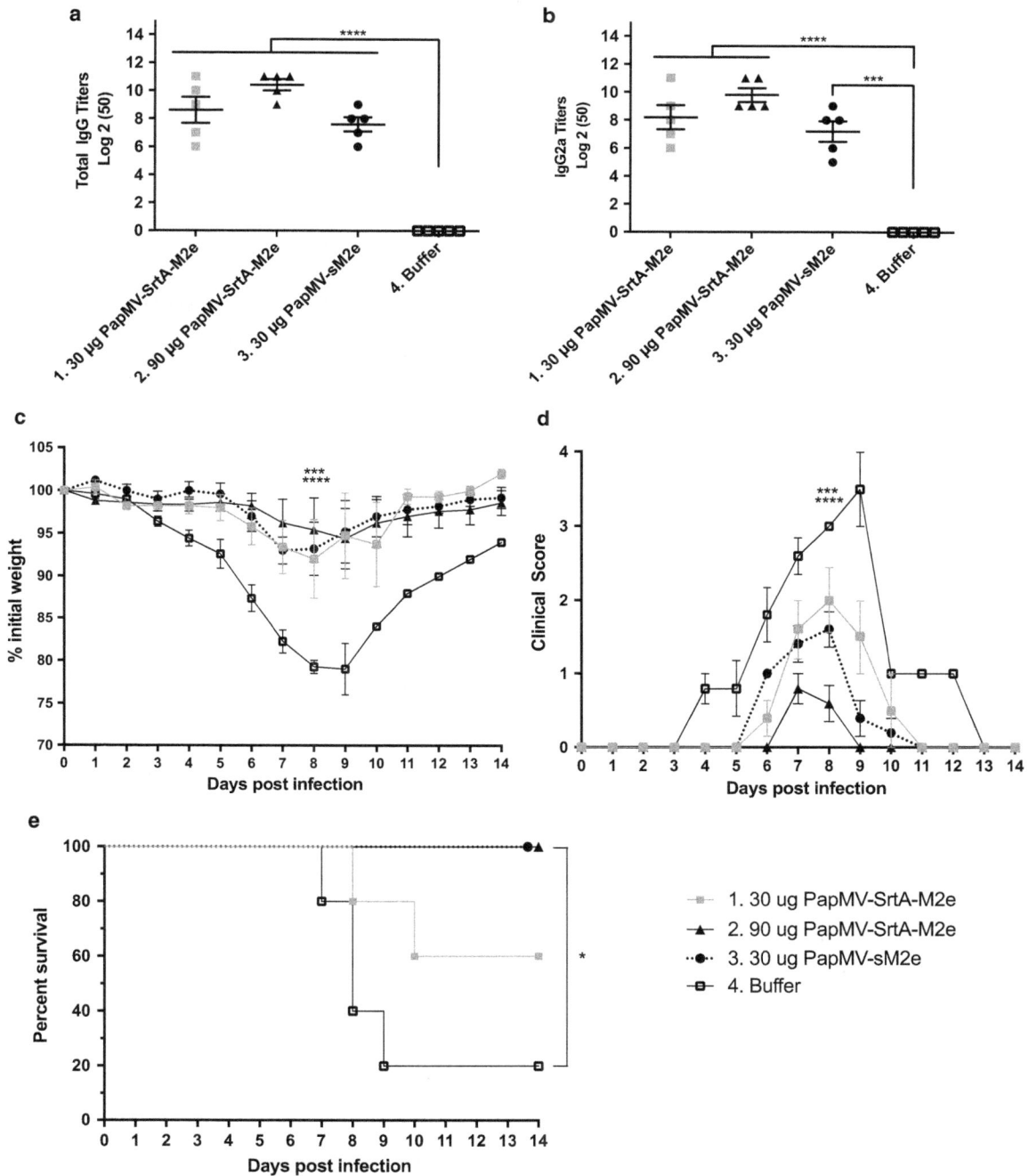

Fig. 5 PapMV-SrtA-M2e induces a specific anti-M2e immune response and protection against an influenza challenge. The M2e peptide has been previously shown to be a highly conserved epitope of influenza A virus that is able to induce protection against an influenza challenge. Therefore, Female Balb/C mice (5 per group) were immunized twice with PapMV nanoparticles coupled to the M2e peptide using the sortase (PapMV-SrtA-M2e), PapMV-sM2e [where a small version of the M2e (sM2e) is fused directly to the N-terminus of the PapMV CP] (positive control), or buffer of the vaccine formulation alone. Blood was taken 13 days following the last immunization, and ELISA assays performed to evaluate levels of anti-M2e total IgG (**a**) or IgG2a (**b**). **P < 0.01 for groups 2 vs 3 total IgG titers and * P< 0.05 for group 2 vs 3 IgG2a titers. To assess the capacity of the PapMV-SrtA-M2e nanoparticles to induce protection to an influenza challenge, immunized mice were infected with 1 × LD80 of influenza A/WSN/33 virus 14 days after the last immunization, and followed for clinical symptoms and survival for 14 days. **c** Mean clinical score of infection signs on a scale of 0 to 4. ***P < 0.001 for group 1 vs 4, and for group 2 vs 3, ****P < 0.0001 for groups 2–3 vs 4, and for group 1 vs 2, all at day 8 post-challenge. **d** Mean weight loss expressed as percentage of initial weight. ***P < 0.001 for group 1 vs 4, and ****P < 0.0001 for groups 2–3 vs 4. **e** Survival of mice expressed as Kaplan–Meier survival curves. *P < 0.5 for groups 2–3 vs 4

pIX or pVIII bacteriophage proteins expressing N-termi-nal glycines but the potential of this approach as a vac-cine platform has not been tested through immunization of animals. Both these studies used SrtA to modify the N-terminus of the CP, while, in contrast, our system con-jugates peptides to the C-terminus of the CP to minimise any modification of the nanoparticle surface that may interfere with the interaction with immune cells. Finally, fusion of the SrtA recognition motif to an internal sur-face-exposed loop of the hepatitis B virus (HBV) core protein led to efficient conjugation of a peptide derived from the enterovirus 71 (EV71) and the AD-4 domain on glycoprotein B (gB) of human cytomegalovirus (HCMV) on HBV icosahedral nanoparticles [50]. These SrtA-mod-ified HBV nanoparticles improved the immune responses directed towards their antigen.

PapMV nanoparticles engineered with the SrtA rec-ognition motif were shown to significantly improve the humoral response triggered towards the attached pep-tides. Although only one-third of the CPs in PapMV-SrtA nanoparticles became coupled to the M2e or T20 peptide, a robust immune response was triggered and PapMV-SrtA-M2e was capable of protecting mice from an influenza challenge. The main differences between the PapMV vaccine platform and the HBV nanoparticle are the shape of the particle and their differential ability to trigger innate immunity: HBV is icosahedral whereas PapMV is rod-shaped. Furthermore, PapMV has the unique capacity to trigger innate immunity through the engagement of TLR7, which results in the induction of IFN-α secretion [15, 17]—a major advantage for the development of an efficient vaccine. The PapMV vaccine platform is also capable of inducing a class switch toward production of IgG2a antibodies, which implies coopera-tion of CD4+ immune cells [51]. IgG2a antibodies con-trol viral infection better than other immunoglobulins due to their higher avidity and their capacity to trigger the antibody-dependent cell-mediated toxicity (ADCC) that is critical to eliminate virus-infected cells with natu-ral killer (NK) cells [52].

Considering the speed at which a candidate vaccine can be developed, and the versatility of this approach, we believe that our SrtA method is advantageous for the presentation of antigens on the PapMV vaccine platform. Sometimes an antigen alone is sufficient to provide pro-tection but, in most cases, this is a very limiting factor.

Conclusion

The covalent attachment of antigens onto PapMV nano-particles using SrtA allows fusion of long peptides (26–39 amino acids) to a vaccine platform. The covalent linkage of the antigen to the platform enhanced significantly the immune response directed to the antigens, and provided

protection to challenge with a viral infection (influenza). This approach is more versatile than fusion of the anti-gen directly to the ORF of the PapMV CP—a route that tolerates fusion of only small peptides. Fusion of longer peptides to the ORF can impair CP structure and its capacity to self-assemble into nanoparticles, and for-mation of nanoparticles is essential to obtain a strong immune response. Thus, the SrtA-based tagging of pep-tide antigens onto already pre-formed PapMV nano-particles allows for fast, flexible and easy nanoparticle modification, representing a promising tool in vaccine design against infectious diseases, as well as in the devel-opment of vaccines in cancer immunotherapy or tailor-made vaccine candidates for personalised medicine [17].

Methods
Production and purification of PapMV nanoparticles
The PapMV nanoparticles harbouring the C-terminal LPETGG SrtA recognition site with (PapMV-SrtA) or without (PapMV-StrA9short) the TSTTR linker, and PapMV-M2e and PapMV nanoparticles were kindly provided by Folia Biotech (Quebec City, Quebec, Can-ada). The nanoparticles were produced and purified as described elsewhere [19]. Briefly, PapMV CPs were pro-duced in E. coli. Purified CPs were assembled in vitro with a non-coding synthetic ssRNA derived from the native sequence of papaya mosaic virus RNA. Levels of contaminant LPS were measured by a Limulus Amebo-cyte Lysate assay (LAL) (Lonza, Walkersville, Maryland, USA). The LPS concentration in the final nanoparticle samples was always below 50 endotoxin units (EU)/mg of protein, and was thus considered negligible. The amount of nanoparticles used to perform each experiment was estimated using the BCA method (Thermo ScientificTM PierceTM BCA Protein Assay), a detergent-compatible formulation based on bicinchoninic acid (BCA) for the colorimetric detection and quantitation of total protein. Since PapMV CP is the main component of the nanopar-ticles (>95% of the total nanoparticle weight), it is a relia-ble method to quantify the amount of nanoparticles used in the experiments conducted in this study.

Expression and purification of recombinant sortase A
Expression and purification of recombinant sortase A followed the method described in [53], with the follow-ing modifications: (1) 2×YT (16 g bacto-tryptone, 10 g yeast extract, 5 g NaCl, pH 7.0) culture medium was used instead of LB medium, (2) IPTG induction was carried out when culture OD_{600nm} reached 1.0 ± 0.1, (3) E. coli was harvested by one 15-min centrifugation at $9000 \times g$ at 4 °C, (4) bacterial pellet from 1 L of culture was sus-pended in 40 mL of cold lysis buffer (50 mM Tris–HCl pH 7.5, 500 mM NaCl, 10 mM imidazole), (5) bacteria

were lysed by three passages through an Emulsiflex C5 homogenizer (Avestin, Ottawa, Ontario, Canada) at 15,000–25,000 psi, (6) 100 U of benzonase (Sigma-Aldrich Inc. Saint-Louis, Missouri, USA) and 5 mM of $MgCl_2$ were added to the cell lysate and incubated for 20 min with gentle agitation before cell lysate clarification by centrifugation at $20,442 \times g$ for 20 min, (7) the clarified supernatant was applied to a 40 mL IMAC column packed with Ni Sepharose 6 Fast Flow Resin (GE Healthcare Life Sciences, Little Chalfont, Buckinghamshire, United Kingdom), and washed with 10 column volumes (CV) of lysis buffer and 5 CV of wash 2 buffer (50 mM Tris–HCl pH 7.5, 150 mM NaCl, 25 mM imidazole). Purification on the IMAC column was done with a ÄKTA purifier 10 FPLC (GE Healthcare Life Sciences). Imidazole was removed using a Sartoflow(R) Slice 200 benchtop crossflow system for diafiltration with a 10 kDa cutoff (Sartorius, Göttingen, Germany). LPS contaminants were removed on an ion exchange membrane bound to quaternary ammonium ligands using the ÄKTA purifier 10, and measured by the LAL assay (Lonza, Basel, Switzerland). LPS levels were under 50 EU/mg of protein and were considered negligible. The yield of SrtA was approximately 100 mg/L of culture.

Peptides

Synthetic peptides, with over 90% purity, used for SrtA labelling, GGG-M2e (GGGSLLTEVETPIRNEWGCRCN DSSD) and GGG-T20 (GGGYTSLIHSLIEESQNQQ EKN EQELLELDKWASLWNWF) were purchased from GenScript (United States) and Biomatik (Canada), respectively.

Evaluation of the structural properties of nanoparticles by dynamic light scattering and transmission electron microscopy

Nanoparticle size was determined by dynamic light scattering (DLS) with a ZetaSizer Nano ZS (Malvern, Malvern, Worcestershire, United Kingdom). PapMV nanoparticles were diluted to 0.12 mg/mL in 10 mM Tris-HCl, pH 8.0, and loaded into a disposable plastic cuvette. Samples were measured four times, and average results generated by cumulative analysis.

Nanoparticle shape was assessed by transmission electron microscopy (TEM). Samples were diluted to 0.02 mg/mL and mixed at a 1:1 ratio with a 3% solution of uranyl acetate for 7 min in the dark; 8 µL of this mixture was then loaded on carbon formvar grids for 5 min in the dark. Excess solution was absorbed and grids were air-dried for at least 2 h before observation. Grids were observed with a FEI-TECNAI-Spirit transmission electron microscope (FEI, Hillsboro, Oregon, USA). DLS and TEM were carried out on the nanoparticles before and after the sortase A transpeptidation reaction to assess the effect of the ligation on the physical properties of the nanoparticles.

Sortase A—mediated transpeptidation reaction

SrtA conjugation reactions were performed in 1X SrtA reaction buffer (50 mM Tris–HCl pH 8.0, 150 mM NaCl, 10 mM $CaCl_2$). Reactions containing 50 µM of SrtA, 50–200 µM of GGG-peptide and 25 µM of different PapMV constructions bearing the SrtA recognition motif LPETGG in nanoparticles were incubated for 2.5 h at room temperature. The molar concentration of nanoparticles was calculated based on the protein concentration of the CP as determined by BCA assay, knowing that 1 kDa = 1 g/mole and 1 µM = 0.000001 mol/L. Reactions were stopped with the addition of EGTA to a final concentration of 10 µM. For the PapMV-SrtA-M2e- and PapMV-SrtA-T20-conjugated samples used in the animal studies, unconjugated peptides and SrtA were removed with an amicon centrifugal filter unit with a cutoff size of 100 kDa (EMD Millipore, Darmstadt, Germany). PapMV-SrtA nanoparticles, being larger than 100 kDa, were retained by the filter unit. Conjugation efficiency was analyzed by 10% Tris-Tricine or 15% Tris–Glycine SDS-PAGE colored with Sypro-Ruby gel stain (Life Technologies Inc. Carlsbad, California, USA). We used 15% Tris–Glycine SDS-PAGE gels for analysis of the conjugation reaction for better resolution of the PapMV-SrtA and PapMV-SrtA coupled to the peptide (Fig. 3; Additional file 2: Figure S2). SrtA reactions seen in Fig. 3 and Additional file 2: Figure S2 were directly analyzed on SDS-PAGE without eliminating the SrtA enzyme excess peptide by amicon 100 kDa.

SDS-PAGE analysis for peptide coupling quantification

Peptide-labelled PapMV-SrtA nanoparticles were diluted to 0.1 µg/µL in cathode buffer for Tris-Tricine gels, or in Tris–Glycine migration buffer supplemented with 30% of SDS loading buffer (50% glycerol, 2% SDS, 0.002% bromophenol blue, 14% 2-mercaptoethanol). Samples were then heated at 95 °C for 10 min and 4 µL of solution (0.4 µg) was loaded on 10% Tris-Tricine SDS-PAGE for the gels shown in Fig. 4 or 15% Tris/Glycine SDS-Page for the gels shown in Fig. 3 and Additional file 2: Figure S2. Gels were colored with Sypro-Ruby gel stain (Life Technologies Inc. Carlsbad, California, USA) following the manufacturer's rapid staining protocol. Sypro-Ruby fluorescence was detected with the 610 nm emission filter following excitation using the green laser (532 nm) of the Typhoon 9200 imager (GE Healthcare Life Sciences). The fluorescence signal resulting from protein bands was quantified using the image analysis software ImageQuant 5.2, and the intensity volumes associated to the different bands, corrected for background signal, were used for the conjugation quantification analysis. The conjugation efficiency was calculated by dividing the signal of coupled PapMV-SrtA protein band by the sum of conjugated and unconjugated PapMV-SrtA signals.

Western blot

After SDS-PAGE migration, proteins were transferred onto PVDF Immobilon®-FL membranes (EMD Millipore, Darmstadt, Germany) and blocked using bløck™-FL buffer (EMD Millipore). Membranes were incubated overnight with the following primary antibodies diluted in the bløck™-FL buffer: rabbit anti-PapMV polyclonal antibody, mouse anti-influenza A M2 (14C2) monoclonal antibody (Santa Cruz Biotechnology Inc. Dallas, Texas, USA) or mouse anti-His$_6$ polyclonal antibody (Roche, Basel, Switzerland). After washing with PBS/0.1% Tween-20, membranes were incubated for 2 h with either Alexa Fluor® 532 coupled goat anti-rabbit IgG (H+L) (Life Technologies Inc. Carlsbad, California, USA) or Alexa Fluor® 488 coupled goat anti-mouse IgG (H+L) (Life Technologies Inc.). After washing, membranes were dried to reduce background, and fluorescent signals were captured using the Typhoon 9200 imager (GE Healthcare Life Sciences).

Animals and immunization

Six- to 8-week-old female BALB/c mice were immunized by intramuscular injection (i.m.) twice, 2 weeks apart, with buffer (50 μL of 10 mM Tris–HCl, pH 8.0) or PapMV-SrtA-M2e, PapMV-sM2e, PapMV-SrtA-T20 or PapMV-SrtA with free T20 peptide. Blood samples were collected before each immunization on days 13 and 27. Serum was separated from the blood by centrifugation in BD Microtainer SST blood collection tubes (BD, East Rutherford, New Jersey, USA) for 2 min at 10,000×g. Total IgG and IgG2a serotype endpoint titers against M2e and T20 peptides in the sera of immunized mice were determined by enzyme linked immunosorbent assay (ELISA) as described below.

Enzyme linked immunosorbent assay

M2e (CSLLTEVETPIRNEWGCRCNDSSD) and T20 (GGGYTSLIHSLIEESQNQQEK NEQELLELDKWA-SLWNWF) peptides were coated overnight at 4 °C in 96-well flat bottom nunc™ MaxiSorp plates (VWR, Radnor, Pennsylvania, USA) at 1 and 2.5 μg/mL, respectively, in 100 μL 0.1 M NaHCO$_3$ buffer, pH 9.6. Plates were blocked with 150 μL per well of PBS/0.1% Tween-20/2% BSA, and washed three times with PBS/0.1% Tween-20, before adding mice sera in twofold serial dilutions starting at 1:50. Plates were incubated for 1.5 h at 37 °C, and washed four times in PBS/0.1% Tween-20 before adding

the peroxidase-conjugated goat anti-mouse IgG or IgG2a secondary antibody (Jackson ImmunoResearch Laboratories, West Grove, Pennsylvania, USA). After washing the plates four times, peptide-specific antibodies were detected with the addition of TMB substrate (Fitzgerald Industries International, Acton, Massachusetts, USA). The reaction was stopped with the addition of 0.18 M H$_2$SO$_4$. Results are expressed as antibody endpoint titers greater than threefold the OD$_{450nm}$ of the background value of the pre-immune sera at the same dilution.

Influenza challenge

Two weeks following the last immunization, immunized mice were infected with 1 LD80 of mouse-adapted influenza A/WSN/33 (H1N1). Mice were lightly anesthetized with isoflurane, and infected with 50 μL of virus by intranasal instillation. Weight loss and clinical signs of disease were monitored for 14 days post-infection. Symptoms were rated from 0 to 4 as follows: (0) no symptoms: (1) lightly spiked fur and curved back; (2) spiked fur and curved back; (3) difficulty moving and slight dehydration; and (4) severe dehydration, lack of reflexes and ocular secretions. Mice reaching a score of 4, or having lost 20% or more of their initial body weight, were euthanized.

Statistical analysis

Differences (antibody titers, weight and symptoms) between immunized groups were measured by Holm–Šídák multiple comparisons test or Tukey's multiple comparisons test. Survival differences were evaluated by Kaplan–Meier survival analysis. P-values <0.05 were considered statistically significant. Statistical analyses were performed using Graph-Pad Prism version 6.0 (GraphPad Software, La Jolla California USA, http://www.graphpad.com).

Additional files

Additional file 1: Figure S1. Engineering of PapMV coat protein carrying the SrtA recognition motif missing the linker. (A) The SrtA recognition motif (LPETGG) of SrtA was inserted into the C-terminus of the PapMV coat protein (CP). (B) The size of the VLPs was assessed by dynamic light scattering (DLS). PapMV-SrtA(short) (65 nm) was showed to be slightly longer than the WT PapMV (54 nm). (C) Transmission electron microscopy (TEM) of WT PapMV (left) and PapMV-SrtA(short) (right) nanoparticles.

Additional file 2: Figure S2. Effect of peptide concentrations on the SrtA coupling reaction. (A) Comparison of the C-terminal sequence of PapMV WT, PapMV-SrtA and PapMV-SrtA(short). In PapMV-SrtA, a linker of 5 amino acids (TSTTR) was added before the SrtA recognition motif LPETGG, while in PapMV-SrtA(short) the recongnition motif was included directly in the native PapMV sequence by deleting two proline residues. (B) SDS-PAGE and western blot of SrtA reactions on PapMV-SrtA and PapMV-SrtA(short) nanoparticles. PapMV nanoparticles (25 μM) were incubated with SrtA (50 μM) and GGG-M2e peptide (50 μM) for 2.5 hours at room temperature. Reactions were stopped with EGTA (10 μM) and passed through a 100 kDa centrifugal filter unit to eliminate excess peptide and contaminating SrtA. PapMV nanoparticles retained by the 100 kDa filter unit were diluted to 0.1 µg/µL in migration buffer supplemented with 30% of SDS loading buffer and 4 µL was loaded onto 15% Tris-Glycine SDS-PAGE. PapMV-SrtA, PapMV-SrtA(short) and SrtA controls correspond to lanes 1, 10 and 11, respectively. Lanes 2-5 and lanes 6-9 represent four experimental replicates of SrtA conjugation on PapMV-SrtA or PapMV-SrtA(short), respectively. Efficient SrtA labelling of GGG-M2e peptide onto PapMV nanoparticles was assessed by SDS-PAGE (top panel), and immunoblotting with a specific antibody against the M2 (bottom panel).

Additional file 3: Figure S3. Optimization of the coupling reaction. SDS-PAGE and Western Blots of SrtA reactions in the presence of PapMV-SrtA nanoparticules and increasing concentrations of GGG-M2e peptide. Western blots were directed against the PapMV CP, the 6xH tag, or the M2e peptide, as indicated on the bottom of each panel. Target products are shown by a red dash. SrtA reactions were diluted to obtain a PapMV-SrtA concentration of 0.1µg/µL (based on molar concentration in the reaction) in migration buffer supplemented with 30% of SDS loading buffer, and 4 µL was loaded onto 10% Tris-Tricine SDS-PAGE for analysis.

Additional file 4: Figure S4. PapMV-SrtA-M2e induces a specific anti-M2e immune response after a single immunization. Female Balb/C mice, 5 per group, were immunized twice with the indicated formulations. Mice were bled 13 days after the first immunization, and levels of anti-M2e total IgG (A) and IgG2a (B) were measured by ELISA. ****$P<0.0001$ for groups 1, 2, 3 vs 4 for total IgG and IgG2a titers.

Additional file 5: Figure S5. PapMV-SrtA-M2e induces a specific anti-M2e immune response and protection against an influenza challenge-(repeat). Female Balb/C mice (10 per group) were immunized twice with 30 µg PapMV-SrtA-M2e, 90 µg PapMV-SrtA-M2e, 30 µg PapMV-sM2e or formulation buffer. At 13 days following the last immunization, mice were bled and ELISA assays performed to evaluate levels of anti-M2e total IgG (A) or IgG2a (B). ****$P<0.0001$ for groups 1, 2, 3 vs 4 total IgG titers and IgG2a titers. Mice were infected with 1 x LD80 of influenza A/WSN/33 virus 14 days after the last immunization, and followed for clinical symptoms and survival for 14 days. (C) Mean weight loss expressed as percentage of initial weight. **$P<0.01$ for group 1 vs 2, ***$P<0.001$ for group 3 vs 4, and for group 2 vs 3, ****$P<0.0001$ for group 2 vs 4, all at day 7 post-challenge. (D) Mean clinical score of infection signs on a scale of 0 to 4. ***$P<0.001$ for group 1 vs 3, and ****$P<0.0001$ for group 1 vs 2 and groups 2, 3 vs 4. (E) Survival of mice expressed as Kaplan-Meier survival curves. *$P<0.5$ for groups 2 vs 4 and **$P<0.01$ for group 3 vs 4.

Authors' contributions

AT performed the final PapMV-SrtA construct screening, the coupling experiments on the PapMV platform, and was involved with experiments with animals. AT also drafted the manuscript. MB performed the structural model of PapMV and helped draft the manuscript. DC and GR helped perform mouse experiments, and to draft the manuscript. LGL contributed to the initial screening of PapMV constructs harbouring the SrtA recognition motif. MB, MELG and PS were responsible for production of batches of PapMV nanoparticles used in this work. DL supervised the entire study, coordinated the involvement of each lab member and finalised the manuscript. All authors read and approved the final manuscript.

Author details
[1] Department of Microbiology, Infectiology and Immunology, Infectious Disease Research Center, Laval University, 2705 Boul. Laurier, Quebec City, PQ G1V 4G2, Canada. [2] Neurosciences, Laval University, 2705 Boul. Laurier, Québec City, PQ G1V 4G2, Canada.

Acknowledgements
We would like to thank the "Plateforme de bio-imagerie du Centre de Recherche en Infectiologie" for the use of the transmission electron microscope. Also, we would like to thank Helen Rothnie for her help in editing the manuscript.

Competing interests
Authors Denis Leclerc and Pierre Savard are shareholders of the company FOLIA BIOTECH INC., a start-up company that has the mandate to exploit this technology commercially to improve currently available vaccines and create new vaccines. This does not alter the authors' adherence to all the journal policies.

Funding
We would like to thank the Natural Sciences and Engineering Research Council of Canada (NSERC Grant RGPIN-2016-05852) and the Canadian Institute of Health Research (CIHR Grant 298143) for funding this research program.

References
1. Chroboczek J, Szurgot I, Szolajska E. Virus-like particles as vaccine. Acta Biochim Pol. 2014;61(3):531–9.
2. Kushnir N, Streatfield SJ, Yusibov V. Virus-like particles as a highly efficient vaccine platform: diversity of targets and production systems and advances in clinical development. Vaccine. 2012;31(1):58–83.
3. Lee KL, Twyman RM, Fiering S, Steinmetz NF. Virus-based nanoparticles as platform technologies for modern vaccines. WIREs Nanomed Nanobiotechnol. 2016;8(4):554–78.
4. Plummer EM, Manchester M. Viral nanoparticles and virus-like particles: platforms for contemporary vaccine design. WIREs Nanomed Nanobiotechnol. 2011;3(2):174–96.
5. Lua LHL, Connors NK, Sainsbury F, Chuan YP, Wibowo N, Middelberg APJ. Bioengineering virus-like particles as vaccines. Biotechnol Bioeng. 2014;111(3):425–40.
6. Noad R, Roy P. Virus-like particles as immunogens. Trends Microbiol. 2003;11(9):438–44.
7. Rodríguez-Limas WA, Sekar K, Tyo KE. Virus-like particles: the future of microbial factories and cell-free systems as platforms for vaccine development. Curr Opin Biotechnol. 2013;24(6):1089–93.
8. Zeltins A. Construction and characterization of virus-like particles. Mol Biotechnol. 2013;53(1):92–107.
9. Zhao L, Seth A, Wibowo N, Zhao CX, Mitter N, Yu C, et al. Nanoparticle vaccines. Vaccine. 2014;32(3):327–37.
10. Roldão A, Mellado MCM, Castilho LR, Carrondo MJT, Alves PM. Virus-like particles in vaccine development. Expert Rev Vaccines. 2010;9(10):1149–76.
11. Fifis T, Gamvrellis A, Crimeen-Irwin B, Pietersz GA, Li J, Mottram PL, et al. Size-dependent immunogenicity: therapeutic and protective properties of nano-vaccines against tumors. J Immunol. 2004;173(5):3148–54.
12. Manolova V, Flace A, Bauer M, Schwarz K, Saudan P, Bachmann MF. Nanoparticles target distinct dendritic cell populations according to their size. Eur J Immunol. 2008;38(5):1404–13.
13. Lebel M-E, Chartrand K, Leclerc D, Lamarre A. Plant viruses as nanoparticle-based vaccines and adjuvants. Vaccines. 2015;3(3):620–37.
14. Mathieu C, Rioux G, Dumas MCC, Leclerc D. Induction of innate immunity in lungs with virus-like nanoparticles leads to protection against influenza and Streptococcus pneumoniae challenge. Nanomedicine. 2013;9(7):839–48.
15. Lebel M-E, Daudelin J-F, Chartrand K, Tarrab E, Kalinke U, Savard P, Leclerc D, Lamarre A. Nanoparticle adjuvant sensing by TLR7 enhances CD8+ T cell-mediated protection from Listeria monocytogenes infection. J Immunol. 2014;192(3):1071–8.
16. Savard C, Guérin A, Drouin K, Bolduc M, Laliberté-Gagné M-E, Dumas M-C, Majeau N, Leclerc D. Improvement of the trivalent inactivated flu vaccine using PapMV nanoparticles. PLoS ONE. 2011;6(6):e21522.
17. Lebel M-E, Chartrand K, Tarrab E, Savard P, Leclerc D, Lamarre A. Potentiating cancer immunotherapy using papaya mosaic virus-derived nanoparticles. Nano Lett. 2016;16(3):1826–32.
18. Leclerc D. Plant viral epitope display systems for vaccine development. Curr Top Microbiol Immunol. 2014;375:47–59.
19. Carignan D, Thérien A, Rioux G, Paquet G, Gagné MÈL, Bolduc M, Savard P, Leclerc D. Engineering of the PapMV vaccine platform with a shortened M2e peptide leads to an effective one dose influenza vaccine. Vaccine. 2015;33(51):7245–53.

20. Rioux G, Babin C, Majeau N, Leclerc D. Engineering of papaya mosaic virus (papmv) nanoparticles through fusion of the HA11 peptide to several putative surface-exposed sites. PLoS ONE. 2012;7(2):5–12.

21. Babin C, Majeau N, Leclerc D. Engineering of papaya mosaic virus PapMV nanoparticles with a CTL epitope derived from influenza NP. J Nanobiotechnol. 2013;11:10.

22. Leclerc D, Beauseigle D, Denis J, Morin H, Pare C, Lamarre A, Lapointe R. Proteasome-independent MHC class I cross-presentation mediated by papaya mosaic virus-like particles leads to the expansion of specific human T cells. J Virol. 2007;81(3):1319–26.

23. Lacasse P, Denis J, Lapointe R, Leclerc D, Lamarre A. Novel plant virus-based vaccine induces protective CTL-mediated antiviral immunity through dendritic cell maturation. J Virol. 2008;82(2):785–94.

24. Hanafi LA, Bolduc M, Laliberté-Gagné ME, Dufour F, Langelier Y, Boulassel MR, Routy JP, Leclerc D, Lapointe R. Two distinct chimeric potexviruses share antigenic cross-presentation properties of MHC class I epitopes. Vaccine. 2010;28(34):5617–26.

25. Clancy KW, Melvin J, McCafferty DG. Sortase transpeptidases: insights into mechanism, substrate specificity, and inhibition. Biopolymers. 2010;94(4):385–96.

26. Ritzefeld M. Sortagging: a robust and efficient chemoenzymatic ligation strategy. Chemistry. 2014;20(28):8516–29.

27. Spirig T, Weiner EM, Clubb RT. Sortase enzymes in Gram-positive bacteria. Mol Microbiol. 2011;82(5):1044–59.

28. Schoonen L, Pille J, Borrmann A, Nolte RJ, van Hest JC. Sortase A-mediated N-terminal modification of cowpea chlorotic mottle virus for highly efficient cargo loading. Bioconjug Chem. 2015;26(12):2429–34.

29. Jiang R, Wang L, Weingart J, Sun XLL. Chemoenzymatic bio-orthogonal chemistry for site-specific double modification of recombinant thrombomodulin. ChemBioChem. 2014;15(1):42–6.

30. Matsumoto T, Sawamoto S, Sakamoto T, Tanaka T, Fukuda H, Kondo A. Site-specific tetrameric streptavidin-protein conjugation using sortase A. J Biotechnol. 2011;152(1–2):37–42.

31. Parthasarathy R, Subramanian S, Boder ET. Sortase A as a novel molecular "stapler" for sequence-specific protein conjugation. Bioconjug Chem. 2007;18(2):469–76.

32. Steinhagen M, Zunker K, Nordsieck K, Beck-Sickinger AG. Large scale modification of biomolecules using immobilized sortase A from Staphylococcus aureus. Bioorganic amp Med Chem. 2013;21(12):3504–10.

33. Sijbrandij T, Cukkemane N, Nazmi K, Veerman EC, Bikker FJ. Sortase A as a tool to functionalize surfaces. Bioconjug Chem. 2013;24(5):828–31.

34. Pritz S, Wolf Y, Kraetke O, Klose J, Bienert M, Beyermann M. Synthesis of biologically active peptide nucleic acid—peptide conjugates by sortase-mediated ligation. J Org Chem. 2007;72(10):3909–12.

35. Antos JM, Miller GM, Grotenbreg GM, Ploegh HL. Lipid modification of proteins through sortase-catalyzed transpeptidation. J Am Chem Soc. 2008;130(48):16338–43.

36. Samantaray S, Marathe U, Dasgupta S, Nandicoori VK, Roy RP. Peptide-sugar ligation catalyzed by transpeptidase sortase: a facile approach to neoglycoconjugate synthesis. J Am Chem Soc. 2008;130(7):2132–3.

37. Levary DA, Parthasarathy R, Boder ET, Ackerman ME. Protein-protein fusion catalyzed by sortase A. PLoS ONE. 2011;6(4):e18342.

38. Madej MP, Coia G, Williams CC, Caine JM, Pearce LA, Attwood R, et al. Engineering of an anti-epidermal growth factor receptor antibody to single chain format and labeling by sortase A-mediated protein ligation. Biotechnol Bioeng. 2012;109(6):1461–70.

39. DiMaio F, Chen C-C, Yu X, Frenz B, Hsu Y-H, Lin N-S, et al. The molecular basis for flexibility in the flexible filamentous plant viruses. Nat Struct Mol Biol. 2015;22(8):642–4.

40. Agirrezabala X, Méndez-López E, Lasso G, Sánchez-Pina MA, Aranda M, Valle M. The near-atomic cryoEM structure of a flexible filamentous plant virus shows homology of its coat protein with nucleoproteins of animal viruses. Elife. 2015;4:e11795.

41. Yang S, Wang T, Bohon J, Gagné ME, Bolduc M, Leclerc D, Li H. Crystal structure of the coat protein of the flexible filamentous papaya mosaic virus. J Mol Biol. 2012;422(2):263–73.

42. Yang J, Yan R, Roy A, Xu D, Poisson J, Zhang Y. The I-TASSER Suite: protein structure and function prediction. Nat Methods. 2015;12(1):7–8.

43. Neirynck S, Deroo T, Saelens X, Vanlandschoot P, Jou WM, Fiers W. A universal influenza A vaccine based on the extracellular domain of the M2 protein. Nat Med. 1999;5(10):1157–63.

44. Denis J, Acosta-Ramirez E, Zhao Y, Hamelin ME, Koukavica I, Baz M, et al. Development of a universal influenza A vaccine based on the M2e peptide fused to the papaya mosaic virus (PapMV) vaccine platform. Vaccine. 2008;26(27–28):3395–403.

45. He Y. Synthesized peptide inhibitors of HIV-1 gp41-dependent membrane fusion. Curr Pharm Des. 2013;19(10):1800–9.

46. Serrano S, Araujo A, Apellániz B, Bryson S, Carravilla P, De La Arada I. Structure and immunogenicity of a peptide vaccine, including the complete HIV-1 gp41 2F5 epitope: implications for antibody recognition mechanism and immunogen design. J Biol Chem. 2014;289(10):6565–80.

47. Rioux G, Mathieu C, Russell A, Bolduc M, Laliberté-Gagné M-E, Savard P, et al. PapMV nanoparticles improve mucosal immune responses to the trivalent inactivated flu vaccine. J Nanobiotechnol. 2014;12:19.

48. Denis J, Majeau N, Elizabeth A-R, Savard C, Bedard MCC, Simard S, et al. Immunogenicity of papaya mosaic virus-like particles fused to a hepatitis C virus epitope: evidence for the critical function of multimerization. Virology. 2007;363(1):59–68.

49. Hess GT, Cragnolini JJ, Popp MW, Allen MA, Dougan SK, Spooner E, et al. M13 bacteriophage display framework that allows sortase-mediated modification of surface-accessible phage proteins. Bioconjug Chem. 2012;23(7):1478–87.

50. Tang S, Xuan B, Ye X, Huang Z, Qian Z. A modular vaccine development platform based on sortase-mediated site-specific tagging of antigens onto virus-like particles. Sci Rep. 2016;12(6):25741.

51. Gamvrellis A, Leong D, Hanley JC, Xiang SD, Mottram P, Plebanski M. Vaccines that facilitate antigen entry into dendritic cells. Immunol Cell Biol. 2004;82(5):506–16.

52. Jegerlehner A, Schmitz N, Storni T, Bachmann MF. Influenza A vaccine based on the extracellular domain of M2: weak protection mediated via antibody-dependent NK cell activity. J Immunol. 2004;172(9):5598–605.

53. Popp MW, Antos JM, Ploegh HL. Site-specific protein labeling via sortase-mediated transpeptidation. Curr Protoc Protein Sci. 2009;15:Unit 15.3.

3

Real-time, label-free monitoring of cell viability based on cell adhesion measurements with an atomic force microscope

Fang Yang[1], René Riedel[1], Pablo del Pino[2], Beatriz Pelaz[2], Alaa Hassan Said[2], Mahmoud Soliman[2], Shashank R. Pinnapireddy[3], Neus Feliu[2], Wolfgang J. Parak[2,4], Udo Bakowsky[3] and Norbert Hampp[1,5*]

Abstract

Background: The adhesion of cells to an oscillating cantilever sensitively influences the oscillation amplitude at a given frequency. Even early stages of cytotoxicity cause a change in the viscosity of the cell membrane and morphology, both affecting their adhesion to the cantilever. We present a generally applicable method for real-time, label free monitoring and fast-screening technique to assess early stages of cytotoxicity recorded in terms of loss of cell adhesion.

Results: We present data taken from gold nanoparticles of different sizes and surface coatings as well as some reference substances like ethanol, cadmium chloride, and staurosporine. Measurements were recorded with two different cell lines, HeLa and MCF7 cells. The results obtained from gold nanoparticles confirm earlier findings and attest the easiness and effectiveness of the method.

Conclusions: The reported method allows to easily adapt virtually every AFM to screen and assess toxicity of compounds in terms of cell adhesion with little modifications as long as a flow cell is available. The sensitivity of the method is good enough indicating that even single cell analysis seems possible.

Keywords: AFM, Cell adhesion, Fast-screening measurement, Gold nanoparticles, Cytotoxicity, Cell viability

Background

The cell membrane is more than just a passive lipid bilayer barrier. Of special relevance, cell membrane proteins are an integral part of the cellular machinery concerning sensing and reacting to what surrounds the cell, through different processes such as signaling, transport and immune response. In particular, cell adhesion molecules and their main function, i.e., cell adhesion, are of prime importance on cell biology and medicine, being a key player on several biological processes such as tumor invasion and metastasis [1], stem-cell fate [2] and cell death and/or growth arrest [3]. Cell detachment, or loss of anchorage in adhesive cells, is a common marker of cell death [4], which could be monitored as a sign of cytotoxicity. For instance, intracellular signals caused by the intracellular accumulation of exogenic agents (e.g. toxins, drugs, nanoparticles, etc.) at toxic concentrations can in general cause cell detachment [5], followed by cell death.

In order to evaluate the safety of a new agent, variety of in vitro cell-based assays is often employed. One feasible strategy to evaluate the potential toxic effects of an unknown compound will be, in the first stage, to evaluate basal cytotoxicity (by using for instance screening assays), and second assess the specific types of toxicity [6] (i.e. to understand the cause cell injury). There are several cell-based assays used to evaluate cytotoxicity, including methods to monitor the function of organelles, cell viability, to track cellular components, etc. Cell viability assays are among the most frequently used methods in all form of cell cultures [7]. There are a variety of cell viability assays that could be used to monitor enzymatic

*Correspondence: hampp@uni-marburg.de
[5] Material Science Center, University of Marburg, Marburg, Germany
Full list of author information is available at the end of the article

activities or general metabolism, some of those assay include the resazurin and tetrazolium reduction, as well as protease activity methods [8].

Most frequently employed standard cytotoxicity methods to assess cell death, including cell viability and proliferation assays, rely on extrinsic labeling or reporter agents which, once internalized, interact with specific cell components providing a signal, typically colorimetric, fluorescent, or bioluminescent. The measured signal can be then related to different cellular parameters that are evaluated and associated in terms of cell viability, such as the activity of mitochondrial enzymes, for instance the succinate dehydrogenase, the intactness of cell membranes, adenosine triphosphate production, etc. [9]. The major limitation of these in vitro methods to evaluate cytotoxicity is that they may be affected by interferences between the compounds and the read-out signal. As example, metallic nanoparticles (NPs) may interact specifically or non-specifically with the reagent or substrate of the assay [10, 11]. Fluorescent NPs may cause crosstalk with fluorescence read-out of the assay. Furthermore, some of the conventional toxicity methodologies are single endpoint assays, i.e., fail to provide real-time continuous monitoring of cell viability, as the assay itself interferes with cell viability [12]. As an alternative to the classic cytotoxicity methods, electrode-impedance-based methods have emerged as a powerful label-free analytical tool to assess cell characteristics [13, 14], including cell viability [15], adhesion, cycle, metastasis, migration, and invasion.

Mass sensors based on micro- and nanomechanical resonators represent a class of ultra-sensitive sensors with enormous potential in the biomedical field [16], with the capability of weighing single cells and single nanoparticles in fluids [17]. Mechanical biosensors have been widely used for ultrasensitive detection of pathogens [18], and also some work has attempted to dynamically inspect living cells [19–24]. There is also some recent work which addresses dynamic (>1 h) qualification of cell viability by a micromechanical mass sensor [25].

Here we report on a micromechanical mass-sensing platform for label-free continuous monitoring (4–5 h) of intoxication in terms of loss in cell adhesion by using the oscillating cantilever of an atomic force microscope (AFM) as probe (more details see in the Additional file 1). AFM is a powerful tool to measure very small forces between a cantilever tip and a surface on the nanoscale, even if the surface to be inspected is soft and submerged in a liquid, e.g., cells in solution. With AFM binding forces between two molecules [26], adhesion of molecules to surfaces [27], adhesion of cells to surfaces [28], or cell to cell adhesion [29] can be recorded. As AFM also allows for lateral resolution also local properties of cell

surfaces can be raster-scanned, such as topography [30], localization of adhesion sites [31], local electro-mechanical signaling [21], or local viscoelastic properties [32, 33]. In the following, an assay will be described, in which cell detachment from the cantilever of an AFM is recorded. Hereby the loss of cell adhesion upon cellular exposure to toxic agents, e.g., NPs or chemicals, is monitored.

In general, normal cells could initiate cell death when lost of cell attachment/contact to the extracellular matrix (ECM) occur [3, 34]. Indeed, it is known that disruption of cell adhesion and cell–matrix interactions with successive detachment of cells may be related to signs of cell death [35]. The importance relays on the fact that cell attachment to ECM plays a central role in cell physiology for instance in cell morphology, proliferation, motility among others [36]. Therefore, in the present study we presented a complementary method to monitor loss of cell detachment, a fast-screening-technique to assess dose dependent toxicity using AFM based methodology. The results obtained with this methodology could possibly be associated to early sign of cell death (before cell death is perceptible). To demonstrate the feasibility of the approach proposed, as control, the results obtained from the cantilever were associated with the results obtained with a common conventional cell viability test, the resazurin assay.

Results and discussion

In our method, a triangular cantilever (SNL-10, $k = 0.12$ N/m, $f_0 = 23$ kHz, Bruker Co) is mounted in a chamber with controlled equilibrated temperature, which can be flushed with different solutions (e.g., NPs or chemical agents in different media and concentrations). Optical images of the cantilever at individual steps of the experiment are shown in Fig. 1. The injection system is schematically shown in Fig. 2a. For a given frequency the cantilever amplitude is highly dependent on the mass of the cantilever or, in our case, on the mass of the cantilever with cells attached (details see in Additional file 1). Because cells attached to different position on the cantilever can have different impact on the deflection, finite elements model can be used for extending the theoretical results of the triangular cantilever [37] (description of finite element model about triangular cantilever in Additional file 1) and in order to control the eventual variations due to cells, the cantilever preparation and characterization are performed for each experiment, i.e., calculation of spring constant before and after experiment, determination of the resonance frequency and deflection sensitivity, to identify that the variance of the deflection is the most appropriate means of analyzing and comparing the data from the different experiments. Figure 2b schematically depicts the method by showing the

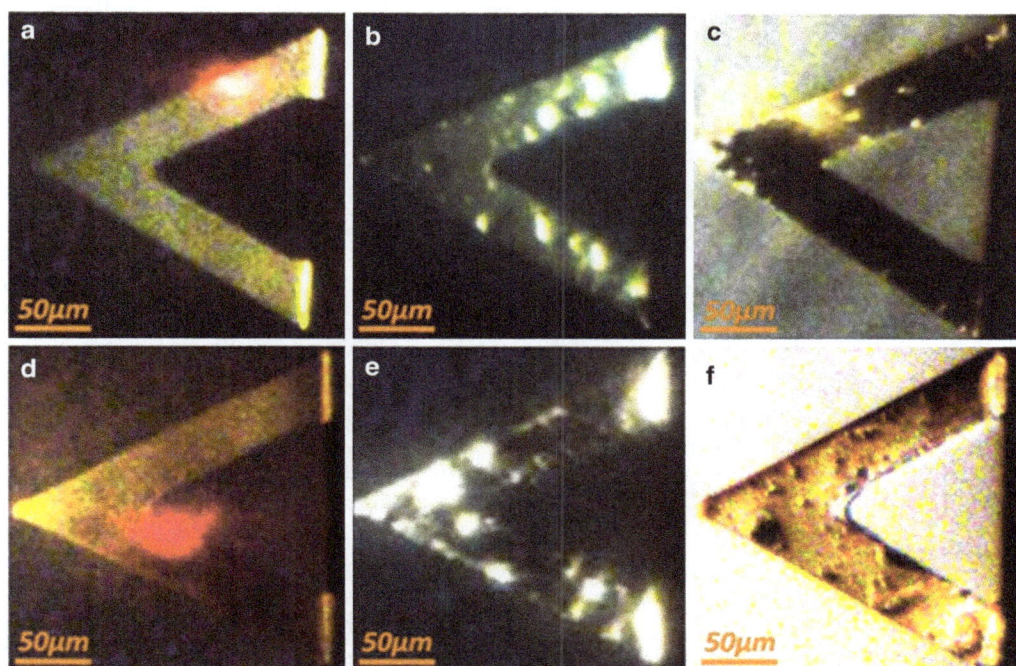

Fig. 1 Optical images of one triangular cantilever. **a** The cantilever is oscillating in air before measurement and the surface is found to be clean and flat. **b** The cantilever is oscillating in solution with cells adsorbed. **c** After washing the cantilever with ethanol and rinsing it with PBS most of the absorbed cells desorbed from the cantilever. **d** The cantilever in water is shown before measurement. **e** After injection of Au NPs the cantilever is oscillating in cell medium (DMEM-HG medium, Capricorn Scientific, Ebsdorfergrund) supplemented with 10% fetal bovine serum (Sigma Aldrich). **f** The cantilever is oscillating with absorbed cells and Au NPs in air after measurement without washing

successive steps through which the cantilever's dynamic deflection was recorded: (1) The readily mounted cantilever started oscillating in air and then flooded with cell medium, meanwhile the deflection will remain at the initial level. (2) A cell suspension (120 μL of a solution of human cervical cancer HeLa cells at 10^5 cells/mL) was injected into the sample chamber and was left for ca. 1 h to allow the cells to sediment and eventually attach to the surface of the cantilever [38]. During this time, the deflection amplitude increased due to the added cell mass. (3) In order to study the effect of chemical agents or NPs on cell adhesion, cells were exposed to these agents/NPs at different concentrations. Upon impairment of cells by these substances, cells could lose contact to the AFM cantilever, and the effects on cell adhesion could be monitored and evaluated. Cell detachment is visible as change of mass of the cantilever-cell system. (4) Finally, the cell is flushed with 70% EtOH and PBS buffer to remove all cells and prepare the system for the next measurement. After the rinsing step, the cantilever is optically inspected to confirm that no rest from the previous experiment were present, which is also confirmed by the reset of the real-time deflection to the initial equilibrium (cf., the deflection of pure medium and PBS in Additional file 1).

As a proof of concept, cells were exposed to differently sized and coated gold nanoparticles (Au NPs), as well as other toxic agents, such as ethanol (70%), $CdCl_2$, and staurosporine (STS) as a common agent typically used to trigger apoptosis [39]. The effects on cell adhesion upon exposure to NPs and chemical agents at different concentration and time points, were evaluated using the above described setup. Au NPs were used as a NP model in this study because they are interesting materials for biomedical application [40] and thus their biocompatibility needs to be further studied. For instance, it has been reported that the metallic surface of Au NPs could trigger catalytic reactions and cause generation of reactive oxygen species (ROS) [41]. Generation of ROS could damage cellular process and induce ROS associate-toxicity through several mechanisms. Indeed, oxidative stress induced for NPs have been shown. Those cell responses need to be taken in consideration when evaluating possible cytotoxicity effects induced by NPs [42, 43].

In the present study, the effects of three different types of Au NPs with different surface coating and size were evaluated. In general, parameters such as the organic coating around the NPs (e.g., intended as result from synthetic surface modification, or non-intended as result

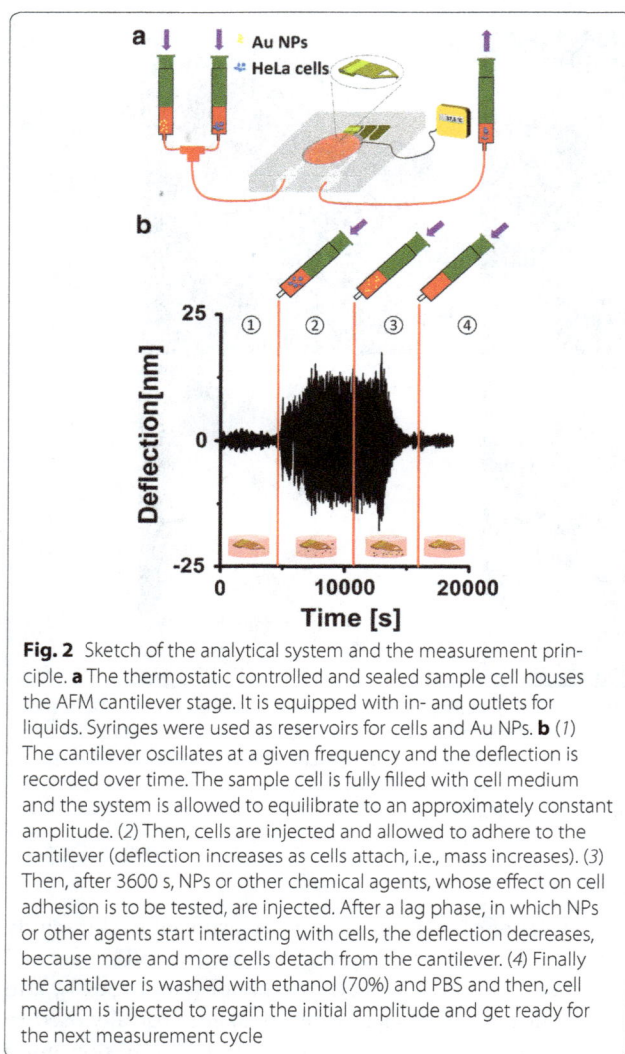

Fig. 2 Sketch of the analytical system and the measurement principle. **a** The thermostatic controlled and sealed sample cell houses the AFM cantilever stage. It is equipped with in- and outlets for liquids. Syringes were used as reservoirs for cells and Au NPs. **b** (*1*) The cantilever oscillates at a given frequency and the deflection is recorded over time. The sample cell is fully filled with cell medium and the system is allowed to equilibrate to an approximately constant amplitude. (*2*) Then, cells are injected and allowed to adhere to the cantilever (deflection increases as cells attach, i.e., mass increases). (*3*) Then, after 3600 s, NPs or other chemical agents, whose effect on cell adhesion is to be tested, are injected. After a lag phase, in which NPs or other agents start interacting with cells, the deflection decreases, because more and more cells detach from the cantilever. (*4*) Finally the cantilever is washed with ethanol (70%) and PBS and then, cell medium is injected to regain the initial amplitude and get ready for the next measurement cycle

from the absorption of macromolecules from the cell media), size, shape, dose, among others, will affect the impact of the NPs on cell function and morphology, typically by impairing metabolic activity, mitochondrial function, as well shaping the degree and pathway(s) of NP internalization by cells [44].

In the present study we used Au NP suspension having varying NP concentrations (3–400 nM). In Fig. 2b, the AFM data after injection of 50 nM of Au NPs into the sealed and temperature-controlled (37.5 °C) sample chamber is shown. Generally, Au NPs are internalized by cells by different mechanisms, one of the most common pathway is endocytosis [45]. After a lag-phase of ca. 1 h, a time that is typically sufficient for internalization of some Au NPs, a diminishing dynamic amplitude in the AFM signal was observed, resulting in loss of cell adhesion which we ascribed to onset of cytotoxicity. In fact, upon

exposure of cells to a potentially dose-dependent toxic agent, cells may change their adhesion properties and be gradually detached from the oscillating cantilever, which would be accompanied by the decrease of the cantilever amplitude, and in this manner recorded. In order to regenerate the cantilever in situ 150 mL of a solution of ethanol (70%) and PBS buffer were injected, respectively. 70% ethanol is known to kill cells. PBS then washed the remaining cell debris away, thus clearing the cantilever, and reduced the amplitude of the cantilever oscillation to the initial value. This process was repeated twice, so that the cantilever will be clearly rinsed and then more measurements may be accomplished during a single session with cells from the same batch. Just before the next measurement and in particular, before cells were injected to the measuring chamber, cell medium was injected again, in order to keep the chamber in conditions suitable for cell culture. The current deflection accompanied with optical images is used for checking the state of the cantilever and chamber.

For a more detailed and comprehensive evaluation of the presented method, the effects of three different types of Au NPs on cells were investigated (details can be found in "Methods"): (i) Au(5)-PMA, i.e., Au NPs having a core diameter of 5 nm which are grafted with poly(isobutylene-alt-maleic anhydride) dodecylamine (in the following referred to as PMA); (ii) Au(13)-PMA, i.e., Au NPs having a core diameter of 13 nm coated with PMA [46]; (iii) Au(13)-PEG, i.e., Au NPs having a core diameter of 13 nm coated with polyethylene glycol (PEG) [47]. Unless otherwise specified a concentration range from 3 to 400 nM (in terms of NP concentration) was tested. For comparison, other common toxic agents were used, such us ethanol (70%), $CdCl_2$, and STS (3 nM–1 μM). Two different cells lines were used for those studies, the human cervical cancer cells (HeLa) and the breast adenocarcinoma cells (MCF7). The dynamic effects on cells caused by the NPs or the chemical agents were monitored by the deflection versus time curves shown in Fig. 3a, b. The agents were injected to the cantilever at different concentrations. Then, after about 1 h exposure, the measurement started (indicated by the red line). After a lag phase, which depended on the agent used, as well as on the dose, cells started to detach (blue line), as indicated by the diminishing deflection amplitude. Notice, that there are no significant changes before NPs have been added, i.e., during the 1 h prior to injection of the agents. After that, the oscillation shows an exponential attenuation, described by a damping coefficient, here referred to as damping value B. The B value is thus an indication for the amplitude damping rate, that is, the damping increases with an increase of B (details about derivation in Additional file 1). The B values extracted

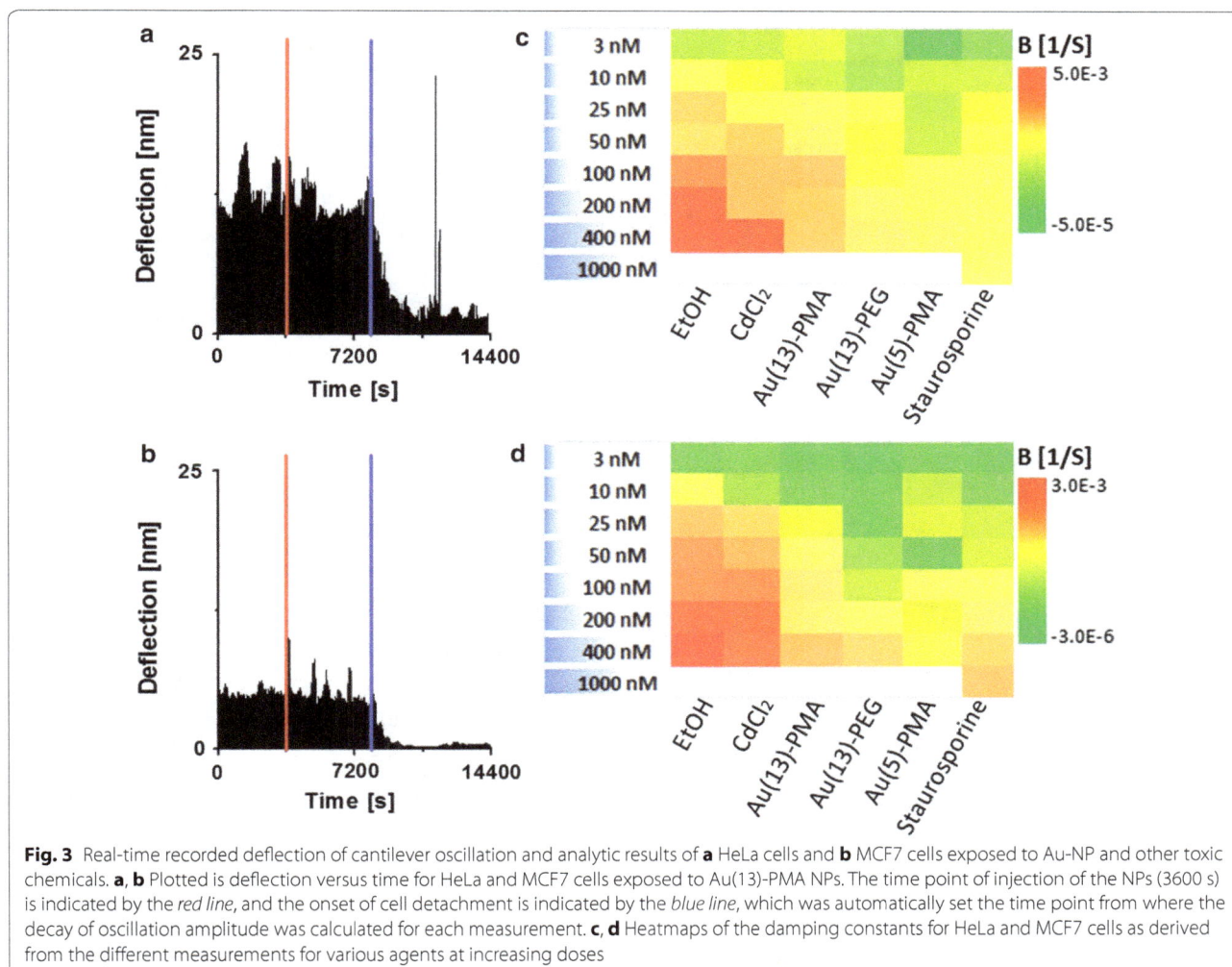

Fig. 3 Real-time recorded deflection of cantilever oscillation and analytic results of **a** HeLa cells and **b** MCF7 cells exposed to Au-NP and other toxic chemicals. **a**, **b** Plotted is deflection versus time for HeLa and MCF7 cells exposed to Au(13)-PMA NPs. The time point of injection of the NPs (3600 s) is indicated by the *red line*, and the onset of cell detachment is indicated by the *blue line*, which was automatically set the time point from where the decay of oscillation amplitude was calculated for each measurement. **c**, **d** Heatmaps of the damping constants for HeLa and MCF7 cells as derived from the different measurements for various agents at increasing doses

from the amplitude decay caused by detachment of cells in each case were quantitatively calculated by a home-made program (details in Additional file 1) (cf., Fig. 3a, b). All original data showing the whole dynamic process for different agents, times and doses are shown in the Additional file 1. Taken these data together, the different B values are condensed into the heatmaps shown in Fig. 3c for HeLa cells and Fig. 3d for MCF7 cells. The following results can be extracted from these heatmaps: (1) In case the same PMA-coating is used, bigger NPs (i.e. diameter of inorganic core of 5 vs. 13 nm), at the same NP concentration, induce a faster onset of cell detachment. (2) In case the NPs had the same diameter of inorganic core (13 nm) and similar surface charge, but different organic coatings (PMA versus PEG) were used, cell detachment is less pronounced for the PEG-coated NPs, probably due to less efficient internalization, as expected from such coatings [48]. (3) Ethanol (necrosis-trigger agent), $CdCl_2$, and staurosporine (apoptosis-trigger agent) were used as

references in order to underline the general applicability of the method, and to demonstrate that it is not limited to detecting cell detachment due to presence of NPs. As expected, ethanol and $CdCl_2$ show early and very fast cell detachment indicative of efficient necrotic agents, while STS shows late and slow cell detachment indicative of apoptosis [49]. The B values versus concentration data points were fitted with logistic curves for both cell lines (cf. Fig. 4), yielding a "half-detachment-dose" value for each agent, so that trends can be extracted. In order to verify that our method could be used to detect and measure toxicity of agents to cells for reference, we evaluated the effects of NPs and compounds exposed to HeLa and MCF7 cells on their cell viability. For that, we used a common cell viability method, the resazurin assay, used to evaluate the metabolic activity of the cells (cf., data in the Additional file 1). This cell viability method was used as a reference control method to compare with the cantilever measurements obtained. As the AFM measurements

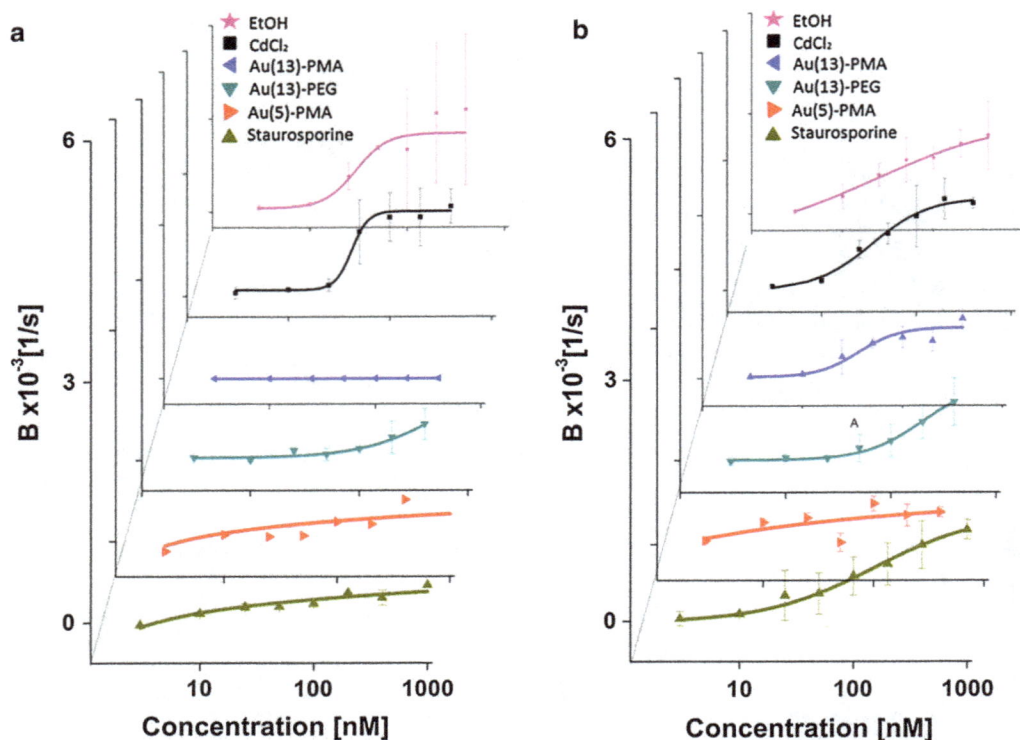

Fig. 4 Damping constants B for different agents (a list of all mean values and standard deviations is presented in the Additional file 1) and the corresponding logistic fit curves. **a** Results for HeLa cells are presented, from which based on the respective logistic fit curves the following "half-detachment-dose" values were extracted: 29 nM (EtOH), 43 nM (CdCl$_2$), 53 nM (Au(13)-PMA), 640 nM (Au(13)-PEG), 98 nM (Au(5)-PMA), and 78 nM (staurosporine). **b** Results for MCF7 cells, from which based on the respective logistic fit curves the following "half-detachment-dose" values were extracted: 21 nM (EtOH), 33 nM (CdCl$_2$), 53 nM (Au(13)-PMA), 190 nM (Au(13)-PEG), 81 nM (Au(5)-PMA), and 150 nM (staurosporine)

were carried out without CO_2 control, the resazurin assays were carried out in the presence or absence of CO_2 (to mimic the conditions of the cantilever, i.e. without CO_2, and standard protocols, i.e. with CO_2). Indeed, the same toxicity trends were observed for the resazurin as for the AFM measurements. There is however one advantage of the AFM assays. In the resazurin assays measurements for different time points have to be carried out separately, while the AFM assays in principle allows for continuous real-time recording.

Conclusions

The results obtained in this work suggest that the presented method is a generally applicable fast-screening-technique based on label-free real-time monitoring tool, which uses cell detachment from an oscillating cantilever to measure cell intoxication. After desired exposure time, the release rate of cells (as quantified in terms of damping values B) from the cantilever was extracted. We speculate that in future, this method may be applied even to single cells or other cell types such as primary cultures.

Methods
Synthesis of Au nanoparticles
Synthesis of 13 nm Au nanoparticles

Citrate-capped Au nanoparticles (NPs) with an average inorganic diameter of 13.5 nm (± 0.8 nm), as determined by transmission electron microscopy (TEM), were synthesized by largely following the protocol reported by Schulz et al. [50]. Briefly, 144 mL of Milli-Q water was added to 250 mL three-necked round-bottomed flask and heated up until boiling with a heating mantle. First, a mixture of sodium citrate (3.5 mL; 60 mM) and citric acid (1.5 mL; 60 mM) was added to the flask and kept under vigorous stirring for 30 min (450 rpm). A condenser was utilized to prevent the evaporation of the solvent. Then 100 μL of ethylene diamine tetraacetic acid (EDTA 30 mM) was added, followed by 1 mL of 25 mM hydrogen tetrachloroaurate (III) aqueous solution. After ca. 70 s the color of the mixture changed from pale yellow to wine-red, which is indicative of the growth of the Au NPs. In this moment the heating was switched off, but not the stirring. When the temperature of the mixture had dropped down to 95 °C, the flask with the NPs was immersed in ice in order to stop the reaction. The absorbance at 450 nm [extinction coefficient $\varepsilon(450) = 1.6 \times 10^8$ M^{-1} cm^{-1}] was used to determine the concentration of the NPs, as previously described by Haiss et al. [51].

Synthesis of 5 nm Au NPs

A modified protocol of the two-phase method published by Brust et al. and Holz et al. was used to produce tetraoctylammonium bromide-capped Au NPs with an inorganic diameter of 5.5 nm (±1.0 nm), as determined by TEM [52, 53]. Briefly, at room temperature, an aqueous solution of hydrogen tetrachloroaurate-(III) (40 mM, 25 mL) and a solution of tetraoctylammonium bromide (TOAB) in toluene (50 mM, 80 mL) were mixed and vigorously shaken (ca. 5 min) in a 500 mL separation funnel. Then, once the $AuCl_4$ ions were fully transferred into the toluene phase, the organic phase was transferred into a 250 mL round bottom flask. Then, a freshly prepared aqueous solution of $NaBH_4$ (350 mM, 25 mL) was added to the solution of gold precursors in toluene under vigorous stirring and kept under stirring for 1 h. The solution was then transferred to a 500 mL separation funnel and 25 mL of 10 mM HCl was added to remove the excess of $NaBH_4$. The mixture was vigorously shaken and the aqueous phase was discarded. Then 25 mL of 10 mM NaOH were added to remove any excess of acid, followed by four washes with Milli-Q water (25 mL). The toluene phase containing the Au NPs was transferred to a 250 mL round bottomed flask. Then, the solution was left under stirring overnight at room temperature. Then, original TOAB coating was exchanged by 1-dodecanethiol, by mixing (65 °C, 3 h) the original NP dispersion in toluene with a solution of 1-dodecanethiol in toluene (4.17 M, 10 mL). Then, the 1-dodecanethiol-capped Au NPs were purified from agglomerates by centrifugation at $1 \times 10^3 g$, whereby the NPs remained in the supernatant. To remove the excess of 1-dodecanethiol, the NPs were precipitated by addition of methanol and collected by centrifugation ($1 \times 10^3 g$). The washing step with methanol was repeated three times to minimize the presence of free surfactant. In order to calculate the concentration of NPs, the absorbance at 520 nm [extinction coefficient $\varepsilon(520) = 8.7 \times 10^6$ M^{-1} cm^{-1}] was used, as previously reported [54].

Surface modification of Au NPs

PEGylation of 13 nm citrate-capped Au NPs

To 150 mL of the as prepared citrate-capped NPs (NP concentration ca. 1.8 nM), 2.7 mg of α-thio-ω-carboxy poly(ethylene glycol) (HS-PEG-COOH, $M_W = 987.19$ Da from Iris Biotech) were added, equivalent 10^4 PEG molecules added per NP. Thus, sufficient PEG was added to ensure full PEG saturation of the NP surface. The

PEGylated Au NPs were purified from PEG excess and re-suspended in deionized water by centrifugal precipitation (three times at $15 \times 10^3 g$, 30 min).

Polymer coating poly(isobutylene-alt-maleic anhydride) dodecylamine-grafted, in the following referred to as PMA of 13 nm citrate-capped Au NPs

Citrate-capped Au NPs were transferred from aqueous media to organic solvent following the protocol of Soliman et al. [46]. Briefly, $3 \cdot 10^4$ PEG molecules ($M_W = 750$ Da; α-methoxy-ω-mercapto-poly(ethylene gylcol) (HS-PEG-CH_3O) from Rapp Polymere) per NP were added and kept under vigorous stirring for 2 h. Then, a 0.4 M solution of dodecylamine (DDA) in chloroform (equal volume as the aqueous solution of NPs) was mixed with the NPs under vigorous stirring, which ultimately allows to transfer the NPs from the aqueous to the chloroform phase. A small amount of NaCl (50 µL 2 M) was added to speed up the NPs' phase transfer. The NPs were then cleaned twice by centrifugal precipitation (8960 g) from excess of PEG and DDA. The precipitated NPs were collected and dispersed in chloroform, in which their concentration was determined by UV/Vis spectroscopy with the molar extinctions coefficients as provided above. Yet, to get PMA-coated Au NPs colloidally stable in aqueous solution, the Au NPs previously coated with PEG/DDA were coated with the amphiphilic polymer PMA by largely following the protocol described by Lin et al. [55]. Briefly, 75% of the anhydride rings of poly(isobutylene-alt-maleic anhydride) were modified with DDA by mixing in tetrahydrofuran (THF) at 65 °C under stirring (12 h). The modified polymer (i.e., PMA) was dried using a Rotavapor at 40 °C under reduced pressure and dispersed in 30 mL chloroform, yielding a stock solution with a final PMA concentration of 0.75 M. Notice that 0.75 M refers to the concentration of the monomers of poly(isobutylene-alt-maleic anhydride). Then, to efficiently achieve PMA-coating of the NPs, a specific volume of PMA, which depends on the total effective surface area (A_{eff}) of the NPs, was used as described by Soliman et al. [46]. Briefly, the NPs were mixed with PMA (0.75 M in terms of monomer units; $R_{p/Area} = 3000$, where $R_{p/Area}$ refers to the number of PMA monomers added per nm^2 of A_{eff}) in a round flask and diluted with chloroform. After 25 min, the chloroform was slowly evaporated at 42 °C under reduced pressure using a Rotavapor, until the solvent was completely evaporated. This procedure was repeated twice. Finally, the dried product was dissolved in sodium borate buffer (SBB, pH = 12), which hydrolyzed the maleic anhydride groups of the PMA, yielding carboxyl groups and thereby providing the NPs with colloidal stability in aqueous solution. The PMA-coated NPs were then filtrated through a syringe membrane filter

(0.22 μm pore size). Finally empty micelles formed by PMA and excess of free PMA were removed by precipitation of the PMA-coated NPs using centrifugation (8960*g*; 40 min, twice) and the buffer was exchanged to water.

PMA coating of 5 nm Au NPs

Equivalently, PMA coating was carried out as described for the 13 nm NPs. A value $R_{p/Area} = 150$ PMA monomers per nm^2 was used instead, which was experimentally optimized to warrant for colloidal stability of NPs with about the same size of inorganic core. The PMA-coated NPs were purified first by gel electrophoresis, as described in previous works (e.g., see Lin et al. [55]) and then by ultracentrifugation ($150 \times 10^3 g$; 60 min, three times).

Characterization of NPs

Transmission electron microscopy (TEM), UV–Vis spectroscopy, dynamic light scattering (DLS) and laser Doppler anemometry (LDA) were used to analyze the colloidal properties of the NPs.

TEM imaging

TEM images of the samples were acquired in a JEM-1230 transmission electron microscope equipped with a LaB6 cathode running at 120 kV and an ORIOUS SC1000 4008×2672 pixels CCD camera (Gatan UK, Abingdon Oxon, UK). UV–Vis spectra were obtained with an Agilent 8453 spectrometer. DLS and LDA measurements were carried out with a Malvern Zetasizer. In Additional file 1, Figure S1a, c show TEM micrographs of PEGylated 13 nm Au NPs and PMA-coated 5 nm Au NPs with negative staining, in which a PEG layer (thickness of ca. 5 nm around cores of 13 nm) and the PMA-coating (thickness of ca. 5 nm around the 5.5 nm cores) are clearly discernible. Additional file 1: Figure S3b shows a TEM micrograph of the PMA-coated NPs (here, only the diameter of the Au core gives contrast).

TEM negative staining

Uranyl acetate was used as negative stain, which allows the formation of a uniform, consistent, and high contrast staining. The sample was prepared on carbon film 400 copper mesh grids purchased from Electron Microscopy Sciences (Hatfield, USA). The specimen grids were exposed to glow-discharge treatment under air plasma for 20 s (2.0×10^{-1} atm. and 35 mA) using a MED 020 modular high vacuum coating system (BAL-TEC AG, Balzers, Liechtenstein). Negatively charged carbon grids were used within 5 min after treatment to ensure hydrophilicity. The on-grid negative staining was performed using a slightly modified single-droplet negative-staining procedure. 1.5 μL sample droplet of NP concentration

ranging from 6 to 15 nM followed by three 2.5 μL droplets of 0.25% weight/volume (w/v) uranyl acetate aqueous solution were placed on a clean Parafilm piece. The treated grid was incubated on the sample droplet for 1 min and then on the staining droplets for 3, 3, and 60 s, respectively. After each incubation step the excess fluid was nearly fully removed by touching the grid edge with Whatman filter paper. Finally, the sample was fully dried for 20 min at 2.0×10^{-1} atm.

UV–Vis absorption spectroscopy

The UV–Vis absorption spectra of the three polymer-coated samples are shown in Addtional file 1: Figure S3d, which clearly show the surface plasmon resonance band of the colloids (ca. 520 nm), more intense in the case of the 13 nm NPs, as expected.

Zetasizer measurements

DLS and ζ-potential values of the three samples are summarized in Table 1. The hydrodynamic diameter (d_h) of the PEGylated Au NPs, as determined by DLS, yielded 22 nm, which matches very well the observations by negative staining TEM. Note however, that the DLS and the negative staining were obtained in aqueous solution and vacuum, respectively. The d_h values of PMA-coated 13 and 5 nm Au colloids were 17 and 11 nm, respectively. Sizes as determined by TEM (inorganic core; d_{TEM}) and DLS (d_h), and ζ-potential values of the polymer coated Au NPs are summarized in Table 1.

Reference assay
Cell culture

HeLa and MCF 7 cells were obtained from the American Type Culture Collection (ATCC, Manassas, VA). Briefly, HeLa cells were cultured in Dulbecco's Modified Eagle's Medium (DMEM) (# D5796) containing 10% Fetal Bovine Serum (FBS) (#S0615), 1% of Penicillin/Streptomycin (P/S) (# 15140-1229) and GlutaMAX™ (#35050-038). MCF7 cells were cultured in Eagle's Minimum Essential Medium (EMEM) (# M5650 supplemented with 10% FBS, 1% of P/S and 0.01 mg/mL human recombinant insulin (# I3536). The cell cultures were kept at 37 °C in a

Table 1 Comparison of diameters taken from TEM and hydrodynamic diameters taken from DLS, as well as ζ-potential values of the examined NP samples

Sample	d_{TEM}/nm	d_h(number)/nm	ζ-potential/mV
Au(13)-PEG	13.5 ± 0.8	21.9 ± 0.3	-23.0 ± 1.9
Au(13)-PMA	13.5 ± 0.8	16.7 ± 0.4	-20.2 ± 0.8
Au(5)-PMA	5.5 ± 1.0	15.3 ± 0.8	-42.7 ± 1.3

DLS and ζ-potential data were recorded in MilliQ water. The hydrodynamic diameter corresponds to the mean value ± standard deviation as obtained from the number distributions humidified atmosphere of 5% CO_2 in air. At confluence, cells were washed with PBS and detached with 0.05% Trypsin EDTA (# 25300-054) solution. Cells then were reseeded in flasks for cell culture or seeded in 96-well plates for the experiments.

Cell viability assay

Cell viability of HeLa and MCF7 cells exposed to Au NPs and chemical agents was evaluated by the Resazurin assay [AlamarBlue® (# 765506) Thermo Fisher, Germany] as previously reported [56–58]. For that, HeLa and MCF7 cells were seeded in 96 black polystyrene plates at the density of 10.000 cells/well in complete cell culture media and were incubated overnight at 37 °C, 5% CO_2. The next day, cells were exposed to NPs and chemical agents at desired concentration for 4 h in the presence or absence of 5% CO_2 at a final volume of 100 µL per well. After the desired time, cells were washed once with PBS, then 100 µL of 10% resazurin solution (in complete cell media) was added to the cells and incubated for 4 h at 37 °C and 5% CO_2. The fluorescence intensity was measured for the presence of resazurin and resorufin with a 96-microwell plate reader connected to a fluorometer (Fluorolog-3, from Horiba Jobin–Yvon, Germany) at an excitation wavelength of 560 nm. The emission was recorded in the range of 570–650 nm, of which an integrated fluorescence intensity was determined. This integrated fluorescence intensity was considered to be proportional to cell viability. Cell viability was normalized to 100% for untreated cells. The results are presented as mean cell viability ± the respective standard deviation (SD), as obtained from three independent experiments (e.g. cell cultures), each one performed in triplicate. Upon incubation with high concentration of toxic agents, cell viability is decreased. All diagrams are presented in Additional file 1.

Authors' contributions

FY designed the method and organized research. RR wrote the software and contributed to the measurements. PDP, BP and MS synthesized, purified and characterized the NPs. AHS, SRP, NF and UB offered the cells and run the cell-assays. PDP, WJP and NH suggested the research plan and experiments. FY, PDP, WJP, NF and NH wrote the manuscript. All authors read and approved the final manuscript.

Author details

[1] Department of Chemistry, University of Marburg, Marburg, Germany. [2] Department of Physics, University of Marburg, Marburg, Germany. [3] Department of Pharmacy, University of Marburg, Marburg, Germany. [4] CIC biomaGUNE, San Sebastián, Spain. [5] Material Science Center, University of Marburg, Marburg, Germany.

Acknowledgements

The help of Dr. H-C. Kim in programming the calculation software is gratefully acknowledged. The authors thank Dr. S. Ashraf for help in cell culture at the initial stages of the project.

Competing interests

The authors declare that they have no competing interests.

Funding

This work was financially supported in part by DFG Grant PA 784/25-1 to WJP.

References

1. Malanchi I, Santamaria-Martinez A, Susanto E, Peng H, Lehr H, Delaloye J, Huelsken J. Interactions between cancer stem cells and their niche govern metastatic colonization. Nature. 2012;481:85–9.
2. Trappmann B, Gautrot JE, Connelly JT, Strange DG, Li Y, Oyen ML, Cohen Stuart MA, Boehm H, Li B, Vogel V, Spatz JP, Watt FM, Huck WT. Extracellular-matrix tethering regulates stem-cell fate. Nat Mater. 2012;11:642–9.
3. Ishikawa F, Ushida K, Mori K, Shibanuma M. Loss of anchorage primarily induces non-apoptotic cell death in a human mammary epithelial cell line under atypical focal adhesion kinase signaling. Cell Death Dis. 2015;6:e1619.
4. Lee MW, Bassiouni R, Sparrow NA, Iketani A, Boohaker RJ, Moskowitz C, Vishnubhotla P, Khaled AS, Oyer J, Copik A, Fernandez-Valle C, Perez JM, Khaled AR. The CT20 peptide causes detachment and death of metastatic breast cancer cells by promoting mitochondrial aggregation and cytoskeletal disruption. Cell Death Dis. 2014;5:e1249.
5. Perillo NL, Marcus ME, Baum LG. Galectins: versatile modulators of cell adhesion, cell proliferation, and cell death. J Mol Med. 1998;76:402–12.
6. Vinken M, Blaauboer BJ. In vitro testing of basal cytotoxicity: establishment of an adverse outcome pathway from chemical insult to cell death. Toxicol In Vitro. 2017;39:104–10.
7. Riss TL, Moravec RA, Niles AL. Cytotoxicity testing: measuring viable cells, dead cells, and detecting mechanism of cell death. Methods Mol Biol. 2011;740:103–14.
8. Riss TL, Moravec RA, Niles AL, Duellman S, Benink HA, Worzella TJ, and Minor L. Cell viability assays. In: Sittampalam GS, Coussens NP, Brimacombe K, et al., editors. Assay guidance manual. Bethesda: Eli Lilly & Company and the National Center for Advancing Translational Sciences; 2004.
9. Soenen S, Rivera-Gil P, Montenegro J, Parak W, De Smedt S, Braeckmans K. Cellular toxicity of inorganic nanoparticles: common aspects and guidelines for improved nanotoxicology evaluation. Nano Today. 2011;6:446–65.
10. Han X, Gelein R, Corson N, Wade-Mercer P, Jiang J, Biswas P, Finkelstein JN, Elder A, Oberdorster G. Validation of an LDH assay for assessing nanoparticle toxicity. Toxicology. 2011;287:99–104.
11. Stone V, Johnston H, Schins RP. Development of in vitro systems for nanotoxicology: methodological considerations. Crit Rev Toxicol. 2009;39:613–26.
12. Kepp O, Galluzzi L, Lipinski M, Yuan J, Kroemer G. Cell death assays for drug discovery. Nat Rev Drug Discov. 2011;10:221–37.
13. Wegener J, Sieber M, Galla HJ. Impedance analysis of epithelial and endothelial cell monolayers cultured on gold surfaces. J Biochem Biophys Methods. 1996;32:151–70.
14. Lo CM, Keese CR, Giaever I. Impedance analysis of MDCK cells measured by electric cell-substrate impedance sensing. Biophys J. 1995;69:2800–7.
15. Tarantola M, Schneider D, Sunnick E, Adam H, Pierrat S, Rosman C, Breus V, Sonnichsen C, Basche T, Wegener J, Janshoff A. Cytotoxicity of metal and semiconductor nanoparticles indicated by cellular micromotility. ACS Nano. 2009;3:213–22.
16. Tamayo J. Mass sensing: optomechanics to the rescue. Nat Nanotechnol. 2015;10:738–9.

17. Burg T, Godin M, Knudsen S, Shen W, Carlson G, Foster J, Babcock K, Manalis S. Weighing of biomolecules, single cells and single nanoparticles in fluid. Nature. 2007;446:1066–9.

18. Arlett JL, Myers EB, Roukes ML. Comparative advantages of mechanical biosensors. Nat Nanotechnol. 2011;6:203–15.

19. Liu Y, Schweizer L, Wang W, Reuben R, Schweizer M, Shu W. Label-free and real-time monitoring of yeast cell growth by the bending of polymer microcantilever biosensors. Sens Actuators B Chem. 2013;178:621–6.

20. Weng Y, Delgado F, Son S, Burg T, Wasserman S, Manalis S. Mass sensors with mechanical traps for weighing single cells in different fluids. Lab Chip. 2011;11:4174–80.

21. Ramos D, Tamayo J, Mertens J, Calleja M, Villanueva LG, Zaballos A. Detection of bacteria based on the thermomechanical noise of a nanomechanical resonator: origin of the response and detection limits. Nanotechnology. 2008;19:035503.

22. Ramos D, Tamayo J, Mertens J, Calleja M, Zaballos A. Origin of the response of nanomechanical resonators to bacteria adsorption. J Appl Phys. 2006;100:10.

23. Davila A, Jang J, Gupta A, Walter T, Aronson A, Bashir R. Microresonator mass sensors for detection of *Bacillus anthracis* sterne spores in air and water. Biosens Bioelectron. 2007;22:3028–35.

24. Domke J, Parak WJ, George M, Gaub HE, Radmacher M. Mapping the mechanical pulse of single cardiomyocytes with the atomic force microscope. Eur Biophys J. 1999;28:179–86.

25. Wu S, Liu X, Zhou X, Liang XM, Gao D, Liu H, Zhao G, Zhang Q, Wu X. Quantification of cell viability and rapid screening anti-cancer drug utilizing nanomechanical fluctuation. Biosens Bioelectron. 2016;77:164–73.

26. Florin E-L, Moy VT, Gaub HE. Adhesion forces between individual ligand-receptor pairs. Science. 1994;264:415–7.

27. Seitz M, Friedsam C, Jostl W, Hugel T, Gaub HE. Probing solid surfaces with single polymers. Chem Phys Chem. 2003;4:986–90.

28. Javier A, Kreft O, Alberola A, Kirchner C, Zebli B, Susha A, Horn E, Kempter S, Skirtach A, Rogach A, Radler J, Sukhorukov G, Benoit M, Parak W. Combined atomic force microscopy and optical microscopy measurements as a method to investigate particle uptake by cells. Small. 2006;2:394–400.

29. Thie M, Rospel R, Dettmann W, Benoit M, Ludwig M, Gaub H, Denker H. Interactions between trophoblast and uterine epithelium: monitoring of adhesive forces. Hum Reprod. 1998;13:3211–9.

30. Radmacher M, Tillamnn RW, Fritz M, Gaub HE. From molecules to cells: imaging soft samples with the atomic force microscope. Science. 1992;257:1900–5.

31. Benoit M, Gabriel D, Gerisch G, Gaub HE. Discrete interactions in cell adhesion measured by single-molecule force spectroscopy. Nat Cell Biol. 2000;2:313–7.

32. Domke J, Dannohl S, Parak WJ, Muller O, Aicher WK, Radmacher M. Substrate dependent differences in morphology and elasticity of living osteoblasts investigated by atomic force microscopy. Colloids Surf B Biointerfaces. 2000;19:367–79.

33. Hofmann UG, Rotsch C, Parak WJ, Radmacher M. Investigating the cytoskeleton of chicken cardiocytes with the atomic force microscope. J Struct Biol. 1997;119:84–91.

34. Chiarugi P, Giannoni E. Anoikis: a necessary death program for anchorage-dependent cells. Biochem Pharmacol. 2008;76:1352–64.

35. Varner JA, Cheresh DA. Integrins and cancer. Curr Opin Cell Biol. 1996;8(5):724–30.

36. Zhao B, Li L, Wang L, Wang CY, Yu J, Guan KL. Cell detachment activates the Hippo pathway via cytoskeleton reorganization to induce anoikis. Genes Dev. 2012;26:54–68.

37. Longo G, Alonso-Sarduy L, Rio LM, Bizzini A, Trampuz A, Notz J, Dietler G, Kasas S. Rapid detection of bacterial resistance to antibiotics using AFM cantilevers as nanomechanical sensors. Nat Nanotechnol. 2013;8:522–6.

38. Voger EA, Bussian RW. Short-term cell-attachment rates: a surface-sensitive test of cell-substrate compatibility. J Biomed Mater Res. 1987;21(10):1197–211.

39. Perez-Hernandez M, del Pino P, Mitchell S, Moros M, Stepien G, Pelaz B, Parak W, Galvez E, Pardo J, de la Fuente J. Dissecting the molecular mechanism of apoptosis during photothermal therapy using gold nanoprisms. ACS Nano. 2015;9:52–61.

40. Bao C, Conde J, Polo E, del Pino P, Moros M, Baptista P, Grazu V, Cui D, de la Fuente J. A promising road with challenges: where are gold nanoparticles in translational research? Nanomedicine. 2014;9:2353–70.

41. Soenen SJ, Manshian B, Montenegro JM, Amin F, Meermann B, Thiron T, Cornelissen M, Vanhaecke F, Doak S, Parak WJ, De Smedt S, Braeckmans K. Cytotoxic effects of gold nanoparticles: a multiparametric study. ACS Nano. 2012;6:5767–83.

42. Kim K, Lee D, Song C, Kang P. Reactive oxygen species-activated nanomaterials as theranostic agents. Nanomedicine. 2015;10:2709–23.

43. Dayem AA, Hossain MK, Lee SB, Kim K, Saha SK, Yang G-M, Choi HY, Cho S-G. The role of reactive oxygen species (ROS) in the biological activities of metallic nanoparticles. Int J Mol Sci. 2017;18(1):120.

44. Pelaz B, Charron G, Pfeiffer C, Zhao Y, de la Fuente J, Liang X, Parak W, del Pino P. Interfacing engineered nanoparticles with biological systems: anticipating adverse nanobio interactions. Small. 2013;9:1573–84.

45. Rothen-Rutishauser B, Kuhn D, Ali Z, Gasser M, Amin F, Parak W, Vanhecke D, Fink A, Gehr P, Brandenberger C. Quantification of gold nanoparticle cell uptake under controlled biological conditions and adequate resolution. Nanomedicine. 2014;9:607–21.

46. Soliman M, Pelaz B, Parak W, del Pino P. Phase transfer and polymer coating methods toward improving the stability of metallic nanoparticles for biological applications. Chem Mater. 2015;27:990–7.

47. del Pino P, Yang F, Pelaz B, Zhang Q, Kantner K, Hartmann R, de Baroja N, Gallego M, Moller M, Manshian B, Soenen S, Riedel R, Hampp N, Parak W. Basic physicochemical properties of polyethylene glycol coated gold nanoparticles that determine their interaction with cells. Angew Chem Int Ed Engl. 2016;55:5483–7.

48. Brandenberger C, Muhlfeld C, Ali Z, Lenz A, Schmid O, Parak W, Gehr P, Rothen-Rutishauser B. Quantitative evaluation of cellular uptake and trafficking of plain and polyethylene glycol-coated gold nanoparticles. Small. 2010;6:1669–78.

49. Kwon H, Lee J, Shin H, Kim J, Choi S. Structural and functional analysis of cell adhesion and nuclear envelope nano-topography in cell death. Sci Rep. 2015;5:15623.

50. Schulz F, Homolka T, Bastus N, Puntes V, Weller H, Vossmeyer T. Little adjustments significantly improve the Turkevich synthesis of gold nanoparticles. Langmuir. 2014;30:10779–84.

51. Haiss W, Thanh NTK, Aveyard J, Fernig DG. Determination of size and concentration of gold nanoparticles from UV–Vis spectra. Anal Chem. 2007;79:4215–21.

52. Brust M, Walker M, Bethell D, Schiffrin D, Whyman R. Synthesis of thiol-derivatized gold nanoparticles in a 2-phase liquid-liquid system. J Chem Soc Chem Commun. 1994;7:801–2.

53. Holz M, Haselmeier R, Mazitov R, Weingartner H. Self-diffusion of neon in water by Ne-21 NMR. J Am Chem Soc. 1994;116:801–2.

54. Kreyling W, Abdelmonem A, Ali Z, Alves F, Geiser M, Haberl N, Hartmann R, Hirn S, de Aberasturi D, Kantner K, Khadem-Saba G, Montenegro JM, Rejman J, Rojo T, de Larramendi IR, Ufartes R, Wenk A, Parak WJ. In vivo integrity of polymer-coated gold nanoparticles. Nat Nanotechnol. 2015;10:619.

55. Lin C, Sperling R, Li J, Yang T, Li P, Zanella M, Chang W, Parak W. Design of an amphiphilic polymer for nanoparticle coating and functionalization. Small. 2008;4:334–41.

56. O'Brien J, Wilson I, Orton T, Pognan F. Investigation of the Alamar Blue (resazurin) fluorescent dye for the assessment of mammalian cell cytotoxicity. Eur J Biochem. 2000;267:5421–6.

57. O'Brien J, Wilson I, Ortaon T, Pognan F. Investigation of the Alamar blue (Resazurin) fluorescent dye for the assessment of mammalian cell cytotoxicity. Toxicology. 2001;164:132.

58. Feliu N, Kohonen P, Ji J, Zhang Y, Karlsson HL, Palmberg L, Nystrom A, Fadeel B. Next-generation sequencing reveals low-dose effects of cationic dendrimers in primary human bronchial epithelial cells. ACS Nano. 2014;9:146–63.

4

Comparative efficacy analysis of anti-microbial peptides, LL-37 and indolicidin upon conjugation with CNT, in human monocytes

Biswaranjan Pradhan, Dipanjan Guha, Krushna Chandra Murmu, Abhinav Sur, Pratikshya Ray, Debashmita Das and Palok Aich*⍟

Abstract

Background: Antimicrobial peptides (AMPs) have the potential to serve as an alternative to antibiotic. AMPs usually exert bactericidal activity via direct killing of microbial pathogens. Reports have proposed that by harnessing innate immune activation, AMPs can regulate pathogen invasion and may control infection. It has been reported that AMPs could be utilized to activate the innate mucosal immune response in order to eliminate pathogenic infections. This way of controlling pathogen infection, by activating host immunity, confers the potential to the select AMPs to alleviate the problem of antibiotic resistance. Among various AMPs tested LL-37 and indolicidin, showed promise to be potential candidates for eliciting enhanced host innate immune responses. LL-37 and indolicidin had exhibited substantial innate immune activation in both human and murine macrophages. Dosage for each of the AMPs, however, was high with adverse side effects.

Results: In this study, we reported that upon conjugation with carbon nanotubes (CNT), each AMP remained biologically functional at a concentration that was 1000-fold less than the dosage required for free AMP to remain active in the cells.

Conclusions: Current study also revealed that while indolicidin induced signalling events mediated through the TNFRSF1A pathway in THP1 cells, followed by activation of NFκB and c-JUN pathways, treatment of cells with LL-37 induced signalling events by activating IL1R, with subsequent activation of NFκB and NFAT2. Thp1 cells, primed with CNT conjugated LL-37 or indolicidin, are protected against *Salmonella typhimurium* infection at 16 h post challenge.

Keywords: Carbon nanotube, Antimicrobial peptides, Cationic peptides, Host defense peptides, Innate immunity

Background

The antimicrobial activity of cationic peptides is mostly elicited via direct interaction with microbes [1, 2]. However, direct attack on microbes to attain anti-microbial effects is not a good strategy because microbes tend to develop resistance against antimicrobial agents over time. An alternative paradigm for prophylactic or therapeutic

success would involve activating the innate immune system of the host through treatment with a sub optimal dosage of antimicrobial agents, rather than a direct attack on the microbes. This methodology could alleviate the possibility of microbes developing a resistance against antimicrobial agents. Keeping this logic in mind, we have compared the antimicrobial activities of two therapeutically potential antimicrobial peptides (AMPs), with potential and proven medicinal properties, LL-37 and indolicidin, in vitro.

LL-37 is a proven and potent AMP. LL-37 is of human origin [3]. LL-37 was first detected in leukocytes and in

*Correspondence: palok.aich@niser.ac.in
School of Biological Sciences, National Institute of Science Education and Research (NISER), HBNI, P.O. Bhimpur-Padanpur, Khurdha, Jatni, Odisha 752050, India

the testis of humans [4]. Subsequently, it was also found inside a large variety of cells, tissues and body fluids. LL-37 was initially recognized for its antimicrobial properties [5–7] against bacteria, fungi and viral pathogens [8, 9]. LL-37 neutralizes lipopolysaccharides [10, 11] because of its high affinity towards LPS [10]. LL-37 also plays a significant role in wound healing, angiogenesis and apoptosis [12]. Most importantly, recent studies suggest that it is also involved in the regulation of cancer [13].

Indolicidin, the other AMP used in the current study, belongs to the cathelicidin class of AMPs. Indolicidin purified from the cytoplasmic granules of bovine neutrophils. Indolicidin is capable of killing gram-negative bacteria by crossing the outer membrane and causing disruption of the cytoplasmic membrane by channel formation [14]. Indolicidin is also active against gram-positive bacteria, fungi, protozoa and enveloped viruses such as HIV-1 [15, 16]. Apart from direct neutralization of microbes, another important function of indolicidin is its ability to modulate the host innate immune system against infectious agents [17, 18]. Indolicidin exerts many immunomodulatory roles, including—but not limited to—chemotaxis, modulation of cytokine and chemokine expression, and leukocyte activation [19, 20]. Instead of utilizing the direct antimicrobial effects of indolicidin, its immunomodulatory properties could be exploited to facilitate pathogen clearance in the host. Interestingly, the concentration of indolicidin required to stimulate the innate immune system is comparable to its antimicrobial concentration of 10–20 µg/ml. This equivalence of concentration, for innate mucosal immunity activation and for antimicrobial activity, is a major concern to develop antimicrobial resistance. It is, therefore, urgently required to have a methodology to reduce the dosage required to modulate host innate immunity.

Both natural and synthetic AMPs have shown promise as 'next generation antibiotics' due to their unique mode of membranolytic action, which minimizes the development of microbial resistance. However, bacteria have evolved the following mechanisms to counteract AMPs: (i) by a transient induction of bacterial signalling systems that help the bacteria to cope with AMPs, and (ii) constitutive resistance as a result of genetic changes. Currently, there are several putative mechanisms known for bacterial resistance to AMPs [2, 21–24]. When AMPs are present at higher concentrations, bacteria modulate their cell surface by making it less negatively charged and less permeable [25–27].

Despite the apparent medical potential of AMPs, their activity is not clinically practical because of weak activity, nonspecific cytotoxicity and proteolytic effect on some host membrane proteins [28]. For example, indolicidin is cytotoxic for rat and human T-lymphocytes [29]. Also,

in vivo studies have confirmed that indolicidin is toxic to erythrocytes [15] at a high concentration (10 µg/ml). Indolicidin's immune modulatory efficacy with respect to concentration needs to be increased in order to avoid damage to the host and development of indolicidin resistance in bacteria. Previous studies have demonstrated that immune modulatory efficacy as well as delivery of CpG is enhanced when conjugated with nanoparticles [30–32]. Additionally, we have recently reported that conjugation of indolicidin with short multi-walled carbon nano-tubes (SM-CNT) enhanced the efficacy of indolicidin by increasing its ability to protect host cells from *Salmonella typhimurium serovar enterica* (ST) MTCC 3232 challenge [1].

In the present study, we have demonstrated that the comparative efficacy and in vitro functioning of LL-37 and indolicidin conjugated with SM-CNTs. We have studied the effects of free and nano-conjugated indolicidin treatment on the human monocyte cell line THP-1 through transcriptomics. We have also selected LL-37 for our current study as it has already been tested for various immune modulatory effects [33]. Our results revealed that following conjugation of LL-37 and indolicidin with SM-CNTs, the immune modulatory efficacy of LL-37 and indolicidin was significantly increased in vitro. Our results revealed that an effective level of activity for the peptides is maintained following CNT–conjugation even at a 1000-fold less dosage than free peptide.

Methods
Synthesis of CNT–indolicidin and CNT–LL-37

LL-37 was obtained from Prof. Bob Hancock, UBC, Canada as a gift and indolicidin was purchased from BR Biochem Lifesciences, India. Both AMPs were obtained as lyophilized powder. LL-37 and indolicidin were conjugated with CNT using EDC-NHS conjugation protocol as described elsewhere [34], which was described in our previous work reported with indolicidin [1]. LL-37 was conjugated using the same protocol 5 mg of LL-37 was suspended in 25 µl of DMSO. The resulting solution was mixed properly followed by further addition of 975 µl of PBS to make a 5 mg/ml peptide solution. This solution was used as the stock peptide solution for our experiment. 400 µl of the 1 mg/ml CNT solution, prepared earlier was put in a clean and sterile microfuge tube. To the above solution, 600 µl of MES buffer (pH = 5.0) used as the appropriate activation buffer was added. This is because activation of the carboxyl groups on the nanotubes using EDC and NHS is most efficient at pH = 4.5–7.2. 5 µl of 0.4 M EDC and 50 µl of 0.1 M NHS was added respectively and the solution was incubated in dark for 45 min at room temperature. Once the activation reaction is complete, 1.4 µl of 2-mercaptoethanol was added to quench the effect of EDC. 960 µl of PB (pH = 7.2) was added to 1 ml of the activated solution. The solution was mixed by gentle pipetting. PBS is used as the conjugation buffer. Therefore, after adding 40 μl of the stock 5 mg/ml peptide solution, the resulting solution was mixed thoroughly and incubated in dark for 2 h at room temperature. In addition, free LL-37 was diluted to the similar extent for proper comparison to the conjugates. Spike was prepared by adding same concentrations of LL-37 to a solution of non-activated CNTs. Free peptides were removed from the conjugate mixture using molecular weight cut-off spin columns (3 MWCO, Millipore, USA). Short multiwalled CNTs were purchased from Cheap tubes with outer diameter 8 nm and inner diameter of 2–5 nm and length between 500 and 2000 nm. Molecular weight of CNTs were calculated based on protocols mentioned before and on the homepage of Hipco [35, 36] assuming standard 0.14 nm of distance between C–C covalent bonds for the circumference and a hexagonal pack distance of 0.283 nm for weak long range interactions. Using these parameters and the value of total surface area as provided by the manufacturer, the average molecular weight determined was 2×10^6. Stock concentration of CNT calculated was 50 μM.

Physical characterization through Fourier transformed infrared (FTIR) spectroscopy

FTIR spectrum of free and conjugated AMPs along with positive controls was collected using Perkin Elmer FTIR model Spectrum RX1 equipment. Purified samples were lyophilized and prepared for FTIR measurement. FTIR measurements were performed at room temperature in the absorbance range from 4000 to 400 cm^{-1} by accumulating 20 scans with a spectral resolution of 1 cm^{-1}. The data was normalized against potassium bromide spectrum, obtained from the same instrument under the same instrumental settings.

Physical characterization through isothermal calorimetry

Isothermal calorimetry (GE Healthcare MicroCalT-MiTC200) was used to investigate the potentiality of peptides to interact with free activated nanoparticles. 1 μl sample was injected at each time point (injection time 5 s) with a gap of 300 s between each injection, 40 such injections were carried out. The baseline setting was at 10 μcal. To get a steady baseline, a 2000 s delay was applied to the system. The resulting thermodynamic parameters related to the binding of the peptide to the carboxyl groups on the nanomaterials were obtained from the signal.

Physical characterization through UV–Visible spectroscopy and isothermal calorimetry (ITC)

UV–Vis spectrophotometry was conducted using NanoDrop 2000 (Thermo-Scientific, USA) in the wavelength range of 200–550 nm. Concentration of peptide was evaluated spectrophotometrically and the stoichiometry of peptide–CNT conjugation was determined by Scatchard plot [1]. Thermodynamic parameters such as changes in free energy, enthalpy and entropy of peptide and CNT binding was determined by titrating the activated CNTs by the peptide using ITC as described before [1].

Physical characterization through binding isotherm

Following activation of the carboxyl groups on free CNTs, carboxylated CNTs were titrated against increasing concentrations of the peptide. Change in absorbance at 260 nm was monitored till saturation of binding was observed. Concentration of CNT used was 5 μM for the titration. Binding isotherm for CNT–LL37 conjugation was determined using Scatchard plot to obtain the association constant and the stoichiometry of binding as described elsewhere for binding of CNT and indolicidin [1]. In addition, free non-activated nanoparticles were also titrated against increasing concentrations of LL-37 to ensure specificity of binding. Dissociation binding constant (K_d) and Stoichiometry (B_{max}) was determined using one, two and multiple site binding isotherm models [37]. All analyses were done using GraphPad Prism 5.01 software, CA, USA.

Peptide and CNT conjugated peptide uptake assay through confocal microscopy

Peptide was labeled with Cy3 (GE HealthScience, USA). Cells were treated with free and conjugated labeled peptide at a concentration of 0.02 μg/ml in terms of peptide for 2 h following fixing of the cells with 2% paraformaldehyde. DAPI (Himedia, India) and cell mask red (Invitrogen, USA) were for nuclear and cell membrane staining. Cells were then mounted using fluoromount G (Southern Biotech, USA) and images were taken by a LSM confocal microscope (Carl Zeiss LSM 780, Germany).

Animal cell culture

The Raw 264.7 murine macrophage and THP-1 human primary monocyte cell lines were obtained from ATCC (Manassas, VA). RPMI-1640 (Himedia, India) supplemented with 4.5 g/l D-glucose, 25 mM HEPES, 0.11 g/l sodium pyruvate, 1.5 g/l sodium bicarbonate, 2 mM L-glutamine and 10% FBS along with 100 units/ml Gentamycin and 100 pg/ml Amphotericin-B was used to maintain both the cell lines. PMA (phorbol-12-myristate-13-acetate) of 100 nM was used for THP-1 cell differentiation into adherent macrophages. Both cell lines were routinely cultured in our laboratory at 37 °C in a humidified atmosphere containing 5% CO_2. The cells were sub-cultured twice a week to maintain an exponential growth state.

Cell treatment

One million macrophage cells per well in 2 ml cell culture media were grown in 6 well plates overnight. Free AMP was administered to the cells at a final concentration of either 0.02 or 20 µg/ml. CNT conjugated AMPs were used at 0.02 µg/ml to treat the cells. Control treatment used free CNT and CNT spiked with AMP. Following treatment, the cells were incubated for 6 h in a CO_2 incubator. Following 6 h of incubation, culture medium was aspirated off and the cells were washed 3 times with PBS. Cells were trypsinized and suspended in cell culture medium and centrifuged at approximately 300g for 5 min at 37 °C. The cell pellet was collected to execute further experiments.

RNA isolation

Total RNA was extracted from treated and un-treated THP-1 cells using RNeasy mini kit (Qiagen #74106, Germany) following manufacturers protocol. In brief, cells were gently lysed with 350 µl RLT buffer by gentle pipetting and then equal volume of 70% ethanol was added to the lysate, mixed by pipetting and passed through RNeasy mini column, which retains the RNA in its silica matrix. The column was then washed once with 750 µl RW1 buffer and twice with 500 µl of RPE buffer to remove unwanted lipid, protein and DNA from the matrix. RNA was then eluted from the matrix with 30 µl nuclease free water and kept in ice. The concentration of extracted RNA was measured using NanoDrop 2000 instrument (Thermo Scientific, USA). RNA integrity was checked in bioanalyzer 2100 (Agilent, USA) using an RNA 6000 Nano kit (Agilent, USA) as per the manufacturer's instruction. RNA integrity number 8.5 or more was considered for downstream experiments, which included quantitative real time polymerase chain reaction (qRT-PCR) and whole genome gene expression microarray.

Complementary DNA (cDNA) synthesis

cDNA was synthesized from total RNA using reverse transcription methodology as described here briefly and detailed protocoled can be obtained from reports published before [38]. 5 µg of total RNA was mixed with the buffer containing affinity script reverse transcriptase and polyT primer. The mixture was kept in the thermo cycler at 45 °C for 45 min to synthesize c-DNA. Next, the temperature was raised to 92 °C for 1 min in order to deactivate the enzyme.

Quantitative real time (qRT-) PCR assay

The qRT-PCR was performed using GoTaq qPCR Kit (Promega #A6002, USA) using the manufacturer protocol and the expression profile of select innate immune genes in terms of fold changes for the treatments with respect to untreated samples was checked. The reaction mixture was 25 µl in each well of a 96-well plate. According to the protocol, 9.4 µl of 2× GoTaq qPCR Master Mix, 12.6 µl of nuclease free water, 100 ng of template cDNA and 1 µM of each of forward and reverse primers (primer details are given in Additional file 1: Table S1) were added in each well. qRT-PCR amplification was performed in a programmable thermos-cycler (Stratagene 3500Mxp, USA) with the following settings: 2 min at 92 °C to activate DNA polymerase for 1 cycle, 15 s at 92 °C for melting and 1 min at 60 °C for primer annealing along with extension of the chain and detection of the florescence for 40 cycles. Cycle threshold (C_t) values were noted, and fold changes of the desired genes were calculated with respect to the control after normalizing with the housekeeping gene, β-actin.

Bacterial protection assay

PMA treated 0.5×10^6 differentiated THP-1 cells were seeded into each well of 12-well plate and incubated at 37 °C for 24 h. Cells were treated with LL-37 and indolicidin separately at a final concentration of 20 and 0.02 µg/ml; whereas conjugated peptides were administered at a lower dose of 0.02 µg/ml. After treatment, plates were incubated for 6 h at 37 °C. Thereafter, THP-1 cells were challenged with ST, a pathogenic bacterium, at a multiplicity of infection (MOI) of 10. Cell viability counts at different time points of 6, 12 and 18 h were determined through trypan blue dye exclusion method.

Microarray and data analysis

Genome wide gene expression study was performed using Agilent Quick-Amp labeling Kit (p/n 5190-0444 Agilent, USA). 500 ng of each RNA samples from the control and treated cells were incubated with reverse transcription mix at 40 °C and converted to cDNA primed by oligodT with a T7 polymerase promoter. cDNA synthesized was used as a template for cRNA generation. cRNA was synthesized by in vitro transcription and the dyes used were Cy3 CTP (to label control sample) and Cy5 CTP (to label test samples). The cDNA synthesis and in vitro transcription steps were carried out at 40 °C. Labeled cRNA was cleaned up and quality assessed for the yields and specific activity. 825 ng each of Cy3 and Cy5 labeled samples were fragmented and hybridized to 4 × 44 k microarray slides. Fragmentation of labeled cRNA and hybridization were conducted using the Gene Expression Hybridization kit of Agilent (Part Number 5188–5242, Agilent, USA). Hybridization was carried out in Agilent's Surehyb Chambers at 65 °C for 17 h. The hybridized slides were washed using Agilent Gene Expression wash buffers (Part Number 5188–5327, Agilent, USA). Slides were scanned using Agilent scan control software and data extraction from images was done using feature extraction software Version 10.7 (Agilent, USA). We further obtained normalized fold change values for all genes present on the microarray slides using Arraypipe (v2.0). Differentially regulated genes at various conditions were functionally clustered using WEB-based GEneSeT AnaLysis Toolkit (Webgestalt).

Graphs and statistical analysis

All graphs and statistical analysis were carried out using GraphPad Prism (V5.04, Prism, USA). Statistical analysis was performed using 2-way ANOVA to calculate levels of significance. One standard deviation was calculated and shown in the graphs.

Results

Characterization of CNT conjugates

The conjugation process and characterization for SM-CNT and indolicidin was already reported elsewhere by the current group [1]. Similar methodologies for conjugation were utilized for LL-37 followed by biophysical characterization using UV–Vis spectroscopy, Binding isotherm, isothermal calorimetry and FT-IR spectroscopy. Conclusive characterization was achieved using FT-IR analysis. Peaks, associated with carbonyl (C=O) stretching at 1680 cm^{-1} and amide (C–N) bending at 1645/cm appeared in CNT–LL-37 conjugated AMPs but not in free AMPs, confirmed formation of a peptide bond between SM-CNT and LL-37 (Fig. 1d). Free AMP when spiked with SM-CNT did not result into any such peak.

Isothermal calorimetry results revealed that SM-CNT and LL-37 conjugation process was exothermic with enthalpy (ΔH) at −290.4 ± 2.2 kcal/mol and free energy change (ΔG) −9.98 ± 2.1 kcal/mol. Free LL-37 when spiked with SM-CNT leads to enthalpy (ΔH) at −62.58 ± 2.6 kcal/mol and free energy change (ΔG) −5.36 ± 2.63 kcal/mol (Fig. 1a). We further established the strength and stoichiometry of binding. Fraction bound with increasing concentration of the peptides was determined by spectrophotometric titration for association of LL-37 with CNT (Fig. 1b, c). Dissociation constant (K_d) for the LL-37 and CNT binding was 4.6 µM with a stoichiometry of binding (B_{max}) value of 1.15 results from binding isotherm analysis (Fig. 1c). The binding constants determined also corroborated with the concentrations used in the experiment. There was no significant association observed when non-activated CNT was mixed with LL-37 (spiked samples). This observation was further validated by thermodynamic parameters described above. The CNT conjugated peptide was internalised by the macrophages (Fig. 1g) whereas the free peptide was diffused all over the macrophage cell membrane (Fig. 1f).

Cell viability of THP-1 cells treated with free and conjugated LL-37 and indolicidin

Before establishing the efficacy of CNT conjugated AMPs, it is important to check viability of the cells following treatment with nano-conjugated peptides. Viability of the human macrophage cell line THP-1 and mice macrophage cell line Raw 264.7 were determined up to 6 h following treatment with free and CNT conjugated AMPS (LL-37 and indolicidin). Macrophage cells were viable for entire 6 h following treatment with free and conjugated AMPs (Fig. 2a, e). Longer than 6 h time point was not chosen since, (a) our goal is to prime early immune response and (b) 4 h was shown as sufficient to exhibit immune modulation in vitro [39]. For free AMPS highest concentration reported was 50 µg/ml while that for conjugated AMP was 2 µg/ml. Effects of untreated as well as free AMPS spiked with equivalent amount of CNTs as control groups of treatment were also evaluated. Experimental data revealed that none of the above treatments were toxic to the macrophage cell lines THP-1 (Fig. 2b, f). Similar results were also observed in Raw 264.7 (data not shown).

Expression of selected innate immune genes

Objective, of the current work, is to understand potential of LL-37 and indolicidin in modulating expression of innate immune genes. Following the establishment of AMP toxicity to cell lines, we, therefore, evaluated the effects of AMP treatment on select innate immune gene expression to understand AMPs potential as an immune stimulant that facilitates antimicrobial activity. Expression kinetics of few select innate immune genes in THP-1 cells following peptide treatment at 20 µg/ml and CNT conjugated peptide at 0.02 µg/ml revealed that, optimal expression of genes occurred at 6 h (Fig. 1h, i). Expression values for a few innate immune genes at the transcriptional level in THP-1 cells following 6 h treatment with free LL-37 at concentrations of 1, 10 and 20 µg/ml were determined by qRT-PCR. Experimental data revealed that with increasing concentration of LL-37, the expression at a transcriptional level for the genes *IFNB1*, *IFNA*, *IL6*, *IL10*, *IL12*, *TNFA* and *NFκB1* increased with respect to time matched untreated controls (Fig. 2c). When the expression of the above-mentioned genes were calculated in THP-1 cells at 6 h following treatment with 0.02 µg/ml of conjugated LL-37, it was observed that the expression of these genes was similar (1-way ANOVA, p ≥ 0.05, n = 3) in comparison to free LL-37 which was treated with a higher dose of 20 µg/ml (Fig. 2c). Similarly, with increasing dose of free indolicidin, the above mentioned innate immune gene expressions increased with respect to time matched untreated control (Fig. 2g). Statistical analysis using 1-way ANOVA (p ≥ 0.05, n = 3) confirms

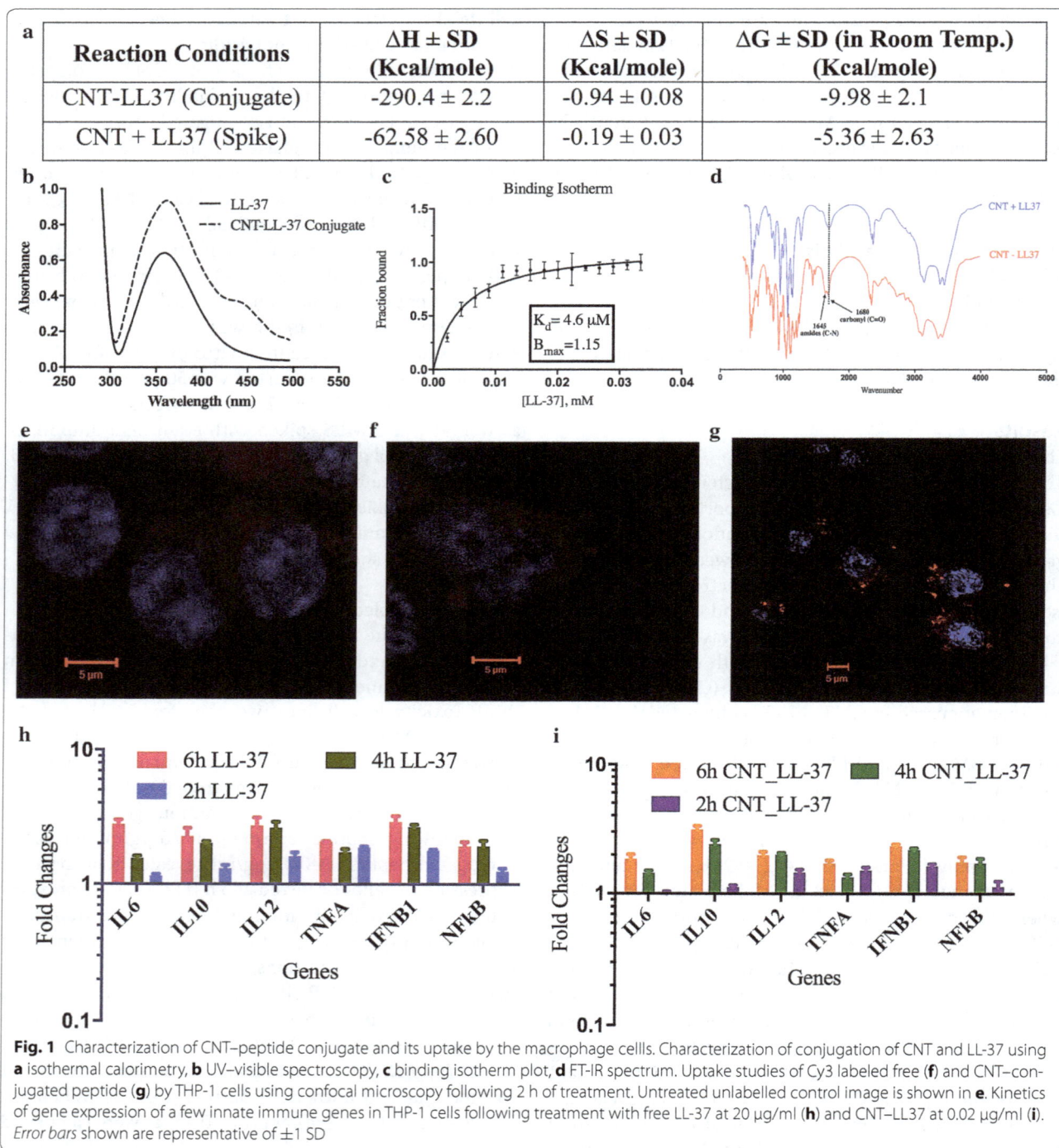

Reaction Conditions	$\Delta H \pm SD$ (Kcal/mole)	$\Delta S \pm SD$ (Kcal/mole)	$\Delta G \pm SD$ (in Room Temp.) (Kcal/mole)
CNT-LL37 (Conjugate)	-290.4 ± 2.2	-0.94 ± 0.08	-9.98 ± 2.1
CNT + LL37 (Spike)	-62.58 ± 2.60	-0.19 ± 0.03	-5.36 ± 2.63

Fig. 1 Characterization of CNT–peptide conjugate and its uptake by the macrophage cellls. Characterization of conjugation of CNT and LL-37 using **a** isothermal calorimetry, **b** UV–visible spectroscopy, **c** binding isotherm plot, **d** FT-IR spectrum. Uptake studies of Cy3 labeled free (**f**) and CNT–conjugated peptide (**g**) by THP-1 cells using confocal microscopy following 2 h of treatment. Untreated unlabelled control image is shown in **e**. Kinetics of gene expression of a few innate immune genes in THP-1 cells following treatment with free LL-37 at 20 μg/ml (**h**) and CNT–LL37 at 0.02 μg/ml (**i**). *Error bars* shown are representative of ±1 SD

that the gene expression pattern was almost similar in 20 μg/ml indolicidin treated cells with respect to 0.02 μg/ml indolicidin conjugated cells (Fig. 2g). It is clear, that, both of the conjugated peptides can induce controlled up-regulation of select innate immune genes at a lower dosage (1000 fold less) than that of the free peptides.

Protection against ST challenge

When innate immune genes are moderately (fold changes up to 5) up-regulated in macrophages, we defined the macrophages as primed [38]. Primed macrophages have the ability to combat bacterial pathogens more efficiently than naive macrophages. To prove this hypothesis,

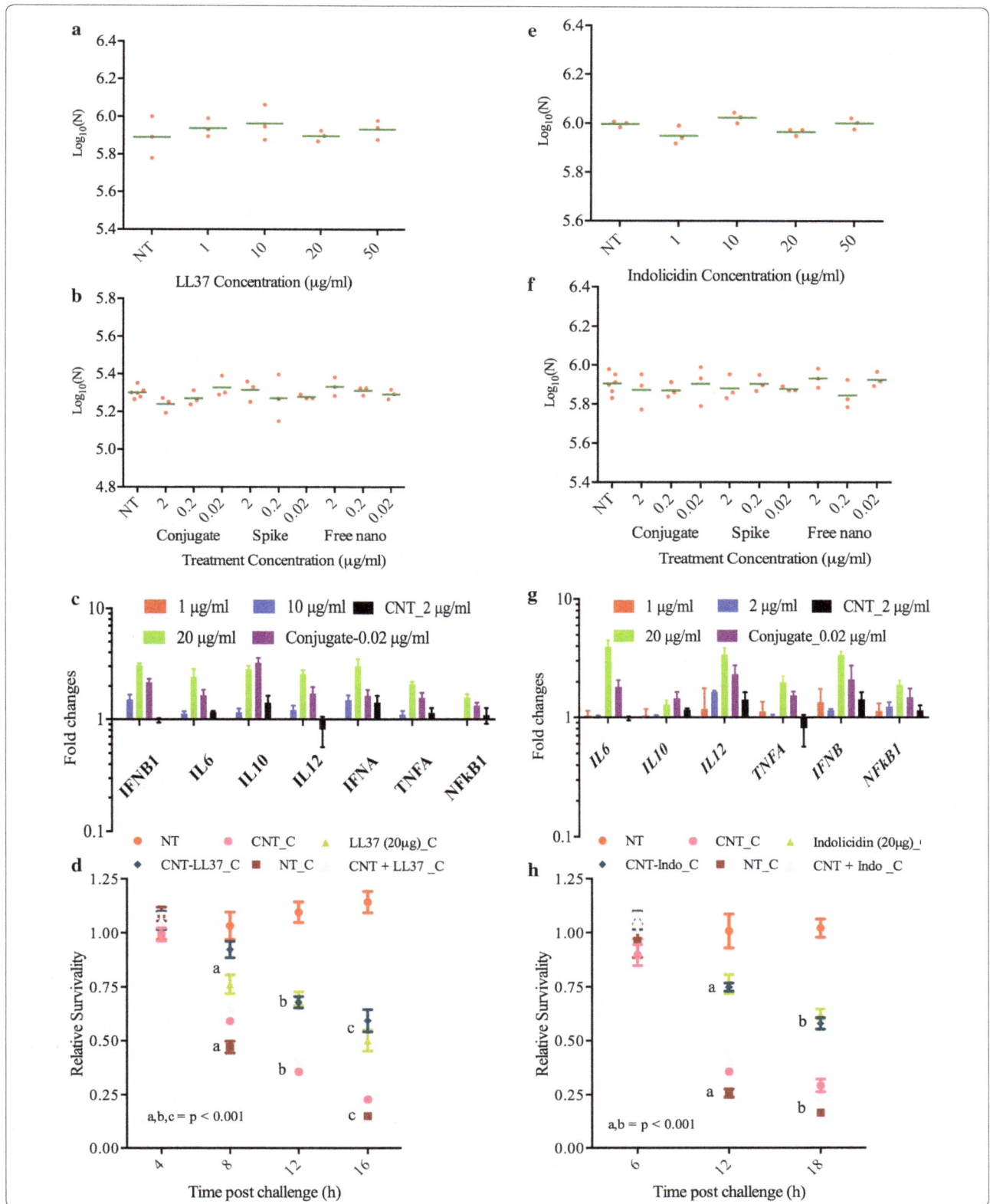

(See figure on previous page.)

Fig. 2 Viability of nano conjugated HDPs followed by gene expression modulation and protection of cells against *Salmonella* challenge. Viability of THP_1 cells following treatment with various concentrations of **a** LL-37, **b** CNT, CNT conjugated LL-37 and LL-37 spiked CNT, **e** indolicidin and **f** CNT, CNT conjugated indolicidin and indolicidin spiked CNT. Expression of select innate immune genes in Thp1 cells following treatment with **c** LL-37 and CNT conjugated LL-37 and **g** indolicidin and CNT–conjugated indolicidin. Relative survivability of THP-1 cells being challenged by *Salmonella* in the absence or presence of **d** free LL-37 at 20 µg/ml or CNT conjugated LL-37 at 0.02 µg/ml and **h** free indolicidin at 20 µg/ml or CNT conjugated indolicidin at 0.02 µg/ml. Significant changes with p ≤ 0.001 at each time point is shown in *letters*

free LL-37 and indolicidin, as well as conjugated AMPs primed THP-1 cells, were exposed to ST at MOI of 10. Results revealed that macrophage cells primed with free LL-37 at 20 µg/ml and CNT–LL-37 at 0.02 µg/ml were significantly (2-way ANOVA, p ≤ 0.001, n = 12) protected against ST challenge with respect to the unprimed cells (Fig. 2d). Primed cell survivability was found to be 80, 65 and 55%, whereas, unprimed cell survival was 48, 36 and 14% at 8, 12 and 16 h post challenge. Similarly, THP-1 cells primed with free indolicidin at 20 µg/ml and CNT–indolicidin at 0.02 µg/ml were significantly (2-way ANOVA, p ≤ 0.001, n = 12) protected against ST infection. The survivability of primed cells was found to be 75 and 55%, whereas, the survival of unprimed cells was 28 and 17% at 12 h and 18 h post challenge (Fig. 2h). The results from our pathogenic challenge study revealed that conjugating LL-37 and indolicidin with SM-CNT increased their immune modulatory efficacy by 1000 folds. However, the exact mechanism through which priming occurs is yet to be elucidated. Therefore, we conducted genome wide transcriptional gene expression microarray studies to understand the plethora of genes that could be responsible for priming.

Genome wide gene expression to elucidate transcriptional pathway biology in vitro

We performed experiments with a view to understanding genome wide transcriptomic profiling by studying gene expression changes in THP-1 cells following treatment with either unconjugated LL-37 or indolicidin at 20 µg/ml or CNT conjugates at 0.02 µg/ml of either conjugate. For control studies, cells were also treated with free CNT or free AMP or AMPs spiked with SM-CNT at their respective conjugate concentrations. The genes were considered to be differentially expressed and statistically significant, if fold changes were ≥1.5 with p ≤ 0.05. There were total of 3784, 1535, 2197, 1563 genes that were differentially expressed in THP-1 cells following 6 h treatment with CNT, CNT + LL-37, CNT–LL-37 and LL-37-20 respectively (Fig. 3a). Out of which 3171, 489, 488, 835 genes were unique in THP-1 cells at 6 h following treatment with CNT, CNT + LL-37, CNT–LL-37 and LL-37-20. Similarly, a total of 3784, 2446, 2221 and 2015 genes were differentially expressed in THP-1 cells at 6 h following treatment with CNT,

CNT + indolicidin, CNT–indolicidin and indolicidin-20 respectively (Fig. 3c). Out of these differentially expressed genes, 1172, 1172, 885 and 1109 were uniquely expressed at 6 h following treatment with CNT, CNT + indolicidin, CNT–indolicidin and indolicidin-20.

The genes were clustered using GeneAnalytics to find out the biological pathways that are populated most. The top 5 pathways related to innate immune signalling and cell cycle regulation are listed in Table 1. Innate immune signalling was enriched with 90 genes in THP-1 cells treated with free indolicidin as well as CNT conjugated indolicidin. There are 46 genes enriched with infectious disease signalling in the THP-1 cells treated with free LL-37 as well as conjugated LL-37. LL-37 also activates TGFβ signaling in THP-1 cells, which indicates feedback suppression of inflammatory pathways. The expression levels of these enriched genes are comparable in free peptide as well as CNT conjugated peptide treatments, but at 1000 fold less concentration (Additional file 2: Table S2, Additional file 3: Table S3). Important genes with their expression level and function are listed in Table 2 (for LL-37) and in Table 3 (for indolicidin). The gene expression profile revealed that conjugated AMPs show similar effects as free AMPs, but at a 1000-fold lower concentration (Tables 2, 3). It was also observed that when indolicidin spiked with CNT, the gene expression profile was better than free peptide at 0.02 µg/ml but we did not find this phenomenon in case of LL-37. This may be due to indolicidin stacking over the CNT surface via π electron cloud overlap of both the substances; increasing the effectiveness of indolicidin delivery into the cell through the added hydrophobicity from CNT. However, more study needs to be done to confirm this phenomenon.

We have tried to populate the pathways with important innate immune genes which were differentially expressed following treatment with both of the conjugates. From the list of differential gene expression, we searched for receptors, adaptors, kinases and transcription factors which are related to immune signaling and match them with the KEGG pathways. The genes of NFκB1 pathway and its downstream genes are up-regulated in THP-1 cells following LL-37 and CNT–LL-37 treatment. It was also observed that interleukin 1 receptor and its subsequent adaptors and kinases such as Myd88, Traf6 and Map3k7 was up-regulated in THP-1 cells following LL-37 and its

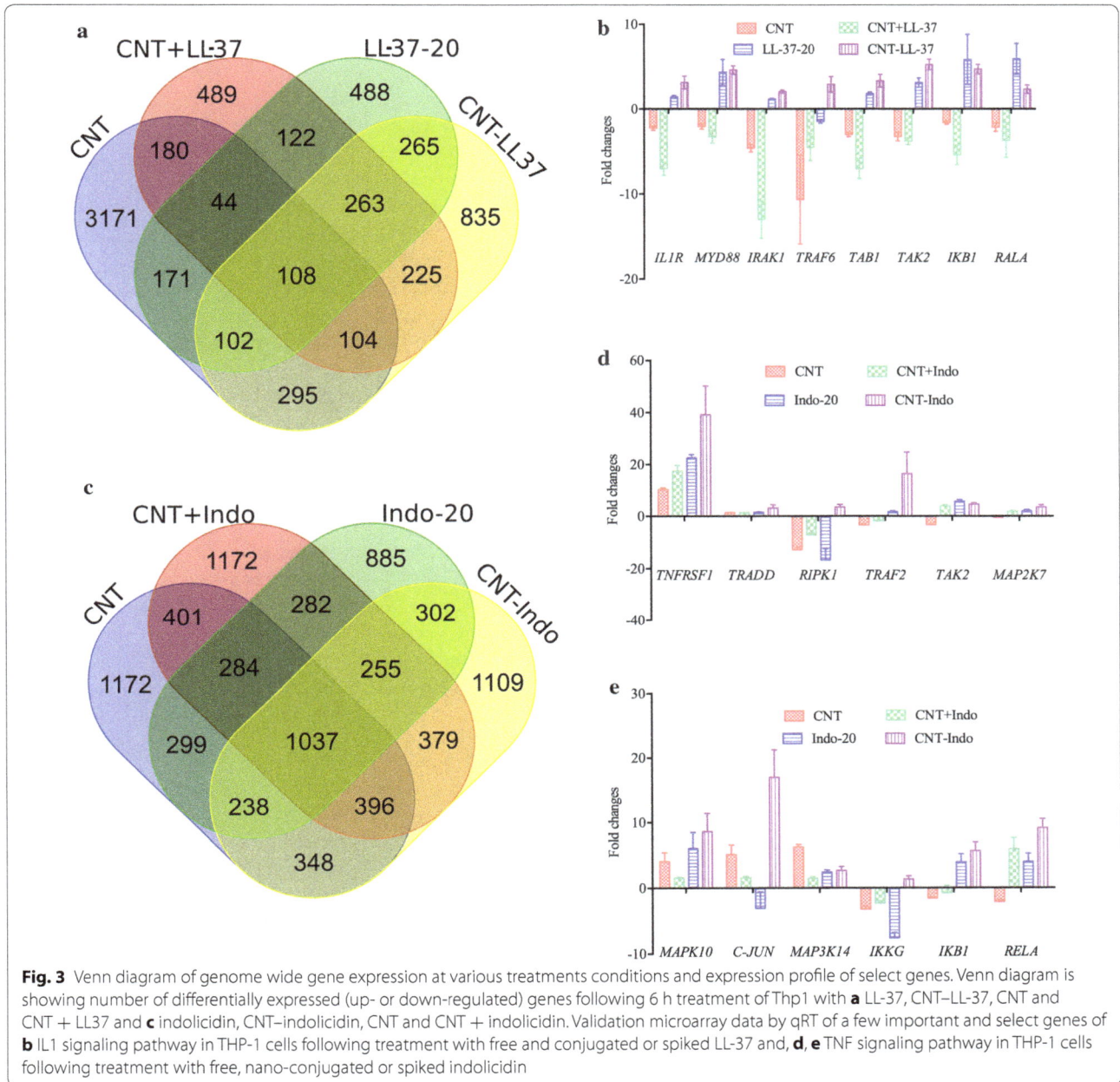

Fig. 3 Venn diagram of genome wide gene expression at various treatments conditions and expression profile of select genes. Venn diagram is showing number of differentially expressed (up- or down-regulated) genes following 6 h treatment of Thp1 with **a** LL-37, CNT–LL-37, CNT and CNT + LL-37 and **c** indolicidin, CNT–indolicidin, CNT and CNT + indolicidin. Validation microarray data by qRT of a few important and select genes of **b** IL1 signaling pathway in THP-1 cells following treatment with free and conjugated or spiked LL-37 and, **d**, **e** TNF signaling pathway in THP-1 cells following treatment with free, nano-conjugated or spiked indolicidin

Table 1 Top 5 enriched pathways in THP-1 following 6 h treatment with LL-37 and indolicidin

LL-37		Indolicidin	
Pathways	Enriched gene number	Pathways	Enriched gene number
AKT signaling	63	Innate immune system	90
Infectious diseases	46	MAPK signaling	54
IGF1R signaling	20	AKT signaling	55
TGFB signaling	20	TNF signaling	21
Cell cycle	22	Chemokine signaling	26

Table 2 Important genes differentially expressed in THP-1 following 6 h treatments with LL-37

Gene symbol	Fold changes WRT NT				Entrez ID	Function
	CNT	LL37-20	CNT–LL37	CNT + LL37		
TSG101	2.1	1.0	6.8	1.0	7251	Acts as a negative growth regulator
GRB2	1.5	1.0	5.2	1.0	2885	Links cell surface GFRs and the Ras signaling pathway
IL9R	1.0	2.6	2.6	2.5	3581	Interleukin-3, 5 and GM-CSF signaling
MAP4K3	1.0	1.7	6.2	1.0	8491	MAPK signaling pathway and TNF signaling
CFLAR	1.0	1.0	6.3	1.0	8837	Acts as an inhibitor of TNFRSF6 mediated apoptosis
ENPP1	3.0	1.0	7.0	1.0	5167	Appears to modulate insulin sensitivity and function
RALBP1	1.0	3.1	4.5	−2.6	10928	Can catalyze transport of glutathione and xenobiotics
SUCLA2	−1.8	4.0	5.6	4.8	8803	Catalyzes succinyl-CoA production
ALOX5	1.0	1.0	6.0	1.0	240	Catalyzes leukotriene biosynthesis and inflammation
MAOB	1.0	4.6	4.3	1.0	4129	Oxidative deamination of biogenic and xenobiotic amines
CCL20	5.5	9.2	10.1	1.0	6364	Chemotactic to lymphocytes and neutrophils. Possesses antibacterial activity *E. coli* and *S. aureus*
IL33	1.0	1.0	3.5	1.0	90865	Activates NF-kappa-B and MAPK signaling pathways
IL36G	1.0	2.9	3.8	2.7	56300	Activates NF-kappa-B and MAPK signaling pathways
SLC2A14	1.0	1.0	4.1	3.3	144195	Facilitative glucose transporter
DEFB105B	1.0	1.0	2.8	1.0	504180	Has antibacterial activity
DEFA5	1.0	1.0	5.0	1.0	1670	Antimicrobial activity against broad spectrum bacteria
EFR3A	−25.4	1.0	7.3	1.0	23167	Signaling through PIP3K and G protein couples receptors
PLA1A	1.0	6.2	5.2	1.0	51365	Stimulate histamine production
INO80B	1.0	1.0	2.8	1.0	83444	Cell cycle arrests at the G1 phase of the cell cycle
TLE1	1.0	3.8	5.8	1.0	7088	Inhibits NF-kappa-B-regulated gene expression
SKP2	1.0	2.3	7.1	1.0	6502	involved in regulation of G1/S transition
NLRC4	1.0	1.0	2.9	1.0	58484	Senses specific proteins from pathogenic bacteria and fungi and responds by assembling an inflammasome complex
BTG3	1.0	5.7	6.1	1.0	10950	Blocks cell cycle at G0/G1 to S phase
SLC22A15	5.6	1.0	9.5	9.0	55356	Probably transports organic cations
IL11RA	1.0	3.6	3.2	1.0	3590	Involved in macrophage proliferation and differentiation
OGFOD2	1.0	4.9	6.9	5.4	79676	Iron ion binding and oxidoreductase activity
IL17RE	1.0	1.0	3.9	3.8	132014	crucial regulator in innate immunity to bacterial pathogens
CCNE2	1.0	4.7	4.5	3.2	9134	Blocks cell cycle at the G1-S phase
IRF1	1.0	1.0	4.7	1.0	3659	Regulation of IFNs against viral and bacterial infections

CNT conjugate treatment. From this it may be inferred that LL-37 signals through the interleukin 1 receptor (IL1R) followed by nuclear factor kappa B1 (NFκB1) translocation to nucleus with subsequent transcription of pro-inflammatory cytokines, chemokines and defensins. We also observed the expression of genes related to cell proliferation and differentiation along with up-regulation of calcium transporter (CACNA1B), protein phosphatase 3 catalytic subunit alpha (PPP3CA) and NFAT2. It may be inferred that calcium release to cytoplasm through CACNA1B activated PPP3CA which in turn dephosphorylate the transcription factor NFAT2 which is subsequently translocated to the nucleus to transcribe genes related to cell proliferation and differentiation. The above pathway is represented in Fig. 4a. The genes involved

in this pathway were also validated through qRT-PCR and represented with their fold changes with respect to untreated time matched controls in Fig. 3b.

It was also observed that many genes related to pro-inflammation, cell proliferation and cell differentiation were significantly up-regulated in THP-1 cells following treatment with indolicidin and CNT conjugated indolicidin. Tumor necrosis factor receptor 1a (TNFSF1A), MAP3K11, MAP3K14, NFκB and c-JUN are up-regulated in THP-1 cells following treatment with indolicidin. From this gene expression data, it may be inferred that TNFRSF1A signals through MAP3K11 and MAP3K14 followed by activation of transcription factors NFκB1 and c-JUN. Eventually the genes related to cell proliferation, differentiation and pro-inflammation get activated in

Table 3 Important genes differentially expressed in THP-1 following 6 h treatments with indolicidin

Gene symbol	Fold changes WRT NT				Entrez Gene ID	Gene functions
	CNT	CNT + Indo	CNT–Indo	Indo-20		
TNFRSF1A	6.5	42.3	29.0	33.2	7132	Activate NFkB, mediated regulator of inflammation
RBCK1	3.4	10.9	6.9	1.0	10616	Activation of canonical NFkB and the JNK signaling
SLC11A1 & SLC5A5	4.4	6.1	6.9	4.3	6556	Transport of glucose and other sugars, bile salts and organic acids, metal ions and amine compounds
	2.2	5.0	8.7	1.8	6528	
ENPP7	5.1	12.1	16.3	8.5	339221	Converts sphingomyelin to ceramide
SOSTDC1	2.8	2.5	5.0	2.5	25928	Enhances Wnt and inhibits TGF-beta signaling
XPR1	1.0	1.6	47.6	−1.9	9213	G-protein coupled receptor activity
S100A5	69.2	67.4	150.4	113.5	6276	Helps in cell cycle progression and differentiation
NAPEPLD	8.0	25.7	22.6	17.7	222236	Responsible for the generation of anandamide, the ligand of cannabinoid and vanilloid receptors
RGS11	2.1	3.0	5.9	1.9	8786	Inhibits signal transduction by G Protein
RGS6	1.0	−1.6	5.9	1.5	9628	Inhibits signal transduction by G protein
RUSC1	1.1	−1.0	5.7	1.6	23623	Activation of the NFkB pathway
IBA57	2.5	4.0	4.9	1.0	200205	Activates iron-sulfur cluster assembly pathway
MYO5B	1.0	−1.3	13.7	−1.0	4645	Vesicular trafficking with the CART complex
AOX1	2.8	2.7	4.7	2.6	316	Regulation of reactive oxygen species homeostasis
RIPK4	3.7	7.3	6.9	1.4	54101	Plays a role in NF-kappa-B activation
FILIP1L	2.0	10.6	14.0	2.2	11259	Leads to inhibition of cell proliferation and migration
NCOA4	2.4	5.9	10.6	5.8	8031	Co-activator of the PPARG
CREB3L3	1.7	7.0	4.5	4.9	84699	Linked to acute inflammatory response
RASSF2	6.5	6.7	12.2	7.3	9770	May promote apoptosis and cell cycle arrest
DAP	2.8	1.7	7.0	2.7	1611	Negative regulator of autophagy
RELA	−1.2	1.7	8.4	−1.2	5970	NFkB pleiotropic transcription factor
CCL20	5.5	1.5	10.9	11.0	6364	Antibacterial activity against *E. coli* and *S. aureus*
LYRM4	3.3	3.4	6.2	2.2	57128	Nuclear and mitochondrial FE-S protein biosynthesis
ZDHHC22	17.0	41.5	44.6	44.4	283576	Feedback regulator of calcium mediated signaling
CYP3A5	3.7	3.6	6.9	3.7	1577	Oxidizes steroids, fatty acids, and xenobiotics

THP-1 cells. This pathway was represented schematically in Fig. 4b and the genes shown were validated through qRT-PCR and plotted with their fold changes with respect to untreated time matched controls in Fig. 3d, e.

Discussion

Cationic AMPs are non-toxic to cells up to a concentration of 50 μg/ml [40]. Our findings corroborate with this pre-established level of AMP toxicity. Our results revealed that following LL-37 and indolicidin treatment up to 50 μg/ml, human and mice macrophage cells show no toxic effects (Fig. 2a, e). The immune modulatory properties of AMPs include modulating pro- and anti-inflammatory responses through [41] various signaling pathways [40] directly [42] or indirectly [43] recruiting effector cells including phagocytes to the site of infection, enhancing intracellular [44] and extracellular [45] bactericidal activity. AMPs also mediate macrophage differentiation [46] which is required for effective clearance of pathogens from the host. AMPs also induce apoptosis

[47] and pyroptosis [48] in the infected cells as a means of clearing pathogens. Despite their effectiveness in pathogen clearance, host defense peptides (HDPs) were not popular for clinical usage because of their high synthesis cost [49]. There is a need to improve the efficacy of AMPs and we tried to accomplish this by conjugating LL-37 and indolicidin to carboxylated CNT. We compared immune modulatory properties of these two peptides in treated macrophages, both in free and CNT conjugated state.

Role of LL-37, indolicidin and their CNT conjugates in modulating pro- and anti-inflammation in THP-1 human macrophage cells

LL-37 activates the canonical NFkB pathway responsible for modulating the expression of various genes involved in the innate immune system [19]. Our results revealed that the expression of pro-inflammatory genes that play a critical role in regulating the NFκB pathway [RELA, TNFRSF1A, TRAF6, ATM and BTRC (Additional file 2: Table S2)] are up-regulated upon CNT–LL-37 treatment;

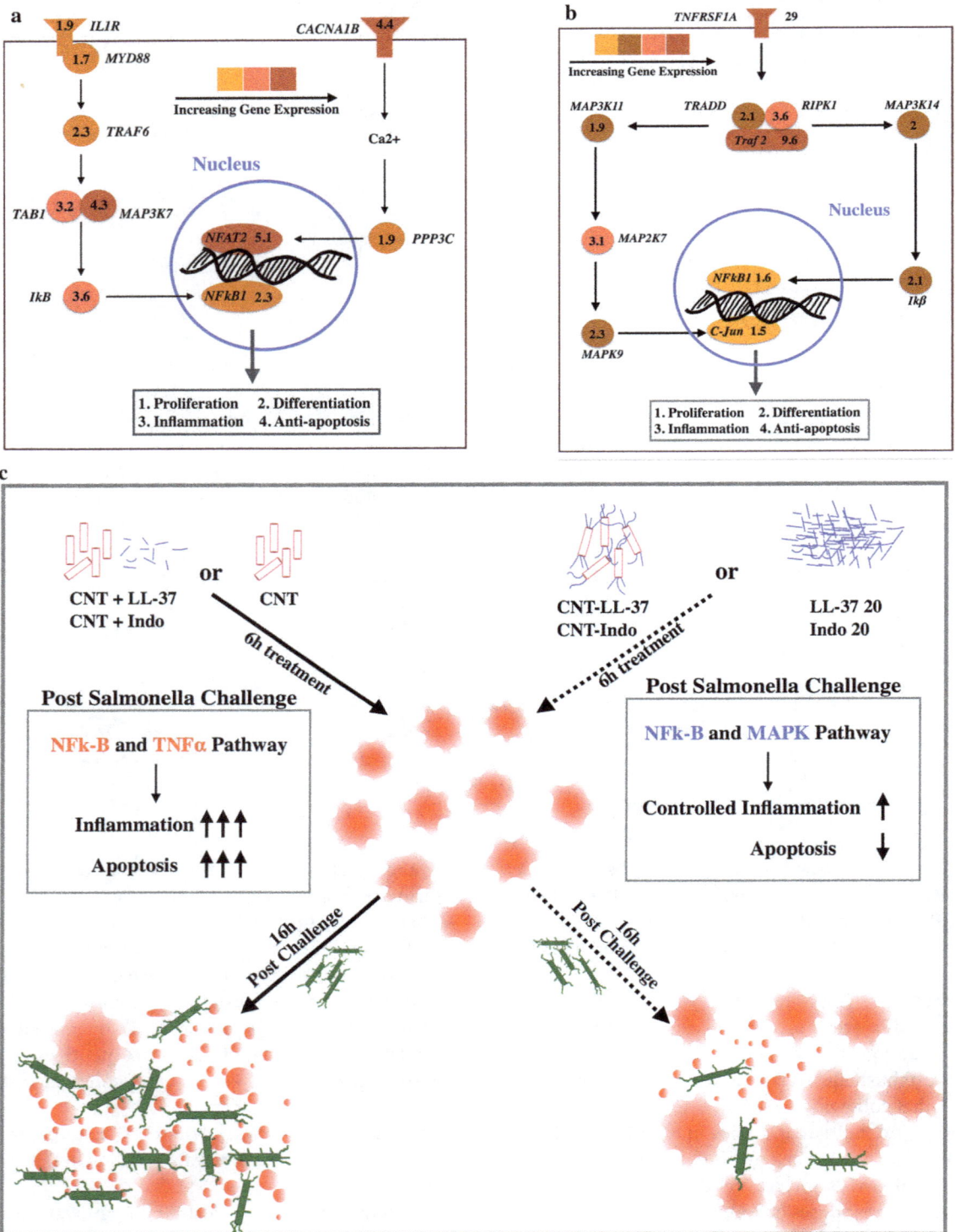

Fig. 4 Pathways that were indicated by genome wide microarray data and validated by qRT-PCR results following treatment with nano-conjugated LL-37 and indolicidin. Activation of **a** IL1 pathway in THP-1 cells following treatment with conjugated LL-37 at 0.02 µg/ml, **b** TNF pathway via TNFRSF1A in THP-1 cells following treatment with conjugated indolicidin at 0.02 µg/ml. **c** Graphical summary of the consolidated schema of the current study depicting how LL-37 and indolicidin primed Thp1 cells to protect against *Salmonella* infection

notably, LL-37 conjugate treated cells also show almost the same expression pattern, but at 1000 fold less concentration than free LL-37. The expression profile of TNFRSF1A, LELA, RIPK4, RUSC1 and RBCK1 genes in CNT–indolicidin as well as free indolicidin treated cells confirms pro-inflammation mediated by the NFκB pathway. LL-37 induced signaling through the P38 MAPK pathway, followed by activation of genes responsible for macrophage differentiation, pro-inflammation and proliferation [50]. Activation of 36 genes (Additional file 2: Table S2) related to the P38 MAPK pathway following LL-37 treatment, and up-regulation of several genes (Additional file 3: Table S3) following indolicidin treatment confirmed that the conjugates stimulate signaling similar to that of the free peptide, but at a 1000-fold lower dose. Through the interaction of phosphoinositide 3-kinase (PI3 K), NFκB and MAPK pathways, LL-37 induced IL-1B, followed by pro-inflammation in monocytes and macrophages [51]. Expression of IL1B (Additional file 2: Table S2, Additional file 3: Table S3) in THP-1 cells following LL-37 and indolicidin conjugate treatment indicates that conjugated AMPs can induce pro-inflammation in macrophages at the same levels stimulated by free peptides, but at a 1000-fold lower concentration. AMPs can also induce the production of IL17 and reactive oxygen species (ROS), enhancing the phagocytic activity of macrophages and capacity for clearing pathogens within the phagosome [52]. Expression of IL17RE (Table 2) in THP-1 cells following treatment with LL-37 as well CNT conjugated LL-37 indicates a similar effect of the conjugate on macrophages. This is a controlled inflammation, indicated by the moderate expression of anti-inflammatory cytokine IL10 along with pro-inflammatory cytokines IL6, IL12, IL1a, IFNa and IFNb (Fig. 2c, g) in the conjugate treatments as well as free LL-37 and indolicidin treatments at 20 µg/ml. This data indicated that LL-37 and indolicidin conjugates are similarly effective in modulating pro-inflammatory pathways as their free peptide forms, but at 1000 fold less concentration.

Role of LL-37, indolicidin and their CNT conjugates in modulating chemokine expression in the THP-1 human macrophage cell line

AMPs are chemo attractants for monocytes, neutrophils, macrophages and T-cells. These molecules can induce various chemokines and chemokine receptors in macrophages to attract these immune cells to the site of infection [43, 53]. Expression of CCL20, CCL4 and CCL19 (Table 2) in CNT–LL-37 and CCL20, CCL19, CCL7, CCL4 (Table 3) in indolicidin conjugate treated THP-1 cells indicates that both conjugates are able to stimulate similar signaling pathways involved in macrophage

chemotaxis. However, this conclusion needs to be verified in vivo.

Additional functions of free and conjugated AMPs in THP-1 cells

AMPs induce autophagy in infected macrophages to facilitate the clearance of intercellular debris, which is controlled through ATG5 gene [54]. Expression of ATG5 in LL-37 and its CNT conjugate treated cells might result in similar autophagic activity. However, CNT itself is a very efficient autophagic inducer, as CNT treatment shows 16 fold up-regulation of ATG5 in THP-1 cells. Apart from enhancing the efficacy of AMPs, CNT itself has this added effect. LL-37 and beta-defensin, induce epidermal growth factor receptor (EGFR) signaling, followed by activation of the PI3 K-AKT and MAPK pathways responsible for cell proliferation during wound healing [55]. THP-1 cells treated with LL-37 and its conjugate also show activation of the PI3 K-AKT and MAPK signaling pathways (Table 1), as well as up-regulation of PDGFRA gene (Additional file 2: Table S2).

It is reported that LL-37 delays apoptosis in monocytes and neutrophils by activating G protein coupled receptor (GPCR) mediated signaling [47]. THP-1 cells treated with LL-37 or its conjugate also appear to exhibit active GPCR signaling as we have recorded the expression of GPR180, GPRC5C, GPR174 and GPR3 (Additional file 2: Table S2); similarly, THP-1 cells treated with indolicidin or its conjugate resulted in GPCR signaling as shown by the expression of GPR135, GPR176, GPR112, GPR110, GPR173 and GPR3 (Additional file 3: Table S3). Compared to LL-37, indolicidin imparts more pro-inflammatory effects in THP-1 cells; however, the regulation of inflammation through TGFβ and IL10 pathway was stronger in LL-37 treated cells, indicated by the genes listed in Table 4. The overall mechanism through which LL-37, indolicidin and their CNT conjugates protect macrophages from *salmonella* induced cytotoxicity could be graphically summarized, as depicted in Fig. 4c.

Data from *Salmonella* challenge studies revealed that, LL-37 or indolicidin primed THP-1 cells can efficiently protect themselves against ST induced cytotoxicity for 16 h post challenge. The genome wide gene expression study shows that pro-inflammatory and anti-apoptotic signaling in THP-1 cells treated with indolicidin may be mediated through *TNFRSF1A*, followed by activation of *NFκB* and *c-JUN*. However, pro-inflammation, cell proliferation and cell differentiation in THP-1 following LL-37 treatment may mediated through *IL1R*, followed by activation *of NFκB* and *NFAT2*. Though immune modulation by LL-37 and indolicidin was partly known before, our data established the complete gene expression and signaling mechanism. The conjugation strategy enhanced

Table 4 LL-37 and indolicidin modulates immune genes and pro-apoptotic genes differently

Gene symbol	CNT	CNT + Indo	CNT–Indo	Indo-20	CNT + LL37	CNT–LL37	LL37-20	Gene description
ANAPC11	−2.4	1.2	1.7	−1.4	1	1	1	Anaphase promoting complex subunit 11
CCL20	5.5	1.5	10.9	11	1	10.1	9.2	Chemokine (C–C motif) ligand 20
DEFA3	1.1	1	1.6	1.1	1	1	1	Defensin, alpha 3
IL31RA	1.4	−6.7	1.6	−1	1	4.4	1	Interleukin 31 receptor A
NFATC2	3.1	2.1	4.3	1.9	1.7	4.9	12.4	Nuclear factor of activated T-cells, cytoplasmic, calcineurin-dependent 2
NFATC2IP	1.3	−1.8	2	1.1	1	1	1	Nuclear factor of activated T-cells, cytoplasmic, calcineurin-dependent 2 interacting protein
PKMYT1	−1	−1.4	1.6	−1.3	1	1	1	Protein kinase, membrane associated tyrosine/threonine 1 (PKMYT1)
RB1	−3.3	−4.1	1.6	1.6	1	1	1	Retinoblastoma 1
SMAD3	−1.3	−2.3	2.5	1.2	1	2.5	1	Mad protein homolog
TGFBR1	1	−1	1.5	−1.1	1	2.5	1.7	Transforming growth factor, beta receptor 1
CHEK1	−3	−1.1	−6.5	1.6	1	−3.9	1	CHK1 checkpoint kinase
CDKN2A	1.5	−1.2	−4.2	−2	1	4.6	4.3	Cyclin-dependent kinase inhibitor 2A
TNFSF4	1.1	1	−3.3	5.5	1	−3.2	−2.9	Tumor necrosis factor (ligand) superfamily, member 4
CDC25C	−1	−1	−2.9	1	−3.6	1	1	Cell division cycle 25 homolog C
CDC14B	−1	−1.3	−2.4	−1.7	−3.8	−7.4	1	CDC14 cell division cycle 14 homolog B
GADD45B	1.8	−1.8	−2.1	−2	1	2.2	1.6	Growth arrest and DNA-damage-inducible, beta
TLR1	−4.4	−1.2	−1.4	−1.8	1	1	1	Toll-like receptor 1
TLR3	1	1	2.3	1	1	2.8	1	Toll-like receptor 3
TNFAIP3	−2.5	−7	−9	1.6	−4.4	1	1	Tumor necrosis factor, alpha-induced protein 3
CCL14	4.3	−1.3	3.9	−1.2	1	1	1	Chemokine (C–C motif) ligand 14
BCL2L2	1.4	2.7	2.5	1.1	1.1	3.2	1.3	BCL2-like 2
Apaf1	4.3	1.1	1.1	1.0	1.2	1.1	1.1	Apoptotic peptidase activating factor 1

the immune modulating efficacy of these two peptides by 1000 fold, which will reduce the cost of these peptides for antimicrobial treatment, thereby increasing treatment access to a wider population of developing countries. Although LL-37 and indolicidin conjugation with CNT shows promise with regards to resisting ST infection in vitro, further trials need to be conducted in vivo for better understanding of its working mechanism.

Conclusions

Present study established an important fact for the usage of nanomaterials in biomedicine is that efficacy of a drug or a bio-agent can be enhanced by conjugating it with a suitable nanomaterial. Our results confirmed that at 1000-fold less concentration of either of the peptide, LL-37 or indolicidin, can be equally effective at 0.02 μg/ml while the same is observed at 20 μg/ml with free peptides. The current report also established the efficacy of the conjugated peptide in protecting macrophage cells against *salmonella* challenge as well as mechanism by which the peptides are protecting is also reported.

Additional files

Additional file 1: Table S1. List of primers used to validate microarray through q-RTPCR.

Additional file 2: Table S2. Gene expression in terms of fold changes following treatment with free LL37, conjugates and other relevant controls.

Additional file 3: Table S3. Expression of genes in terms of fold changes following treatment free and conjugated Indolicidin and other conditions.

Abbreviations
AMPs: antimicrobial peptides; CNT: carbon nanotubes; SM-CNT: short multi-walled carbon nano-tubes; ST: *Salmonella typhimurium serovar enterica*; PMA: phorbol-12-myristate-13-acetate; qRT-PCR: quantitative real time polymerase chain reaction; MOI: multiplicity of infection; cDNA: complementary DNA; Webgestalt: WEB-based GEneSeT AnaLysis Toolkit.

Authors' contributions
BP executed most of the experiments and analyzed data. DG performed some experiments. BP and DG also assisted in writing the draft manuscript. AS and KM assisted in performing a few experiments. PR and DD helped in analyzing transcriptomic data. LL-37 was a gift from Professor REW Hancock of Hancock Lan, UBC, BC, Canada. PA conceptualized, supervised and finalized the work. All authors read and approved the final manuscript.

Acknowledgements
Authors acknowledge the facility and infrastructure provided by NISER to execute the work.

Competing interests
The authors declare that they have no competing interests.

Funding
Funding to support part of the work was approved by the Department of Science and Technology, Ministry of Science and Technology, Govt. of India with Grant # SR/NM/NS-58/2010 to Palok Aich, PhD.

References

1. Sur A, Pradhan B, Banerjee A, Aich P. Immune activation efficacy of indolicidin is enhanced upon conjugation with carbon nanotubes and gold nanoparticles. PLoS ONE. 2015;10:e0123905.
2. Tzeng Y-L, Ambrose KD, Zughaier S, Zhou X, Miller YK, Shafer WM, Stephens DS. Cationic antimicrobial peptide resistance in Neisseria meningitidis. J Bacteriol. 2005;187:5387–96.
3. Dürr UHN, Sudheendra US, Ramamoorthy A. LL-37, the only human member of the cathelicidin family of antimicrobial peptides. Biochim et Biophys Acta Biomembr. 2006;1758:1408–25.
4. Agerberth B, Gunne H, Odeberg J, Kogner P, Boman HG, Gudmundsson GH. FALL-39, a putative human peptide antibiotic, is cysteine-free and expressed in bone marrow and testis. Proc Natl Acad Sci. 1995;92:195–9.
5. Johansson J, Gudmundsson GH, Rottenberg ME, Berndt KD, Agerberth B. Conformation-dependent antibacterial activity of the naturally occurring human peptide LL-37. J Biol Chem. 1998;273:3718–24.
6. Turner J, Cho Y, Dinh N-N, Waring AJ, Lehrer RI. Activities of LL-37, a cathelin-associated antimicrobial peptide of human neutrophils. Antimicrob Agents Chemother. 1998;42:2206–14.
7. Larrick JW, Hirata M, Zhong J, Wright SC. Anti-microbial activity of human CAP18 peptides. Immunotechnology. 1995;1:65–72.
8. Sorensen O, Arnljots K, Cowland JB, Bainton DF, Borregaard N. The human antibacterial cathelicidin, hCAP-18, is synthesized in myelocytes and metamyelocytes and localized to specific granules in neutrophils. Blood. 1997;90:2796–803.
9. Nilsson MF, Sandstedt B, Sørensen O, Weber G, Borregaard N, Ståhle-Bäckdahl M. The human cationic antimicrobial protein (hCAP18), a peptide antibiotic, is widely expressed in human squamous epithelia and colocalizes with interleukin-6. Infect Immun. 1999;67:2561–6.
10. Larrick JW, Hirata M, Balint RF, Lee J, Zhong J, Wright SC. Human CAP18: a novel antimicrobial lipopolysaccharide-binding protein. Infect Immun. 1995;63:1291–7.
11. Cowland JB, Johnsen AH, Borregaard N. hCAP-18, a cathelin/pro-bactenecin-like protein of human neutrophil specific granules. FEBS Lett. 1995;368:173–6.
12. Bandurska K, Berdowska A, Barczynska Felusiak R, Krupa P. Unique features of human cathelicidin LL-37. BioFactors. 2015;41:289–300.
13. Kuroda K, Okumura K, Isogai H, Isogai E. The human cathelicidin antimicrobial peptide LL-37 and mimics are potential anticancer drugs. Front oncol. 2015;5:144.
14. Falla TJ, Karunaratne DN, Hancock REW. Mode of action of the antimicrobial peptide indolicidin. J Biol Chem. 1996;271:19298–303.
15. Ahmad I, Perkins WR, Lupan DM, Selsted ME, Janoff AS. Liposomal entrapment of the neutrophil-derived peptide indolicidin endows it with in vivo antifungal activity. Biochim et Biophys Acta Biomembr. 1995;1237:109–14.
16. Robinson W, McDougall B, Tran D, Selsted ME. Anti-HIV-1 activity of indolicidin, an antimicrobial peptide from neutrophils. J Leukoc Biol. 1998;63:94–100.
17. Choi KY, Chow LN, Mookherjee N. Cationic host defence peptides: multi-faceted role in immune modulation and inflammation. J Innate Immun. 2012;4:361–70.
18. Hancock REW, Sahl H-G. Antimicrobial and host-defense peptides as new anti-infective therapeutic strategies. Nat Biotechnol. 2006;24:1551–7.
19. Mansour SC, Pena OM, Hancock RE. Host defense peptides: front-line immunomodulators. Trends Immunol. 2014;35:443–50.
20. Zerfas BL, Gao J. Recent advances in peptide immunomodulators. Curr Top Med Chem. 2015;16:187–205.
21. Nawrocki KL, Crispell EK, McBride SM. Antimicrobial peptide resistance mechanisms of gram-positive bacteria. Antibiotics. 2014;3:461–92.
22. Gunn JS, Lim KB, Krueger J, Kim K, Guo L, Hackett M, Miller SI. PmrA–PmrB-regulated genes necessary for 4-aminoarabinose lipid A modification and polymyxin resistance. Mol Microbiol. 1998;27:1171–82.
23. Nizet V. Antimicrobial peptide resistance mechanisms of human bacterial pathogens. Curr Issues Mol Biol. 2006;8:11.
24. Henderson JC, Fage CD, Cannon JR, Brodbelt JS, Keatinge-Clay AT, Trent MS. Antimicrobial peptide resistance of Vibrio cholerae results from an lps modification pathway related to nonribosomal peptide synthetases. ACS Chem Biol. 2014;9:2382–92.
25. Tran AX, Whittimore JD, Wyrick PB, McGrath SC, Cotter RJ, Trent MS. The lipid A 1-phosphatase of Helicobacter pylori is required for resistance to the antimicrobial peptide polymyxin. J Bacteriol. 2006;188:4531–41.
26. McPhee JB, Bains M, Winsor G, Lewenza S, Kwasnicka A, Brazas MD, Brinkman FSL, Hancock REW. Contribution of the PhoP–PhoQ and PmrA–PmrB two-component regulatory systems to Mg²⁺-induced gene regulation in Pseudomonas aeruginosa. J Bacteriol. 2006;188:3995–4006.
27. McPhee JB, Lewenza S, Hancock REW. Cationic antimicrobial peptides activate a two-component regulatory system, PmrA–PmrB, that regulates resistance to polymyxin B and cationic antimicrobial peptides in Pseudomonas aeruginosa. Mol Microbiol. 2003;50:205–17.
28. Vaara M. New approaches in peptide antibiotics. Curr Opin Pharmacol. 2009;9:571–6.
29. Schluesener H, Radermacher S, Melms A, Jung S. Leukocytic antimicrobial peptides kill autoimmune T cells. J Neuroimmunol. 1993;47:199–202.
30. de Titta A, Ballester M, Julier Z, Nembrini C, Jeanbart L, van der Vlies AJ, Swartz MA, Hubbell JA. Nanoparticle conjugation of CpG enhances adjuvancy for cellular immunity and memory recall at low dose. Proc Natl Acad Sci USA. 2013;110:19902–7.
31. Wei M, Chen N, Li J, Yin M, Liang L, He Y, Song H, Fan C, Huang Q. Polyvalent immunostimulatory nanoagents with self-assembled CpG oligonucleotide-conjugated gold nanoparticles. Angew Chem Int Ed Engl. 2011;51:1202–6.
32. Ballester M, Nembrini C, Dhar N, de Titta A, de Piano C, Pasquier M, Simeoni E, van der Vlies AJ, McKinney JD, Hubbell JA, Swartz MA. Nanoparticle conjugation and pulmonary delivery enhance the protective efficacy of Ag85B and CpG against tuberculosis. Vaccine. 2011;29:6959–66.
33. Scott MG, Davidson DJ, Gold MR, Bowdish D, Hancock REW. The human antimicrobial peptide LL-37 is a multifunctional modulator of innate immune responses. J Immunol. 2002;169:3883–91.
34. Chen RJ, Zhang Y, Wang D, Dai H. Noncovalent sidewall functionalization of single-walled carbon nanotubes for protein immobilization. J Am Chem Soc. 2001;123:3838–9.
35. Adler AH, Schreiber N, Sorkin A, Sorkin S, Wagner G. Visualization of MD and MC simulation for atomistic modeling. Comput Phys Commun. 2002;147:665–9.
36. Dresselhaus MSD, Avouris G. Carbon nanotubes: synthesis, structure, properties, and applications. Berlin: Springer; 2001.
37. Sanders CR. Biomolecular ligand-receptor binding studies: theory, practice, and analysis. Nashville: Vanderbilt University; 2010.
38. Pradhan B, Guha D, Ray P, Das D, Aich P. Comparative analysis of the effects of two probiotic bacterial strains on metabolism and innate immunity in the RAW 264.7 murine macrophage cell line. Probiotics Antimicrob Proteins. 2016;8:73–84.

39. Awad S, Hassan AN, Muthukumarappan K. Application of exopolysac-charide-producing cultures in reduced-fat Cheddar cheese: texture and melting properties. J Dairy Sci. 2005;88:4204–13.

40. Mookherjee N, Brown KL, Bowdish DME, Doria S, Falsafi R, Hokamp K, Roche FM, Mu R, Doho GH, Pistolic J. Modulation of the TLR-mediated inflammatory response by the endogenous human host defense peptide LL-37. J Immunol. 2006;176:2455–64.

41. Abraham P, George S, Kumar KS. Novel antibacterial peptides from the skin secretion of the Indian bicoloured frog *Clinotarsus curtipes*. Bio-chimie. 2014;97:144–51.

42. Tjabringa GS, Ninaber DK, Drijfhout JW, Rabe KF, Hiemstra PS. Human cathelicidin LL-37 is a chemoattractant for eosinophils and neutro-phils that acts via formyl-peptide receptors. Int Arch Allergy Immunol. 2006;140:103–12.

43. Nijnik A, Pistolic J, Filewod NCJ, Hancock REW. Signaling pathways medi-ating chemokine induction in keratinocytes by cathelicidin LL-37 and flagellin. J Innate Immun. 2012;4:377–86.

44. Niyonsaba F, Madera L, Afacan N, Okumura K, Ogawa H, Hancock REW. The innate defense regulator peptides IDR-HH2, IDR-1002, and IDR-1018 modulate human neutrophil functions. J Leukoc Biol. 2013;94:159–70.

45. Mantovani A, Cassatella MA, Costantini C, Jaillon S. Neutrophils in the activation and regulation of innate and adaptive immunity. Nat Rev Immunol. 2011;11:519–31.

46. Pena OM, Afacan N, Pistolic J, Chen C, Madera L, Falsafi R, Fjell CD, Hancock REW. Synthetic cationic peptide IDR-1018 modulates human macrophage differentiation. PLoS ONE. 2013;8:e52449.

47. Barlow PG, Beaumont PE, Cosseau C, Mackellar A, Wilkinson TS, Hancock REW, Haslett C, Govan JRW, Simpson AJ, Davidson DJ. The human cathelicidin LL-37 preferentially promotes apoptosis of infected airway epithelium. Am J Respir Cell Mol Biol. 2010;43:692–702.

48. Hu Z, Murakami T, Suzuki K, Tamura H, Kuwahara-Arai K, Iba T, Nagaoka I. Antimicrobial cathelicidin peptide LL-37 inhibits the LPS/ATP-induced pyroptosis of macrophages by dual mechanism. PLoS ONE. 2014;9:e85765.

49. Afacan NJ, Yeung AT, Pena OM, Hancock RE. Therapeutic potential of host defense peptides in antibiotic-resistant infections. Curr Pharm Des. 2012;18:807–19.

50. Mookherjee N, Lippert DND, Hamill P, Falsafi R, Nijnik A, Kindrachuk J, Pistolic J, Gardy J, Miri P, Naseer M. Intracellular receptor for human host defense peptide LL-37 in monocytes. J Immunol. 2009;183:2688–96.

51. Yu J, Mookherjee N, Wee K, Bowdish DME, Pistolic J, Li Y, Rehaume L, Hancock REW. Host defense peptide LL-37, in synergy with inflammatory mediator IL-1β^2, augments immune responses by multiple pathways. J Immunol. 2007;179:7684–91.

52. Van Der Does AM, Joosten SA, Vroomans E, Bogaards SJP, Van Meij-gaarden KE, Ottenhoff THM, Van Dissel JT, Nibbering PH. The antimi-crobial peptide hLF1–11 drives monocyte-dendritic cell differentiation toward dendritic cells that promote antifungal responses and enhance Th17 polarization. J Innate Immun. 2012;4:284–92.

53. Mookherjee N, Hancock REW. Cationic host defence peptides: innate immune regulatory peptides as a novel approach for treating infections. Cell Mol Life Sci. 2007;64:922–33.

54. Yuk J-M, Shin D-M, Lee H-M, Yang C-S, Jin HS, Kim K-K, Lee Z-W, Lee S-H, Kim J-M, Jo E-K. Vitamin D3 induces autophagy in human monocytes/macrophages via cathelicidin. Cell Host Microbe. 2009;6:231–43.

55. Niyonsaba F, Ushio H, Nakano N, Ng W, Sayama K, Hashimoto K, Nagaoka I, Okumura K, Ogawa H. Antimicrobial peptides human β-defensins stimulate epidermal keratinocyte migration, proliferation and produc-tion of proinflammatory cytokines and chemokines. J Investig Dermatol. 2007;127:594–604.

5

Enzyme adsorption-induced activity changes: a quantitative study on TiO$_2$ model agglomerates

Augusto Márquez[1], Krisztina Kocsis[1], Gregor Zickler[1], Gilles R. Bourret[1], Andrea Feinle[1], Nicola Hüsing[1], Martin Himly[2]*, Albert Duschl[2], Thomas Berger[1]* ⓘ and Oliver Diwald[1]

Abstract

Background: Activity retention upon enzyme adsorption on inorganic nanostructures depends on different system parameters such as structure and composition of the support, composition of the medium as well as enzyme loading. Qualitative and quantitative characterization work, which aims at an elucidation of the microscopic details governing enzymatic activity, requires well-defined model systems.

Results: Vapor phase-grown and thermally processed anatase TiO$_2$ nanoparticle powders were transformed into aqueous particle dispersions and characterized by dynamic light scattering and laser Doppler electrophoresis. Addition of β-galactosidase (β-gal) to these dispersions leads to complete enzyme adsorption and the generation of β-gal/TiO$_2$ heteroaggregates. For low enzyme loadings (~4% of the theoretical monolayer coverage) we observed a dramatic activity loss in enzymatic activity by a factor of 60–100 in comparison to that of the free enzyme in solution. Parallel ATR-IR-spectroscopic characterization of β-gal/TiO$_2$ heteroaggregates reveals an adsorption-induced decrease of the β-sheet content and the formation of random structures leading to the deterioration of the active site.

Conclusions: The study underlines that robust qualitative and quantitative statements about enzyme adsorption and activity retention require the use of model systems such as anatase TiO$_2$ nanoparticle agglomerates featuring well-defined structural and compositional properties.

Keywords: Nanoparticles, Agglomerates, TiO$_2$, β-galactosidase, Adsorption, Enzymatic activity, IR spectroscopy

Background

A huge variety of synthetic approaches towards mesoporous materials with well-defined and tunable pore structures has become available only recently [1]. This has provided a promising new field to perform structure–activity studies at the microscopic level and for the development of novel hybrid materials for biocatalysis [2, 3] or biosensing [4, 5]. The functional properties of such materials are determined by the physicochemical parameters characterizing the interaction of biomolecules with the porous inorganic structure. The knowledge-based manipulation and optimization of these interactions requires well-defined mesoporous structures. At the same time, related model systems are also needed for insights into the interaction between biomolecules and engineered nanomaterials [6]. This applies for less defined and more dynamic structures such as particle agglomerates and/or aggregates, which are ubiquitous both in nature and in technology and which play an important role in emerging fields such as nanotoxicology or nanomedicine. Particle agglomerates and aggregates constitute complex structures and typically experience transformations in response to minute changes of the surrounding environment. For this reason, it is extremely difficult to establish model systems that mimic such a behavior sufficiently.

The properties of colloidal inorganic nanoparticles in biological media are subject to the adsorption of

---

*Correspondence: Martin.Himly@sbg.ac.at; Thomas.Berger@sbg.ac.at
[1] Department of Chemistry and Physics of Materials, Paris Lodron University of Salzburg, Jakob-Haringer-Strasse 2a, 5020 Salzburg, Austria
[2] Department of Molecular Biology, Paris Lodron University of Salzburg, Hellbrunnerstrasse 34/III, 5020 Salzburg, Austria

biomolecules and the formation of a protein corona around the particles [7, 8]. The generation of protein/particle composites affects the structure both of the protein (by adsorption-induced changes in protein conformation) and of the particles (by changes of the agglomeration and dispersion state) [9]. Since protein- and particle-related changes are strongly sensitive to a variety of chemical, physical and biological factors governing the interaction of the biomolecules with the inorganic material, enormous efforts have been devoted to the elucidation of related microscopic details [10–15]. Also for this purpose more systematic and quantitative studies involving reference systems are needed [13].

Vapor phase-grown particle powders of isolated anatase TiO_2 nanocrystals and decontaminated particle surfaces have proven to be an appropriate model system for the study of hydration-induced microstructural and electronic property changes [16–20]. Only recently, we have employed such particle powders as model systems for protein adsorption studies and attained a reproducible qualitative and quantitative assessment of the interaction of a model serum protein (bovine serum albumin, BSA) with compositionally and structurally well-defined particle agglomerates. Such nanoparticle agglomerates are used in the present study to evaluate the impact of enzyme adsorption on its catalytic activity. For this purpose we selected β-galactosidase (β-gal) which has been used in previous model studies on enzyme immobilization at metal oxide nanomaterials [21–26]. Here we report the first qualitative and quantitative results obtained for enzyme adsorption using FT-IR spectroscopy and light scattering techniques and discuss challenges and pitfalls which typically arise during the evaluation of enzymatic activity changes between free and adsorbed proteins.

Methods
Chemicals
Na_2CO_3 (Sigma Aldrich, ACS reagent, ≥99.5%), Na_2HPO_4 (Sigma Aldrich, ACS Reagent ≥99%), citric acid (Sigma Aldrich, ≥99.5%), o-Nitrophenyl-β-D-galactopyranoside (ONPG, Sigma Aldrich, >99% HPLC), $MgCl_2$ (Sigma Aldrich, ACS Reagent >99%) and D_2O (Aldrich, 99 atom% D) were used as received. β-gal was produced in house as a highly purified recombinant protein according to protocols previously described in detail [27]. All H_2O solutions were prepared using water with a resistivity of 18 M Ω cm (Millipore, Milli-Q).

Nanoparticle synthesis
Anatase TiO_2 nanocrystals were prepared by metal organic chemical vapor synthesis (MOCVS) based on the decomposition of titanium (IV) isopropoxide at $T = 1073$ K in a hot wall reactor system [28, 29]. For purification, the

obtained powder samples were subjected to thermal treatment under high vacuum conditions ($p < 10^{-5}$ mbar). First, the powder sample was heated to $T = 873$ K using a rate of $r \leq 5$ K min^{-1}. Subsequent oxidation with O_2 at this temperature was applied to remove organic remnants from the precursor material and to guarantee the stoichiometric composition of the oxide. Such post-synthesis treatment eliminates organic remnants as evidenced by IR spectroscopy [29]. With transmission electron microscopy (TEM) we found that the majority of crystallites display a highly irregular but approximately equidimensional shape (Additional file 1: Figure S1) [28, 29]. The average diameter of these nanocrystals corresponds to 13 nm. The specific surface area deduced from the TEM mean diameter value corresponds to 121 m^2 g^{-1} and is in good agreement with results from nitrogen sorption measurements (130 ± 13 m^2 g^{-1}) [28].

Preparation and characterization of particle dispersions
The size distribution and zeta potential of agglomerates was determined for aqueous TiO_2 particle dispersions with a concentration of 0.1 mg mL^{-1}. To ensure sufficient particle dispersion an ultrasonic finger (amplitude 25%, 30 min, UP200St, Ti-sonotrode 2 mm, Hielscher Ultrasonics GmbH) was used while the dispersion was cooled in an ice bath to prevent heating via mechanical sample agitation. Protein adsorption was performed at room temperature by mixing 15 mL of TiO_2 particle dispersion with 5 mL of protein solution (final particle concentration: $[TiO_2] = 0.1$ mg mL^{-1}, [β-gal] = 50 μg mL^{-1}). The dispersion was then mechanically stirred for 12 h.

We will refer to secondary particles, which are built up from TiO_2 nanocrystals in enzyme-free water, as agglomerates. Nanocrystals in agglomerates are held together by weak physical interactions. On the other hand we refer to β-gal/TiO_2 composites forming after the addition of enzyme to preformed agglomerates as heteroaggregates. Protein adsorption onto the oxide is expected to involve not only weak physical interactions, but also chemical bonding.

A Zetasizer Nano ZSP ZEN5600 (Malvern Instruments) was used to determine the size distribution of dispersed TiO_2 particles, proteins and protein/TiO_2 heteroaggregates by dynamic light scattering (DLS) as well as the zeta potentials by laser Doppler electrophoresis. To obtain size distribution functions from DLS measurements 70 size classes were used for TiO_2 dispersions and 300 size classes for enzyme solutions. To transform the intensity distribution of hydrodynamic diameters into a number distribution several assumptions have to be made, including that all particles are spherical and homogeneous. To support data obtained from DLS we therefore performed an additional electron microscopic study of TiO_2 agglomerates and β-gal/TiO_2 heteroaggregates.

Electron microscopic characterization of TiO$_2$ agglomerates and β-gal/TiO$_2$ heteroaggregates

After DLS analysis small volumes of aqueous dispersions containing TiO$_2$ agglomerates or β-gal/TiO$_2$ heteroaggregates were dropped onto a carbon film on a Cu grid. The immobilized samples were then dried at room temperature. The samples were imaged with a field emission gun SEM Ultra Plus 55 (Zeiss) at an accelerating voltage of 5 kV with the Gemini In-Lens secondary electron detector. For related TEM measurements a TECHNAI F20 microscope equipped with a field emission gun and S-twin objective lens was used.

Activity of free and immobilized β-gal-enzymatic assay

The activity of β-gal was studied at 5.5 < pH < 8.5 in McIlvaine's buffer (0.1 M citric acid and 0.2 M Na$_2$HPO$_4$ at different ratios) both in the free state ([β-gal] = 0.5 µg mL^{-1}) and when adsorbed on TiO$_2$ nanoparticle agglomerates ([TiO$_2$] = 1 mg mL^{-1}, [β-gal] = 8.5 µg mL^{-1}). MgCl$_2$ (1 mM) and ONPG (0.5 mM) were used as cofactor and substrate, respectively.

An aqueous TiO$_2$ particle dispersion was first treated ultrasonically and then mixed with aqueous β-gal solution. The dispersion was mechanically stirred (VWR VMS-C4, 600 rpm) at room temperature for 12 h. Finally the sample was centrifuged (EBA 20 centrifuge, Hettich, 6000 rpm). Bradford analysis confirmed that enzyme adsorption was complete and leads to complete enzyme elimination from the supernatant. The β-gal/TiO$_2$ heteroaggregates were washed first with water and then with the respective buffer solution. Finally, the enzymatic reaction was started by adding ONPG to the buffer and agitation of the mixture. The reaction progress was monitored as a function of time by withdrawing 0.5 mL of the reaction mixture and mixing with 0.7 mL of 1 M Na$_2$CO$_3$ aqueous solution to stop the reaction. The resulting dispersion was centrifuged (Biofuge fresco, Heraeus, 13,000 rpm) and the concentration of the reaction product (o-nitrophenol, ONP) was determined photometrically at a wavelength λ = 420 nm (molar extinction coefficient, ε (420 nm) = 2.13·10^4 M^{-1} cm^{-1} at pH 10.2) [30]. The specific enzyme activity is represented in U/mg, where one enzyme unit (U) corresponds to the amount of enzyme catalysing the reaction of 1 µmol of ONP per minute.

Study of protein conformation—ATR-FTIR-spectroscopy

For IR measurements an attenuated total reflection (ATR) unit (PIKE Technologies, Veemax II) was attached to a Bruker Vertex 70 FTIR spectrometer equipped with a MCT detector. The measurements were performed at an incident angle of 55° using a hemispherical ZnSe prism. Spectra were obtained by averaging 100 scans at a resolution of 4 cm^{-1} and are represented as −log (R/R$_0$),

where R and R$_0$ are the reflectance values corresponding to the single beam spectra recorded for sample and reference, respectively.

Proteins in solution

First, the protein solution was applied to the ATR prism and spectra were recorded until the establishment of adsorption equilibrium. Afterwards, the prism was rinsed several times with pure water. A spectrum of the hydrated protein layer which remains at the prism surface serves as the reference. Finally, the pure water was replaced again by the respective protein solution and the sample spectrum was measured. This procedure assures that the detected signals result exclusively from proteins in solution and not from the protein layer at the surface of the ATR prism.

Proteins adsorbed on a porous TiO$_2$ nanoparticle film

A TiO$_2$ film was immobilized on the ATR prism by doctor blade deposition using Scotch tape as spacer. First, 40 µL of a 0.4 M TiO$_2$ nanoparticle dispersion was applied per cm^2 of prism surface and spread over it. The resulting film was then dried in a nitrogen stream. Afterwards, the IR cell was assembled by pressing a glass cell against the pre-coated prism using a Teflon ring as the junction. Finally, the cell was filled with water or aqueous protein solution to measure the background or sample spectrum, respectively.

Spectrum fitting

The amide I and II region of protein spectra (1700–1480 cm^{-1}) was fitted by a set of Gaussian-shaped bands. From the second derivative of an experimental spectrum we determined the minima and defined related values as band positions [31, 32]. Prior to spectrum fitting a horizontal baseline, which was extrapolated from the flat spectral region between 2000 and 1800 cm^{-1}, was subtracted from the original spectrum. All band positions and band widths were kept constant regardless of whether the protein was free in aqueous (H$_2$O or D$_2$O) solution or adsorbed on TiO$_2$. Spectra were fitted by optimizing the contribution of single bands using iterative Chi square minimization (Levenberg–Marquardt algorithm). For visualization of the protein structure of β-gal the UCSF Chimera software tool [33] was used.

Results and discussion

β-gal adsorption-induced colloidal property changes of vapor phase-grown TiO$_2$ nanoparticles in aqueous dispersion

Anatase TiO$_2$ nanoparticle powders, which were grown in the gas phase and processed in vacuum, are characterized by ensembles of equidimensional nanoparticles (primary particle size: 13 nm) with low concentrations of solid–solid interfaces and contaminant-free surfaces (Additional file 1: Figure S1) [16]. After ultrasonic treatment particles form agglomerates in aqueous dispersion, which display agglomerate size distributions peaking at ~80 nm (Fig. 1). The hydrodynamic diameter of β-gal in aqueous solution was determined to be ~13 nm (Fig. 1). As tracked by DLS, β-gal adsorption onto TiO_2 agglomerates increases the agglomerate size. Final particle size distributions with a maximum at ~140 nm (Fig. 1) suggest the formation of β-gal/TiO_2 heteroaggregates, where an enzyme corona forms at the outer surface of TiO_2 agglomerates. TiO_2 de-agglomeration by enzyme adsorption followed by a re-agglomeration upon the formation of larger β-gal/TiO_2 heteroaggregates represents an alternative scenario. However, only the protein free dispersions of vapor-phase grown TiO_2 particles were subjected to an ultrasonic treatment in order to achieve model agglomerates with reproducible particle size distribution. After enzyme addition, we minimized further input of mechanical energy and stirred the dispersions only mildly until an adsorption equilibrium was reached. We expect that further breakup of the TiO_2 agglomerates and subsequent protein inclusion during re-agglomeration is rather unlikely under such conditions. Adsorption-induced changes of the agglomerates' surface charge provide strong evidence for protein corona formation, i.e. the generation of a ß-gal shell around TiO_2 agglomerates. The zeta potential of TiO_2 agglomerates in aqueous suspension decreases from its starting value related to pure TiO_2 dispersions (−3.0 mV) to that of β-gal—containing

dispersions (−27 mV) corresponding to the zeta potential of β-gal in solution (−29 mV).

To complement data obtained from DLS we performed an electron microscopic study of TiO_2 agglomerates and β-gal/TiO_2 heteroaggregates. For this purpose we used conventional scanning electron microscopy (SEM, Fig. 2) and transmission electron microscopy (TEM, Additional file 1: Figure S1). It has to be emphasized that the immobilization of colloidal particles onto a sample grid is associated with a change of the particle surrounding from a condensed liquid phase to a high vacuum environment. It is well known that such a transformation may be associated with a significant modification of the microstructure. Furthermore, care has to be taken when deducing from a limited number of microscopically resolved and analyzed entities generalized conclusions about the structural properties of a large particle ensemble. Indeed we observe on the same sample grid different microstructures. On the one hand, extended almost two-dimensional structures are observed, which can conveniently be studied by TEM (Additional file 1: Figure S1). We tentatively associate the formation of these structures with a drying-induced spreading of primary particles on the substrate both for TiO_2 agglomerates and for β-gal/TiO_2 heteroaggregates. Importantly, the primary particle size does not change upon enzyme adsorption (Additional file 1: Figure S1). On the other hand, we observe sample spots characterized by highly agglomerated or aggregated secondary particles, which feature however a quite narrow size distribution. The resulting structures seem to be three-dimensional thus escaping TEM investigation. However, from scanning electron micrographs we determined the number-weighted size distribution of β-gal/TiO_2 heteroaggregates (Fig. 2). Remarkably, the size distribution of β-gal/TiO_2 heteroaggregates resembles quite closely the size distribution of TiO_2 agglomerates in dispersion. Whereas the adsorbed protein layer escapes detection by SEM, these results suggest that TiO_2 agglomerates serve as substrates for enzyme adsorption upon conservation of their original size. On the other hand, this observation suggests that a high local concentration of agglomerates is beneficial for avoiding drying artefacts upon sample preparation thus preserving the original secondary particle size in dispersion. At the moment, however, we do not fully control the parameters governing sample deposition on the microscopy grid. Further work in this direction is underway.

To estimate the number of β-gal molecules being adsorbed per TiO_2 agglomerate we used the Bradford assay. The supernatant solution confirms complete enzyme adsorption. Moreover, we neglected enzyme adsorption at the walls of the glass beaker and considered its adsorption on TiO_2 agglomerates with a size of 80 nm

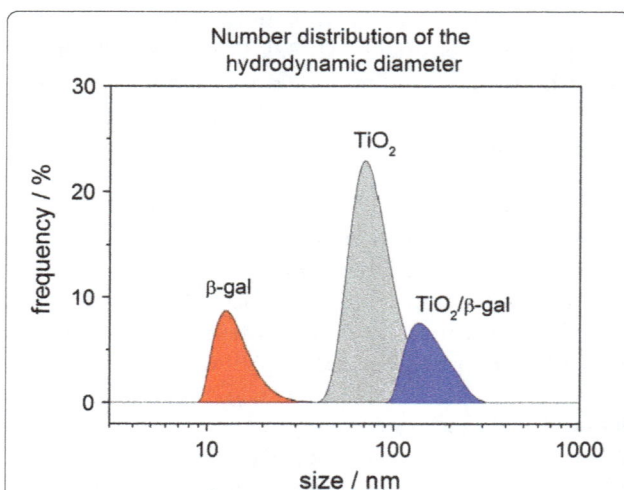

Fig. 1 Particle size distribution as related to β-gal, TiO_2 agglomerates and mixtures of β-gal and TiO_2 agglomerates in aqueous suspension. A size increment from ~80 to ~140 nm reveals the adsorption of β-gal and the formation of β-gal/TiO_2 composites. [β-gal] = 50 µg mL^{-1}; [TiO_2] = 0.1 mg mL^{-1}

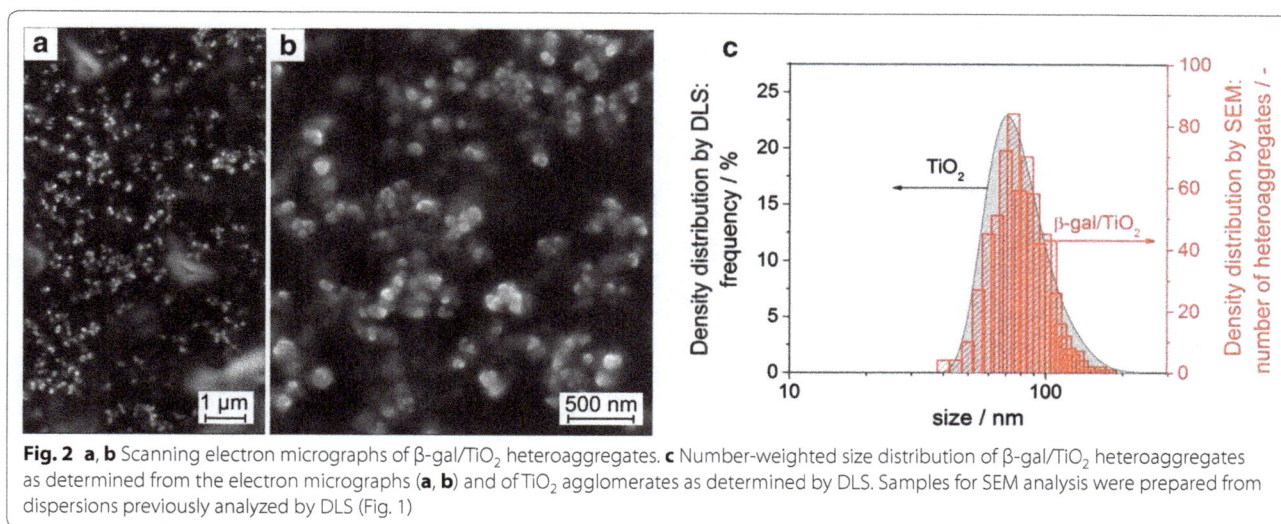

Fig. 2 a, b Scanning electron micrographs of β-gal/TiO$_2$ heteroaggregates. **c** Number-weighted size distribution of β-gal/TiO$_2$ heteroaggregates as determined from the electron micrographs (**a, b**) and of TiO$_2$ agglomerates as determined by DLS. Samples for SEM analysis were prepared from dispersions previously analyzed by DLS (Fig. 1)

and a solid fraction of 0.5 (i.e. 50% porosity) [16]. On this basis we estimated the number of 325 β-gal molecules per agglomerate. A rough estimate[1] of the number of β-gal molecules which would correspond to a protein monolayer covering the external surface of TiO$_2$ agglomerates yields ~150 β-gal molecules per TiO$_2$ agglomerate. A comparison with the experimental value (325 β-gal molecules per TiO$_2$ agglomerate) led us to the conclusion that β-gal may form multilayers at the outer surface of TiO$_2$ agglomerates. This is different to BSA adsorption on comparable TiO$_2$ agglomerate structures where we observed saturation at the level of one monolayer equivalent as reported only recently [34].

Adsorption of β-gal and the impact on enzymatic activity

β-gal (molecular weight: 465,000 g mol^{-1}) hydrolyzes lactose and other β-galactosides into monosaccharides. For the following experiments we used heteroaggregated β-gal/TiO$_2$ systems and explored how enzyme adsorption on TiO$_2$ agglomerates impacts on the enzyme's biological activity. More specifically, we focus on the interaction of isolated enzyme molecules with the inorganic substrate. To exclude significant protein–protein interactions such as protein aggregation and multilayer formation at the oxide surface, agglomerate coverage was kept below one monolayer equivalent.

After an adsorption time of 12 h, Bradford analysis clearly reveals that the enzyme ([β-gal] = 8.5 µg mL^{-1})

was completely removed from the aqueous solution by the dispersed TiO$_2$ agglomerates ([TiO$_2$] = 1 mg mL^{-1}). This uptake corresponds to ~6 β-gal molecules per TiO$_2$ agglomerate, or ~4% of a theoretical monolayer. The β-gal/TiO$_2$ heteroaggregates were first washed with water and then treated in the respective buffer solution.

In comparison to the protein in free form ([β-gal] = 0.5 µg mL^{-1}) the enzymatic activity of the protein adsorbed on TiO$_2$ agglomerates (Fig. 3; Additional file 1: Figure S2) was studied in the range 5.5 < pH < 8.5 using a McIlvaine's buffer (0.1 M citric acid and 0.2 M Na$_2$HPO$_4$ at different ratios). Consistent with previous studies [26] the activity of free β-gal was observed to peak at ~pH 7. Upon adsorption the enzymatic activity decreases drastically by a factor of 60 at pH ≤ 7. At pH 6.5 the activity of adsorbed β-gal was also tested in pure water (i.e. in the absence of the buffer) yielding a comparable activity. For the regime pH > 7 we observed an even more pronounced decrease—by a factor of ~100—of the apparent enzyme activity.

Buffer species in solution are needed for the stabilization of the pH during the activity measurements. Their potential competition with proteins for adsorption sites needs to be included in the present discussion. HPO$_4^{2-}$ bind as bidentate species and—in principle—can also induce protein desorption [35]. At this point, we hypothesize that the observed deviation of the activity trend (Fig. 3) above pH > 7 is due to buffer-induced desorption of β-gal, since at pH values between 7.2 and 8.5 HPO$_4^{2-}$ is the most abundant phosphate species (H$_3$PO$_4$: pK$_{a1}$ = 2.15, pK$_{a2}$ = 7.2, pK$_{a3}$ = 12.4). To test this hypothesis we studied the effect of HPO$_4^{2-}$ on the concentration of adsorbed β-gal using ATR-IR spectroscopy on a porous film of vapor phase grown TiO$_2$ nanoparticles

---

[1] For this purpose, agglomerates were modeled by spheres with a diameter of 80 nm (i.e. the hydrodynamic diameter determined by DLS, Fig. 1). The required adsorption area per β-gal molecule was approximated by the projection area of a spherical molecule with a diameter of 13 nm (i.e. hydrodynamic diameter of β-gal in solution, Fig. 1) to a planar surface. This approximation yields a monolayer coverage of 150 β-gal molecules per TiO$_2$ agglomerate.

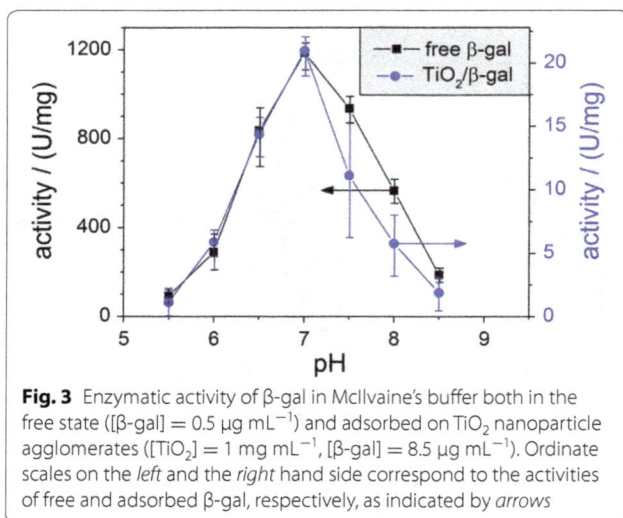

Fig. 3 Enzymatic activity of β-gal in McIlvaine's buffer both in the free state ([β-gal] = 0.5 µg mL^{-1}) and adsorbed on TiO$_2$ nanoparticle agglomerates ([TiO$_2$] = 1 mg mL^{-1}, [β-gal] = 8.5 µg mL^{-1}). Ordinate scales on the *left* and the *right* hand side correspond to the activities of free and adsorbed β-gal, respectively, as indicated by *arrows*

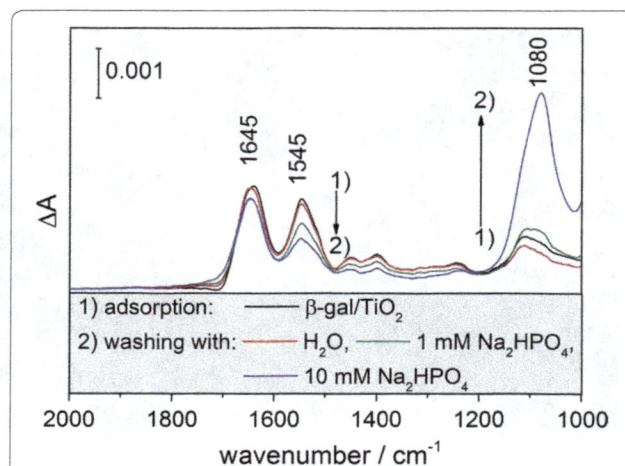

Fig. 4 ATR-IR spectra of β-gal adsorbed on a porous film of TiO$_2$ nanoparticles ([β-gal] = 150 µg mL^{-1}, [TiO$_2$] = 1.3 mg mL^{-1}, adsorption time: 8 h), background spectrum: TiO$_2$ film in contact with water. Following protein adsorption the film was extensively rinsed first with water and then with Na$_2$HPO$_4$ solutions ([Na$_2$HPO$_4$] = 1 and 10 mM, respectively)

(Fig. 4).[2] Upon β-gal adsorption protein-specific amide bands at 1645 (amide I) and 1545 cm^{-1} (amide II) appear in the IR spectrum, the intensities of which remain constant after removal of the protein solution and extensive washing with pure H$_2$O. However, washing the film with Na$_2$HPO$_4$ solutions of increasing concentration clearly depletes the intensities of the protein-specific bands and generates a new band at 1080 cm^{-1} characteristic of HPO$_4^{2-}$ [36]. Buffer induced protein desorption was also observed when using HEPES (4-(2-hydroxyethyl)-1-piperazineethanesulfonic acid) instead of McIlvaine's buffer (not shown).

Several aspects have to be considered when comparing the activity of enzymes in free and adsorbed form. First of all, the number of enzyme molecules involved has to be known. While this type of information can be obtained in a straightforward manner for proteins in solution, related efforts are challenging with regard to adsorbed proteins. The total number of adsorbed molecules may change significantly upon minute changes of solution composition as shown above (Fig. 4). Furthermore, mass transport and, thus, local gradients of reaction educts and products are significantly influenced by (a) the distribution of the enzyme within pores or at the oxide surface (sub-monolayer, monolayer, multilayers) and (b) the distribution of enzyme/oxide heteroaggregates within the reaction volume (i.e. their dispersion state). The dispersion state in turn depends on protein coverage and solution composition and pH.

Finally, the protein/particle ratio and, more precisely, the coverage of the surface available on single particles, agglomerates or aggregates need to be specified. For instance, it has been shown that the degree of conformational change of adsorbed proteins depends on the protein/particle surface ratio [37]. In the following we use ATR-IR spectroscopy to probe the structural changes β-gal experiences during adsorption.

Experiments on the formation of the protein corona focus for individual proteins often on the formation of a complete protein shell where the entire particle surface is covered [38, 39]. In the present study, only 4% of the surface was covered with protein, which is more realistic for any individual protein when nanoparticles have entered the human body. Of note, buffering compounds will be present in such a situation.

Structural changes of β-gal adsorbed on porous films of vapor phase-grown TiO$_2$ nanoparticles

The ATR-IR spectra of β-gal adsorbed from aqueous solution on porous films of vapor phase-grown TiO$_2$ nanoparticles (Fig. 4; Additional file 1: Figure S3) are characterized by two characteristic protein-specific bands at 1645 cm^{-1} (amide I) and 1545 cm^{-1} (amide II) [31, 32, 40, 41]. A detailed analysis of ATR-IR spectra was performed to explore adsorption-induced changes in protein conformation. The amide I band intensity exhibits sensitive dependence on the protein backbone, since different secondary structures contribute to the amide I band in a narrow wavenumber range [31, 32, 42]. β-gal is a tetramer, which is mainly composed of α-helical and

---

[2] Due to sensitivity reasons we had to use for ATR-IR experiments a [β-gal]/[TiO$_2$] ratio, which is about 14 times higher than the one used in the enzymatic assay. However, using as the basis for enzyme quantification the same considerations as in Footnote 1, this concentration corresponds to ~50% of a theoretical monolayer.

β-sheet segments and which has smaller contributions from random coil and β-turns [43]. Every tetramer contains four functional active sites. The active site is formed primarily by an α/β barrel structure of each monomer, however, it includes also critical catalytic residues of other monomers [44].

IR spectra of β-gal both in aqueous (H_2O or D_2O) solution and adsorbed on a TiO_2 film are shown in the Additional file 1: Figure S3. The spectral range between 1700 and 1480 cm^{-1} which features the amide I and amide II bands (Additional file 1: Figures S4, S5) was subjected to a band deconvolution procedure. Additional file 1: Figure S6 shows the fitting result for the amide I band of β-gal (free and adsorbed) in an H_2O environment. Upon adsorption a significant decrease of the component at ~1630 cm^{-1} (attributed to β-sheet structures) and an increase of the component at ~1655 cm^{-1} (attributed to α-helix and random structures) is observed (Additional file 1: Figure S6, Table S1). With regard to the fitting of the amide I band the overlap of α-helix and random structures represents a fundamental problem in H_2O. This overlap can be greatly reduced in D_2O due to small band shifts of the amide I components upon a hydrogen/deuterium exchange [31, 32]. Furthermore, interference from the H_2O bending mode, which overlaps with the amide I band, can be eliminated in D_2O. The experimental IR spectrum and the corresponding fitting results for β-gal (free and adsorbed) in a D_2O environment are provided in Fig. 5 and in the Additional file 1: Figures S3 and S5. The second derivative of the spectrum corresponding to adsorbed β-gal (Fig. 5b) features a contribution at 1641 cm^{-1}, which is absent in the spectrum of the free protein (Fig. 5a) and which is attributed to random structures. From the fitting results it can be deduced that this contribution grows on the expense of the β-sheet content, which is reduced by ~30% as compared to the free protein, while the structure of the α-helices remains essentially undisturbed. The adsorption induced deterioration of β-sheets is intimately related to the structure of β-gal: [44] the central domain forming the α/β barrel structure (indicated by the black arrow in Fig. 5c) constitutes the core of each monomer and is surrounded by four domains containing mainly β-sheet structures (indicated by yellow arrows in Fig. 5c). These outer parts of the protein preferentially interact with the oxide surface upon adsorption. The associated conversion of β-sheets into random structures and a modification of the active sites are expected to significantly contribute to the dramatic decrease of the enzymatic activity upon adsorption. The restricted accessibility of reactive centers upon enzyme adsorption may additionally contribute to the observed activity loss (Table 1).

The results presented here describe a situation where the local environment of the enzyme is strongly influenced by the inorganic structure of the TiO_2 agglomerates due to a high enzyme dispersion at the agglomerate surface. At low surface coverages, this leads to a significant modification of the protein conformation and, consequently, to a dramatic loss in enzyme activity. As the extent of conformational change of a protein adsorbed onto nanostructured materials may critically depend on coverage [37], the present situation corresponds to the worst case with regard to the conservation of the enzyme activity upon immobilization. Immobilization strategies which involve enzyme adsorption may involve enzyme distortion-related activity losses [46]. While at low enzyme loadings the contact area between the enzyme and the substrate may be maximized, contact area minimization does occur at higher loadings [46]. For low enzyme coverages the co-adsorption of proteins (e.g. BSA) or polymers has been found to improve the activity of enzymes immobilized on 2D or microstructured substrates [46, 47]. Similar effects have been reported for materials which are structured at lower length scales, i.e. for nanoparticulate systems [48].

The systematic investigation of coverage-dependent adsorption processes including associated protein conformation changes and, as a result therefrom, competitive processes require rigorous control of both the structural properties of nanoparticle agglomerates and of the enzyme loading. We believe that the characterization approach presented here, i.e. the combination of systematic and quantitative adsorption studies and molecular insight from spectroscopy, can only be performed on structurally and compositionally well-defined model agglomerates. Related reference systems may serve as a valuable tool for the future investigation of more complex bio/nano hybrid systems containing commercially available components.

Conclusions

Enzymatic activity changes of β-gal when adsorbed at low coverages (~4% of the theoretical monolayer coverage) on compositionally well-defined anatase TiO_2 agglomerates were linked to results from complementary FT-IR measurements. Substantial loss in biological activity was observed over the entire pH range investigated, i.e. 5.5 < pH < 8.5, and corresponds to a decrease by a factor of 60–100. Band analysis of the FT-IR active amide modes clearly revealed the adsorption-induced deterioration of β-sheet structures and—as a result—the origin of the deactivation of the catalytic site. The here presented combination between surface science-inspired and molecular spectroscopy based adsorption studies on oxide nanomaterials also include the determination of

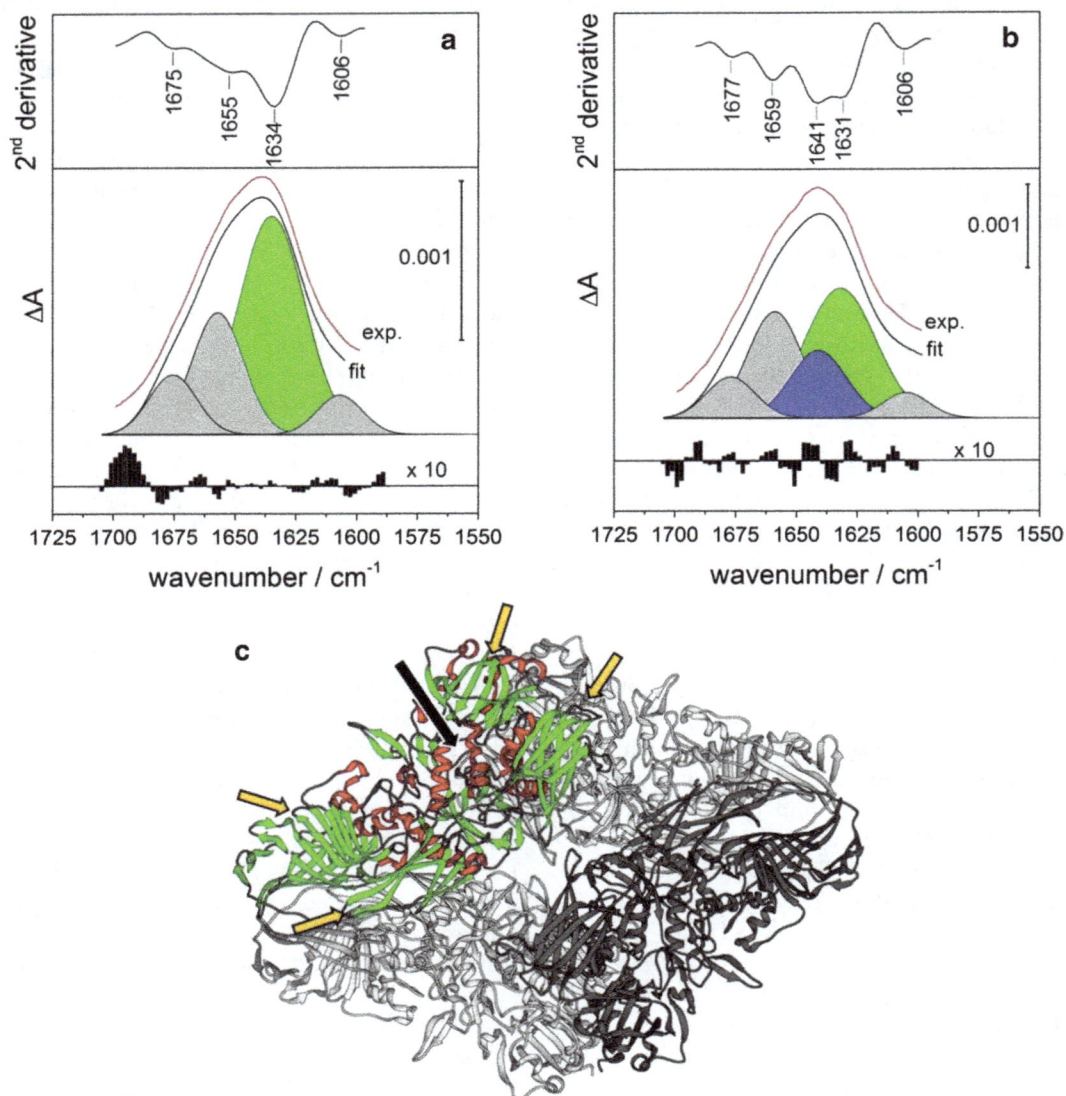

Fig. 5 Second derivative of the amide I band, fitting results and corresponding residuals for **a** free β-gal in D$_2$O ([β-gal] = 150 μg mL^{-1}) and **b** β-gal adsorbed on a porous film of TiO$_2$ nanoparticles ([β-gal] = 150 μg mL^{-1}, [TiO$_2$] = 1.3 mg mL^{-1}, adsorption time: 8 h), background spectrum: TiO$_2$ film in contact with D$_2$O. The band parameters of the deconvoluted single components are listed in Table 1. **c** Graphical representation of the protein structure of β-gal retrieved from the Protein Databank (pdb entry: 1f4a) [45] with the α/β barrel carrying the active center of one of the four monomers (indicated by a *black arrow*) surrounded by four domains rich in β-sheet structures (marked by *yellow arrows*). α-helices and β-sheets of one monomer are highlighted in *red* and *green*, respectively

Table 1 Band parameters of the deconvoluted single components contributing to the amide I band of free β-gal in D$_2$O and β-gal adsorbed on TiO$_2$ (corresponding to the fitting results represented in Fig. 5)

Structure	Free β-gal			Adsorbed β-gal		
	Peak position (cm^{-1})	FWHM (cm^{-1})	Area (%)	Peak position (cm^{-1})	FWHM (cm^{-1})	Area (%)
Inter β-sheet	1606	22	7	1606	22	5
β-Sheet	1634	32	58	1631	34	42
Random	–	–	–	1641	27	17
α-Helix	1655	25	25	1659	26	27
Turn	1675	23	10	1677	23	9

size and surface charge of the metal oxide-based agglomerates and requires well-defined particulate model systems with pre-processed and bare oxide surfaces. Robust qualitative and quantitative statements about enzyme adsorption-induced activity changes become accessible only on such systems. Moreover, they allow for insights into cooperative adsorption effects involving different types of proteins or other molecules from the buffer solution.

Abbreviations
ATR: attenuated total reflection; BSA: bovine serum albumin; DLS: dynamic light scattering; HEPES: 4-(2-attenuated)-1-piperazineethanesulfonic acid; MOCVS: metal organic chemical vapor synthesis; ONP: o-nitrophenol; ONPG: o-nitrophenyl-β-D-galactopyranoside; SEM: scanning electron microscopy; TEM: transmission electron microscopy; β-gal: β-galactosidase.

Authors' contributions
AM, KK and GZ performed the experimental work. AM, KK, MH and TB contributed to the analysis and representation of the data. TB wrote the manuscript. All authors participated in the design of the study and in the interpretation of the data. All authors revised the manuscript and approved the final version of the paper.

Acknowledgements
We thank Johannes Bernardi from the Technical University of Vienna for TEM measurements. We gratefully acknowledge assistance by Dr. Markus Wiederstein when using the UCSF Chimera package.

Competing interests
The authors declare that they have no competing interests.

Funding
This work was financially supported by the University of Salzburg within the Allergy-Cancer-BioNano (ACBN) Research initiative.

References
1. Feinle A, Elsässer MS, Hüsing N. Sol–gel synthesis of monolithic materials with hierarchical porosity. Chem Soc Rev. 2016;45:3377–99.
2. Zhou Z, Hartmann M. Progress in enzyme immobilization in ordered mesoporous materials and related applications. Chem Soc Rev. 2013;42:3894–912.
3. Fried DI, Brieler FJ, Fröba M. Designing inorganic porous materials for enzyme adsorption and applications in biocatalysis. ChemCatChem. 2013;5:862–84.
4. Bhakta SA, Evans E, Benavidez TE, Garcia CD. Protein adsorption onto nanomaterials for the development of biosensors and analytical devices: a review. Anal Chim Acta. 2015;872:7–25.
5. Topoglidis E, Cass A, O'Regan B, Durrant JR. Immobilisation and bioelectrochemistry of proteins on nanoporous TiO$_2$ and ZnO films. J Electroanal Chem. 2001;517:20–7.
6. Monopoli MP, Pitek AS, Lynch I, Dawson KA. Formation and characterization of the nanoparticle-protein corona. Methods Mol Biol. 2013;1025:137.
7. Lynch I, Cedervall T, Lundqvist M, Cabaleiro-Lago C, Linse S, Dawson KA. The nanoparticle-protein complex as a biological entity; a complex fluids and surface science challenge for the 21st century. Adv Colloid Interf Sci. 2007;134–135:167–74.
8. Vidic J, Haque F, Guigner JM, Vidy A, Chevalier C, Stankic S. Effects of water and cell culture media on the physicochemical properties of ZnMgO nanoparticles and their toxicity toward mammalian cells. Langmuir. 2014;30:11366–74.
9. Schulze C, Kroll A, Lehr CM, Schäfer UF, Becker K, Schnekenburger J, Schulze Isfort C, Landsiedel R, Wohlleben W. Not ready to use-overcoming pitfalls when dispersing nanoparticles in physiological media. Nanotoxicology. 2008;2:51–61.
10. Nel AE, Mädler L, Velegol D, Xia T, Hoek E, Somasundaran P, Klaessig F, Castranova V, Thompson M. Understanding biophysicochemical interactions at the nano-bio interface. Nat Mater. 2009;8:543–57.
11. Mahmoudi M, Lynch I, Ejtehadi MR, Monopoli MP, Bombelli FB, Laurent S. Protein-nanoparticle interactions: opportunities and challenges. Chem Rev. 2011;111:5610–37.
12. Walkey CD, Chan W. Understanding and controlling the interaction of nanomaterials with proteins in a physiological environment. Chem Soc Rev. 2012;41:2780–99.
13. Pino PD, Pelaz B, Zhang Q, Maffre P, Nienhaus GU, Parak WJ. Protein corona formation around nanoparticles-from the past to the future. Mater Horizons. 2014;1:301–13.
14. Kumar A, Das S, Munusamy P, Self W, Baer DR, Sayle DC, Seal S. Behavior of nanoceria in biologically-relevant environments. Environ Sci Nano. 2014;1:516–32.
15. Mudunkotuwa IA, Grassian VH. Biological and environmental media control oxide nanoparticle surface composition: the roles of biological components (proteins and amino acids), inorganic oxyanions and humic acid. Environ Sci Nano. 2015;2:429–39.
16. Elser MJ, Berger T, Brandhuber D, Bernardi J, Diwald O, Knözinger E. Particles coming together: electron centers in adjoined TiO$_2$ nanocrystals. J Phys Chem B. 2006;110:7605–8.
17. Siedl N, Elser MJ, Halwax E, Bernardi J, Diwald O. When fewer photons do more: a comparative O$_2$ photoadsorption study on vapor-deposited TiO$_2$ and ZrO$_2$ nanocrystal ensembles. J Phys Chem C. 2009;113:9175–81.
18. Baumann SO, Elser MJ, Auer M, Bernardi J, Hüsing N, Diwald O. Solid–solid interface formation in TiO$_2$ nanoparticle networks. Langmuir. 2011;27:1946–53.
19. Elser MJ, Diwald O. Facilitated lattice oxygen depletion in consolidated TiO$_2$ nanocrystal ensembles: a quantitative spectroscopic O$_2$ adsorption study. J Phys Chem C. 2012;116:2896–903.
20. Berger T, Diwald O. Defects in metal oxide nanoparticle powders. In: Jupille J, Thornton J, editors. Defects at oxide surfaces. Berlin: Springer; 2015.
21. Di Serio M, Maturo C, de Alteriis E, Parascandola P, Tesser R, Santacesaria E. Lactose hydrolysis by immobilized β-galactosidase: the effect of the supports and the kinetics. Catal Today. 2003;79–80:333–9.
22. Satyawali Y, Roy SV, Roevens A, Meynen V, Mullens S, Jochems P, Doyen W, Cauwenberghs L, Dejonghe W. Characterization and analysis of the adsorption immobilization mechanism of β-galactosidase on metal oxide powders. RSC Adv. 2013;3:24054–62.
23. Bernal C, Sierra L, Mesa M. Application of hierarchical porous silica with a stable large porosity for β-galactosidase immobilization. ChemCatChem. 2011;3:1948–54.
24. Bernal C, Sierra L, Mesa M. Improvement of thermal stability of β-galactosidase from Bacillus circulans by multipoint covalent immobilization in hierarchical macro-mesoporous silica. J Mol Catal B. 2012;84:166–72.
25. Verma ML, Barrow CJ, Kennedy JF, Puri M. Immobilization of β-d-galactosidase from Kluyveromyces lactis on functionalized silicon dioxide

nanoparticles: characterization and lactose hydrolysis. Int J Biol Macromol. 2012;50:432–7.

26. Biró E, Budugan D, Todea A, Péter F, Klébert S, Feczkó T. Recyclable solid-phase biocatalyst with improved stability by sol-gel entrapment of β-d-galactosidase. J Mol Catal B. 2016;123:81–90.

27. Deressa T, Stöcklinger A, Wallner M, Himly M, Kofler S, Hainz K, Brandstetter H, Thalhamer J, Hammerl P. Structural integrity of the antigen is a determinant for the induction of T-helper type-1 immunity in mice by gene gun vaccines against *E. coli* beta-galactosidase. PLoS ONE. 2014;9:e102280.

28. Berger T, Sterrer M, Diwald O, Knözinger E, Panayotov D, Thompson TL, Yates JT Jr. Light-induced charge separation in anatase TiO_2 particles. J Phys Chem B. 2005;109:6061–8.

29. Berger T, Sterrer M, Diwald O, Knözinger E. Charge trapping and photoadsorption of O_2 on dehydroxylated TiO_2 nanocrystals-an electron paramagnetic resonance study. ChemPhysChem. 2005;6:2104–12.

30. Wallenfels K, Malhotra OP. Galactosidases. Adv Carbohydr Chem. 1962;16:239.

31. Barth A, Zscherp C. What vibrations tell us about proteins. Q Rev Biophys. 2002;35:369–430.

32. Barth A. Infrared spectroscopy of proteins. Biochim Biophys Acta. 2007;1767:1073–101.

33. Pettersen EF, Goddard TD, Huang CC, Couch GS, Greenblatt DM, Meng EC, Ferrin TE. UCSF chimera-a visualization system for exploratory research and analysis. J Comput Chem. 2004;25:1605–12.

34. Márquez A, Berger T, Feinle A, Hüsing N, Himly M, Duschl A, Diwald O. Langmuir. 2017;33:2551–8.

35. Moulton SE, Barisci JN, McQuillan AJ, Wallace GG. ATR-IR spectroscopic studies of the influence of phosphate buffer on adsorption of immunoglobulin G to TiO_2. Colloids Surf A. 2003;220:159–67.

36. Wei T, Kaewtathip S, Shing K. Buffer effect on protein adsorption at liquid/solid interface. J Phys Chem C. 2009;113:2053–62.

37. Mahmoudi M, Shokrgozar MA, Sardari S, Moghadam MK, Vali H, Laurent S, Stroeve P. Irreversible changes in protein conformation due to interaction with superparamagnetic iron oxide nanoparticles. Nanoscale. 2011;3:1127–38.

38. Röcker C, Pötzl M, Zhang F, Parak WJ, Nienhaus GU. A quantitative fluorescence study of protein monolayer formation on colloidal nanoparticles. Nat Nanotechnol. 2009;4:577–80.

39. Casals E, Pfaller T, Duschl A, Oostingh GJ, Puntes V. Time evolution of the nanoparticle protein corona. ACS Nano. 2010;4:3623–32.

40. Glassford SE, Byrne B, Kazarian SG. Recent applications of ATR FTIR spectroscopy and imaging to proteins. Biochim Biophys Acta. 2013;1834:2849–58.

41. Mudunkotuwa IA, Minshid AA, Grassian VH. ATR-FTIR spectroscopy as a tool to probe surface adsorption on nanoparticles at the liquid-solid interface in environmentally and biologically relevant media. Analyst. 2014;139:870–81.

42. Arrondo J, Muga A, Castresana J, Goñi FM. Quantitative studies of the structure of proteins in solution by fourier-transform infrared spectroscopy. Prog Biophys Mol Biol. 1993;59:23–56.

43. Arrondo J, Muga A, Castresana J, Bernabeu C, Goñi FM. An infrared spectroscopic study of β-galactosidase structure in aqueous solutions. FEBS Lett. 1989;252:118–20.

44. Juers DH, Matthews BW, Huber RE. LacZ β-galactosidase: structure and function of an enzyme of historical and molecular biological importance. Protein Sci. 2012;21:1792–807.

45. Juers DH, Jacobson RH, Wigley D, Zhang XJ, Huber RE, Tronrud DE, Matthews BW. High resolution refinement of β-galactosidase in a new crystal form reveals multiple metal-binding sites and provides a structural basis for α-complementation. Protein Sci. 2000;9:1685–99.

46. Cao L. Carrier-bound immobilized enzymes: principles, application and design. New York: Wiley; 2006.

47. Wehtje E, Adlercreutz P, Mattiasson B. Improved activity retention of enzymes deposited on solid supports. Biotechnol Bioeng. 1993;41:171–8.

48. Ni Y, Li J, Huang Z, He K, Zhuang J, Yang W. Improved activity of immobilized horseradish peroxidase on gold nanoparticles in the presence of bovine serum albumin. J Nanopart Res. 2013;15:2038.

6

A novel covalent approach to bio-conjugate silver coated single walled carbon nanotubes with antimicrobial peptide

Atul A. Chaudhari[1], D'andrea Ashmore[1], Subrata deb Nath[2], Kunal Kate[2], Vida Dennis[1], Shree R. Singh[1], Don R. Owen[3], Chris Palazzo[3], Robert D. Arnold[4], Michael E. Miller[5] and Shreekumar R. Pillai[1*]

Abstract

Background: Due to increasing antibiotic resistance, the use of silver coated single walled carbon nanotubes (SWCNTs-Ag) and antimicrobial peptides (APs) is becoming popular due to their antimicrobial properties against a wide range of pathogens. However, stability against various conditions and toxicity in human cells are some of the major drawbacks of APs and SWCNTs-Ag, respectively. Therefore, we hypothesized that APs-functionalized SWCNTs-Ag could act synergistically. Various covalent functionalization protocols described previously involve harsh treatment of carbon nanotubes for carboxylation (first step in covalent functionalization) and the non-covalently functionalized SWCNTs are not satisfactory.

Methods: The present study is the first report wherein SWCNTs-Ag were first carboxylated using Tri sodium citrate (TSC) at 37 °C and then subsequently functionalized covalently with an effective antimicrobial peptide from Therapeutic Inc., TP359 (FSWCNTs-Ag). SWCNTs-Ag were also non covalently functionalized with TP359 by simple mixing (SWCNTs-Ag-M) and both, the FSWCNTs-Ag (covalent) and SWCNTs-Ag-M (non-covalent), were characterized by Fourier transform infrared spectroscopy (FT-IR), Ultraviolet visualization (UV–VIS) and transmission electron microscopy (TEM). Further the antibacterial activity of both and TP359 were investigated against two gram positive (*Staphylococcus aureus* and *Streptococcus pyogenes*) and two gram negative (*Salmonella enterica* serovar Typhimurium and *Escherichia coli*) pathogens and the cellular toxicity of TP359 and FSWCNTs-Ag was compared with plain SWCNTs-Ag using murine macrophages and lung carcinoma cells.

Results: FT-IR analysis revealed that treatment with TSC successfully resulted in carboxylation of SWCNTs-Ag and the peptide was indeed attached to the SWCNTs-Ag evidenced by TEM images. More importantly, the present study results further showed that the minimum inhibitory concentration (MIC) of FSWCNTs-Ag were much lower (~7.8–3.9 μg/ml with IC50: ~4–5 μg/ml) compared to SWCNTs-Ag-M and plain SWCNTs-Ag (both 62.6 μg/ml, IC50: ~31–35 μg/ml), suggesting that the covalent conjugation of TP359 with SWCNTs-Ag was very effective on their counterparts. Additionally, FSWCNTs-Ag are non-toxic to the eukaryotic cells at their MIC concentrations (5–2.5 μg/ml) compared to SWCNTs-Ag (62.5 μg/ml).

Conclusion: In conclusion, we demonstrated that covalent functionalization of SWCNTs-Ag and TP359 exhibited an additive antibacterial activity. This study described a novel approach to prepare SWCNT-Ag bio-conjugates without loss of antimicrobial activity and reduced toxicity, and this strategy will aid in the development of novel and biologically important nanomaterials.

Keywords: Carbon nanotubes, Antimicrobial, Peptide, Bacteria, Cytotoxicity, Bio-conjugation

*Correspondence: spillai@alasu.edu
[1] Center for Nanobiotechnology Research, Alabama State University, Montgomery, AL, USA
Full list of author information is available at the end of the article

Background

Carbon nanotubes are well known for their wide range of applications in diverse fields, including biomedicine [1, 2]. Of relevance, single-walled carbon nanotubes (SWNTs) have been used for biomedical molecular imaging and effective drug delivery in vivo as well as in vitro [3–9]. Additionally, metallic nanocomposites of SWCNTs, especially silver coated SWCNTs (SWCNTs-Ag) have shown a remarkable antibacterial activity against gram positive as well as gram negative pathogens over the past few years [10–12]. The results from these studies are promising as there is an urgent need of developing novel antimicrobial strategies due to increasing resistance to several broad spectrum antibiotics [13–15]. However, the application of SWCNTs-Ag has been limited due to several known mechanisms of toxicity to eukaryotic cells [16–18]. Functionalization strategies have been developed and reported to lessen their toxicity, such as pegylation or surface modification using biological entities like DNA/RNA or protein [11, 12, 16–18]. In our previous study we demonstrated that pegylation of SWCNTs-Ag reduced toxicity to different eukaryotic cell lines without reducing their anti-bacterial activity [11]. Although surface modification of SWCNTs-Ag reduced their toxic effects on eukaryotic cells, the functionalization was not stable and the dosage required for the antibacterial activity remained high (62.5 μg/ml).

Functionalization using antimicrobial peptides (APs) may reduce the dosage required for the antibacterial activity of SWCNTs-Ag due to the antibacterial activity of both the components. The application of host derived or synthetic APs are becoming popular due to their effectiveness and broad range of antibacterial activity [19–23]. Several in vitro as well as in vivo models have successfully demonstrated that APs exhibit effective antimicrobial activity [23–29]. Despite these advantages, the use of APs has several disadvantages, such as enzymatic degradation leading to loss of activity, expensive to produce and instability in solution. However, a suitable delivery vehicle can mitigate these challenges by enhancing delivery of the peptides to the infected site and minimizing degradation. Carbon nanotubes have been used effectively as a formulation platform for the targeted delivery of anticancer agents, DNAs, RNAs and proteins [30–35]. Thus, functionalization of SWCNTs to produce nanotube bioconjugates for the desired application is a rational approach. In general, functionalization can be achieved by either covalent bonding or non-covalent interactions to SWCNTs [30, 31, 33, 34]. However, various covalent functionalization protocols involves harsh treatment of carbon nanotubes such as oxidation of nanotubes, addition of 1,3-dipolar cycloaddition on the nanotube sidewalls or treatment with highly concentrated acids such as HNO_3 and H_2SO_4 [33, 36–38]. Although covalent chemical reactions offer stable

functionalized SWCNTs, the sidewalls of the nanotubes get severely damaged in the process. In contrast, non-covalent functionalization by amphiphilic molecules either through passive absorption or by coating of the nanotube surfaces maintains the structure and optical properties of SWCNTs [33]. However, non-covalently functionalized SWCNTs are of limited use due to poor stability and issues associated with biocompatibility. Although, SWCNTs have been functionalized earlier using anti-cancer drugs, DNA, RNA or proteins [30, 32–35], the metallic nanocomposites of SWCNTs such as SWCNTs-Ag have yet to be successfully functionalized with biological molecules.

Therefore a unique strategy is required which will result in minimal damage to the structure of SWCNTs-Ag, provide maximum biocompatibility and optimum antibacterial activity with high potency (i.e., relatively low dosage). In the present study, we developed a novel approach to functionalize SWCNTs-Ag covalently using an effective antimicrobial peptide from Therapeutic Inc, TP359. The functionalized SWCNTs-Ag were further characterized by Fourier transform infrared spectroscopy (FT-IR) and transmission electron microscopy (TEM) and compared with the non-covalently functionalized SWCNTs-Ag. Further, the antibacterial activity of functionalized SWCNT-Ag (covalent as well as non-covalent) was investigated against two gram positive (*Staphylococcus aureus* and *Streptococcus pyogenes*) and two gram negative pathogens (*Salmonella enterica* serovar Typhimurium and *Escherichia coli*).

Results

Carboxylation of silver coated-single walled carbon nanotubes (SWCNTs-Ag)

Carboxylation on the surface of SWCNTs-Ag was confirmed by FT-IR. FT-IR spectrum of pure TSC, SWCNTs-Ag and different ratios of TSC to SWCNTs-Ag are shown in Fig. 1a–f. The characteristic absorption bands of –COO at around 1394 (symmetric stretching) and 1589 cm^{-1} (asymmetric stretching), and –COOH bands at around 3322 cm^{-1} were observed on pure TSC whereas the pure SWCNTs-Ag did not show any of these characteristic peaks on their surfaces (Fig. 1e, f). The similar peaks corresponding to the –COO and –COOH stretching were observed on TSC treated SWCNTs-Ag at different ratios (Fig. 1a–d). The presence of –COOH and –COO stretching on the TSC-SWCNTs-Ag confirmed the carboxylation of SWCNTs-Ag with the treatment of TSC.

Covalent and non-covalent functionalization of SWCNTs-Ag with an antimicrobial peptide TP359

Figure 2 shows the FT-IR pattern of the plain SWCNTs-Ag, TP359, FSWCNTs-Ag (covalent) and SWCNTs-Ag-M (non-covalent). FSWCNTs-Ag showed presence of

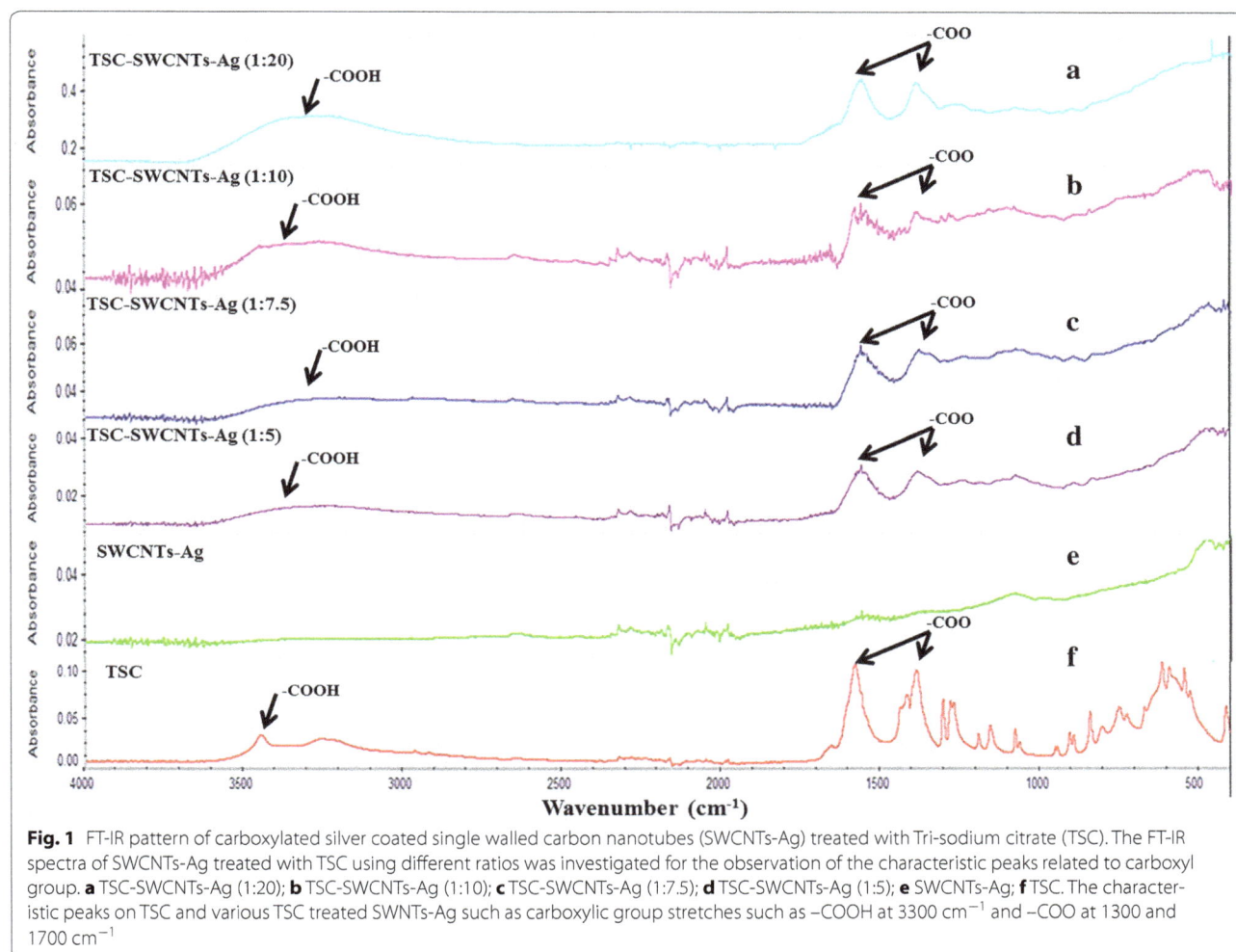

Fig. 1 FT-IR pattern of carboxylated silver coated single walled carbon nanotubes (SWCNTs-Ag) treated with Tri-sodium citrate (TSC). The FT-IR spectra of SWCNTs-Ag treated with TSC using different ratios was investigated for the observation of the characteristic peaks related to carboxyl group. **a** TSC-SWCNTs-Ag (1:20); **b** TSC-SWCNTs-Ag (1:10); **c** TSC-SWCNTs-Ag (1:7.5); **d** TSC-SWCNTs-Ag (1:5); **e** SWCNTs-Ag; **f** TSC. The characteristic peaks on TSC and various TSC treated SWNTs-Ag such as carboxylic group stretches such as –COOH at 3300 cm^{-1} and –COO at 1300 and 1700 cm^{-1}

functional groups for alcohols and phenols (–O–H) at 3400–3600 cm^{-1}, carbonyl (–C=O) at 1760 cm^{-1} and amine (N–H) stretches at 1580 cm^{-1} similar to that of TP359, whereas SWCNTs-Ag did not exhibit any of these peaks (Fig. 2a–c). Conversely, the non-covalent bonding of SWCNTs-Ag and the peptide (SWCNTs-Ag-M) showed the presence of alkane (–C–H) at 3000 cm^{-1} and amine stretch at 1580 cm^{-1} similar to that of TP359 however the –O–H stretch was not apparent (Fig. 2d). The FT-IR spectra thus showed the non-covalent and covalent functionalization of SWCNTs-Ag with TP359, later being more apparent (Fig. 2). Similarly, the UV–VIS pattern of FSWCNTs-Ag, SWCNTs-Ag and SWCNTs-Ag-M showed a characteristic peak only on FSWCNTs-Ag at ~250 nm validating the functionalization (Fig. 3).

Further, TEM imaging of SWCNTs-Ag, TSC-SWCNTs-Ag, TP359, SWCNTs-Ag-M and FSWCNTs-Ag was performed and the electronmicrographs are presented in Fig. 4a–e. The silver coating on SWCNTs-Ag and TSC-SWCNTs-Ag can be visualized clearly (Fig. 4a, b, indicated by white dotted arrows) indicating that TSC treatment did not alter the silver coating on the SWCNTs. TEM images of SWCNTs-Ag-M showed that the peptide adhered onto the surface of the SWCNTs-Ag resulting in the coating of SWCNTs-Ag surface with the peptide (Fig. 4d, indicated by a yellow solid arrow). While in the FSWCNTs-Ag samples, the peptide appeared to be attached to SWCNTs-Ag (Fig. 4e, indicated by red solid arrow). Similarly, AFM analysis of FSWCNTs-Ag and plain SWCNTs-Ag was performed to further validate the above findings. Figure 5 shows the surface characteristics of plain SWCNTs-Ag and FSWCNTs-Ag. SWCNTs-Ag appears to be homogeneously distributed on the surface and there was no significant elevation observed on their surface (Fig. 5a–d). In contrast, peptide attachment appears to be evident on FSWCNTs-Ag characterized by significant elevation on the surface (Fig. 5e–h).

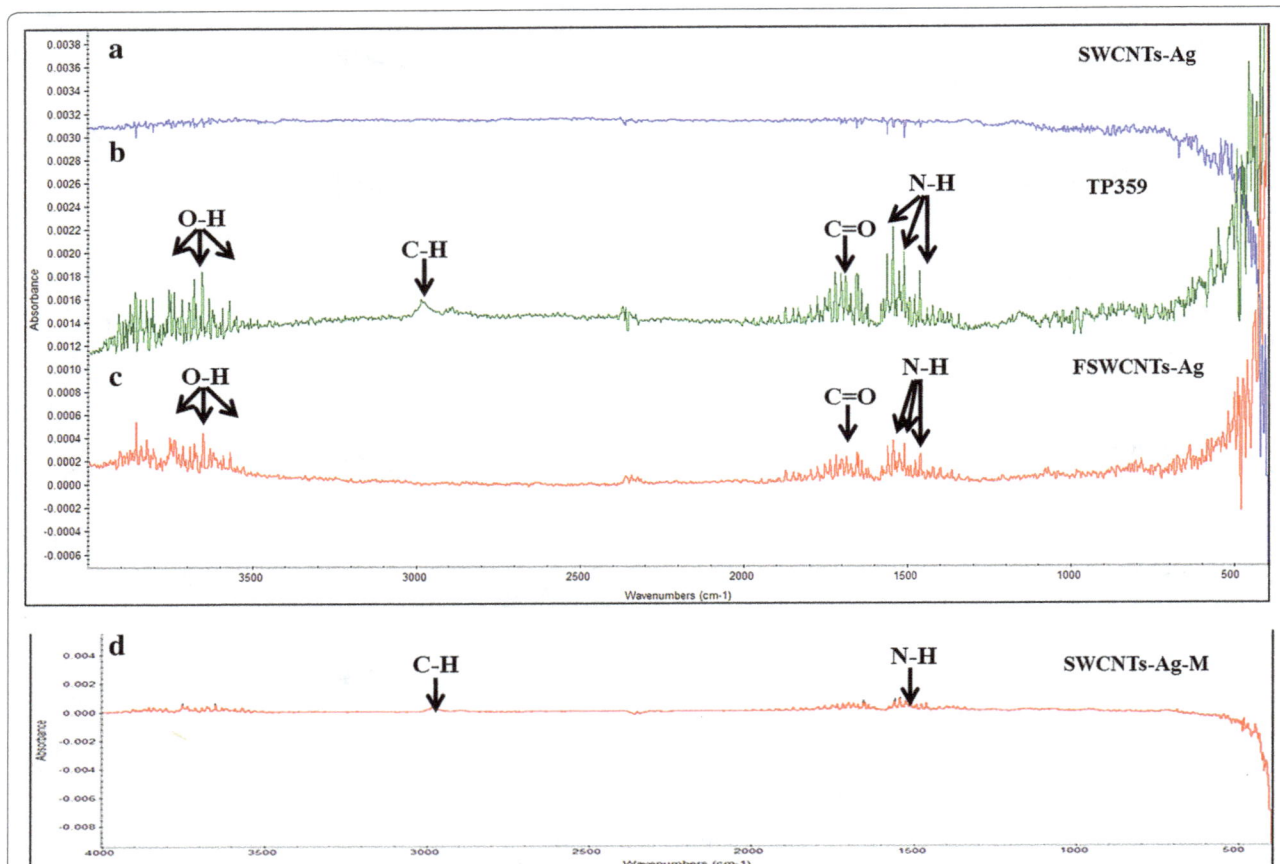

Fig. 2 FT-IR pattern of silver coated single walled carbon nanotubes (SWCNTs-Ag) functionalized with the antimicrobial peptide TP359. SWCNTs-Ag were either carboxylated using TSC and subsequently functionalized with TP359 (FSWCNTs-Ag) or simply mixed with TP359 (SWCNTs-Ag-M) and FT-IR spectra of both the formulations was investigated. **a** SWCNTs-Ag; **b** TP359; **c** FSWCNTs-Ag; **d** SWCNTs-Ag-M. The characteristic peaks for amine groups (N–H) at 1580–1650 cm^{-1}, alkane group (C–H) at 3100 cm^{-1}, carbonyl group (C=O) at 1760 cm^{-1} and carboxylic group (O–H) at 3300 cm^{-1} was observed on TP359, FSWCNTs-Ag and SWCNTs-Ag-M

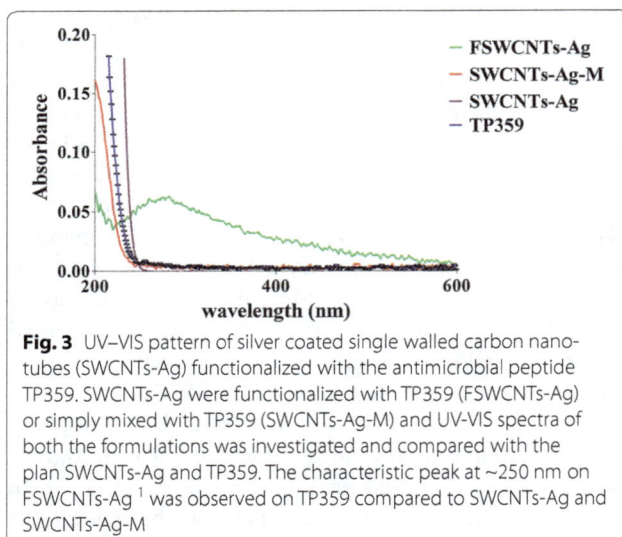

Fig. 3 UV–VIS pattern of silver coated single walled carbon nanotubes (SWCNTs-Ag) functionalized with the antimicrobial peptide TP359. SWCNTs-Ag were functionalized with TP359 (FSWCNTs-Ag) or simply mixed with TP359 (SWCNTs-Ag-M) and UV-VIS spectra of both the formulations was investigated and compared with the plan SWCNTs-Ag and TP359. The characteristic peak at ~250 nm on FSWCNTs-Ag [1] was observed on TP359 compared to SWCNTs-Ag and SWCNTs-Ag-M

Antibacterial activity of FSWCNTs-Ag compared to SWCNTs-Ag-M, SWCNTs-Ag, TP359 and TSC-SWCNTs-Ag

The MICs of FSWCNTs-AG, SWCNTs-Ag-M, SWC-NTs-Ag, TP359 and TSC-SWCNTs-Ag were investigated against two gram negative and two gram positive bacteria. Also the quantitative (cfu/ml) analysis of bacteria exposed to these nanocomposites was determines using plate counting assay. The MICs of FSWCNTs-AG, SWC-NTs-Ag-M, SWCNTs-Ag, TP359 and TSC-SWCNTs-Ag are shown in Additional file 1: Figure S1. Table 1 represents the MICs, bactericidal and bacteriostatic concentrations. The MICs for TP359 and FSWCNTs-Ag were lower compared to SWCNTs-Ag, SWCNTs-Ag-M and TSC-SWCNTs-Ag (Additional file 1: Figure S1; Table 1). TP359 exhibited strong antibacterial activity against all four pathogens and the MIC values were in between 7.8–3.9 μg/ml (*Escherichia coli*); 1.9–0.9 μg/ml (*Salmonella*

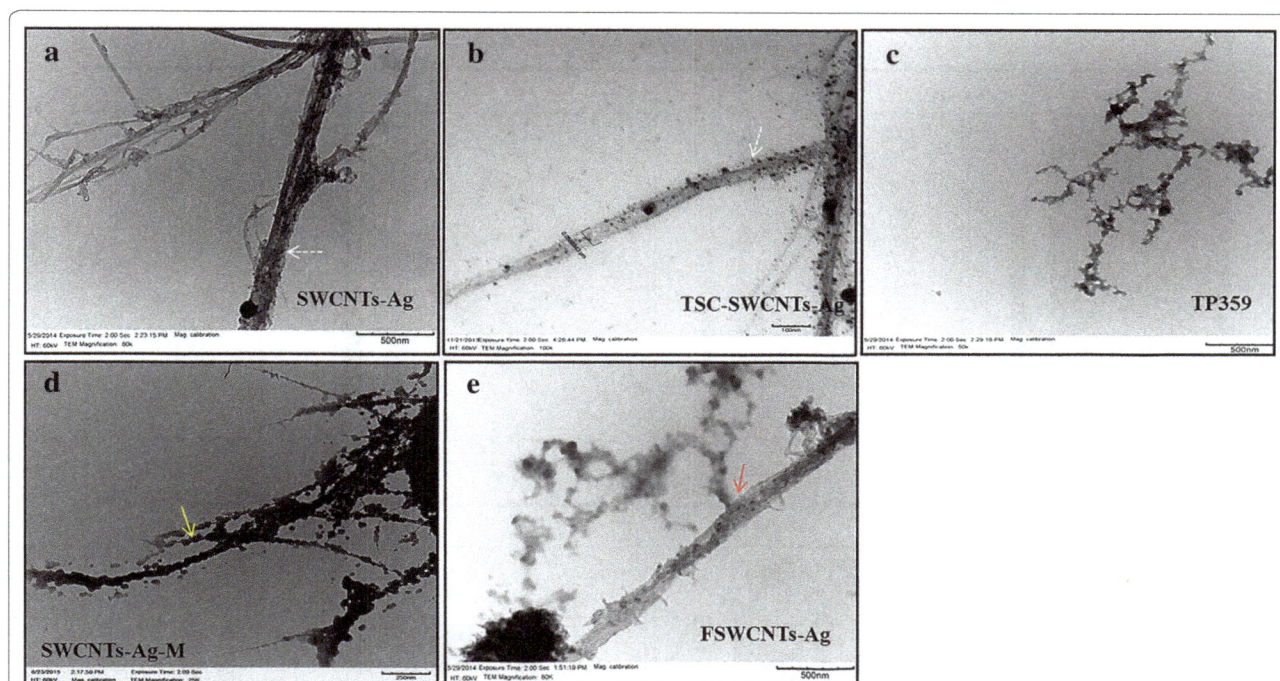

Fig. 4 Transmission electron microscopy of silver coated single walled carbon nanotubes (SWCNTs-Ag) functionalized with the antimicrobial peptide TP359. Carboxylated SWCNTs-Ag were functionalized with TP359 and the functionalization of the peptide was confirmed by TEM imaging. **a** SWCNTs-Ag; **b** TSC-SWCNTs-Ag; **c** TP359; **d** SWCNTs-Ag-M; **e** FSWCNTs-Ag. The attachment of the peptide to the SWCNTs-Ag is seen clearly

Typhimurium) and 3.9–1.9 μg/ml (for *Staphylococcus aureus* and *Streptococcus pyogenes*). On the other hand, the MIC for SWCNTs-Ag was in between 62.5–31.3 μg/ml and 125–62.5 μg/ml against gram −ve (*Escherichia coli* and *Salmonella* Typhimurium) and gram +ve (*Staphylococcus aureus* and *Streptococcus pyogenes*) bacteria, respectively (Table 1). Compared to SWCNTs-Ag, FSWCNTs-Ag exhibited stronger antibacterial activity at much lower concentration such as 7.8–3.9 μg/ml (against *Escherichia coli, Staphylococcus aureus* and *Streptococcus pyogenes*, TP359 concentration is 0.3 μg/ml) and 1.9–0.9 μg/ml (*Salmonella* Typhimurium, TP359 concentration is 0.08 μg/ml). On the contrary for SWCNTs-Ag-M (non-covalent functionalization strategy), the MIC values were still greater than FSWCNTs-Ag and TP359 against *Escherichia coli* and *Salmonella* Typhimurium (31.3–15.6 μg/ml); *Staphylococcus aureus* (62.5–31.3 μg/ml) and *Streptococcus pyogenes* (125–62.5 μg/ml). MIC values of TSC-SWCNTs-Ag against all four bacterial pathogens were similar to that of MICs of SWCNTs-Ag (Additional file 1: Figure S1; Table 1). Additionally, the quantitative analysis of bacteria exposed to these concentrations showed logarithmic decrease in all four bacterial pathogens (Fig. 6a–d). The half maximal inhibitory concentrations (IC50) values (based on the quantitative bacterial growth vs concentrations) for FSWCNTs-Ag (range

for all four 1.3–5 μg/ml) were ~tenfolds lower than plain SWCNTs-Ag (range for all four 23–35 μg/ml) (Additional file 1: Table S1). Further, the KB assay results, presented as Table 2 [(*Escherichia coli* and *Salmonella* Typhimurium, Additional file 1: Figure S2) and Additional file 1: Figure S3 (*Staphylococcus aureus* and *Streptococcus pyogenes*)], confirmed the antibacterial activity of FSWC-NTs-Ag and TP359 compared to SWCNTs-Ag. The zone of inhibition for TP359 and FSWCNTs-Ag against *E coli* was observed at 20, 10 and 5 μg/ml concentrations whereas SWCNTs-Ag showed slight zone of inhibition only at 20 μg/ml (Table 2, Additional file 1: Figure S2a–c). When *Salmonella* Typhimurium were treated with TP359 and FSWCNTs-Ag at different concentrations, a clear zone of inhibition was observed for all the concentrations compared to SWCNTs-Ag (Table 2; Additional file 1: Figure S2a–c). Similarly, TP359 and FSWCNTs-Ag inhibited the growth of gram positive bacteria at 20, 10 and 5 μg/ml and showed clear zone of inhibition whereas SWCNTs-Ag did not show inhibition at any concentrations (Table 2; Additional file 1: Figure S3a–c).

In vitro cell toxicity of TP359, SWCNTs-Ag and FSWCNTs-Ag
Next, we explored the cell toxicity of TP359, FSWC-NTs-Ag and SWCNTs-Ag at four different concentrations. Figure 7a–f represents cell toxicity of TP359,

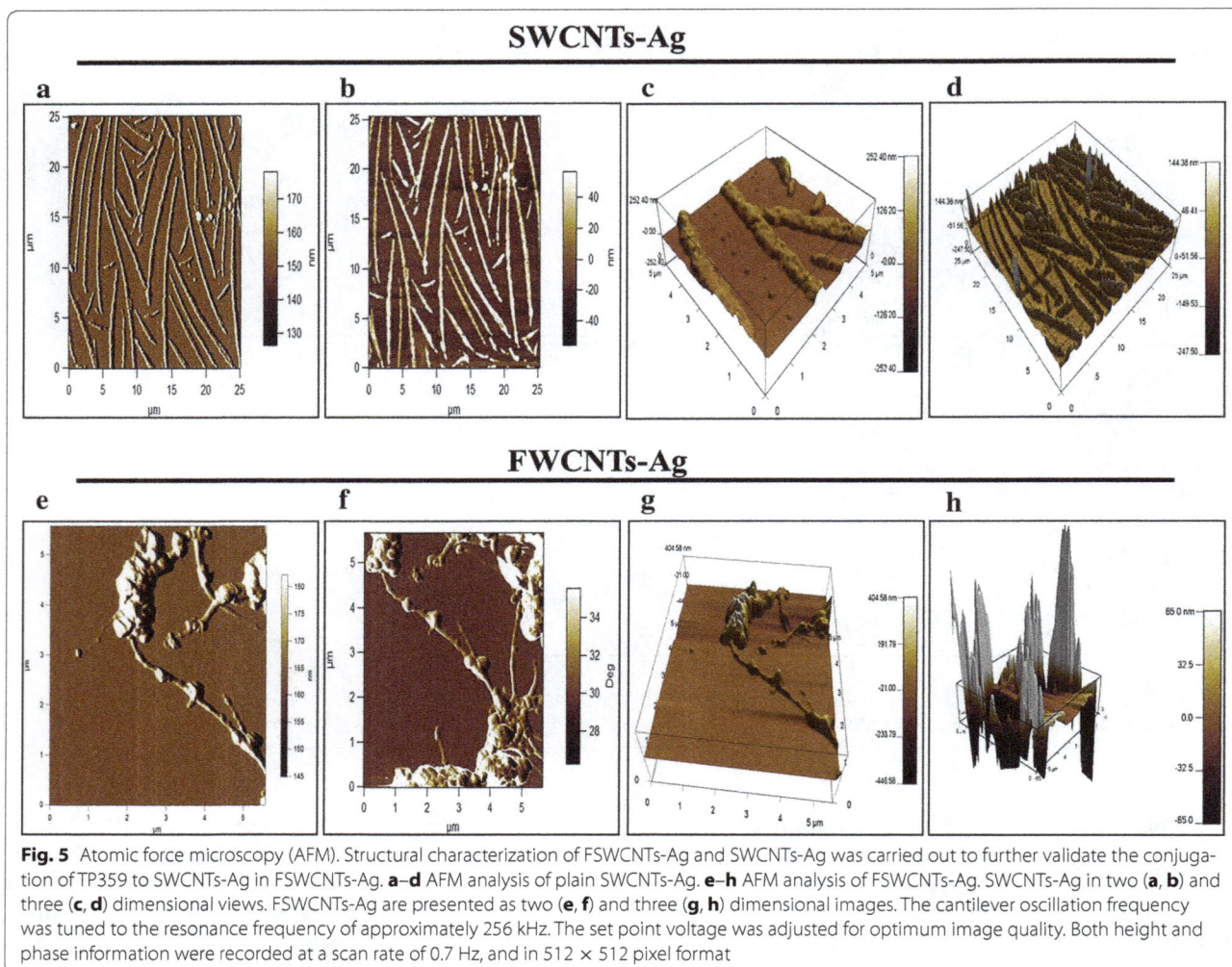

Fig. 5 Atomic force microscopy (AFM). Structural characterization of FSWCNTs-Ag and SWCNTs-Ag was carried out to further validate the conjugation of TP359 to SWCNTs-Ag in FSWCNTs-Ag. **a–d** AFM analysis of plain SWCNTs-Ag. **e–h** AFM analysis of FSWCNTs-Ag. SWCNTs-Ag in two (**a**, **b**) and three (**c**, **d**) dimensional views. FSWCNTs-Ag are presented as two (**e**, **f**) and three (**g**, **h**) dimensional images. The cantilever oscillation frequency was tuned to the resonance frequency of approximately 256 kHz. The set point voltage was adjusted for optimum image quality. Both height and phase information were recorded at a scan rate of 0.7 Hz, and in 512 × 512 pixel format

SWCNTs-Ag and FSWCNTs-Ag in A549 (Fig. 7a, c, e) and J774 cells (Fig. 7b, d, f). TP359 was not toxic to A549 cells or J774 cells at any concentration tested (Fig. 7a, b). On the other hand, FSWCNTs-Ag and SWCNTs-Ag were significantly toxic at 20 and 10 µg/ml (with approximately 30 % cell staining) and were relatively non-toxic at 5 and 2.5 µg/ml concentrations and almost 90 % cells relative to control (Fig. 7c–f).

Discussion

The present study reports a novel strategy for covalent attachment of antimicrobial peptide to SWCNTs-Ag and antibacterial activity of covalently functionalized SWCNTs-Ag compared to non-covalently functionalized SWCNTs-Ag. In this approach, SWCNTs-Ag were successfully carboxylated using Tri sodium citrate (TSC) and did not involve the use of extreme treatments of SWC-NTs-Ag for carboxylation. TSC is a well-known reducing agent and have been engaged in the size controlled

synthesis of gold nanoparticles (GNPs) [39, 40]. Besides its role in synthesis of nanoparticles, TSC have also been reported to carboxylate nanoparticles such as GNPs and superparamagnetic iron oxide nanoparticles [41–43]. Congruent with these previous findings, our data showed that treatment of SWCNTs-Ag with TSC resulted in carboxylation of SWCNTs-Ag. However, the striking difference between our study and the previously reported findings is that we used a range of relatively less concentrations of TSC [approximately 0.15 (1:20 ratio) to 0.60 mM (1:5 ratio)] for carboxylation of SWCNTs-Ag compared to 40 mM of TSC that was used to carboxylate the gold nanoparticles [43]. As reported previously [42, 43], the nanoparticles were treated with TSC at high temperatures (90 °C) whereas in this study SWCNTs-Ag were treated with TSC at 37 °C. Carboxylation of SWC-NTs generally involves harsh acidic treatments such as combination of H_2SO_4 and HNO_3 at extreme temperatures that has been associated with physical damage to

Table 1 MICs, bactericidal and bacteriostatic concentrations

Material	Concentration (μg/ml)			
	Escherichia coli	*Salmonella* Typhimurium	*Staphylococcus aureus*	*Streptococcus pyogenes*
TP359				
MICs	7.8–3.9	1.9–0.9	3.9–1.9	3.9–1.9
Bactericidal	7.8*	1.9*	3.9*	3.9*
Bacteriostatic	3.9*	0.9*	1.9*	1.9*
SWCNTs-Ag				
MICs	62.5–31.2	62.5–31.2	125–62.5	125–62.5
Bactericidal	62.5	62.5	125	125
Bacteriostatic	31.2	31.2	62.5	62.5
TSC-SWCNTs-Ag				
MICs	62.5–31.2	62.5–31.2	125–62.5	125–62.5
Bactericidal	62.5	62.5	125	125
Bacteriostatic	31.2	31.2	62.5	62.5
FSWCNTs-Ag				
MICs	7.8–3.9**	1.9–0.9**	7.8–3.9*	7.8–3.9**
Bactericidal	7.8*	1.9**	7.8**	7.8**
Bacteriostatic	3.9	0.9	3.9	3.9
SWCNTs-Ag-M				
MICs	31.2–15.6	31.2–15.6	62.5–31.2	125–62.5
Bactericidal	31.2	31.2	62.5	125
Bacteriostatic	15.6	15.6	31.2	62.5

The p ≤ 0.05 indicating significant * differences, or p ≤ 0.01 indicating highly significant ** differences

carbon nanotubes [36, 44, 45]. Our results demonstrated that SWCNTs-Ag can be carboxylated without exposure to extreme temperatures or high concentration of chemicals that are associated with degradation.

Further, the carboxylated SWCNTs-Ag were covalently functionalized with the antimicrobial peptide TP359, and non-covalent functionalization was achieved by simply mixing TP359 and plain SWCNTs-Ag. Non-covalent functionalization (SWCNTs-Ag-M) showed adherence of the peptide onto the surface of the SWCNTs-Ag whereas covalent functionalization (FSWCNTs-Ag) showed the attachment of the peptide to SWCNTs-Ag. It has been established that covalent functionalization of nanoparticles exhibit greater stability and improved activity [36–38]. In contrast, non-covalent functionalization is highly unstable and less effective than the covalent bonding [33]. The protocol described herein provides a relatively simple and stable approach to the covalent functionalization of the SWCNTs-Ag that neither involves extreme chemical treatment, nor causes damage to the carbon nanotubes due to exposure to high temperatures.

Evaluation of antibacterial activity of FSWCNTs-AG, SWCNTs-Ag-M, SWCNTs-Ag, TP359 and TSC-SWC-NTs-Ag revealed that the antibacterial concentrations of FSWCNTs-Ag were much lower compared to SWC-NTs-Ag-M and TSC-SWCNTs-Ag, suggesting that

the covalent conjugation of TP259 with SWCNTs-Ag showed increased efficacy compared to their counterparts as far as the antibacterial activity was concerned. It is therefore interesting to examine the mechanism of antibacterial action of the FSWCNTs-Ag compared to the peptide alone and SWCNTs-Ag. The present study thus defines a novel protocol to functionalize the silver coated single walled carbon nanotubes with the peptide which can be used to further evaluate the exact mechanism for antibacterial action for peptide functionalized SWCNTs-Ag compared with peptide and plain SWCNTs-Ag alone. The functionalization of SWCNTs with DNA, RNA or chemotherapeutic drugs has shown remarkable success in achieving the desired targets for the action of these biological molecules [30, 32–35]. However, it is known that the SWCNTs exhibit less potent antibacterial activity compared to SWCNTs-Ag (MIC concentrations 62.5 μg/ml). Therefore in this study, we attempted the functionalization of SWCNTs-Ag (vs. un-functionalized SWNCTs) with a known antimicrobial peptide in order to reduce the effective antimicrobial dosage of SWCNTs-Ag. This may has potential to reduce off the toxicity of SWCNTs-Ag to human cells and may reduce the emergence of drug resistance phenotypes. Our results suggested that covalently conjugating the peptide to SWCNTs-Ag effectively reduced

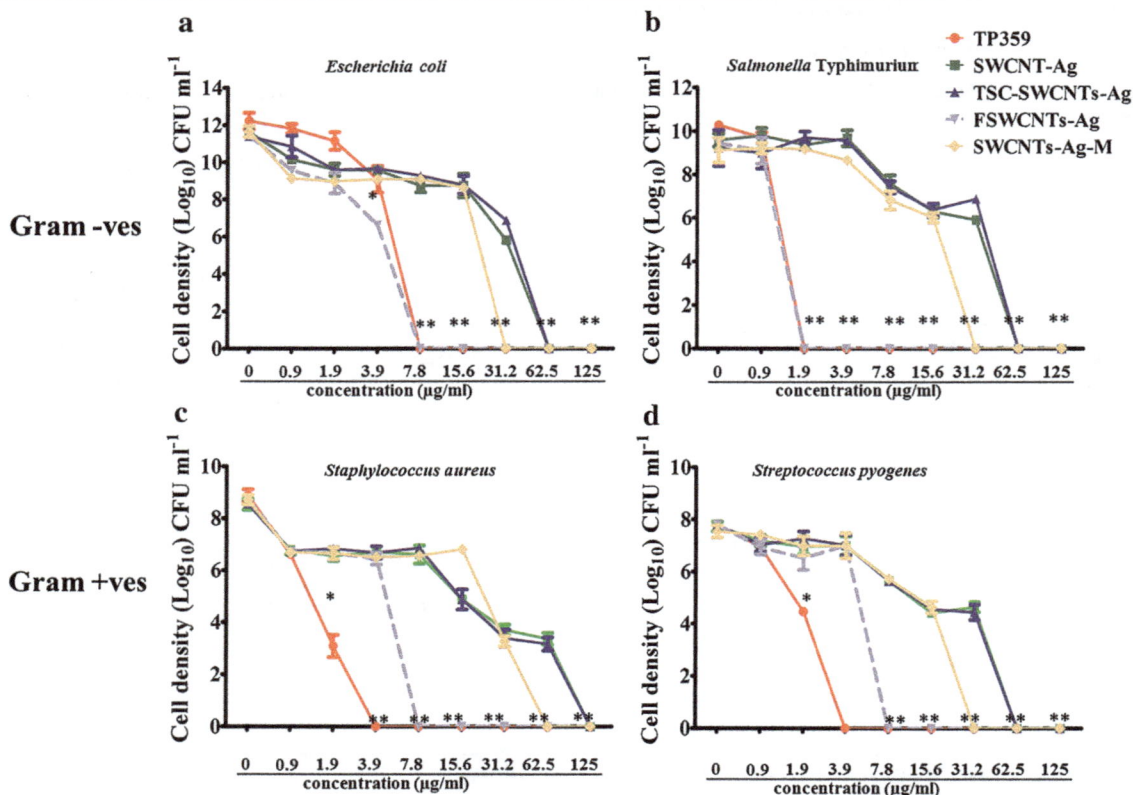

Fig. 6 Quantitative analysis of gram positive and gram negative bacteria exposed to various concentrations of nanocomposites. **a** log cfu/ml of surviving *Escherichia coli* upon exposure to peptide and nanocomposites. **b** Quantification of *Salmonella* Typhimurium upon exposure to peptide and nanocomposites. **c** *Staphylococcus aureus* exposed to various nanocomposites. **d** Quantitative analysis of *Streptococcus pyogenes*. Bacteria were grown in LB broth containing various concentrations of nanocomposites and all the cultures were incubated at 37 °C with shaking at 250 rpm and the cfu/ml counts were done at 24 h. Statistical differences were indicates as * when p ≤ 0.05, or ** when value was p ≤ 0.01. *Error bars* represent standard deviations determined from at least four replicates

the antibacterial dosage of the SWCNTs-Ag as well as TP359. We also demonstrated that noncovalent functionalization did not show any significant antibacterial activity compared to SWCNTS-Ag. It should be noted that FSWCNTs-Ag and SWCNTs-Ag-M, were washed after incubation and resulted in the loss of unbound (in case of FSWCNTs-Ag) and non-adherent (in case of SWCNTs-Ag-M) TP359. The loss of the TP359 in the supernatant of SWCNTs-Ag-M was greater than the supernatant of FSWCNTs-Ag (Table 3). As expected, this suggests that non-covalent binding of the peptide (SWCNTs-Ag-M) is relatively unstable and is easily removed in the supernatant and therefore SWCNTs-Ag-M did not exhibit any significant antibacterial activity as good as FSWCNTs-Ag. Although the supernatant of FSWCNTs-Ag lost most (almost 80 %) of the unbound TP359, FSWCNTs-Ag still exhibited effective antibacterial properties which indicate that the covalent bonding of the peptide to SWCNTs-Ag is stable. Our results also indicated that treatment of SWCNTs-Ag with TSC did not affect the antibacterial

activity of SWCNTs-Ag. TSC is used widely as a reducing agent. It is possible that the treatment with TSC may reduce the silver coating on the SWCNTs and could result into the loss of antibacterial activity. However, based on TEM and the antibacterial activity assay we did not observe loss of the silver coating on the SWCNTs-Ag. This may be a result of using lower concentrations of TSC for carboxylation of SWCNTs-Ag. The results from the present study are interesting as the novel FSWCNTs-Ag exhibited excellent antibacterial activity. However, it is important to further evaluate the antibacterial activity in detail using advanced methodology such as scanning electron microscopy, transmission electron microscopy, qRTPCR studies to evaluate gene expression, haemolytic activity etc.

Finally the cytotoxicity of FSWCNTs-AG, SWCNTs-Ag and TP359 was evaluated. It has been reported that non covalent functionalization of SWCNTs or SWCNTs-Ag using non-toxic polymers such as poly ethylene glycol reduces the toxicity in eukaryotic cells [11, 18, 41,

Table 2 Diameters of zones of inhibition for four bacterial pathogens measured by the Kirby–Bauer disc diffusion assay

Bacteria	Zones of inhibition (mm)												Amoxicillin-clavulanic acid (n = 3)
	Concentration (µg)												
	TP359 (n = 3)				FSWCNTs-Ag (n = 3)				SWCNTs-Ag (n = 3)				
	20	10	5	2.5	20	10	5	2.5	20	10	5	2.5	
Salmonella Typhimurium	12 ± 1.5**	10.0 ± 1.0*	6.0 ± 2.0*	5.0 ± 2.0*	11.6 ± 2.0*	11.3 ± 1.2**	8.3 ± 1.5*	7.3 ± 1.5*	0.1 ± 0.1	0.13 ± 0.04	0.0 ± 0.0	0.0 ± 0.0	22.0 ± 0.5
E. coli	10.0 ± 1.0**	7.0 ± 1.0*	6.6 ± 1.0*	0.0 ± 0.0	10.3 ± 1.5**	7.6 ± 1.5*	6.0 ± 1.7*	0.0 ± 0.0	0.27 ± 1.06	0.15 ± 1.0	0.0 ± 0.0	0.0 ± 0.0	18.0 ± 1.5
Streptococcus pyogenes	10.0 ± 1.0**	7.3 ± 1.5*	4.6 ± 0.6*	0.0 ± 0.0	11.0 ± 1.0**	8.3 ± 0.5*	7.0 ± 10*	0.0 ± 0.0	0.12 ± 0.1	0.13 ± 0.15	0.0 ± 0.0	0.0 ± 0.0	18.6 ± 1.5
Staphylococcus aureus	10.0 ± 1.0**	7.0 ± 1.0*	5.0 ± 1.0*	0.0 ± 0.0	11.0 ± 1.0**	9.6 ± 2.1*	5.6 ± 1.5*	0.0 ± 0.0	0.1 ± 0.1	0.03 ± 0.05	0.0 ± 0.0	0.0 ± 0.0	17.3 ± 0.5

The $p \leq 0.05$ indicating significant * differences, or $p \leq 0.01$ indicating highly significant ** differences

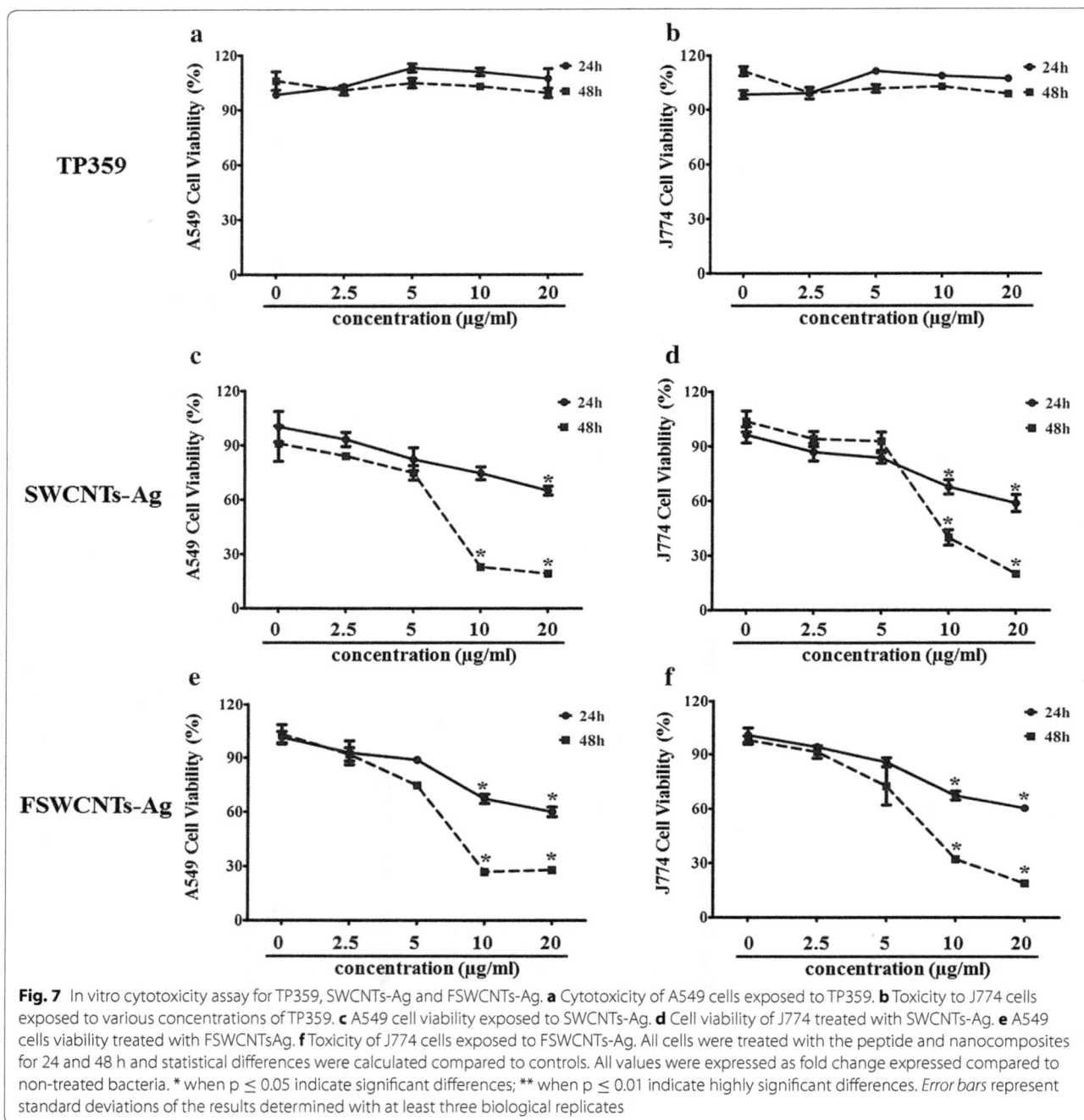

Fig. 7 In vitro cytotoxicity assay for TP359, SWCNTs-Ag and FSWCNTs-Ag. **a** Cytotoxicity of A549 cells exposed to TP359. **b** Toxicity to J774 cells exposed to various concentrations of TP359. **c** A549 cell viability exposed to SWCNTs-Ag. **d** Cell viability of J774 treated with SWCNTs-Ag. **e** A549 cells viability treated with FSWCNTsAg. **f** Toxicity of J774 cells exposed to FSWCNTs-Ag. All cells were treated with the peptide and nanocomposites for 24 and 48 h and statistical differences were calculated compared to controls. All values were expressed as fold change expressed compared to non-treated bacteria. * when $p \leq 0.05$ indicate significant differences; ** when $p \leq 0.01$ indicate highly significant differences. *Error bars* represent standard deviations of the results determined with at least three biological replicates

44, 45]. In our previous report we have shown successfully that non-covalently pegylated SWCNTs-Ag are less toxic compared to non-pegylated SWCNTs-Ag at their bactericidal concentrations [11]. In the present study our results suggest that covalent functionalization with peptide did not reduce the toxicity of SWCNTs-Ag, as SWCNTS-Ag and FSWCNTs-Ag were toxic at higher concentrations (20 and 10 µg/ml) and the toxicity was

reduced further at lower concentrations (Fig. 7). However, the important difference between the two formulations is that the FSWCNTs-Ag retained antibacterial activity at lower concentrations whereas non-functionalized SWCNTs-Ag (although being non-toxic at lower concentrations) did not exhibit antibacterial activity at equal concentrations. In particular, MIC concentrations of the FSWCNTs-Ag against *Salmonella* Typhimurium

Table 3 TP359 peptide concentration measure by BSA assay

Nanocomposite[a]	TP359 concentration (µg/ml)					
	Initial[b]	W_1^c	W_2^d	W_3^e	W_4^f	Final [(initial − (W1 + W2 + W3)][g]
FSWCNTs-Ag	1000	731	48.7	12.7	0	208
SWCNTs-Ag-M	1000	818	76.2	29.7	0	76.1

[a] Silver coated single walled carbon nanotubes (SWCNTs-Ag) were functionalized with TP359 either covalently (FSWCNTs-Ag) or non-covalently (SWCNTs-Ag-M)

[b] Starting concentration of TP359 used for functionalization

[c] First wash after centrifugation

[d] Second wash after centrifugation

[e] Third wash after centrifugation

[f] Fourth wash after centrifugation

[g] Final concentration of TP359 that was present on the functionalized nanocomposites

are much lower than its not-toxic concentrations to eukaryotic cells. It is still not clear if the conjugation efficiency can be enhanced by increasing the peptide concentration or how different conditions, such as pH variation, temperature etc., affect the stability of the FSWCNTs-Ag; these factors are under investigation.

Conclusions

In conclusion, the present study describes a novel approach for covalent bio-conjugation of silver coated SWCNTs with an antimicrobial peptide. Functionalization exhibited an "additive effect" relationship between the SWCNTs-Ag and TP359 as far as antibacterial activity is concerned. This additive effect is evidenced by the requirement of much lower concentrations of the TP359 (0.3–0.08 µg/ml) and SWCNTs-Ag (7.8–1.9 µg/ml) in the FSWCNTs-Ag to obtain the desired antibacterial effect against all bacterial pathogens compared to the plain TP359 (MIC: 7.8–1.9 µg/ml) and SWCNTs-Ag (MIC: 125–62.5 µg/ml). Also the IC50 values of FSWCNTs-Ag (~4–5 µg/ml) were much lower as opposed to plain SWCNTs-Ag IC50: ~31–35 µg/ml) More importantly, our results showed that the FSWCNTs-Ag are non-toxic to murine macrophages and Hep2 cells at their MIC concentrations (5–2.5 µg/ml). The protocol described in the present study provides a beneficial approach to bioconjugate SWCNTs-Ag and will aid into development of novel and biologically important nanomaterials.

Methods

Preparation of nanomaterial and antimicrobial peptide TP359

Silver (typically 40–50 wt%) coated single walled carbon nanotubes (SWCNTs-Ag), were purchased from NanoLab, Inc. Waltham, MA, USA. One milligrams of SWCNTs-Ag were dispersed in 1 ml of sterile distilled water to produce its aqueous dispersion. The dispersion was immediately sonicated for 1–2 h to obtain the SWCNTs-Ag dispersion (1 mg/ml). Similarly, 1 mg of a novel antimicrobial peptide from Therapeutic peptides Inc. (Baton Rouge, LA, USA), TP359 of 5 amino acids, was solubilized in sterile water. TP359 is a lipidated cationic oligopeptide (LCOP), myristoyl-KKALK$_d$ amide (US Patent 8431523).

Carboxylation of SWCNTs-Ag using tri-sodium citrate (TSC)

SWCNTs-Ag were carboxylated using Trisodium citrate dihydrate (TSC) (Alfa Aesar, Ward Hill, MA, USA) as described previously with fewer modifications [42, 43, 46]. Briefly, SWCNTs-Ag were mixed with TSC at different ratios of TSC to SWCNTs-Ag (1 mg/ml) such as 1:20 (1 part of TSC to 20 parts of SWCNTs-Ag), 1:10, 1:7.5 and 1:5. The mixture was subjected to continuous stirring at 37 °C for 4 h. The samples were designated as TSC-SWCNTs-Ag, vacuum dried and the coating of carboxyl group on SWCNTs-Ag surface was further analyzed using FT-IR as described elsewhere [11].

Bio-conjugation of TP359 to carboxylated SWCNTs-Ag using covalent bonding

The carboxylated SWCNTs-Ag were covalently conjugated with TP359 using a 1-Ethyl-3-(3-dimethylaminopropyl) carbodiimide (EDC, Sigma Aldrich, St. Louis, MO, USA) and N-Hydroxysuccinimide (NHS, Sigma Aldrich, St. Louis, MO, USA). Briefly, 1 mg of TSC-treated SWCNTs-Ag (1:20) were mixed with 0.5 mg of EDC and 0.25 mg of NHS and subjected to continuous stirring for 1 h, followed by addition of 1 mg of TP359 and stirring of the mixture for 4 h. Unbound peptide was removed by repeated centrifugation (at least four times) at 40,000×g for 50 min and the functionalized SWCNTs-Ag (FSWCNTs-Ag) were finally re-suspended in sterile nuclease-free water 1 ml. Each wash after centrifugation was collected to measure the peptide concentration using BCA™ protein assay kit (Thermo scientific, Rockford, IL, USA). Functionalization was further assessed by using FT-IR spectroscopy, UV–VIS spectra, atomic force microscopy (AFM) and transmission electron microscopy (TEM).

Non-covalent functionalization of TP359 and SWCNTs-Ag

Non covalent functionalization of SWCNTs-Ag with TP359 was carried out to compare with the covalent functionalization strategy. For this purpose, 1 mg of SWCNTs-Ag were mixed with 1 mg of TP359 and kept for stirring of the mixture for 4 h. The samples were subjected to repeat centrifugation (at least four times) at 40,000×g for 50 min and the sample was finally re-suspended in 1 ml of sterile nuclease-free water and denoted as SWCNTs-Ag-M. Each wash after centrifugation was collected to measure the peptide concentration using BCA™ protein assay kit (Thermo scientific, Rockford, IL, USA). The concentration of TP359 in the covalently and non-covalently attached SWCNTs-Ag is presented in Table 3.

Characterization of TSC-treated SWCNTs-Ag, FSWCNTs-Ag, SWCNTs-Ag-M and TP359 using FT-IR, TEM

The nanocomposite formulations were characterized using FT-IR analysis and TEM imaging to confirm the presence of carboxyl group in TSC-SWCNTs-Ag and covalent attachment of the peptide to SWCNTs-Ag. In order to confirm the carboxylation of SWCNTs-Ag, FTIR spectra of different ratio of TSC to SWCNTs-Ag (such as 1:20, 1:10, 1:7.5 and 1:5) were recorded and compared to only TSC and SWCNTs-Ag. The FT-IR pattern of FSWCNTs-Ag, SWCNTs-Ag-M, TP359 and SWCNTs-Ag were compared to confirm the functionalization. The spectra were recorded in attenuated total reflectance (ATR) mode using an infrared (IR) spectrophotometer (Nicolet 380 FT-IR; Thermo Fisher Scientific) as 64 scans per sample, ranging from 400 to 4000 cm^{-1} and a resolution of 4 cm^{-1}. The sample chamber was purged with dry N$_2$ gas.

TEM (TEM, Zeiss EM 10C 10CR, Carl Zeiss Meditec, Oberkochen, Germany) was used to examine the covalent attachment of TP359 to SWCNTs-Ag and compared with TSC-SWCNTs-Ag, SWCNTs-Ag-M, TP359 and SWCNTs-Ag. The samples for TEM were prepared as described previously [12]. Briefly, the samples were sonicated, diluted ten times and placed on the carbon-coated copper grid (200 mesh) and observed by TEM.

AFM

Approximately 50 µl of SWCNTs-Ag and FSWCNTs-Ag solution were dropped on a fresh cleaved glass slide and air-dried for overnight under clean bench. The dried samples were used to conduct atomic force microscopy (AFM; Asylum MF3D Conductive AFM, Santa Barbara, California). Air contact mode and standard silicon cantilevers (Budget Sensors, Sofia, Bulgaria) 300 µm in length and 30 µm in width were used for imaging. The cantilever oscillation frequency was tuned to the resonance frequency of approximately 256 kHz. The set point voltage was adjusted for optimum image quality. Both height and phase information were recorded at a scan rate of 0.7 Hz, and in 512 × 512 pixel format. AFM images in a large scanning area were processed using Asylum software (Asylum Research).

Ultraviolet visualization (UV–Vis)

UV–Vis was employed to ascertain the functionalization of the peptide TP359 on SWCNTs-Ag. SWCNTs-Ag, FSWCNTs-Ag and SWCNTs-Ag-M nanoparticles were each diluted in deionized water and their absorbance and spectral wavelengths were assessed using the DU 800UV/Vis spectrophotometer (Beckman Coulter, Fullerton, CA).

Bacterial experiments

Salmonella enterica serovar Typhimurium (ATCC® 13311™), *Escherichia coli* (ATCC® 25922™), *Staphylococcus aureus* (ATCC® 9144™) and *Streptococcus pyogenes* (ATCC® 8135™) were purchased from American Type Culture Collection (ATCC, VA USA). Bacteria were grown at 37 °C in Luria–Bertani (LB) broth (Difco, Sparks, MD, USA) with continuous shaking until the optical density (OD) was 0.6–0.8 (at 600 nm). Bactericidal activity of SWCNTs-Ag, TP359, TSC-SWCNTs-Ag, FSWCNTs-Ag and SWCNTs-Ag-M was investigated using the parameters such as minimum inhibitory concentrations (MIC), quantitative growth analysis and the Kirby-Bauer (KB) disc diffusion assay against all four pathogens.

Determination of minimum inhibitory concentration (MIC)

The MIC values for SWCNTs-Ag, TP359, TSC-SWCNTs-Ag, FSWCNTs-Ag and SWCNTs-Ag-M were evaluated in quadruplet wells of sterile 96-well microtiter plates using the broth microdilution assay as previously described broth microdilution procedure [47, 48]. Briefly, 1 × 10^5 cfu/ml of the bacteria were exposed to doubling concentrations of the samples starting at 0.9 µg/ml. For this purpose, the optical density of bacterial culture was measured at 600 nm and the cfu/ml was determined using standard curve equation. One milliliter of media containing 1 × 10^9 cfu/ml of the bacteria were diluted tenfolds (10^8, 10^7, 10^6, 10^5) subsequently to obtain 1 × 10^5 cfu/ml of the bacteria. All plates were sealed lightly (with ventilation) and then incubated at

37 °C for 24 h. Each plate consisted of eight dilutions of the samples, one negative control (no sample or no bacterial culture) and one positive control (only bacterial culture without samples). The concentration of the first well with no turbidity was considered as the MIC. The inhibition of bacterial growth was determined by measuring absorbance at 600 nm with a TECAN Sunrise™ enzyme-linked immunosorbent assay (ELISA) plate reader (Tecan US, Inc Morrisville, NC, USA). To avoid the background absorbance of the nanoparticles, the plates were read at 0 h before keeping in the incubator. Then the 0 h readings were subtracted from the 24 h readings to obtain the actual absorbance values at OD600. All experiments were repeated at least three times. To determine whether there is a synergistic or additive effect between the TP359 and SWCNTs-Ag in the FSWCNTs-Ag nanocomposite, we determined the fractional inhibitory concentration (FIC) test as described earlier [49]. The combined antibacterial effect of nanoparticles A and B (where A is TP359, B is SWCNTs-Ag, and AB is FSWCNTs-Ag) was calculated using the following formula as:

$$\text{FIC index} = \left[\text{MIC(AB)}/\text{MIC(A)}\right] + \left[\text{MIC(AB)}/\text{MIC(B)}\right].$$

The results were indicated as: FIC index values below 0.5 indicate synergistic effect, above 2 indicate antagonistic effects and values between 0.5 and 2.0 indicate additive effects.

Quantitative growth analysis of bacteria

Growth of all four pathogens was quantified at 24 h post-exposure to SWCNTs-Ag, TP359, TSC-SWCNTs-Ag, FSWCNTs-Ag and SWCNTs-Ag-M. 1×10^5 cfu/ml of the bacteria were exposed to doubling concentrations of the samples starting at 0.9 µg/ml. The cultures were incubated at 37 °C with shaking at 250 rpm for 24 h. Post incubation, 1 ml aliquots of bacterial culture were collected, subjected to serial tenfold dilution in sterile LB broth, and appropriate dilution was then spread on PCA to determine the cfu/ml. Each sample was analyzed in quadruplet. The bacterial cfu/ml for each sample was counted and the logarithmic decrease in bacterial growth was expressed as $\log_{10}$ cfu/ml value of each sample. The IC50 values for plain SWCNTs-Ag and FSWCNTs-Ag were calculated using % inhibition on Y axis and concentrations on X axis.

KB assay

Based on the MIC findings, the antibacterial activity of TP359 and FSWCNTs-Ag at four concentrations such as 20 (4× of MIC), 10 (2× of MIC), 5 (MIC), 2.5 (0.5× of MIC) µg/ml was further tested using the KB assay against all four pathogens as described earlier with fewer modifications and was compared with similar concentrations

of SWCNTs-Ag [50]. Bacterial suspensions of each bacterial strain (10^5 cfu/ml) were swabbed on the surface of Mueller–Hinton agar plates and filter paper discs (Fisher Scientific, MO) containing different concentrations of SWCNTs-Ag, TP359 and FSWCNTs-Ag, were placed on the plate. A broad spectrum combination of amoxicillin and clavulanic acid (30 µg) was used as a positive control (BD, BBL™, USA). Plates were incubated at 37 °C overnight and 'zones of inhibition' were observed.

In vitro cell toxicity assay

The cell toxicity to SWCNTs-Ag, TP359 and FSWCNTs-Ag was determined using Cell Titer 96® Non-Radioactive cell proliferation kit (Promega, Madison, WI). A549 (human lung carcinoma) and J774 (murine macrophages) cell lines were used for the cytotoxicity assay. Cytotoxicity was determined using a colorimetric assay based on the reduction of tetrazolium dye MTT [3-(4,5-dimethylthiazol-2yl)-2,5-diphenyltetrazolium bromide). As per the manufacturer's protocol, 1×10^4 cells/well in 100 µl of minimum essential medium-10 (MEM-10, Gibco, Life technologies, Grand Island, NY) were seeded into a 96-well plate. After overnight incubation at 37 °C in 5 % CO_2 humidified atmosphere, the media from the 96-well plate were replaced with the MEM-10 containing 20, 10, 5 and 2.5 µg/ml of SWCNTs-Ag, TP359 and FSWCNTs-Ag. The treated cells were further incubated at 37 °C and 5 % CO2 for 24 and 48 h. At the end of the corresponding incubation time, 15 µl of MTT dye was added into the each well, the plates were sealed using aluminum foil and was allowed to incubate again for the next 4 h. The reaction was then stopped with 100 µl of stop solution. The absorbance of the plate was measured at 570 nm on a TECAN Sunrise™ enzyme-linked immunosorbent assay (ELISA) plate reader (Tecan US, Inc., Morrisville, NC, USA). Non-treated cells, in growth media, were used as a control.

Statistical analyses

All data are expressed as the mean ± standard deviation (SD) unless otherwise specified. Analyses were performed with GraphPad Prism Version 4 software (GraphPad Software, Inc., La Jolla, CA). Statistical differences for MICs and cytotoxicity assay were evaluated by using post hoc pairwise comparison of two-way ANOVA. Differences were considered to be statistically significant when the p values were ≤0.05 or 0.01.

Abbreviations

APs: antimicrobial peptides; ANOVA: analysis of variance; ATR: attenuated total reflectance; ELISA: enzyme-linked immunosorbent assay; FIC: fractional inhibitory concentration; FSWCNTs-Ag: covalently functionalized silver coated single walled carbon nanotubes; FT-IR: fourier transform infrared spectroscopy; IC50: the half maximal inhibitory concentration; IR: infrared; MIC: minimum inhibitory concentrations; MTT: 3-(4,5-dimethylthiazol-2yl)-2,5-diphenyltetrazolium bromide; OD: optical density; SWCNTs: single walled carbon nanotubes; SWCNTs-Ag: silver coated single walled carbon nanotubes; SWCNTS-Ag-M: non-covalently functionalized silver coated single walled carbon nanotubes; TEM: transmission electron microscopy; TP359: therapeutic peptide 359; TSC: tri-sodium citrate; TSC-SWCNTs-Ag: tri-sodium treated silver coated single walled carbon nanotubes; UV-VIS: ultraviolet visualization.

Authors' contributions
AC and SP conceived the concept. AC conducted most of the experiments, and SDN, DA, KK, DO, CP, RA and MM, conducted part of the experiments. AC, VD, SP, SDN, DA, KK, SS and RA, analyzed the data and co-wrote the paper. The manuscript was written through contributions of all authors. All authors read and approved the final manuscript.

Author details
[1] Center for Nanobiotechnology Research, Alabama State University, Montgomery, AL, USA. [2] Department of Mechanical Engineering, University of Louisville, Louisville, KY, USA. [3] Therapeutic Peptides Inc., 7053 Revenue Drive, Baton Rouge, LA 70809, USA. [4] Department of Drug Discovery and Development, Auburn University, Auburn, AL, USA. [5] Research Instrumentation Facility, Auburn University, Auburn, AL, USA.

Acknowledgements
This research was supported by grants from the National Science Foundation-CREST (HRD-1241701), NSF-HBCU-UP (HRD-1135863) and National Institutes of Health-MBRS-RISE (1R25GM106995-01).

Competing interests
The authors declare that they have no competing interests.

Funding
We have declared the funding sources. There is no role played by the funding body in the design of the study and collection, analysis, and interpretation of data and in writing the manuscript.

References
1. Bianco A, Kostarelos K, Partidos CD, Prato M. Biomedical applications of functionalised carbon nanotubes. Chem Commun (Camb). 2005;5:571–7.
2. Liu Z, Tabakman S, Welsher K, Dai H. Carbon nanotubes in biology and medicine: in vitro and in vivo detection, imaging and drug delivery. Nano Res. 2009;2:85–120.
3. Chen ZTS, Goodwin AP, Kattah MG, Daranciang D, Wang X, Zhang G, Li X, Liu Z, Utz PJ, Jiang K, Fan S, Dai H. Protein microarrays with carbon nanotubes as multicolor Raman labels. Nat Biotechnol. 2008;26:1285–92.
4. De la Zerda A, Zavaleta C, Keren S, Vaithilingam S, Bodapati S, Liu Z, Levi J, Smith BR, Ma TJ, Oralkan O, et al. Carbon nanotubes as photoacoustic molecular imaging agents in living mice. Nat Nanotechnol. 2008;3:557–62.
5. Heller DABSET, Strano MS. Single-walled carbon nanotube spectroscopy in live cells: towards long-term labels and optical sensors. Adv Mater. 2005;17:2793–9.
6. Leeuw TK, Reith RM, Simonette RA, Harden ME, Cherukuri P, Tsyboulski DA, Beckingham KM, Weisman RB. Single-walled carbon nanotubes in the intact organism: near-IR imaging and biocompatibility studies in Drosophila. Nano Lett. 2007;7:2650–4.
7. Liu ZLX, Tabakman SM, Jiang K, Fan S, Dai H. Multiplexed multicolor Raman imaging of live cells with isotopically modified single walled carbon nanotubes. J Am Chem Soc. 2008;130:13540–1.
8. Welsher K, Liu Z, Daranciang D, Dai H. Selective probing and imaging of cells with single walled carbon nanotubes as near-infrared fluorescent molecules. Nano Lett. 2008;8:586–90.
9. Zavaleta C, de la Zerda A, Liu Z, Keren S, Cheng Z, Schipper M, Chen X, Dai H, Gambhir SS. Noninvasive Raman spectroscopy in living mice for evaluation of tumor targeting with carbon nanotubes. Nano Lett. 2008;8:2800–5.
10. Brahmachari S, Mandal SK, Das PK. Fabrication of SWCNT-Ag nanoparticle hybrid included self-assemblies for antibacterial applications. PLoS One. 2014;9:e106775.
11. Chaudhari AA, Jasper SL, Dosunmu E, Miller ME, Arnold RD, Singh SR, Pillai S. Novel pegylated silver coated carbon nanotubes kill Salmonella but they are non-toxic to eukaryotic cells. J Nanobiotechnol. 2015;13:23.
12. Rangari VK, Mohammad GM, Jeelani S, Hundley A, Vig K, Singh SR, Pillai S. Synthesis of Ag/CNT hybrid nanoparticles and fabrication of their nylon-6 polymer nanocomposite fibers for antimicrobial applications. Nanotechnology. 2010;21:095102.
13. El-Sayed A, El-Shannat S, Kamel M, Castaneda-Vazquez MA, Castaneda-Vazquez H. Molecular epidemiology of Mycobacterium bovis in humans and cattle. Zoonoses Public Health. 2015. doi:10.1111/zph.12242.
14. Friedman ND, Temkin E, Carmeli Y. The negative impact of antibiotic resistance. Clin Microbiol Infect. 2015;S1198-743X(15):1028–9.
15. Liu YY, Wang Y, Walsh TR, Yi LX, Zhang R, Spencer J, Doi Y, Tian G, Dong B, Huang X, et al. Emergence of plasmid-mediated colistin resistance mechanism MCR-1 in animals and human beings in China: a microbiological and molecular biological study. Lancet Infect Dis. 2015;16:161–8.
16. Brahmachari S, Das D, Shome A, Das PK. Single-walled nanotube/amphiphile hybrids for efficacious protein delivery: rational modification of dispersing agents. Angew Chem Int Ed Engl. 2011;50:11243–7.
17. Brahmachari S, Ghosh M, Dutta S, Das PK. Biotinylated amphiphile-single walled carbon nanotube conjugate for target-specific delivery to cancer cells. J Mater Chem B. 2014;2:1160–73.
18. Vardharajula S, Ali SZ, Tiwari PM, Eroglu E, Vig K, Dennis VA, Singh SR. Functionalized carbon nanotubes: biomedical applications. Int J Nanomed. 2012;7:5361–74.
19. Brogden KA, Ackermann M, McCray PB Jr, Tack BF. Antimicrobial peptides in animals and their role in host defences. Int J Antimicrob Agents. 2003;22:465–78.
20. Brogden NK, Brogden KA. Will new generations of modified antimicrobial peptides improve their potential as pharmaceuticals? Int J Antimicrob Agents. 2011;38:217–25.
21. Hancock RE. Peptide antibiotics. Lancet. 1997;349:418–22.
22. Lakshmaiah Narayana J, Chen JY. Antimicrobial peptides: possible anti-infective agents. Peptides. 2015;72:88–94.
23. Nuri RST, Shai Y. Defensive remodeling: How bacterial surface properties and biofilm formation promote resistance to antimicrobial peptides. Biochim Biophys Acta. 2015;1848:3089–100.
24. da Silva A Jr, Teschke O. Effects of the antimicrobial peptide PGLa on live Escherichia coli. Biochim Biophys Acta. 2003;1643:95–103.
25. Maher S, McClean S. Investigation of the cytotoxicity of eukaryotic and prokaryotic antimicrobial peptides in intestinal epithelial cells in vitro. Biochem Pharmacol. 2006;71:1289–98.
26. Nekhotiaeva N, Elmquist A, Rajarao GK, Hallbrink M, Langel U, Good L. Cell entry and antimicrobial properties of eukaryotic cell-penetrating peptides. FASEB J. 2004;18:394–6.
27. van Berkel PH, Welling MM, Geerts M, van Veen HA, Ravensbergen B, Salaheddine M, Pauwels EK, Pieper F, Nuijens JH, Nibbering PH. Large scale production of recombinant human lactoferrin in the milk of transgenic cows. Nat Biotechnol. 2002;20:484–7.
28. Zagulski TLP, Zagulska A, Broniek S, Jarzabek Z. Lactoferrin can protect mice against a lethal dose of Escherichia coli in experimental infection in vivo. Br J Exp Pathol. 1989;70:697–704.
29. Zhang L, Parente J, Harris SM, Woods DE, Hancock RE, Falla TJ. Antimicrobial peptide therapeutics for cystic fibrosis. Antimicrob Agents Chemother. 2005;49:2921–7.
30. Liu Z, Cai W, He L, Nakayama N, Chen K, Sun X, Chen X, Dai H. In vivo biodistribution and highly efficient tumour targeting of carbon nanotubes in mice. Nat Nanotechnol. 2007;2:47–52.
31. Liu Z, Davis C, Cai W, He L, Chen X, Dai H. Circulation and long-term fate of functionalized, biocompatible single-walled carbon nanotubes

in mice probed by Raman spectroscopy. Proc Natl Acad Sci USA. 2008;105:1410–5.

32. Liu Z, Sun X, Nakayama-Ratchford N, Dai H. Supramolecular chemistry on water-soluble carbon nanotubes for drug loading and delivery. ACS Nano. 2007;1:50–6.

33. Liu Z, Tabakman SM, Chen Z, Dai H. Preparation of carbon nanotube bioconjugates for biomedical applications. Nat Protoc. 2009;4:1372–82.

34. Liu ZWMHM, Dai HJ. siRNA delivery into human T cells and primary cells with carbon-nanotube transporters. Angew Chem Int Ed Engl. 2007;46:2023–7.

35. Zhang Z, Yang X, Zhang Y, Zeng B, Wang S, Zhu T, Roden RB, Chen Y, Yang R. Delivery of telomerase reverse transcriptase small interfering RNA in complex with positively charged single-walled carbon nanotubes suppresses tumor growth. Clin Cancer Res. 2006;12:4933–9.

36. Niyogi S, Hamon MA, Hu H, Zhao B, Bhowmik P, Sen R, Itkis ME, Haddon RC. Chemistry of single-walled carbon nanotubes. Acc Chem Res. 2002;35:1105–13.

37. Pm TN. Functionalization of carbon nanotubes via 1,3-dipolar cycloadditions. J Mater Chem B. 2004;14:437–9.

38. Rosca IDWFUM, Akaska T. Oxidation of multiwalled carbon nanotubes by nitric acid. Carbon. 2005;43:3124–31.

39. Hu J, Zhang Y, Liu B, Liu J, Zhou H, Xu Y, Jiang Y, Yang Z, Tian ZQ. Synthesis and properties of tadpole-shaped gold nanoparticles. J Am Chem Soc. 2004;126:9470–1.

40. Schneider G, Decher G. Functional core/shell nanoparticles via layer-by-layer assembly. Investigation of the experimental parameters for controlling particle aggregation and for enhancing dispersion stability. Langmuir. 2008;24:1778–89.

41. Dumortier H, Lacotte S, Pastorin G, Marega R, Wu W, Bonifazi D, Briand JP, Prato M, Muller S, Bianco A. Functionalized carbon nanotubes are non-cytotoxic and preserve the functionality of primary immune cells. Nano Lett. 2006;6:1522–8.

42. Hinterwirth H, Lindner W, Lammerhofer M. Bioconjugation of trypsin onto gold nanoparticles: effect of surface chemistry on bioactivity. Anal Chim Acta. 2012;733:90–7.

43. Zou X, Ying E, Dong S. Seed-mediated synthesis of branched gold nanoparticles with the assistance of citrate and their surface-enhanced Raman scattering properties. Nanotechnology. 2006;17:4758–64.

44. Haberl N, Hirn S, Wenk A, Diendorf J, Epple M, Johnston BD, Krombach F, Kreyling WG, Schleh C. Cytotoxic and proinflammatory effects of PVP-coated silver nanoparticles after intratracheal instillation in rats. Beilstein J Nanotechnol. 2013;4:933–40.

45. Schipper ML, Nakayama-Ratchford N, Davis CR, Kam NW, Chu P, Liu Z, Sun X, Dai H, Gambhir SS. A pilot toxicology study of single-walled carbon nanotubes in a small sample of mice. Nat Nanotechnol. 2008;3:216–21.

46. Okoli CFAQJ, Toprak MS, Dalhammar G, Muhammed M, Rajarao G. Characterization of supermagnetic iron oxide nanoparticles and its application in protein purification. J NanoSci Nanotechnol. 2011;11:1–6.

47. Palomino JC, Martin A, Camacho M, Guerra H, Swings J, Portaels F. Resazurin microtiter assay plate: simple and inexpensive method for detection of drug resistance in *Mycobacterium tuberculosis*. Antimicrob Agents Chemother. 2002;46:2720–2.

48. Soehnlen MK, Kunze ME, Karunathilake KE, Henwood BM, Kariyawasam S, Wolfgang DR, Jayarao BM. In vitro antimicrobial inhibition of *Mycoplasma bovis* isolates submitted to the Pennsylvania Animal Diagnostic Laboratory using flow cytometry and a broth microdilution method. J Vet Diagn Invest. 2011;23:547–51.

49. Khaled HB-RD, Gonzalez-Felicianob JA, Villalobos-Santosb JC, Makarova VI, et al. Synergistic antibacterial activity of PEGylated silver–graphene quantum dots nanocomposites. Appl Mater Today. 2015;1:80–7.

50. Bauer AW, Kirby WM, Sherris JC, Turck M. Antibiotic susceptibility testing by a standardized single disk method. Am J Clin Pathol. 1966;45:493–6.

7

Evaluation of the antibacterial power and biocompatibility of zinc oxide nanorods decorated graphene nanoplatelets: new perspectives for antibiodeteriorative approaches

Elena Zanni[1,2]*, Erika Bruni[1], Chandrakanth Reddy Chandraiahgari[2,3], Giovanni De Bellis[2,3], Maria Grazia Santangelo[4], Maurizio Leone[4], Agnese Bregnocchi[2,3], Patrizia Mancini[5], Maria Sabrina Sarto[2,3] and Daniela Uccelletti[1,2]

Abstract

Background: Nanotechnologies are currently revolutionizing the world around us, improving the quality of our lives thanks to a multitude of applications in several areas including the environmental preservation, with the biodeterioration *phenomenon* representing one of the major concerns.

Results: In this study, an innovative nanomaterial consisting of graphene nanoplatelets decorated by zinc oxide nanorods (ZNGs) was tested for the ability to inhibit two different pathogens belonging to bacterial *genera* frequently associated with nosocomial infections as well as biodeterioration phenomenon: the Gram-positive *Staphylococcus aureus* and the Gram-negative *Pseudomonas aeruginosa*. A time- and dose-dependent bactericidal effect in cell viability was highlighted against both bacteria, demonstrating a strong antimicrobial potential of ZNGs. Furthermore, the analysis of bacterial surfaces through Field emission scanning electron microscopy (FESEM) revealed ZNGs mechanical interaction at cell wall level. ZNGs induced in those bacteria deep physical damages not compatible with life as a result of nanoneedle-like action of this nanomaterial together with its nanoblade effect. Cell injuries were confirmed by Fourier transform infrared spectroscopy, revealing that ZNGs antimicrobial effect was related to protein and phospholipid changes as well as a decrease in extracellular polymeric substances; this was also supported by a reduction in biofilm formation of both bacteria. The antibacterial properties of ZNGs applied on building-related materials make them a promising tool for the conservation of indoor/outdoor surfaces. Finally, ZNGs nanotoxicity was assessed in vivo by exploiting the soil free living nematode *Caenorhabditis elegans*. Notably, no harmful effects of ZNGs on larval development, lifespan, fertility as well as neuromuscular functionality were highlighted in this excellent model for environmental nanotoxicology.

Conclusions: Overall, ZNGs represent a promising candidate for developing biocompatible materials that can be exploitable in antimicrobial applications without releasing toxic compounds, harmful to the environment.

*Correspondence: elena.zanni@uniroma1.it
[1] Department of Biology and Biotechnology C. Darwin, Sapienza University of Rome, Piazzale Aldo Moro 5, Rome, Italy
Full list of author information is available at the end of the article

Background

Nowadays, an ever-growing interest is focused on nanoscience that works with and/or creates promising materials characterized by nanostructured dimensions. Nanotechnologies have extensively been developed in the last years, expanding more and more the range of possible applications. At the present, nanotechnology has implications in a plethora of areas including medicine, food industry and environmental field, depending on specific nanomaterial features such as mechanical, thermal and chemical properties as well as large surface area [1–3]. Nevertheless, the antimicrobial power together with optical/light properties make some nanostructures particularly helpful in applications involved in the conservation of cultural heritage and/or building construction. In fact, historic buildings need to be preserved avoiding the risk of biodeterioration. Such process lead to unpleasant alteration of the material determined by the metabolism of bacteria, fungi, algae and lichens [4, 5]. Biodeteriorative activities determine severe damages to architectural surfaces, church frescoes or wall paintings that are found in catacombs and caverns. Among the bacterial isolates derived from wall paintings, *Pseudomonas* and *Staphylococcus* genera are the most predominant together with *Bacillus*, *Streptomyces* and *Mycobacterium* [6]. The formation of bacterial biofilm on construction material plays a key role in the possible occurrence of pathogen infections in nosocomial environments as well as in building biodeterioration [7, 8]. In fact, bacterial growth on wall surface as well as on medical devices represents a severe concern in the health care system, taking into account that bacteria are becoming multiresistant to antibiotics. From this perspective, the development of surfaces able to kill or inhibit bacterial growth without the use of antibiotics/drugs is attracting a great interest, and new wall paint and coatings, containing nanoparticles that possess antimicrobial activity, represent an emerging approach in order to prevent both the spread of nosocomial infections and biodeteriorative activity [9].

Among nanomaterials, great interest is currently addressed to the synthesis and development of graphene-based nanocomposites as reported in [10–13]. In particular, decoration of graphene with metal oxide offers unique properties that extensively broaden its application in chemical, medical and pharmaceutical fields [14, 15].

Several studies reported impressive antimicrobial power for metal oxide-based nanoparticles [16, 17]. Moreover, it has been possible to grow ZnO nanostructures onto graphene, so that decoration or functionalization was typically achieved only over the exposed surface of graphene. In our recent study, the synthesis of ZnO nanorods (ZnO-NRs) with controlled shape and density onto unsupported multilayer graphene flakes (also known as graphene nanoplatelets GNPs) was reported [18].

These zinc oxide nanorods-decorated graphene nanoplatelets (ZNGs) were characterized by the ability to kill the bacterium causing dental caries, namely *Streptococcus mutans*. ZNGs were found to efficiently kill and to control *S. mutans* cells by inhibiting both planktonic and biofilm growth [19]. This hybrid nanomaterial combines the remarkable electrical and antimicrobial properties offered by GNPs together with optical features and the highly effective killer action against both Gram-positive and Gram-negative bacteria of ZnO-NRs. Moreover, the characteristic grey color of graphene based nanomaterials is mitigated by ZnO whitening effect, making this hybrid nanostructure a promising candidate for the development of novel nanofiller-based wall paint in the field of building construction and cultural heritage.

Herein, ZNGs were used to inhibit two pathogens belonging to genera frequently associated to biodeterioration: the Gram-positive *Staphylococcus aureus* and the Gram-negative *Pseudomonas aeruginosa*; a mechanical mode of action against both bacteria has been suggested. Environmental nanotoxicity was assessed through the soil free-living nematode *Caenorhabditis elegans*.

Methods

Production of nanostructures and suspensions

Zinc oxide nanorods decorated graphene nanoplatelets were produced by a simple hydrothermal method as described in Chandraiahgari et al. [18]. Briefly, GNPs were derived through a solvothermal exfoliation process as described previously in Rago et al. [20]. Thereafter, ZnO-NRs with high density were grown directly over unsupported GNPs suspended in aqueous solution [18]. The morphology of the produced nanomaterials was investigated through high-resolution field emission scanning electron microscopy (FE-SEM) (Fig. 1). Figure 1a shows the pristine GNPs having thickness in the range of 2–10 nm and average lateral dimensions in the range of 1–10 μm. Figure 1b shows the hybrid ZNG nanomaterial composed by GNPs and in situ grown rod shaped ZnO-NRs. ZnO-NRs having average diameter of ~36 nm and length in the range of 300–400 nm are directly grown over the planar shaped pristine GNPs with high density. Superior crystallinity and chemical purity of these nanomaterials were systematically investigated and results are found to be identical to our earlier works [18]. It resulted out that the ZNGs are composed by hexagonal wurtzite crystalline ZnO and crystalline graphitic carbon compounds. No other impurities were detected thus ensuring the purity of the produced ZNG nanostructures. Aqueous colloidal suspensions of ZNGs were prepared through the dispersion of ZNG powder in ultrapure and

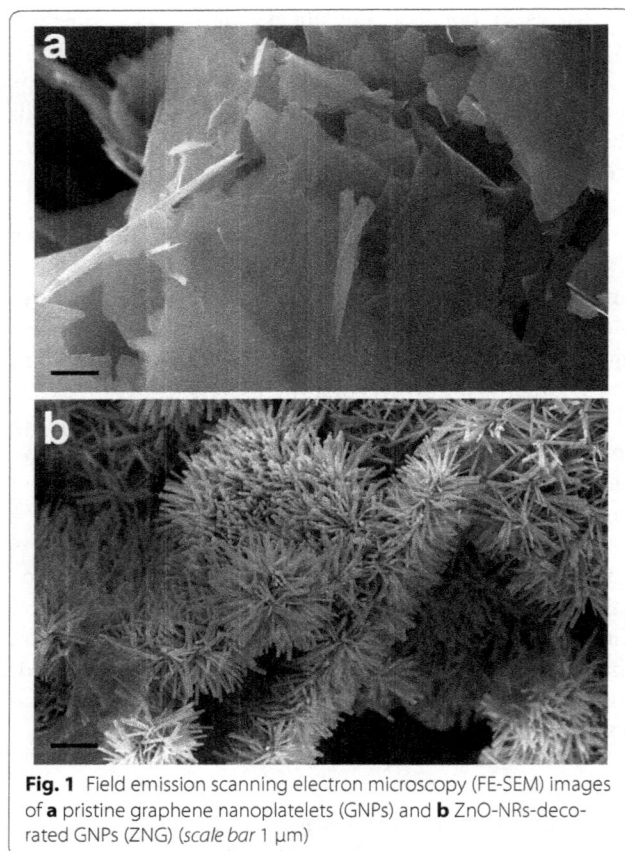

Fig. 1 Field emission scanning electron microscopy (FE-SEM) images of **a** pristine graphene nanoplatelets (GNPs) and **b** ZnO-NRs-decorated GNPs (ZNG) (*scale bar* 1 μm)

sterilized deionized water using probe ultrasonication. The homogenous suspensions were then readily transferred to 50 mL sterilized centrifuge tubes.

Bacterial strains and media

Pseudomonas aeruginosa ATCC 15692 and *Staphylococcus aureus* ATCC 25923 were the bacterial strains used in this study. They were grown in LB (Luria–Bertani) broth at 37 °C.

Cells viability test

Viability was evaluated in both suspensions and solid substrates. For liquid assay, bacteria were incubated at 37 °C under gentle shaking in H_2O_{dd} suspensions of ZNGs at various concentrations (ranging from 0.1 to 50 μg/mL). The bacterial concentration inoculated was 5×10^7 cells/mL. Both microbial strains were exposed to increasing concentrations of ZNGs and compared to the respective untreated controls. The experiments were carried out at 2 and 24 h of treatment.

In the case of antimicrobial test on solid surfaces, ZNGs applied on plywood samples (2.5 cm × 2.5 cm) covered or not by a commercial paint were drop casted with 150 μL of a ZNG suspension (250 μg/mL) and

air-dried. After a 30 min of UV-sterilization, 200 μL of *S. aureus* suspension (6×10^5 cell/mL) were spotted onto the plywood surfaces. Cells were extracted at the initial time of contamination (t0) and after 4 h of incubation at 25 °C by washing plywood substrates in a sterile bag with 10 mL of sterile H_2O_{dd}.

The ability of bacterial survival was assessed by the colony count method (Colony Forming Unit, CFU) for both types of tests, by spreading the diluted samples onto LB agar plates.

Evaluation of biofilm formation

The biofilm growth in 96-well microtiter plate was estimated by using the Crystal Violet (CV) assay. In the case of *S. aureus*, each well was inoculated with 200 μL of a suspension containing *S. aureus* cells (final concentration 1×10^7 cell/mL), the Tryptic Soy Broth medium (TSB, Becton–Dickinson and Company, Franklin Lakes, NJ, USA) with 2% glucose (to stimulate biofilm formation) and ZNGs, present or not at various concentrations (in triplicate). For *P. aeruginosa*, 100 μL of a suspension of LB broth and ZNGs inoculated with bacterial aliquot (0.5 OD_{600}) were placed in every well. After incubation of the plates under stirring (25 rpm) at 37° C for 24 h, the culture medium was removed and the wells were washed twice with H_2O_{dd} with the purpose to remove cells not adhered. Plates were then kept at 65 °C for 20 min. Finally, every well was stained with 0.3% Crystal Violet (Sigma-Aldrich) and incubated at RT for 15 min. After several washes with H_2O_{dd}, plates were left to dry and wells were then treated with 200 μL of 96% EtOH for CV elution. Absorbance at 600 nm was then measured by using a multiplate reader (Promega, GloMax multi+ detection system).

Pyocyanin assay in *P. aeruginosa*

For this test, 12-well microtiter plates were used. Each well was filled with 900 μL of LB broth containing or not different concentration of ZNGs (in triplicate) and inoculated with *P. aeruginosa* cells at a final concentration of 5×10^7 cell/mL from an overnight growth culture, reaching a final volume of 1.5 mL (adding sterile H_2O_{dd}). Plates were incubated at 37 °C overnight without agitation. Next, ON cultures were centrifuged and the supernatant absorbance was measured at 380 nm.

Preparation of bacterial cells for FE-SEM imaging

Treated and untreated cells of *P. aeruginosa* were incubated at 37 °C for 1 h, while *S. aureus* ones for 30 min. Short treatment times were chosen to obtain images in which the effects of ZNGs on bacterial cells were clearly visible. The tested concentration of ZNGs was 50 μg/mL in 1 mL of sterile water. The protocol for samples preparation was performed as described in Olivi et al. [21]. Imaging was performed using a Zeiss Auriga FE-SEM, operated at an accelerating voltage of 5 kV.

FTIR

To investigate the antimicrobial properties of ZNGs, Fourier Transform Infrared (FTIR) spectroscopy was used. The comparison of the FTIR spectra of untreated bacterial cells and of bacterial cells treated with this nanocomposite allowed to assess whether the treatment induced alterations of the bacterial cell structure and surface components. Briefly, about 5×10^8 cell/ mL of overnight grown cultures of *P. aeruginosa* and *S. aureus* were incubated in 1 mL of sterile H_2O_{dd} at 37° C for 90 min under gentle agitation, with or without ZNGs (10 µg/mL). Both ZNGs concentration and time of exposure were chosen in order to have a high cellular survival and to appreciate the early structural changes in treated bacteria. Cells were withdrawn and then fixed with 1 mL of a freshly prepared 4% (v/v) formaldehyde solution. After incubation for 1 h in the dark, the samples were washed three times and the cells were initially suspended in 20 µL of H_2O (water suspension) or of D_2O (deuterium oxide suspension). FTIR spectra have been collected either on dried samples or on liquid samples. Dried samples were prepared by drop-casting 20 µL of a bacterial suspension onto a CaF_2 window and then leaving the liquid suspension to air-drying. Measurements on liquids were performed by placing 50 µL of a bacterial suspension in deuterated water between two CaF_2 windows separated by a 50 µm Teflon spacer. In both cases, FTIR spectra of untreated bacterial samples and of treated bacterial samples have been acquired and then analyzed. FTIR measurements were carried out with a Bruker Vertex 70 spectrometer equipped with a DTGS (doped triglycine sulfate) detector. During data collection the sample was at room temperature and the sample compartment was under continuum purging with dry N_2 gas. Each spectrum is an average over 256 scans and has a spectral resolution of 2 cm^{-1}. In the case of dried samples, the intensity transmitted by the CaF_2 substrate was used as a reference to obtain the sample absorbance. In the case of liquid samples, absorbance was calculated using the intensity transmitted by the CaF_2 cell filled with pure D_2O as a reference.

Method of cultivation for *C. elegans*

In this study the *C. elegans* wild type strain N_2 was used. It was maintained at 16 °C on agar plates of the Nematode Growth Medium (NGM) covered by a layer of bacterial suspension of *Escherichia coli* OP50 as feeding source [22].

Lifespan analysis

Nematode treatment with ZNGs was performed starting on 1-day adults or newly hatched L1 larvae, resulting from synchronized cultures that were transferred to NGM-OP50 plates with ZNGs at the indicated concentrations. Every day nematodes were placed onto freshly prepared plates and 100 µL of ZNG suspensions were distributed before worms seeding. The nematodes were monitored daily for their survival with respect to untreated nematodes, and were considered dead when there was no response to the delicate touch of a platinum wire. At least 60 nematodes per condition were used in each experiment.

Brood size

OP50-NGM plates containing or not ZNGs were seeded with adult worms (in triplicate) and were incubated at 16 °C, allowing embryos laying. Next, each animal was transferred onto a fresh plate every day, and the number of progeny was recorded for 4 days until the worm stopped laying eggs.

Body length analysis

Nematode larvae exposed to ZNGs starting from embryos hatching, were photographed at the indicated time points by using a Leica MZ10F stereomicroscope with a Jenoptik CCD camera. Length of worm body was determined by using the Delta Sistemi IAS software. An average of 30 nematodes were imaged on at least three independent experiments.

Pumping rate measurements

The pharyngeal pumping rate was measured in *C. elegans* individuals exposed or not to ZNGs starting from their larval development as described in lifespan assay. About 10 worms for each experimental condition were analyzed for the number of their pharyngeal contractions during a time interval of 30 s. This analysis was repeated at the indicated time points.

Body bending evaluation

The locomotion behavior of nematodes, treated with ZNGs starting from embryos hatching, was analyzed by body bending counting at the indicated time points. After several washes in M9 buffer to remove bacteria, nematodes were placed in 10 µL of M9 buffer allowing them to swim freely. About 10 worms for each experimental condition were monitored for the number of head thrashes within a minute.

Statistical analysis

All experiments were performed at least in triplicate. Data are presented as mean ± SD. The statistical significance was determined by Student's t test or one-way ANOVA analysis coupled with a Bonferroni post test (GraphPad Prism 5.0 software, GraphPad Software Inc., La Jolla, CA, USA), and defined as $*p < 0.05$, $**p < 0.01$, and $***p < 0.001$.

Results and discussion

In this study, the antimicrobial effects exerted by ZNGs were investigated against bacteria belonging to *Staphylococcus* and *Pseudomonas* genera, frequently found in wall paintings as biodeteriorative agents. In particular, strains of *Staphylococcus aureus* and *Pseudomonas aeruginosa* were employed to confirm the antimicrobial activity of this nanomaterial on both Gram-positive and Gram-negative bacteria, respectively. After just 2 h of treatment with ZNGs, *P. aeruginosa* revealed significant differences in bacterial survival with respect to the control when concentration of 50 µg/mL was used (Fig. 2a). Indeed, in this case only 23% of survival was observed, in contrast to a slight stimulation of cell growth showed by treatment with 1 µg/mL (Fig. 2a). Similar to short-term treatment, the 24 h-exposure to low amounts of ZNGs (0.1 and 1 µg/mL) induced a higher cell growth in *P. aeruginosa* with respect to untreated cells. By contrast, a noteworthy bacterial mortality was highlighted when ZNGs concentration was increased; exposure to 10 and 50 µg/mL led to 78 and 99.8% reduction in bacterial survival, respectively (Fig. 2b). Notably, ZNGs exerted the highest antimicrobial power against *S. aureus*, resulting already effective after 2 h of treatment at 10 µg/mL, and pointing out a 96% mortality rate at the maximum tested concentration (50 µg/mL) (Fig. 3a). In the case of long-term exposure, a strikingly elevated killer action of ZNGs was revealed even at extremely low concentrations; a mortality rate of 99.3% was indeed observed with just 1 µg/mL (Fig. 3b). In our previous studies on pristine GNPs and ZnO nanorods, the two constituents of this hybrid nanomaterial, showed a strikingly high antimicrobial power against both Gram-positive and Gram-negative bacteria including *P. aeruginosa*, *S. mutans*, *S. aureus* and *Bacillus subtilis* [20, 23, 24]. Although graphene represents an attractive material for various applications due to its unique and antimicrobial, electrical and mechanical properties (reviewed in Zhu et al. [25]), it is visibly black in colour as other 2D carbon-based materials. This reduces the efficiency of its use in applications where aesthetical dimension matters. The characteristic darkness of GNPs is here softened by ZnO whitening (as demonstrated in Zanni et al. [19]) and ZnO decoration also prevents GNPs aggregation, consenting, thus, its exploitation in the development of novel nanofiller-based wall paints in the field of building construction and cultural heritage. Our results are in line with the notion that a

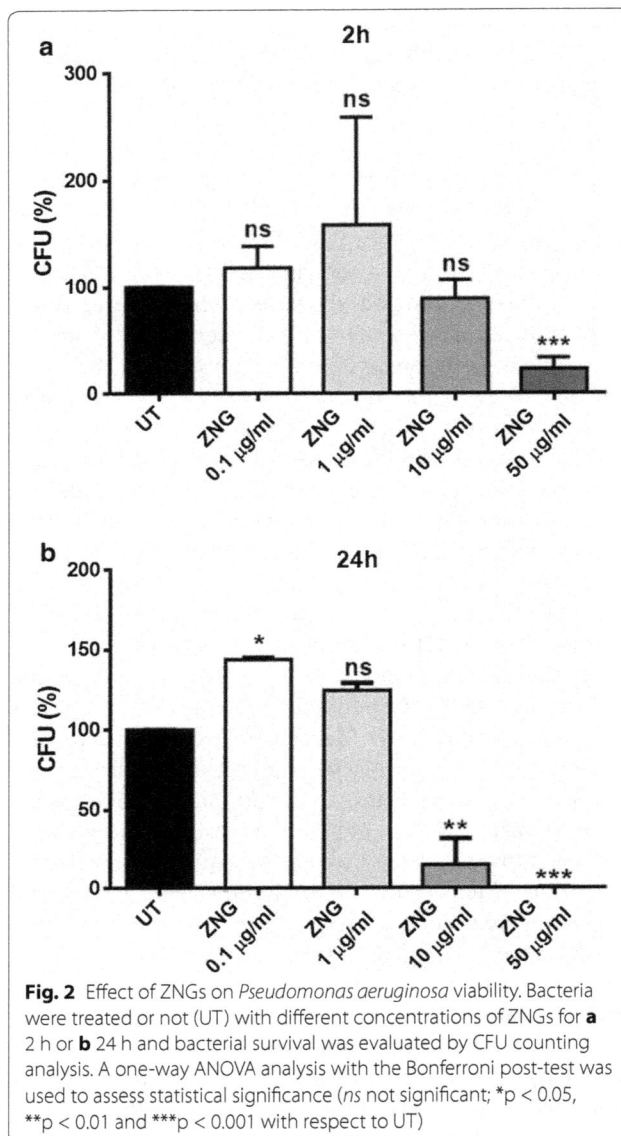

Fig. 2 Effect of ZNGs on *Pseudomonas aeruginosa* viability. Bacteria were treated or not (UT) with different concentrations of ZNGs for **a** 2 h or **b** 24 h and bacterial survival was evaluated by CFU counting analysis. A one-way ANOVA analysis with the Bonferroni post-test was used to assess statistical significance (*ns* not significant; $*p < 0.05$, $**p < 0.01$ and $***p < 0.001$ with respect to UT)

higher inhibition activity of ZnO nanoparticles has been reported against Gram-positive bacteria with respect to Gram-negative bacteria [26, 27]. After just 2 h, ZNGs treatment (10 µg/mL) resulted in 70% more antibacterial activity against the Gram-positive *S. aureus* than the Gram-negative *P. aeruginosa*. Overall, our results indicate a time- and dose-dependent bactericidal action of ZNGs against the planktonic forms of two representative Gram-positive and Gram-negative bacteria.

Next, a FE-SEM analysis was performed in order to examine the interactions between bacterial cells and ZNGs. The untreated cells of *S. aureus* and *P. aeruginosa* resulted to be intact with their round and rod shaped morphology, respectively (Fig. 4a, c). Conversely, in

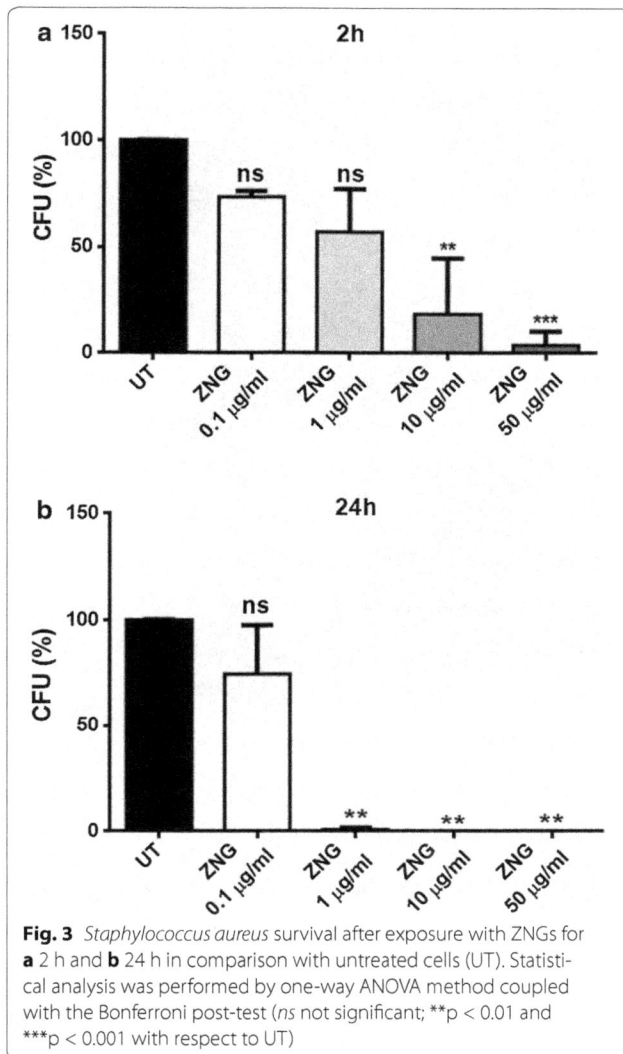

Fig. 3 *Staphylococcus aureus* survival after exposure with ZNGs for **a** 2 h and **b** 24 h in comparison with untreated cells (UT). Statistical analysis was performed by one-way ANOVA method coupled with the Bonferroni post-test (*ns* not significant; **p < 0.01 and ***p < 0.001 with respect to UT)

treated cells the bacterial surface showed mechanical injuries caused by direct contact with ZNGs (Fig. 4b, d), which perforated the cell wall as a result of ZnO-NRs that protrude from the sheets of GNPs. Severe membrane disruption and cytoplasm leakage were observed in bacterial cells treated with ZNGs, in contrast to some cells that maintained their membrane integrity, but showing a poor living state. It can be hypothesized that nanorods adhere to cells and then act as a network of nanoneedles that pierce the bacterial wall and trap the cells, thereby inducing severe mechanical damage. On the other hand, nanosheets of GNPs offer large surface area, providing a preferred growth orientation for the ZnO-NRs over the GNP surface. Indeed, the adhesion of the nanostructures to the cell wall resulted to be improved, enhancing the penetration of the ZnO-NRs through the cell membrane. Because of their large lateral dimensions and their very

sharp edges, GNPs work together with ZnO NRs to provoke mechanical injuries by acting as nanoknives, as suggested in different studies on graphene-based materials [28, 29].

Furthermore, the production of pyocyanin, a virulence factor secreted by *P. aeruginosa* cells, was evaluated in bacteria exposed or not to increasing concentrations of ZNGs. The production of this bacterial blue-green pigment decreased when *P. aeruginosa* cells interacted with ZNGs; exposure to 50 and 100 µg/mL led to 50 and 70% reduction in the virulence factor secretion compared to the control, respectively (Additional file 1: Figure S1). Lee et al. demonstrated in *P. aeruginosa* that ZnO nanoparticles inhibited biofilm formation as well as the production of several virulence factors including pyocyanin [30]. Biofilm formation plays a key role as a detrimental effect in the environment in terms of biodeterioration and the spread of hospital-acquired infections. The biofilm inhibitory activity of ZNGs was thus investigated in both *P. aeruginosa* and *S. aureus* cells after 24 h of treatment. Crystal violet assay highlighted a considerable inhibitory activity of ZNGs against *P. aeruginosa* biofilm; a 13% reduction of biofilm development was observed already starting from the treatment with 10 µg/mL compared to untreated sample. Notably, the ZNG anti-biofilm activity became more evident when *P. aeruginosa* cells were exposed to 50 µg/mL, and even higher (50%) in bacteria treated with the maximum tested concentration (100 µg/mL) (Fig. 5a). Those results demonstrated the effectiveness of this nanomaterial in controlling biofilm growth of *P. aeruginosa*, one of the most wide-spread Gram-negative bacteria. Remarkably, ZNGs resulted to be stronger inhibitors of biofilm formation for the Gram-positive *S. aureus* with respect to *P. aeruginosa*. Indeed, *S. aureus* cells treated for 24 h with 10 µg/mL ZNGs, already showed a 36% decrease of biofilm formation in comparison to the untreated bacteria (Fig. 5b). Such decrease became more evident with 50 and 100 µg/mL ZNGs concentrations, showing 84 and 95% reduction in *S. aureus* biofilm development, respectively (Fig. 5b). Surface moisture is one of the main features that allow bacterial biofilm growth on different types of substrates including also architectural surfaces. Several treatments have been developed to avoid biofilm formation although biofilm removal from contaminated surfaces resulted to be not very effective [31]. Our results demonstrate that biofilm production is reduced by ZNGs treatment and that for its inhibition higher concentrations are required. This is in agreement with the notion that bacteria from biofilms are more resistant to antibacterial agents than their planktonic form [32, 33]. Recently we demonstrated that ZNGs resulted to be effective in inhibiting both the growth and the biofilm

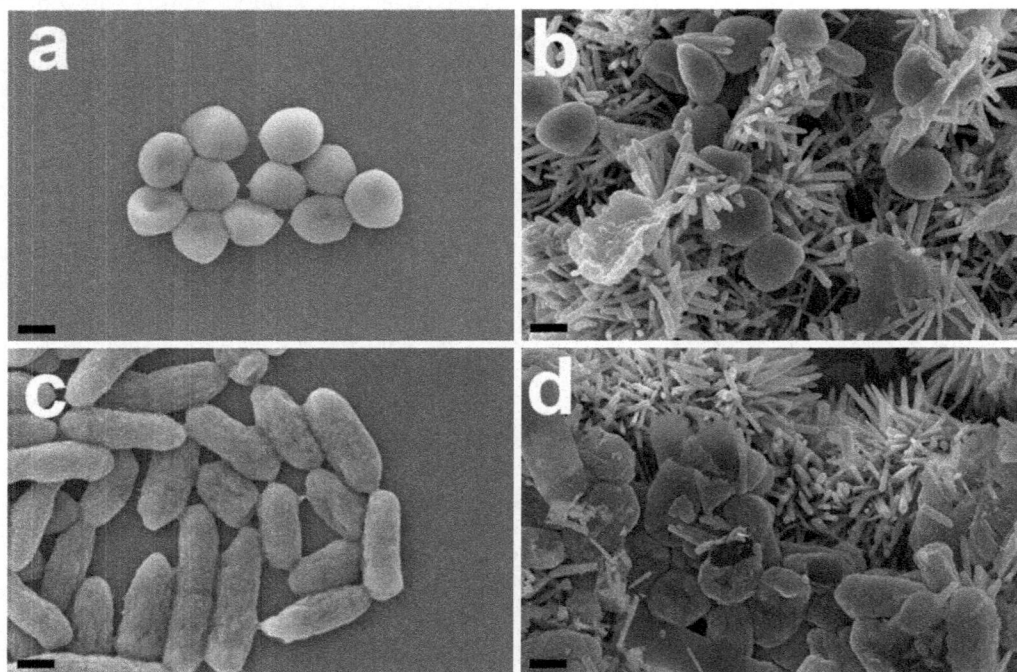

Fig. 4 FE-SEM micrographs of bacterial cells after exposure to zinc oxide nanorods-decorated GNPs. *S. aureus* cells incubated with **a** H_2O_{dd} or **b** ZNGs suspension (50 µg/mL). *P. aeruginosa* **c** untreated cells are shown in comparison to **d** the same bacteria exposed to ZNGs (50 µg/mL) (*scale bar* 400 nm)

formation of *S. mutans*, the Gram-positive bacterium responsible for dental caries [19].

To assess whether the treatment induced alterations of the bacterial cell structure and surface components, the FTIR spectroscopy was employed. Figure 6 shows the FTIR spectra of *P. aeruginosa* bacteria treated or not with ZNGs for 90 min. Such choice of experimental time point was taken in order to detect the early changes in the bacterial cells due to the treatment. Several absorption bands related to dynamic properties of different functional groups of proteins, fatty acids and polysaccharides of bacterial cells are clearly visible in the FTIR spectra. In the case of dried samples (Fig. 6a) spectra in the 1800–1300 cm^{-1} is dominated by Amide I (~1656 cm^{-1}) and Amide II (~1543 cm^{-1}) bands, which give quantitative information on protein secondary structure [34–36]. The 1300–900 cm^{-1} spectral region (see Fig. 6a inset) contains the absorption band of phosphodiester and free phosphate functional groups (~1239 cm^{-1}) and a band associated to various polysaccharides (~1090 cm^{-1}) [37]. However, spectra from dried samples resulted to be characterized by low signal-to-noise ratio in the 1800–1300 cm^{-1} region. In order to put in evidence the different contributions to the line shape of the Amide I band, FTIR spectra in D$_2$O solution samples were performed and are shown in Fig. 6b. For these samples, since

data below 1300 cm^{-1} are dominated by D$_2$O absorption bands, only the 1800–1300 cm^{-1} spectral region is reported. In this spectral range, the most prominent absorption bands are the Amide I′ (~1656 cm^{-1}) and Amide II′ (which is shifted at ~1450 cm^{-1} in D$_2$O) [34]. The exposure to ZnO NRs-decorated GNPs indeed clearly affects the shape and intensity of several absorption bands including the >C=O stretching of esters and of carbonic acid (1760–1700 cm^{-1}) in both the dried and liquid samples indicating a modification of lipids and fatty acids content [38]. For both kind of sample preparations, it is also evident a decrease of the signal in the spectral region 1694–1675 cm^{-1} associated to "β-turns" e "antiparallel pleated β-sheets" of proteins [35, 39], suggesting changes in the secondary structure of proteins. In the case of liquid samples (Fig. 6b), we observed also a decrease in the intensity of Amide I′ band confirming a modification at the level of proteins secondary structure. The same spectral change is not detected when comparing the FTIR spectra of the two dried samples (Fig. 6a) probably due to the presence of residual water content in the treated sample. In the Amide II (Amide II′) region essentially no change due to the treatment is detected, except for a small growth in the high wavenumbers tail of the band which could be affected by changes in the amino acid environment around the carboxylate (COO$^-$) group

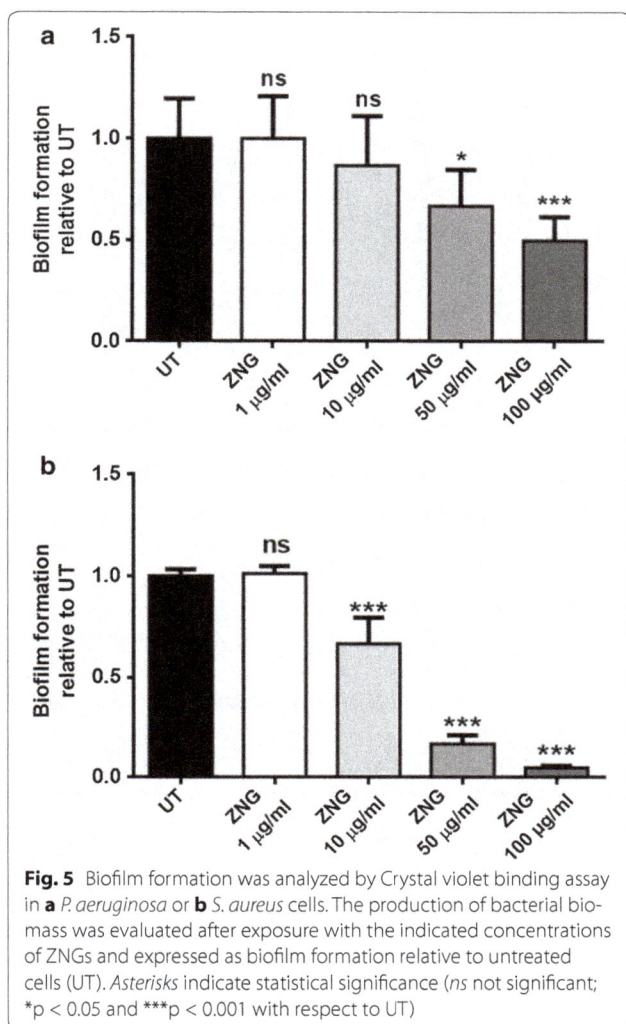

Fig. 5 Biofilm formation was analyzed by Crystal violet binding assay in **a** *P. aeruginosa* or **b** *S. aureus* cells. The production of bacterial biomass was evaluated after exposure with the indicated concentrations of ZNGs and expressed as biofilm formation relative to untreated cells (UT). *Asterisks* indicate statistical significance (*ns* not significant; *p < 0.05 and ***p < 0.001 with respect to UT)

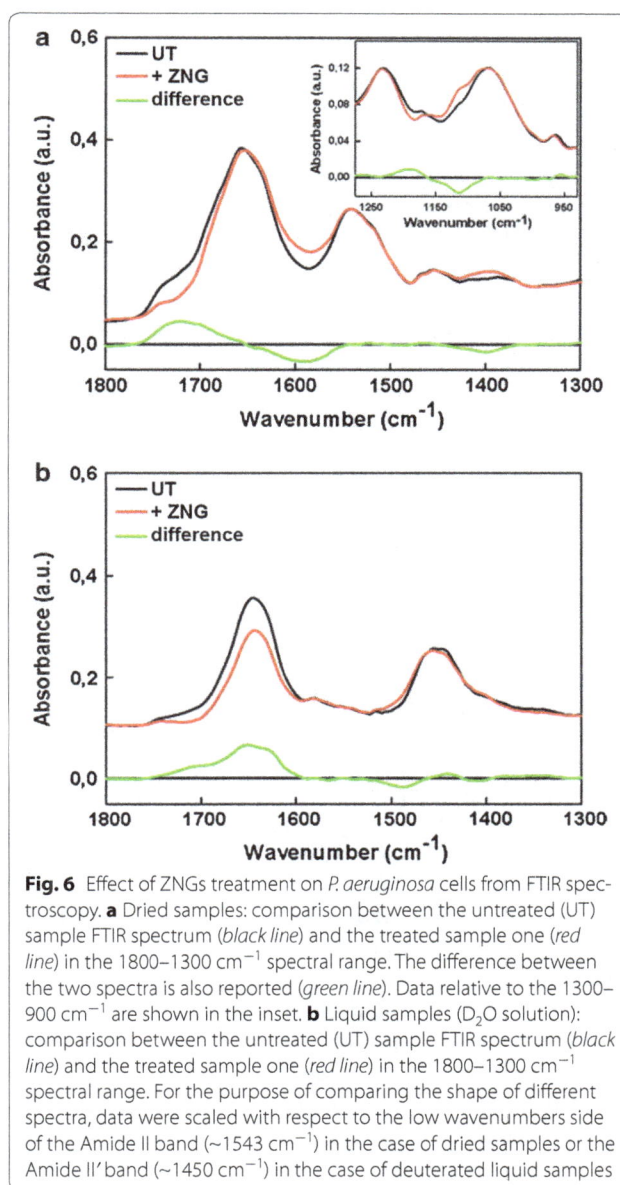

Fig. 6 Effect of ZNGs treatment on *P. aeruginosa* cells from FTIR spectroscopy. **a** Dried samples: comparison between the untreated (UT) sample FTIR spectrum (*black line*) and the treated sample one (*red line*) in the 1800–1300 cm^{-1} spectral range. The difference between the two spectra is also reported (*green line*). Data relative to the 1300–900 cm^{-1} are shown in the inset. **b** Liquid samples (D$_2$O solution): comparison between the untreated (UT) sample FTIR spectrum (*black line*) and the treated sample one (*red line*) in the 1800–1300 cm^{-1} spectral range. For the purpose of comparing the shape of different spectra, data were scaled with respect to the low wavenumbers side of the Amide II band (~1543 cm^{-1}) in the case of dried samples or the Amide II' band (~1450 cm^{-1}) in the case of deuterated liquid samples

(~1574 cm^{-1} and ~1560 cm^{-1}) of aspartates and glutamates [40]. In the case of dried samples, a further evidence of such change is the increase of the signal of the carboxylate C = O symmetric stretching (~1397 cm^{-1}).

In the spectral region between 1300 and 900 cm^{-1} (inset of Fig. 6a), the band at about 1234 cm^{-1}, attributed to phosphodiester functional groups of DNA/RNA polysaccharide backbone structures, is essentially unaffected by ZNGs treatment. On the other hand, the band at ~1069 cm^{-1}, attributed to the symmetric stretching vibration of PO$_2^-$ groups in nucleic acids and to C–O–C and C–O–P stretching vibrations of various oligo- and poly-saccharides, becomes wider because of the appearance of a component at 1114 cm^{-1}. This observation suggests that an alteration in bacterial polysaccharide structures (Extracellular polymeric substances, EPS) results from the interaction of the bacterial cell surface with ZNGs, in agreement with the biofilm results. Indeed, cells forming a biofilm are surrounded by EPS, which represents the immediate environment of these cells, thus playing a relevant role in nutrient acquisition and in the protection of the bacterial cells from environment and mechanical stresses. Consistent with this, Wang et al. suggested a protective role for bacterial EPS against ZnO nanoparticles killer action, via nanostructures sequestering [41]. We can hypothesize that ZNGs act by lowering EPS production and thus by inhibiting cellular barrier mechanisms.

Fourier transform infrared spectroscopy data indicate that EPS reduction is more pronounced in *S. aureus* than in *P. aeruginosa* bacteria. This observation is in line with biofilm growth inhibition results (Fig. 5), confirming a stronger antimicrobial effect of ZnO NRs-decorated GNPs on *S. aureus*. Moreover, the FTIR spectrum of the treated *S. aureus* bacteria was, in all repeated experiments, always noisier with respect to the spectrum of the untreated ones. This result could be a further indication of the intensive interaction between the ZNGs and the external structure of the *S. aureus* bacteria. Similar overall changes (mainly alterations in the structure of proteins and polysaccharides) induced by the treatment with ZNGs are observed in the case of *S. aureus* (Additional file 1: Figure S2). However, in the 1300–900 cm^{-1} region, the band associated to saccharide structures underwent a bigger modification. Indeed, in the case of *S. aureus*, the appearance of two components, one at ~1119 cm^{-1} and the other at ~998 cm^{-1}, was also observed. The FTIR results support the hypothesis that ZNGs exposure produces cell damages. In particular, the FTIR analysis suggests that the antimicrobial effect-related changes are associated with protein and phospholipid damages. This is consistent with the previously observed results demonstrating modifications in protein structures as well as membrane injuries in *S. aureus* cells treated with ZnO NRs [23]. Moreover, several studies highlighted both partial protein unfolding and changes in phospholipids as a meaning of the interaction between cell wall biomolecules and nanomaterials surface [42–44]. Cell surface proteins play important roles in cellular physiological activities, including DNA stability and replication, which in turn may lead to DNA damages.

The possible exploitation of this nanomaterial in indoor/outdoor applications prompted us to evaluate their antimicrobial properties by contaminating ZNG-decorated surfaces with *S. aureus* cells. To this aim, ZNGs were drop casted on building-related materials such as plywood sheets or samples covered by a commercial paint. ZNGs-treated surfaces induced about 95% of mortality in *S. aureus* cells already after 4 h from the contamination (Fig. 7). Overall our results demonstrate that this hybrid nanomaterial may represent a promising approach to overcome/reduce microbial growth in different application fields ranging from historical and cultural heritage to nosocomial environments and wearable medical devices.

The ever-growing demand for nanomaterials raise the need to understand if their environmental release could impact negatively on human healthiness. Therefore, nanotoxicity and environmental risk assessment have gained much attention in the last decade. *Caenorhabditis elegans* is a worm that has been often found in soil and

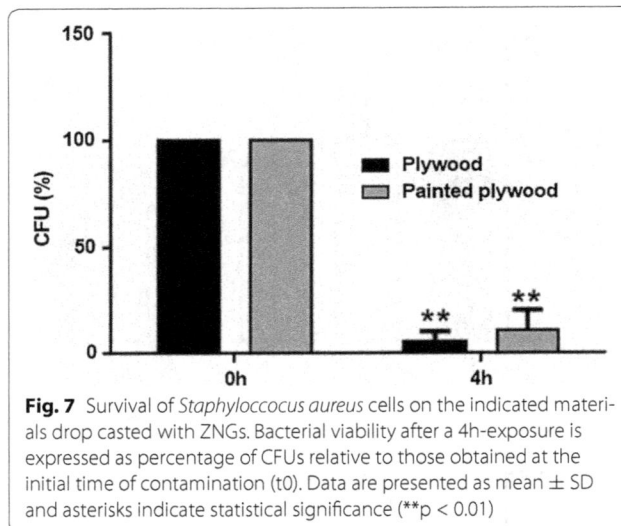

Fig. 7 Survival of *Staphyloccocus aureus* cells on the indicated materials drop casted with ZNGs. Bacterial viability after a 4h-exposure is expressed as percentage of CFUs relative to those obtained at the initial time of contamination (t0). Data are presented as mean ± SD and asterisks indicate statistical significance (**p < 0.01)

leaf-litter environments, and it is emerging as a powerful model for studying neurobiology, developmental biology as well as environmental toxicology. This nematode has been extensively used to study nanotoxicity of different nanomaterials due to several features including its short lifecycle, compact genome as well as ease of maintenance (as reviewed in Gonzalez-Moragas et al. [45]). Starting from this, the biocompatibility of ZnO NRs-decorated GNPs were evaluated in the animal model *C. elegans*. Indeed, to study the effects of ZNGs on the physiology of an entire organism, several analyses were performed in worms treated with ZNGs. First, the lifespan of adult worms exposed or not to ZNGs was investigated. As shown in Fig. 8a, no significant differences were highlighted between the longevity curves of animals treated with several concentrations of ZNGs and that one relative to the control; in all experimental conditions a 50% reduction of the nematode viability was obtained around the 10th day of adulthood similarly to untreated worms, demonstrating the lack of acute toxicity in vivo.

Notably, also the longevity curve of worms exposed to ZNGs starting from egg hatching resulted to be similar to that one of untreated worms (Fig. 8b), indicating that larval stage exposure to those nanoparticles did not impact negatively on *C. elegans* healthiness. Next, we evaluated the fertility rate of the ZNGs-fed nematodes as an indicator for chronic toxicity. The reproductive potential of *C. elegans* was not affected by ZnO-NRs-decorated GNPs administration to adult animals, which were able to produce an average of ~300 embryos as in the case of untreated worms (Fig. 8c); similar results were obtained when nematodes were exposed to ZNGs all along their larval development (Additional file 1: Figure

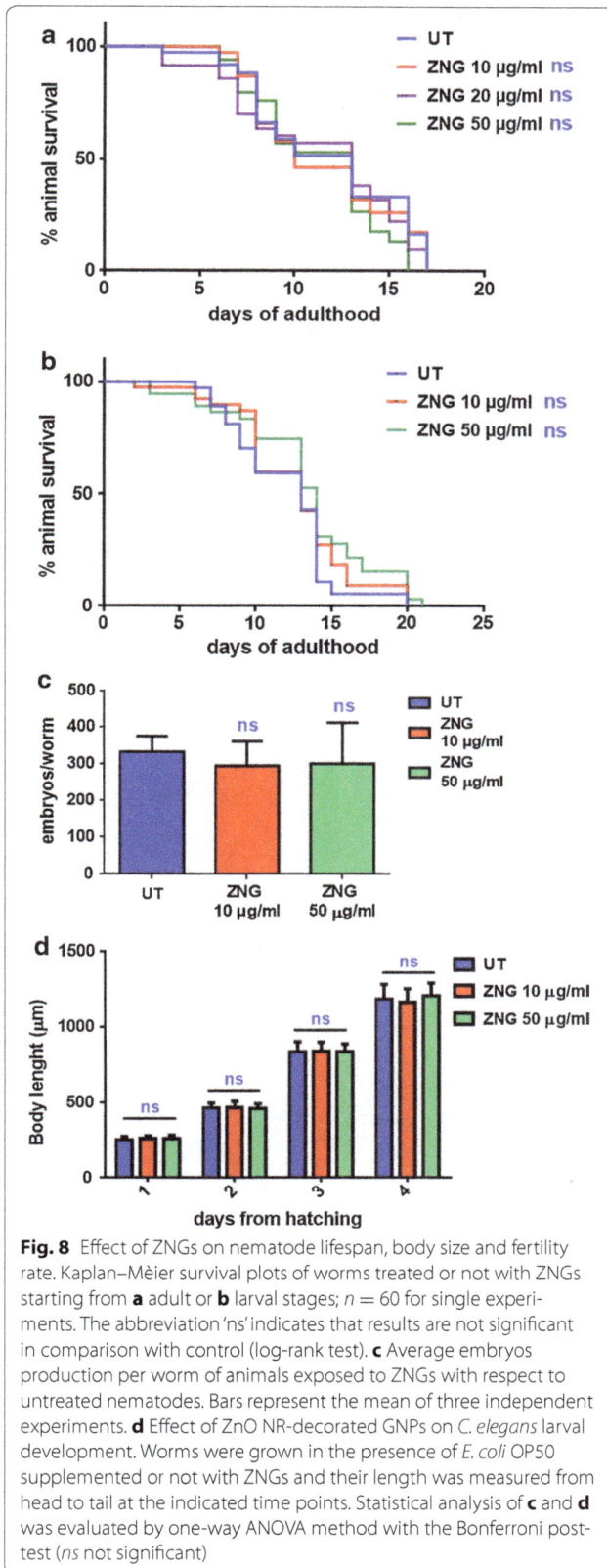

Fig. 8 Effect of ZNGs on nematode lifespan, body size and fertility rate. Kaplan–Mèier survival plots of worms treated or not with ZNGs starting from **a** adult or **b** larval stages; n = 60 for single experiments. The abbreviation 'ns' indicates that results are not significant in comparison with control (log-rank test). **c** Average embryos production per worm of animals exposed to ZNGs with respect to untreated nematodes. Bars represent the mean of three independent experiments. **d** Effect of ZnO NR-decorated GNPs on *C. elegans* larval development. Worms were grown in the presence of *E. coli* OP50 supplemented or not with ZNGs and their length was measured from head to tail at the indicated time points. Statistical analysis of **c** and **d** was evaluated by one-way ANOVA method with the Bonferroni post-test (*ns* not significant)

S3). Then, the effect of different concentrations of ZnO-based hybrid nanomaterial was tested on *C. elegans* larval development. Indeed, after egg hatching, larvae were exposed to ZNGs and their body length was monitored every day. Even in this case, the size of treated worms did not significantly change with respect to the control at each analyzed time-point (Fig. 8d). Finally, the neuromuscular functionality of nematodes was investigated by measuring the contractions of the pharynx, a neuromuscular pump, to assess if ZNGs exposure could affect *C. elegans* swallowing ability. Even in this case, pharyngeal pumping rates were not decreased when ZNGs were administered to nematodes (Fig. 9a). In parallel, the analysis of locomotion behavior was performed to determine the impact of ZNGs exposure on *C. elegans* muscles and neurons. To this aim, the estimation of head thrashes was analyzed and resulted to be not influenced by ZNGs treatment during the first days of adulthood as well as along senescence (Fig. 9b), indicating that the exposure

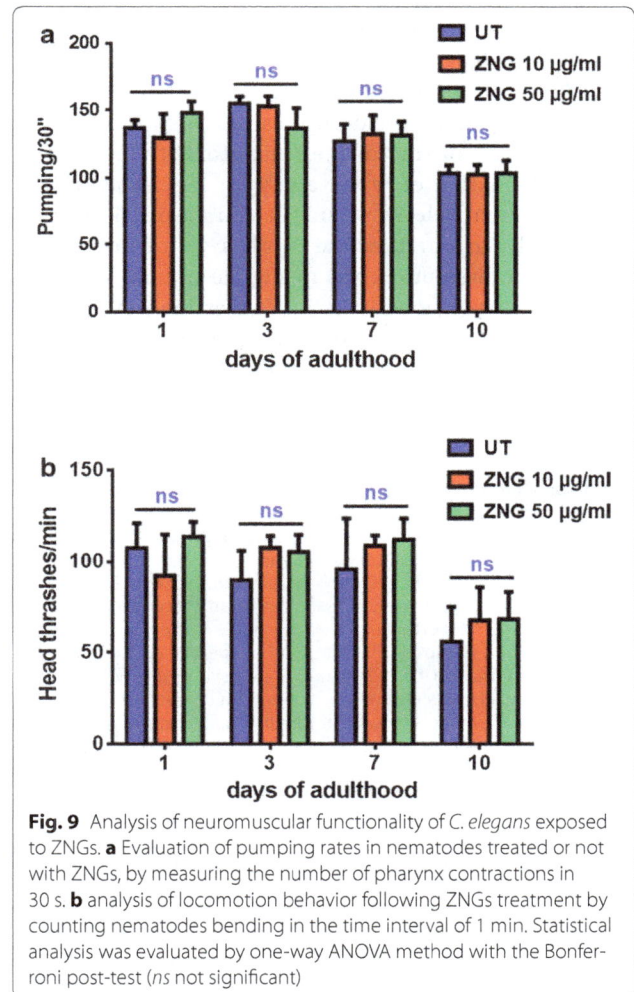

Fig. 9 Analysis of neuromuscular functionality of *C. elegans* exposed to ZNGs. **a** Evaluation of pumping rates in nematodes treated or not with ZNGs, by measuring the number of pharynx contractions in 30 s. **b** analysis of locomotion behavior following ZNGs treatment by counting nematodes bending in the time interval of 1 min. Statistical analysis was evaluated by one-way ANOVA method with the Bonferroni post-test (*ns* not significant)

to ZNG did not determine negative effects on the nervous system of nematodes and hence on their motility.

We have already demonstrated that the two components of ZNGs, namely ZnO NRs and GNPs, showed non-toxic effect in *C. elegans* [23, 24]. Moreover, the lack of cytotoxicity of ZnO nanorods has been assessed in different human cell lines [38]. However, it has been reported that ZnO nanoparticles resulted to be toxic to different model systems including also *C. elegans* [39–41] and that the main components of their nanotoxicity resulted to be the reactive oxygen species production and the consequent release of zinc ion in suspensions [42]. In our previous studies we demonstrated that Zn ion dissolution was negligible in ZnO NRs as well as in ZNGs [10, 15]. We can thus speculate that the lack of ZNGs toxicity in *C. elegans* could be ascribed to the low concentration of bioavailable Zn^{2+}.

Our data are consistent with the observation that the two components of ZNGs, namely ZnO NRs and GNPs, showed no harmful effects in *C. elegans* [15, 16], and that ZnO NRs resulted to be not cytotoxic in different human cell lines [46]. Although it has been reported that ZnO nanoparticles induced toxic effect in different model systems including also *C. elegans* [47–49], the main components of their nanotoxicity resulted to be the production of reactive oxygen species as well as the consequent release of zinc ion in suspensions [50]. Remarkably, in our previous studies we demonstrated that Zn ion dissolution was negligible in both ZnO NR and ZNG suspensions [19, 23], suggesting that the low concentration of bioavailable Zn^{2+} may account for lack of harmful effects in *C. elegans* exerted by of ZNGs.

Additional file

Additional file 1: Figure S1. Pyocyanin production in *Pseudomonas aeruginosa* treated or not (UT) with ZNGs for 24 h. Results are the mean of three independent experiments and error bars represent standard deviation. Statistical significance is defined as *p < 0.5 and ***p < 0.001 with respect to UT, while its absence is indicated with the abbreviation 'ns'. **Figure S2.** Effect of graphene-ZnO nanorods treatment on *Staphylococcus aureus* from FTIR spectroscopy. **a** Dried samples: comparison between the untreated (UT) sample FTIR spectrum (black line) and the treated sample one (red line) in the 1800–1300 cm^{-1} spectral range. The difference between the two spectra is also reported (green line). Data relative to the 1300–900 cm^{-1} range are shown in the inset. **b** Liquid samples (D_2O solution): comparison between the untreated (UT) sample FTIR spectrum (black line) and the treated sample one (red line) in the 1800–1300 cm^{-1} spectral range. For the purpose of comparing the shape of different spectra, data were scaled with respect to the low wavenumbers side of the Amide II band (~1543 cm^{-1}) in the case of dried samples or the Amide II band (~1450 cm^{-1}) in the case of deuterated liquid samples. **Figure S3.** Average embryos production per worm of animals exposed to ZNGs starting from L1 larval stage with respect to untreated nematodes. Bars represent the mean of three independent experiments. Statistical analysis was evaluated by one-way ANOVA method with the Bonferroni post-test (ns not significant).

Authors' contributions

Conceived and designed the experiments: DU, MSS. Wrote the paper: EZ, DU, EB. Critical revision of manuscript: PM. Performed the nanomaterial fabrication: CRC, AB. Did bacterial and nematode experiments/treatments: EB. Performed SEM analysis: GDB. Performed FTIR experiments: MGS. Analyzed and supervised FTIR data: ML. All authors read and approved the final manuscript.

Author details

[1] Department of Biology and Biotechnology C. Darwin, Sapienza University of Rome, Piazzale Aldo Moro 5, Rome, Italy. [2] Research Center on Nanotechnology Applied to Engineering of Sapienza (CNIS), SNNLab, Sapienza University of Rome, Piazzale Aldo Moro 5, Rome, Italy. [3] Department of Astronautical, Electrical and Energy Engineering, Sapienza University of Rome, Via Eudossiana 18, Rome, Italy. [4] Department of Physics and Chemistry, University of Palermo, Palermo, Italy. [5] Department of Experimental Medicine, Sapienza University of Rome, Viale Regina Elena 324, Rome, Italy.

Acknowledgements

We thank Dr. Domenico Cavallini for helpful support.

Competing interests

The authors declare that they have no competing interests.

Funding

The authors wish to thank Italian MIUR for funding by the PON R&C 2007–2013 program with the Project PON03PE_00214_1 Nanotechnologies and Nanomaterials for Cultural Heritages (TECLA, B62F14000560005).

References

1. Eleftheriadou M, Pyrgiotakis G, Demokritou P. Nanotechnology to the rescue: using nano-enabled approaches in microbiological food safety and quality. Curr Opin Biotech. 2016;44:87–93.
2. Wang C, Yu C. Detection of chemical pollutants in water using gold nanoparticles as sensors: a review. Rev Anal Chem. 2013;32:1–14.
3. Huang X, Yin Z, Wu S, Qi X, He Q, Zhang Q, Yan Q, Boey F, Zhang H. Graphene-based materials: synthesis, characterization, properties, and applications. Small. 2011;7:1876–902.
4. Cappitelli F, Principi P, Pedrazzani R, Toniolo L, Sorlini C. Bacterial and fungal deterioration of the Milan Cathedral marble treated with protective synthetic resins. Sci Total Environ. 2007;385:172–81.
5. Ragon M, Fontaine MC, Moreira D, Lopez-Garcia P. Different biogeographic patterns of prokaryotes and microbial eukaryotes in epilithic biofilms. Mol Ecol. 2012;21:3852–68.
6. Sterflinger K, Pinar G. Microbial deterioration of cultural heritage and works of art—tilting at windmills? Appl Microbiol Biot. 2013;97:9637–46.
7. Gaylarde CC, Morton LHG. Deteriogenic biofilms on buildings and their control: a review. Biofouling. 1999;14:59–74.
8. Taylor E, Webster TJ. Reducing infections through nanotechnology and nanoparticles. Int J Nanomed. 2011;6:1463–73.

9. Chelazzi D, Poggi G, Jaidar Y, Toccafondi N, Giorgi R, Baglioni P. Hydroxide nanoparticles for cultural heritage: consolidation and protection of wall paintings and carbonate materials. J Colloid Interface Sci. 2013;392:42–9.

10. Wang J, Wang H, Wang Y, Li J, Su Z, Wei G. Alternate layer-by-layer assembly of graphene oxide nanosheets and fibrinogen nanofibers on a silicon substrate for a biomimetic three-dimensional hydroxyapatite scaffold. J Mater Chem B. 2014;2:7360–8.

11. Zhao X, Zhang P, Chen Y, Su Z, Wei G. Recent advances in the fabrication and structure-specific applications of graphene-based inorganic hybrid membranes. Nanoscale. 2015;7:5080–93.

12. Li D, Zhang W, Yu X, Wang Z, Su Z, Wei G. When biomolecules meet graphene: from molecular level interactions to material design and applications. Nanoscale. 2016;8:19491–509.

13. Yu X, Zhang W, Zhang P, Su Z. Fabrication technologies and sensing applications of graphene-based composite films: advances and challenges. Biosens Bioelectron. 2017;89:72–84.

14. Zhang P, Wang H, Zhang X, Xu W, Li Y, Li Q, Wei G, Su Z. Graphene film doped with silver nanoparticles: self-assembly formation, structural characterizations, antibacterial ability, and biocompatibility. Biomater Sci. 2015;3:852–60.

15. Ding J, Zhu S, Zhu T, Sun W, Li Q, Wei G, Su Z. Hydrothermal synthesis of zinc oxide-reduced graphene oxide nanocomposites for an electrochemical hydrazine sensor. RSC Adv. 2015;5:22935–42.

16. Kaviyarasu K, Geetha N, Kanimozhi K, Maria Magdalane C, Sivaranjani S, Ayeshamariam A, Kennedy J, Maaza M. In vitro cytotoxicity effect and antibacterial performance of human lung epithelial cells A549 activity of zinc oxide doped TiO$_2$ nanocrystals: investigation of bio-medical application by chemical method. Mater Sci Eng C Mater Biol Appl. 2017;74:325–33.

17. Maria Magdalane C, Kaviyarasu K, Judith Vijaya J, Siddhardha B, Jeyaraj B. Facile synthesis of heterostructured cerium oxide/yttrium oxide nanocomposite in UV light induced photocatalytic degradation and catalytic reduction: synergistic effect of antimicrobial studies. J Photochem Photobiol B. 2017;173:23–34.

18. Chandraiahgari CR, De Bellis G, Balijepalli SK, Kaciulis S, Ballirano P, Migliori A, Morandi V, Caneve L, Sarto F, Sarto MS. Control of the size and density of ZnO-nanorods grown onto graphene nanoplatelets in aqueous suspensions. Rsc Adv. 2016;6:83217–25.

19. Zanni E, Chandraiahgari CR, De Bellis G, Montereali MR, Armiento G, Ballirano P, Polimeni A, Sarto MS, Uccelletti D. Zinc oxide nanorods-decorated graphene nanoplatelets: a promising antimicrobial agent against the cariogenic bacterium Streptococcus mutans. Nanomaterials. 2016;6:179.

20. Rago I, Bregnocchi A, Zanni E, D'Aloia AG, De Angelis F, Bossu M, De Bellis G, Polimeni A, Uccelletti D, Sarto MS et al. Antimicrobial activity of graphene nanoplatelets against Streptococcus mutans. IEEE Nano 2015; pp. 9–12.

21. Olivi M, Zanni E, De Bellis G, Talora C, Sarto MS, Palleschi C, Flahaut E, Monthioux M, Rapino S, Uccelletti D, et al. Inhibition of microbial growth by carbon nanotube networks. Nanoscale. 2013;5:9023–9.

22. Stiernagle T. Maintenance of C. elegans (February 11, 2006), WormBook, ed. The C. elegans Research Community, WormBook, doi/10.1895/wormbook. 1.101. 1. 2006.

23. Rago I, Chandraiahgari CR, Bracciale MP, De Bellis G, Zanni E, Guidi MC, Sali D, Broggi A, Palleschi C, Sarto MS, et al. Zinc oxide microrods and nanorods: different antibacterial activity and their mode of action against Gram-positive bacteria. Rsc Adv. 2014;4:56031–40.

24. Zanni E, De Bellis G, Bracciale MP, Broggi A, Santarelli ML, Sarto MS, Palleschi C, Uccelletti D. Graphite Nanoplatelets and Caenorhabditis elegans: insights from an in vivo Model. Nano Lett. 2012;12:2740–4.

25. Zhu Y, Murali S, Cai W, Li X, Suk JW, Potts JR, Ruoff RS. Graphene and graphene oxide: synthesis, properties, and applications. Adv Mater. 2010;22:3906–24.

26. Premanathan M, Karthikeyan K, Jeyasubramanian K, Manivannan G. Selective toxicity of ZnO nanoparticles toward Gram-positive bacteria and cancer cells by apoptosis through lipid peroxidation. Nanomedicine. 2011;7:184–92.

27. Reddy KM, Feris K, Bell J, Wingett DG, Hanley C, Punnoose A. Selective toxicity of zinc oxide nanoparticles to prokaryotic and eukaryotic systems. Appl Phys Lett. 2007;90:2139021–3.

28. Liu S, Hu M, Zeng TH, Wu R, Jiang R, Wei J, Wang L, Kong J, Chen Y. Lateral dimension-dependent antibacterial activity of graphene oxide sheets. Langmuir. 2012;28:12364–72.

29. Hui L, Piao J-G, Auletta J, Hu K, Zhu Y, Meyer T, Liu H, Yang L. Availability of the basal planes of graphene oxide determines whether it is antibacterial. ACS Appl Mater Interfaces. 2014;6:13183–90.

30. Lee J-H, Kim Y-G, Cho MH, Lee J. ZnO nanoparticles inhibit Pseudomonas aeruginosa biofilm formation and virulence factor production. Microbiol Res. 2014;169:888–96.

31. Chen X, Stewart PS. Biofilm removal caused by chemical treatments. Water Res. 2000;34:4229–33.

32. Mah TFC, O'Toole GA. Mechanisms of biofilm resistance to antimicrobial agents. Trends Microbiol. 2001;9:34–9.

33. Berlutti F, Catizone A, Ricci G, Frioni A, Natalizi T, Valenti P, Polimeni A. Streptococcus mutans and Streptococcus sobrinus are able to adhere and invade human gingival fibroblast cell line. Int J Immunopathol Pharmacol. 2010;23:1253–60.

34. Barth A. Infrared spectroscopy of proteins. BBA. Bioenergetics. 2007;1767:1073–101.

35. Militello V, Casarino C, Emanuele A, Giostra A, Pullara F, Leone M. Aggregation kinetics of bovine serum albumin studied by FTIR spectroscopy and light scattering. Biophys Chem. 2004;107:175–87.

36. Navarra G, Tinti A, Leone M, Militello V, Torreggiani A. Influence of metal ions on thermal aggregation of bovine serum albumin: aggregation kinetics and structural changes. J Inorg Biochem. 2009;103:1729–38.

37. Maquelin K, Kirschner C, Choo-Smith LP, van den Braak N, Naumann D, Puppels GJ. Identification of medically relevant microorganisms by vibrational spectroscopy. J Microbiol Methods. 2002;51:255–71.

38. Kansiz M, Heraud P, Wood B, Burden F, Beardall J, McNaughton D. Fourier Transform Infrared microspectroscopy and chemometrics as a tool for the discrimination of cyanobacterial strains. Phytochemistry. 1999;52:407–17.

39. Jackson M, Mantsch HH, Chapman D. Infrared spectroscopy of biomolecules. New York: Wiley; 1996. p. 314–6.

40. Barth A. The infrared absorption of amino acid side chains. Prog Biophys Mol Biol. 2000;74:141–73.

41. Wang Q, Kang FX, Gao YZ, Mao XW, Hu XJ. Sequestration of nanoparticles by an EPS matrix reduces the particle-specific bactericidal activity. Sci Rep. 2016;6:21379.

42. Li HY, Gao YC, Li CX, Ma G, Shang YL, Sun Y. A comparative study of the antibacterial mechanisms of silver ion and silver nanoparticles by Fourier transform infrared spectroscopy. Vib Spectrosc. 2016;85:112–21.

43. Wei X, Yu J, Ding L, Hu J, Jiang W. Effect of oxide nanoparticles on the morphology and fluidity of phospholipid membranes and the role of hydrogen bonds. J Environ Sci. 2017;57:221–30.

44. Faghihzadeh F, Anaya NM, Schifman LA, Oyanedel-Craver V. Fourier transform infrared spectroscopy to assess molecular-level changes in microorganisms exposed to nanoparticles. Nanotechnol Environ Eng. 2016;1:1.

45. Gonzalez-Moragas L, Roig A, Laromaine A. C. elegans as a tool for in vivo nanoparticle assessment. Adv Colloid Interface Sci. 2015;219:10–26.

46. Zanni E, De Palma S, Chandraiahgari CR, De Bellis G, Cialfi S, Talora C, Palleschi C, Sarto MS, Uccelletti D, Mancini P. In vitro toxicity studies of zinc oxide nano- and microrods on mammalian cells: a comparative analysis. Mater Lett. 2016;179:90–4.

47. Aruoja V, Dubourguier HC, Kasemets K, Kahru A. Toxicity of nanoparticles of CuO, ZnO and TiO$_2$ to microalgae Pseudokirchneriella subcapitata. Sci Total Environ. 2009;407:1461–8.

48. Kasemets K, Ivask A, Dubourguier HC, Kahru A. Toxicity of nanoparticles of ZnO, CuO and TiO$_2$ to yeast Saccharomyces cerevisiae. Toxicol Vitro. 2009;23:1116–22.

49. Wu QL, Nouara A, Li YP, Zhang M, Wang W, Tang M, Ye BP, Ding JD, Wang DY. Comparison of toxicities from three metal oxide nanoparticles at environmental relevant concentrations in nematode Caenorhabditis elegans. Chemosphere. 2013;90:1123–31.

50. Song WH, Zhang JY, Guo J, Zhang JH, Ding F, Li LY, Sun ZT. Role of the dissolved zinc ion and reactive oxygen species in cytotoxicity of ZnO nanoparticles. Toxicol Lett. 2010;199:389–97.

8

The role of intracellular trafficking of CdSe/ZnS QDs on their consequent toxicity profile

Bella B. Manshian[1,2]*[ID], Thomas F. Martens[3,4], Karsten Kantner[5], Kevin Braeckmans[3,4], Stefaan C. De Smedt[3], Jo Demeester[3], Gareth J. S. Jenkins[2], Wolfgang J. Parak[5,6], Beatriz Pelaz[5], Shareen H. Doak[2], Uwe Himmelreich[1] and Stefaan J. Soenen[1]

Abstract

Background: Nanoparticle interactions with cellular membranes and the kinetics of their transport and localization are important determinants of their functionality and their biological consequences. Understanding these phenomena is fundamental for the translation of such NPs from in vitro to in vivo systems for bioimaging and medical applications. Two CdSe/ZnS quantum dots (QD) with differing surface functionality (NH_2 or COOH moieties) were used here for investigating the intracellular uptake and transport kinetics of these QDs.

Results: In water, the COOH- and NH_2-QDs were negatively and positively charged, respectively, while in serum-containing medium the NH_2-QDs were agglomerated, whereas the COOH-QDs remained dispersed. Though intracellular levels of NH_2- and COOH-QDs were very similar after 24 h exposure, COOH-QDs appeared to be continuously internalised and transported by endosomes and lysosomes, while NH_2-QDs mainly remained in the lysosomes. The results of (intra)cellular QD trafficking were correlated to their toxicity profiles investigating levels of reactive oxygen species (ROS), mitochondrial ROS, autophagy, changes to cellular morphology and alterations in genes involved in cellular stress, toxicity and cytoskeletal integrity. The continuous flux of COOH-QDs perhaps explains their higher toxicity compared to the NH_2-QDs, mainly resulting in mitochondrial ROS and cytoskeletal remodelling which are phenomena that occur early during cellular exposure.

Conclusions: Together, these data reveal that although cellular QD levels were similar after 24 h, differences in the nature and extent of their cellular trafficking resulted in differences in consequent gene alterations and toxicological effects.

Keywords: Quantum dot NPs, Intracellular localization, Endosomal uptake, Gene alterations, Nanotoxicity

Background

The scope of the use of nanomaterials (NMs) not only for technological, but also in biomedical and clinical applications is still increasing, where mainly imaging purposes and more recently therapeutic purposes are being explored to greater depth. This is driven by the high number of unique physical and chemical properties that many materials possess when downsized to the nanoscale. One such type of NM are quantum dots (QDs), which are small colloidal semiconductor nanoparticles (NPs) that possess remarkable photophysical properties, including high photostability and brightness, along with very narrow and size-tunable emission spectra [1, 2]. These properties have enabled the real-time tracking of surface-located receptors in live cells over longer time periods [3, 4], as well as intracellular tracking of single molecules and protein [5–7]. QDs also have potential as probes for in vivo fluorescence imaging [8]. They are being explored as therapeutic agents [9], such as in photodynamic

*Correspondence: bella.manshian@kuleuven.be
[1] Biomedical NMR Unit/MoSAIC, KU Leuven Campus Gasthuisberg, Herestraat 49, 3000 Louvain, Belgium
Full list of author information is available at the end of the article

therapy where the QDs could be used to eradicate cancer cells [10]. Despite alternative materials, the predominantly used QDs are based on II/VI group semiconductor materials, and thus typically comprise Cd. Given their chemical composition and the presence of highly toxic elements such as Cd^{2+} [11, 12], the use of QDs in live cells, tissues, and clinical applications has remained limited. Despite various strategies being explored to reduce their toxicity (e.g. Cd^{2+}-free QDs, dual polymer-silica coated QDs), their practical use in biomedical applications remains moderate. This is in part due to the absence of sufficient information about the precise mechanisms and kinetics involved in the interaction of QDs with biological entities. Some recent studies have tackled this topic [13–15] yet more research is required to understand the effects of specific physico-chemical differences in NPs on their toxicity profiles [16]. Additionally, one inherent issue with the field of nanosafety research is the near endless number of potential interactions of NPs with biological components, of which only a selected few can be examined in every single study for a selected in vitro or in vivo model [17]. As most studies will focus on key mechanisms, such as the induction of reactive oxygen species (ROS) or gross cell viability studies, more subtle effects are often overlooked and differences between the various in vitro and in vivo models used can drastically alter the outcome of any study [18, 19]. As such, several key questions regarding the potential toxic effects of QDs remain thus far not fully answered.

In the present work, two different types of QDs (bearing negative and positive surface charge) are being used to examine cyto- and genotoxic effects on cultured human cells. Continuing on the results obtained in a previous work with the same QDs [20], further investigations are performed here to evaluate the kinetics of their cellular uptake, intracellular localization, and the alterations they induce to the cellular homeostasis in an effort to attaining a better understanding of the observed differences in their toxicity profiles. Intracellular cadmium levels are quantified and correlated to changes in cellular homeostasis. One major aim of this study is therefore to link the differences in physicochemical parameters with the kinetics of cellular processing and their toxicity levels. A second aim is to further elucidate upon the mechanisms by which the different QDs exert their toxicity. For this purpose, the effect of the intracellular environment on QD functionality and chemical stability are investigated. Additionally, detailed gene expression studies are performed and activation of important cytoskeletal regulator and stress and toxicity signalling pathways are examined. Finally, all results are combined and analysed together, in order to evaluate whether the differences in physicochemical properties of the QDs are linked to their

respective uptake kinetics and levels, and whether their intracellular processing also influences QD behaviour and their mechanism of toxicity.

Therefore, this study is a more comprehensive investigation and exploration of the processes responsible for the differences in the cellular and NP interactions that was previously published using the same QDs [20].

Results

Properties of the QDs

The COOH- and NH_2-CdSe/ZnS QDs were purchased from different vendors but both NPs had the same core and surface coating. The diameter of the inorganic part (i.e. the CdSe core and the ZnS shell, excluding the organic surface coating) of the QDs was determined as 4.6 ± 0.5 nm for the carboxyl, COOH-QDs (QD−) and 6.9 ± 0.9 nm for the amine, NH_2-QDs (QD+) (for details please see Additional file 1: Figure S1). The emission spectra were different for the different QDs where COOH-QDs had their first excitation peak at 585 and 664 nm for the NH_2-QDs. At the same QD concentration and at 450 nm excitation the COOH-QDs were much brighter than the NH_2-QDs. In water, the COOH- and NH_2-QDs were negatively and positively charged, respectively. In serum-containing medium the NH_2-QDs were agglomerated (as indicated by the largely increased hydrodynamic diameter), whereas the COOH-QDs remained dispersed (for more QD characterisation information please see the supporting information, a summary is given in Additional file 1: Table S7).

Cellular uptake by confocal microscopy

Quantum dots internalisation by HFF-1 cells following 4 and 24 h exposure was examined by confocal microscopy of tubulin stained cells. QDs were confirmed to be in the cells by 3D imaging. Both, NH_2- and COOH-QDs were readily taken up by the cells, as observed from the images (Fig. 1). However, upon semi-quantification of cellular QD levels, clear differences in fluorescence levels were observed after 4 h, where COOH-QDs resulted in higher cellular fluorescence. After 24 h, fluorescence had however dropped significantly, reaching the same level as the NH_2-QDs. The NH_2-QDs did not show any significant differences between 4 and 24 h exposure and appeared to rapidly reach maximum intracellular fluorescence levels.

Alterations to QD properties with varying pH conditions

The effects of altering pH levels on the QD properties were tested here by dissolving the NPs in citrate containing PBS the pH of which was adjusted to 4.5, 5.5, and 7.4 (please see Additional file 1: Figure S11). Our results showed that the fluorescence of the COOH-QDs is indeed quenched after 48 h at all pH levels and was

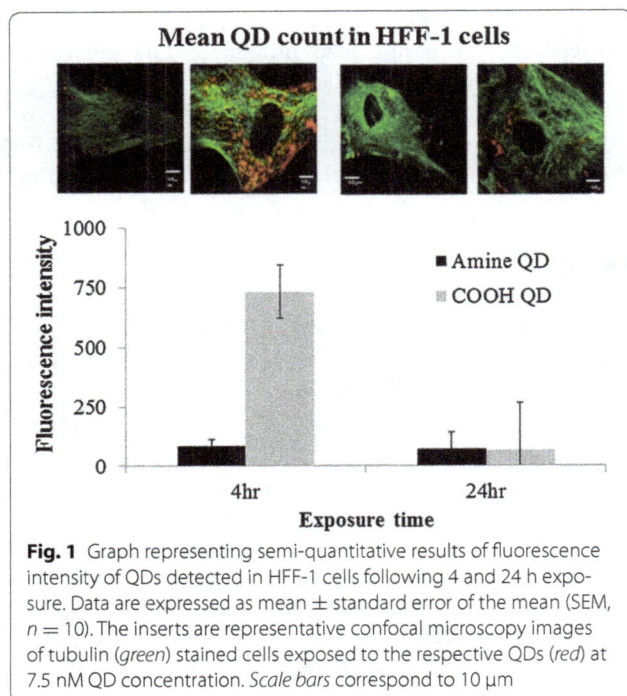

Fig. 1 Graph representing semi-quantitative results of fluorescence intensity of QDs detected in HFF-1 cells following 4 and 24 h exposure. Data are expressed as mean ± standard error of the mean (SEM, $n = 10$). The inserts are representative confocal microscopy images of tubulin (*green*) stained cells exposed to the respective QDs (*red*) at 7.5 nM QD concentration. *Scale bars* correspond to 10 μm

most prominent at the lower pH levels, but the overall effects were strong in all conditions (Additional file 1: Figure S11A). In contrast, NH_2-QDs showed no degradation effects of fluorescence following incubation with the three solutions for up to 4 days (Additional file 1: Figure S11B).

Determination of QD properties upon cellular internalization

To evaluate whether the semiconductor part of the QDs dissolved in the cellular environment, the Measure-IT (Invitrogen Ltd. UK) commercially available kit was used to assess free cadmium ion content in HFF-1 cells treated with COOH- or NH_2-QDs for 24 h. In order to assess the effect of the low endosomal pH on the QDs, non-proliferating HFF-1 cells were used, as highly proliferative cells would have complicated this analysis by the continuous dilution of both intracellular QDs and intracellular free ions [21]. Data (Additional file 1: Figure S11) revealed significant increases in cellular free Cd^{2+} levels in COOH- and NH_2-QD treated cells at all the tested time points, starting at day 2, for COOH-QDs and starting from day 3 for NH_2-QDs.

Evaluation of cellular QD trafficking
Confocal microscopy
Cellular interaction of the two QDs was studied using confocal microscopy based analysis of cells expressing

green fluorescent protein (GFP)-tagged Lamp1 (lysosomal marker) or EEA1 (marker for early endosomes). Colocalization between either QD and Lamp1 or EEA1 was determined from the acquired images using the ImageJ analysis tool. Lysosomes or endosomes were considered as colocalized with the QDs when their respective intensities were higher than the threshold of their individual channels and if their ratio of intensity was more than the ratio setting value [22]. Figure 2 reveals that after 24 h incubation, there is a clear difference between the two types of QDs, where NH_2-QDs result in much higher levels of QDs colocalizing with lysosomes. In contrast, COOH-QDs result in much higher levels of QDs colocalizing with early endosomes.

Fluorescence single particle tracking
In this analysis, the two QDs showed different profiles of uptake and localization in the intracellular environment at the different time points (Fig. 3). NH_2-QDs were taken up by the Rab5a-positive early endosomes with endosomal colocalization increasing with time until 1 h post exposure. They then appeared to be immediately transported into the LAMP1-containing organelles, being mainly lysosomes (Fig. 3a), where the majority of these QDs remained until 6 h post exposure. The QDs that were not found to be colocalized with Rab5a or LAMP1 (at 120–180 min time points) were likely present in intermediary organelles such as late endosomes [23]. After

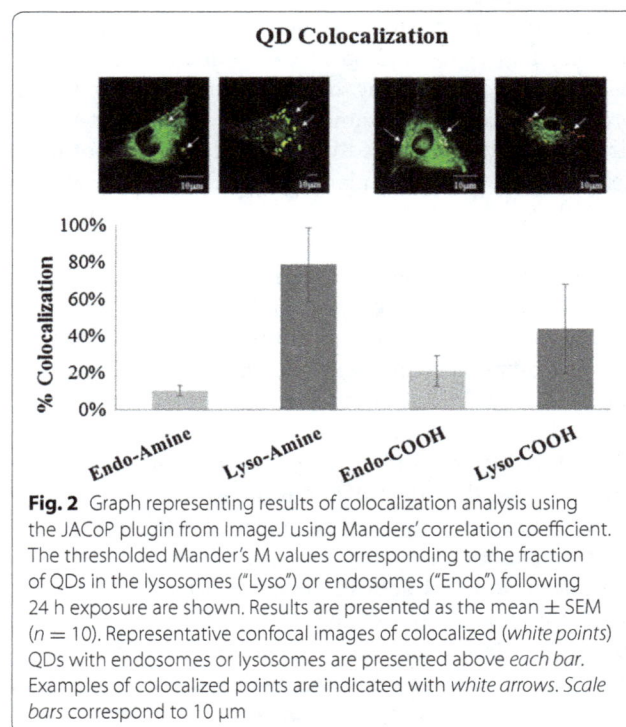

Fig. 2 Graph representing results of colocalization analysis using the JACoP plugin from ImageJ using Manders' correlation coefficient. The thresholded Mander's M values corresponding to the fraction of QDs in the lysosomes ("Lyso") or endosomes ("Endo") following 24 h exposure are shown. Results are presented as the mean ± SEM ($n = 10$). Representative confocal images of colocalized (*white points*) QDs with endosomes or lysosomes are presented above *each bar*. Examples of colocalized points are indicated with *white arrows*. *Scale bars* correspond to 10 μm

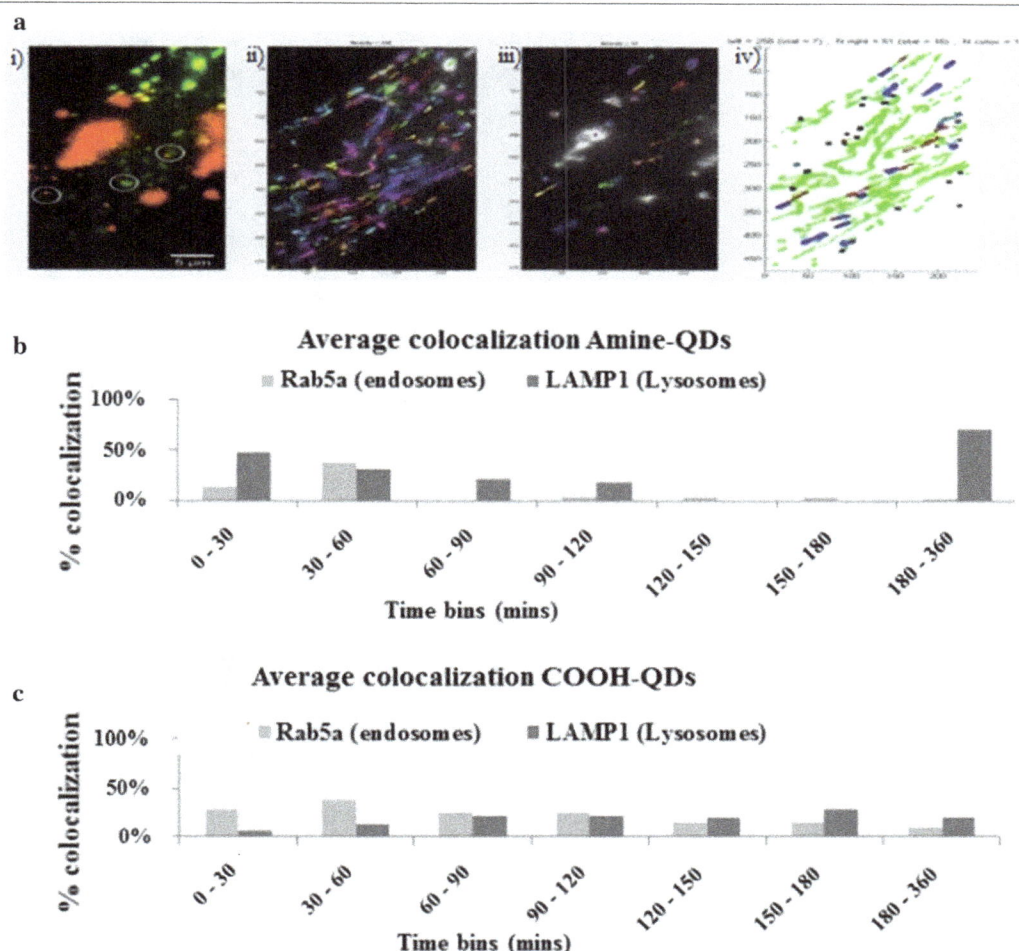

Fig. 3 Plots from the intracellular trafficking profile of QDs in HFF-1 cells using early endosomes-GFP, and lysosomes-GFP. **a** Images showing an example of the (*i*) overlay of NH$_2$-QDs with the lysosomal marker, (*ii*) the tracks for the *green* lysosomal channel, (*iii*) tracks for the QD channel, and (*iv*) colocalization of *green* (lysosomal) and *red* (QD) tracks. The *scale bar* corresponds to 5 μm. **b, c** Graphs represent trajectory-based dynamic colocalization of fluorescent NH$_2$-QDs with the endosomal marker (Rab5a) and lysosomal marker (LAMP1) that was calculated and *plotted* as a function of time. *Each dot* corresponds to 1 min movie recording that was taken in different cells at that specific time point

3–6 h, the majority of the detected QDs were found in the lysosomal compartment. The COOH-QDs displayed a completely different profile, where up to 6 h only a low number of QDs were present in the lysosomal compartment (Fig. 3b).

Exocytosis investigation with ICP-MS

The results of this analysis (Fig. 4) revealed clear differences in the cellular release of Cd^{2+} ions by HFF-1 cells, depending on the type of QD. Figure 4a shows that after 4 h the cell culture media contained sixfold higher amounts of Cd^{+2} ions following exposure to the NH$_2$-QDs as compared to the COOH-QDs. Higher number of exocytosed NH$_2$-QDs as compared to the COOH-QDs. COOH-QDs demonstrated dose and time

dependent increase in the level of QDs released into the cell culture media (Fig. 4c). On the other hand, exocytosis of NH$_2$-QDs did not have a time dependent pattern however there was a dose dependency in the 4 h treatments up to 6 h post removal of the NPs from the culture media. This pattern has disappeared in the 24 h experiments (Fig. 4b).

Evaluation of QD induced cellular stress

Next, the toxic effects of the QDs were evaluated following 4 and 24 h exposure, using an already validated high-content imaging approach [14], where a few parameters were selected at sub-cytotoxic concentrations, which were defined in another work [24]. These were the levels of reactive oxygen species (ROS), mitochondrial ROS,

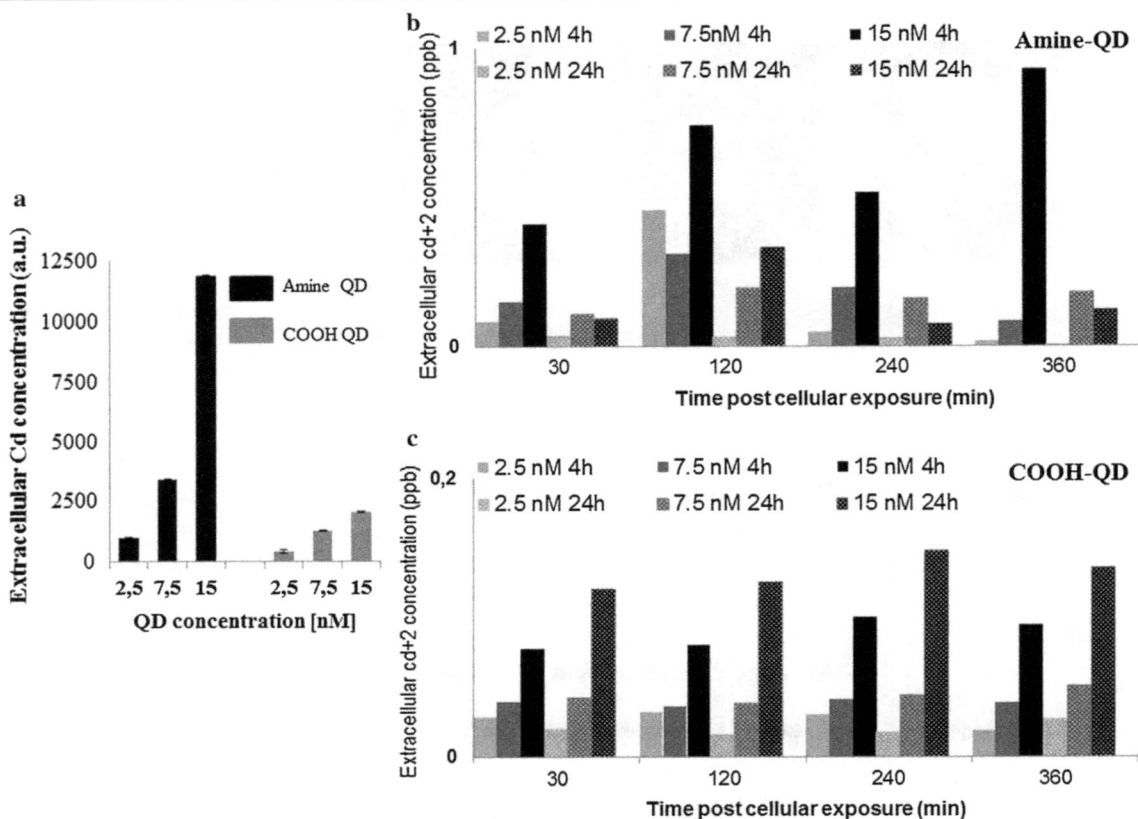

Fig. 4 a Graph representing the amount of elemental Cd remaining in the cell culture medium for each exposure concentration following 4 h incubation. **b, c** Figures showing the *number* of elemental Cd, relative to the control, detected in the cell culture media at each concentration after 4 h (*solid filled bars*) and 24 h (*dotted bars*) incubation, which was followed by immediate washing of the cells. Samples were collected and measured with ICP-MS at the different time points. Please note the difference in the *y*-scale between graphs **b** and **c**

induction of autophagy, and alterations to cell morphology. A heat-map is used here to compare toxicity profiles between both types of QDs. For this analysis, the control values were all normalised to 100%. The data show no major effect of the NH_2-QDs at any of the parameters tested (Fig. 5, for more detailed results and images see Additional file 1: Figures S12–14). The COOH-QDs however resulted in induction of mitochondrial ROS (Additional file 1: Figure S14) and reduction in cell area after 24 h exposure (Additional file 1: Figure S13). Neither of the two QD types resulted in a significant effect on cellular autophagy. Neither of the QDs tested here resulted in significant induction of ROS. However, one should keep in mind that there are different types of ROS that can be generated by various processes in different cellular compartments. Here, the mitochondrial-specific probe did indicate induction of mitochondrial ROS, even at the lowest concentration of COOH-QDs. Interestingly, increasing the QD concentration did not correlate with higher levels of mitochondrial ROS, as under the conditions used, a near-constant high level of mitochondrial

ROS was observed, when cells were exposed to the COOH-QDs.

Gene expression studies

In the arrays investigating genes involved in cellular stress and toxicity different sets of genes were found to be up- or down-regulated following 24 h exposure to NH_2- or COOH-QDs. NH_2-QD exposure resulted in an increase in CCL2, IL1A, IL1B, IL6, IL8, and TNFα genes (Fig. 6a). In contrast, exposure to the higher concentrations (7.5 and 15 nM) of COOH-QDs resulted in the downregulation of several genes mainly involved in the hypoxic processes (Fig. 6b). The most significant of these genes included CFTR, AQP4, and ADM, all of which demonstrated no notable changes from exposure to the NH_2-QDs. Significant changes were defined as at least twofold changes as compared to untreated control levels. Another important difference between the two QDs was the significant downregulation of VEGFA recorded with the COOH-QDs which was absent from exposure to the NH_2-QDs. On the other hand, like its counterpart,

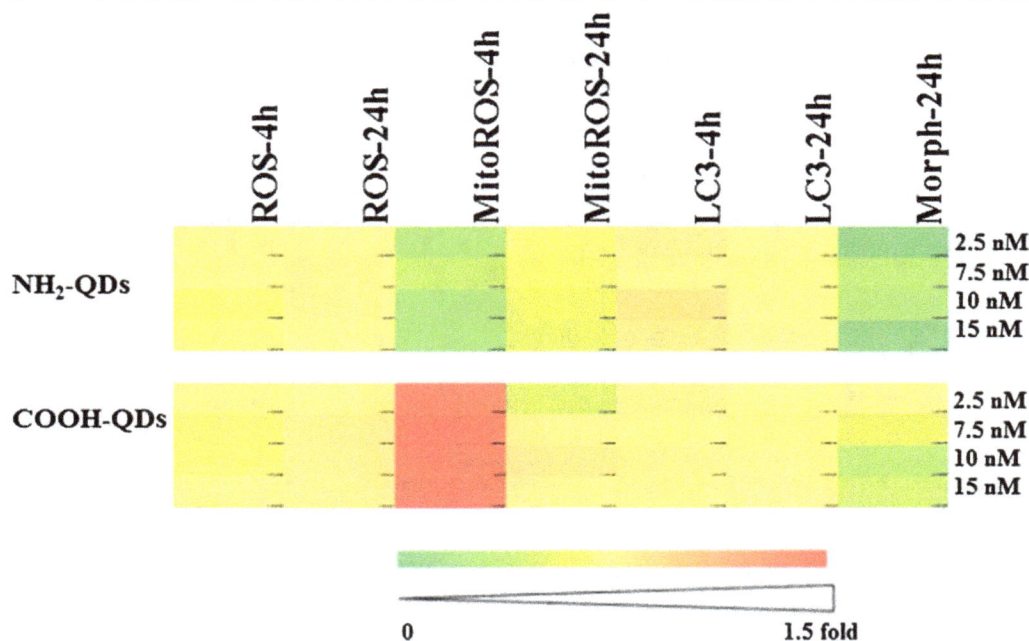

Fig. 5 A heat map of the level of toxicity detected with the different toxicity screening assays upon exposure of the cells to the NH_2-QDs or COOH-QDs at 2.5, 7.5, 10, and 15 nM concentrations

COOH-QD resulted in the upregulation of TNF gene up to 7.5 nM concentration.

The results of the PCR arrays exploring changes to cytoskeletal regulators showed that exposure of the HFF-1 cells to NH_2-QDs caused an increase in CCNA1, CDK5R1, IQGAP2, MYLK2, and WAS genes at all tested concentrations. However, the effects were more significant at the two highest concentrations tested (7.5 and 15 nM) (Fig. 6c).

For COOH-QDs, a set of eleven genes were found to be significantly affected, two of which; MYLK2 and WAS, were also found to be upregulated from exposure to the NH_2-QDs (Fig. 6d). For the cells exposed to COOH-QDs, these genes were significantly affected along with others, such as ARAP1, CDC42BPA, and CDC42EP2, which were all significantly downregulated at all the tested concentrations.

Discussion
Cellular uptake
Though characterisation studies demonstrated a clear difference in fluorescence intensity between the NH_2- and COOH-QDs, with the latter being much brighter, yet the comparison here is not simply the difference in the uptake level between the two QDs. For the interpretation of the results of confocal microscopy shown in Fig. 1 two different effects need to be discussed. First, a possible difference in cellular internalization of these QDs, as

previously demonstrated [20], where negatively charged well dispersed COOH-QDs were shown to be more readily taken up in different cell types compared to the positively charged agglomerated NH_2-QDs. This however would be in contrast to another study, where positively charged ZnO NPs, prone to agglomeration, were found to be internalized to a higher extent than negatively charged, well dispersed, polymer-coated ZnO NPs [25]. Second, the possibility that the fluorescence properties of the QDs may have changed, leading to changes in intracellular signal not due to changes in QD concentration, but due to fluorescence loss of the QDs upon being localized in acidic endosomes/lysosomes. According to Additional file 1: Figure S11 this effect is stronger for the COOH-QDs than for the NH_2-QDs, which would explain the loss of intracellular fluorescence over time of cells incubated with COOH-QDs. Fluorescence loss in acidic pH itself may be caused by different mechanisms. Low pH can cause the generation of trap states by partial loss of the ligands shell, which quenches fluorescence. The latter was tested and described in the next paragraph. Third, some of the QDs may have been exocytosed after being endocytosed [26], which also would lead to a decline in the intracellular fluorescence detected over time.

pH effect on QD stability
Generally, cellular internalization of NPs occurs through endocytic processes [27], during which NP stability may

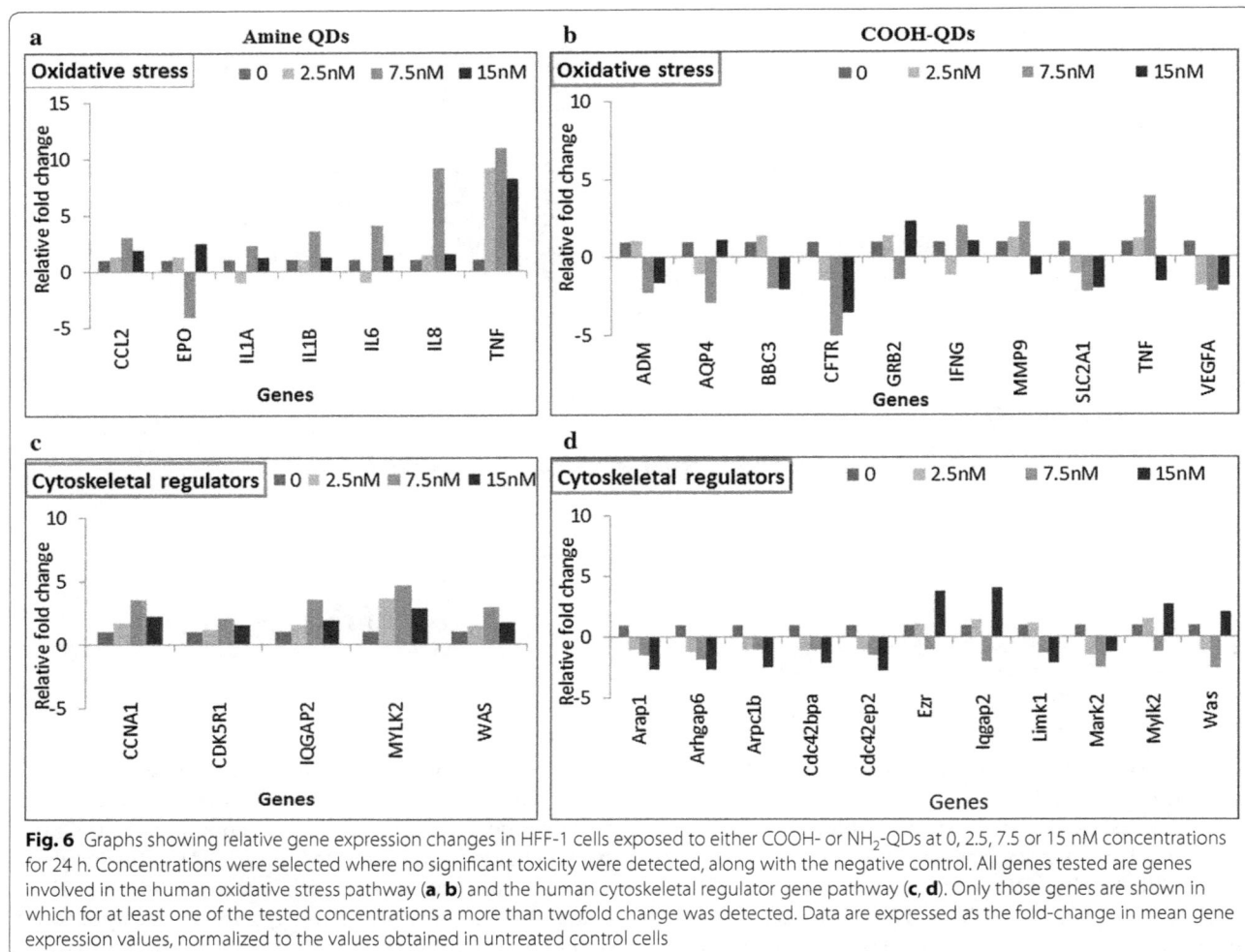

Fig. 6 Graphs showing relative gene expression changes in HFF-1 cells exposed to either COOH- or NH$_2$-QDs at 0, 2.5, 7.5 or 15 nM concentrations for 24 h. Concentrations were selected where no significant toxicity were detected, along with the negative control. All genes tested are genes involved in the human oxidative stress pathway (**a**, **b**) and the human cytoskeletal regulator gene pathway (**c**, **d**). Only those genes are shown in which for at least one of the tested concentrations a more than twofold change was detected. Data are expressed as the fold-change in mean gene expression values, normalized to the values obtained in untreated control cells

be affected due to changes in the surrounding pH conditions ranging from pH 7.4 representing the extracellular environment, pH 5–6 for late and early endosomes, consecutively to the more acidic pH 4.5 of the lysosomal milieu [28]. It has been shown that various NPs, including QDs, are sensitive to the acidic degrading environment of the lysosomes, resulting in a gradual dissolution of the NPs and release of metal ions, which in this case, would be amongst others, highly toxic Cd^{2+} ions [29]. Therefore, the effect of changing pH levels on NP stability were tested. The results of this test showed no degradation effects for the NH$_2$-QDs up to 4 days in all three solutions while the COOH-QDs were found quenched starting from day 2. It is not clear if this observation is due to the high chemical stability of the NH$_2$-QDs. One possibility is that these QDs formed large aggregates (as seen in the characterisation results in Additional file 1: Figure S8), which could have sedimented to the bottom of the wells, resulting in an absence of significant signal

alteration. Concerning the photophysical properties of the two different types of QDs, within the first day of incubation, there was a significant loss in fluorescence of the (initially bright) COOH-QDs (Additional file 1: Figure S11A) but not for the (already initially weakly fluorescent) NH$_2$-QDs (Additional file 1: Figure S11B). The loss in intracellular QD fluorescence from 4 to 24 h after exposure for the COOH-QDs thus could be explained by a possible fluorescence quenching. Partial loss of the ligand shell may have caused the reduction in the fluorescence of the COOH-QDs. However, the NH$_2$-QDs were already initially agglomerated, further loss of ligands is thus less likely and thus the initially already weak fluorescence does not decrease further upon incubation.

Alternatively, low pH may lead to corrosion of the QDs, i.e. in their dissolution, leading to the release of free Cd^{2+} ions. In order to investigate the last point, intracellular levels of free Cd^{2+} ions (i.e. Cd^{2+} ions released from internalized QDs) were detected (Additional file 1: Figure S11C), as explained in the next section.

Changes to QD properties following intracellular uptake

In the intracellular environment the two QDs appear to have undergone some degradation as evidenced with the free Cd^{2+} ions detected. Both QDs were significantly degraded in the cell starting at day 2 for the COOH-QDs and day 3 for the NH_2-QDs. These data are in line with earlier studies on QDs, where degradation of the QDs typically displays a lag time of one to several days, after which there is a gradual increase in cellular Cd^{2+} levels [21, 30]. Slower release of Cd^{2+} from the NH_2-QDs may be explained by the fact that they are agglomerated, thus their surface is less accessible. In addition, there is indication that the ZnS shell around the CdSe core is thicker for the NH_2-QDs than for COOH-QDs (Additional file 1: Table S1), which also may account for the slower release of Cd.

The absolute amount of released Cd^{2+} ions correlates to the number of internalized QDs. However, due to loss in QD fluorescence upon potential partial loss of the ligand shell the data shown in Fig. 1 do not allow us to make a statement about the absolute amount of QDs that has been incorporated by cells. Additional file 1: Figure S11C shows clear release of intracellular Cd^{2+} from internalized QDs. Cadmium is a heavy metal that has been shown to be highly toxic in mammalian cells [31]. Free cadmium ions have also been correlated with toxicity detected in cells exposed to cadmium based QDs [32]. However, Cd^{2+}-mediated toxicity would depend on the balance between cell cycle kinetics and degradation kinetics (i.e. the release of Cd^{2+}), where toxicity will only occur when the cellular Cd^{2+} concentration exceeds a certain toxic threshold, which may not be the case for highly proliferating cells (i.e. in the limiting case, if cell division were faster than release of Cd^{2+} from internalized QDs) then no Cd^{2+}-mediated effects would occur). More subtle, sub-cytotoxic effects will be more easily detected in non-proliferating cells, as they should occur at lower Cd^{2+} levels. This assumption is supported by a previous publication [20], where significant chromosomal damage was detected in HFF-1 cells exposed to 7.5 nM QD concentration of either the NH_2- or COOH-QDs, conditions under which no acute cytotoxicity was observed.

Intracellular trafficking of the QDs: confocal microscopy, fSPT, and ICP-MS

In order to gain insight into the kinetics of the uptake of these QDs into HFF-1 cells and to better understand differences in their cellular interaction and consequent effects on cellular homeostasis, we performed confocal microscopy based analysis of cells expressing Lamp1-lysosomal marker and EEA1-early endosomal marker. The positively charged NH_2-QDs were mainly localized in lysosomal compartments while the COOH-QDs were mostly found to reside in the early endosomes. These results are in good agreement with another study, in which positively charged FeO_x NPs with moderate colloidal stability localized only with lysosome, whereas negatively charged FeO_x NPs with good colloidal stability first were found in endosomes and the later also in lysosomes [33]. These experiments, however, suffer from the high number of endosomes and lysosomes, which requires high lateral resolution in imaging to delineate all the different cellular organelles. Additionally, all these organelles are dynamic and are in constant movement, where in case of not sufficient lateral resolution some colocalization observed might be accidental due to the close proximity of one passing QD (agglomerate) and cellular organelles. Therefore, in order to overcome these limitations these tests were followed with more precise kinetic studies, which involved live tracking of QDs with a dual colour fluorescence single particle tracking (fSPT) system.

The dynamic, trajectory-based colocalization of the QDs with the stained endosomes or lysosomes was analysed using motion trajectories acquired using the fSPT system via the recorded movies of the identified green and red objects. Algorithms in custom built MatLab software were utilised to perform calculations. The dynamic colocalization coefficient, which detected correlated movement between the two objects, was thus the fraction of trajectories of one fluorescence channel that showed correlated movement with trajectories from the second channel. The fSPT system allows for recording of movies of both, the stained organelles and the QDs under investigation, thereby providing time-dependent, live event information regarding the true colocalization of both components [34]. The results obtained from these experiments were in line with our confocal microscopy results explained above. Results were also in agreement with a study in which (polyethylene-coated) gold (Au) NPs, of high colloidal stability, were passed from small vesicles (<150 nm, such as endosomes) to bigger vesicles (>1000 nm, such as lysosomes), whereas agglomerated Au NPs had their peak inside small vesicles at intermediate incubation times (4 h) [35]. In this cited study both Au NPs were negatively charged.

It has been reported that functionalized NPs are prone to exocytosis [36] which is an important parameter to investigate with NPs that are to be used as imaging contrasts, especially that previous studies on Au NPs have demonstrated differences in intracellular NPs due to exocytosis thus highlighting the importance of duration and concentration of NP exposure for their optimal use for cell labelling [37]. In order to investigate this parameter and to better understand the results of the confocal microscopy analysis (Fig. 1) from this work, where depletion in fluorescence was noted between 4 and 24 h time points following exposure of HFF-1 cells to the COOH-QDs, ICP-MS was conducted. For this purpose, cells were exposed to the QDs for 4 and 24 h, after which the incubation media were removed. Cells were then extensively washed and given fresh QD-free media, after which samples were collected at 0, 30, 60, 120, 240, and 360 min time points to evaluate the presence of free Cd^{2+} ions in the extracellular medium. In the following analysis we are assuming that the detected Cd in the extracellular medium originates from exocytosed QDs (note that ICP-MS measures the elemental amount of Cd, regardless of whether it originates from Cd-based QDs or Cd ions). Also, it is important to note that this assumption excludes QDs sticking to the extracellular membrane, which might not have been removed by the thorough washing steps [38].

The amount of exocytosed QDs should scale with the amount of QDs that have been incorporated by cells before the washing procedure. The higher the exposure concentration of QDs to cells, the higher, thus, the number of exocytosed QDs should be, which was true in these experiments (Fig. 4a). In case one assumes that positively charged agglomerated QDs are internalized to a higher extent than negatively charged well dispersed QDs [25], then the higher number of exocytosed NH_2-QDs as compared to the COOH-QDs could be understood. It is also worthwhile noting the difference in the size of the COOH- and NH_2-QDs where the latter is slightly larger where one would assume that some of the additional Cd detected with these NPs is due to the additional Cd atoms present. However, the extent of released Cd ions cannot be justified with only this parameter which makes us assume that there is an effect of NP trafficking also involved in the observed difference. Time dependence of exocytosis of the NH_2-QDs at low and short exposure condition (2.5 nM, 4 h) follows the trend of average colocalization of these QDs with endosomes (Fig. 4b versus Fig. 3b). As in the colocalization experiments time dependence of exocytosis of NH_2-QDs does not follow a linear pattern.

Cytotoxicity studies

The results of the cytotoxicity studies showed a difference in the toxicity profile of the NH_2-QDs compared to the COOH-QDs where the latter induced more cytotoxicity especially in the form of mitochondrial ROS. Autophagy has been linked to a great variety of NPs [39, 40], and has been associated with different types of QDs

in various studies [41, 42], yet no autophagy was found to be induced in these studies. The lack of a clear induction of autophagy is therefore somewhat surprising, but may be due to the nature of the cell type used in the present study. Generally, nanomaterial-induced autophagy is primarily associated with cancer cell types, where in comparative studies, it has been shown that healthy, noncancerous cell types (such as the ones used in this study) display lower levels of autophagy induction [40, 43]. Similarly, even though ROS has been considered to be a key player in toxicological profile of several types of NPs [44], however, some recent studies have suggested that this view may have been exaggerated, in part due to interactions of the NPs with the most common assays used for ROS detection [45]. In particular for imaging-based experiments, the induction of ROS has not been shown to be clearly predominant with many different types of NPs [46]. Moreover, in some other work we have seen that it is mitochondrial oxidative stress that is associated with the NP induced cellular damage [47].

The lack of any significant effect with the NH_2-QDs suggests that the mitochondrial ROS induction might be due to the internalization process itself where uptake with these QDs appeared to be less than the COOH-QDs and much more NH_2-QDs were found to be exocytosed by the cells compared to the COOH-QDs.

Gene alterations following QD exposure

To support the observations obtained above and to gain more insight into the molecular mechanisms involved in the alterations to the cellular homeostasis and to correlate this to the different trafficking mechanism of the two QDs, the gene expression levels of two key cellular pathways were investigated. The first pathway focuses on genes involved in cellular stress and toxicity, and can be seen as an overview of cellular homeostasis. Different sets of genes were up- or down-regulated following 24 h exposure to NH_2- or COOH-QDs. Cell exposed to NH_2-QD resulted the upregulation of CCL2, IL1A, IL1B, IL6, IL8, and TNFα genes, all of which are involved in the induction of inflammatory responses [48]. Similar effects have been reported following exposure to various NPs. For example, exposure of leukocytes, monocytes, and macrophages isolated from human blood, to polystyrene NPs, resulted in an increase in phagocytosis due to the presence of the NPs [49]. In contrast, exposure to COOH-QDs resulted in a decrease in an array of genes mainly involved in cellular hypoxia. This finding is in line with earlier findings, where the involvement of genes linked to hypoxia have been associated with cellular NP toxicity [50]. The induction of high levels of mitochondrial metabolism, as indicated by the induction of mitochondrial ROS, may result in an artificial hypoxia-like scenario. Although the level of available oxygen is sufficient for basal cellular metabolism, the persistent higher metabolism results in higher energy demands, which may not always be met by the overproducing mitochondria. This *"lack of energy"* therefore will be highly similar to the typical scenario of low oxygen consumption, resulting in alterations to the expression levels of genes typically associated with hypoxia. The occurrence of hypoxia-like processes is interesting, because hypoxia is a main feature of tumor cells resulting in resistance to cancer therapeutic agents [51]. It has been reported that some of the primary adaptive responses to hypoxia include the expression of genes involved in angiogenesis, such as the vascular endothelial growth factor A (VEGFA) gene and the SLC2A1 gene responsible for the metabolic adaption of cells [52]. Previous reports have suggested that the inhibition of these genes could lead to killing of tumor cells or the suppression of resistance to cancer therapeutic agents [51]. This raises the question of whether such NPs could be used for therapeutic applications. Interestingly, no changes were noted in this array from both QDs for the genes involved in oxidative stress which is consistent with our ROS results presented above.

Next to cellular stress and toxicity responses, we also looked into analysis of the cytoskeletal regulator pathway genes. Array results showed that exposure of cells to NH_2-QDs induced an increase in the levels of a few genes that are involved in cell mobility and migration. Some of these genes, such as IQGAP2, which effects cellular morphology by regulating the actin cytoskeleton by interacting with cytoskeletal components, cell adhesion, and cell signaling molecules [53–55]. This gene has been implicated in invasion and metastasis of cancer cells [55]. MYLK2, and WAS genes, which were also significantly upregulated in NH_2-QD treatments, are involved in the trafficking of molecules into the cell [56, 57]. Exposure to the COOH-QDs resulted in an array of upregulated and downregulated genes. The most notable ones were ARAP1, CDC42BPA, and CDC42EP2 which are genes involved in forming cell projections and their downregulation is in line with the high-content imaging studies, where at higher COOH-QD concentrations cells were less spread resulting in a lower cell surface area. The deformation of cellular cytoskeleton networks by various NPs is also in line with various other reports, where, in particular at higher NP concentrations, clear deformations of actin and tubulin cytoskeleton have been observed, which could result in secondary effects like altered cellular mobility and migration capacities [58].

Conclusions

Most NP studies consider physico-chemical properties and their correlation to either kinetics, or toxicity. The present work reveals the importance of understanding how the cell interacts with NPs from a kinetic and mechanistic point of view and then how to interpret these observations to NP properties in an effort to elucidate the differences observed in toxicity and gene alteration results between different NPs. Upon exposing human fibroblasts to two types of QDs, one with COOH moieties, which was well dispersed, and the other with NH_2 moieties, which was agglomerated, the toxicological profile for these QDs was different. The state of agglomeration turned out to be a very relevant physico-chemical parameter describing the difference between both types of QDs. The latter clearly had an effect on the process by which the cells trafficked these NPs thus resulting in different effects on cellular homeostasis. The cellular uptake was studied at different time points, where clear differences were observed. NH_2-QDs were taken up by the cells rather quickly, but soon resulted in a steady-state level, after which no additional uptake was observed, and were eventually transferred to the lysosomal compartment (Fig. 3b). COOH-QDs followed a different pathway, where they were internalized at a high rate which persisted over at least 6 h (Fig. 3c). There was only a minimal transfer of COOH-QDs to the lysosomal compartment. Generally, both QDs but more so with the NH_2-QDs perinuclear localization of the NPs was noted which could be due to the residence of more acidic lysosomes that are performing degradation process in that region [59]. Although no acute cytotoxicity was observed for either of the two QD types under the conditions used, the differences in cellular internalization however resulted in variations in their stress response profile, where high-content imaging and gene expression studies revealed the induction of mitochondrial ROS, cytoskeletal remodelling, and hypoxia-like cellular responses from exposure to the COOH-QDs, which could all be linked to higher energy demands. A hypothetical sketch is shown in Fig. 7.

Together, these data reveal that though the two QDs differed in physico-chemical properties they were internalised by the cells to a similar extent. Differences in their uptake kinetics, however, appear to be accountable for the significant changes discovered in their toxicity and gene expression profiles.

Methods
Cell culture
Human foreskin fibroblasts HFF-1 (ATCC Manassas, VA) cells were cultured in Dulbecco's Modified Eagle's Medium (DMEM) in the presence of 15% foetal bovine

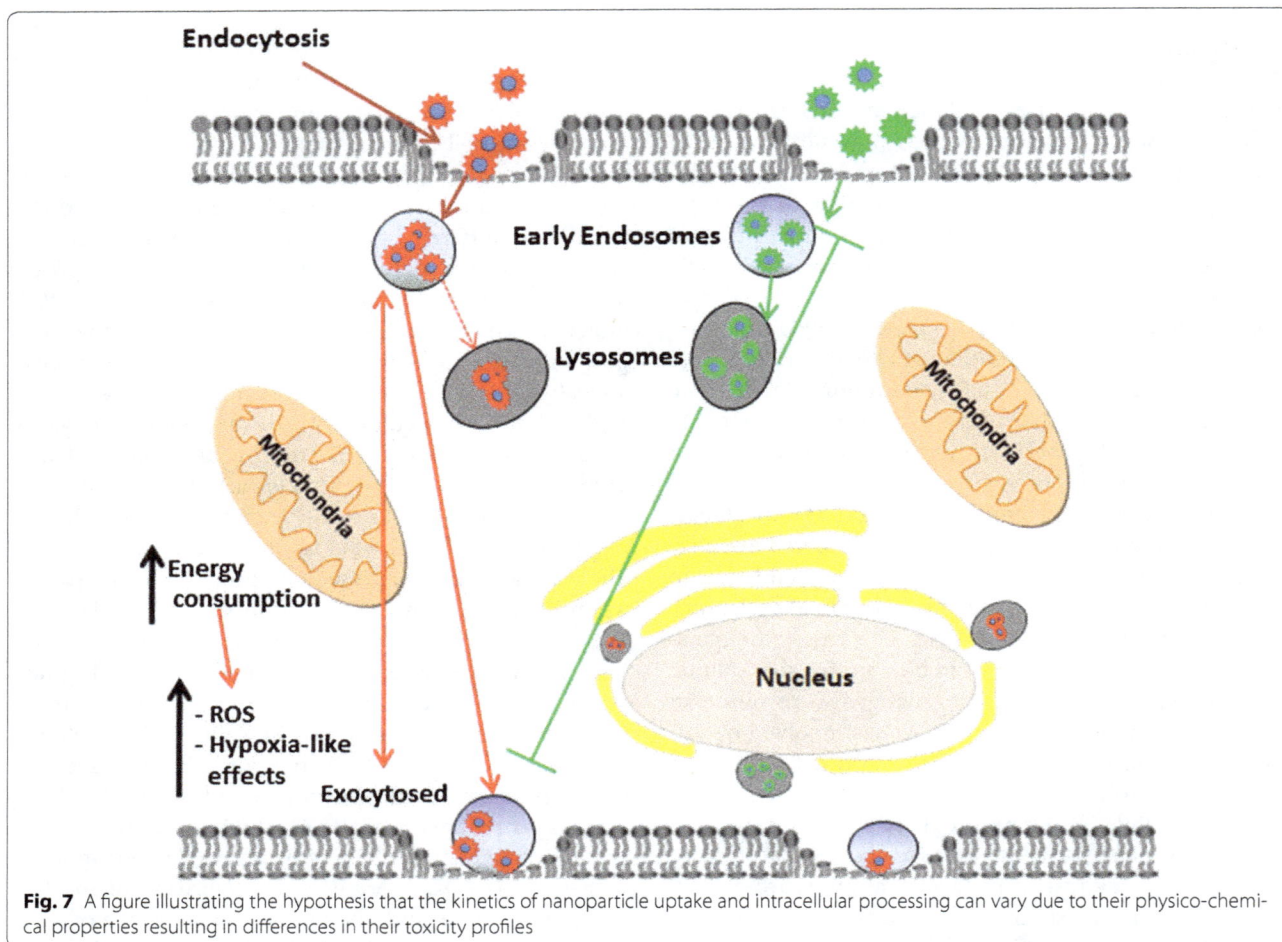

Fig. 7 A figure illustrating the hypothesis that the kinetics of nanoparticle uptake and intracellular processing can vary due to their physico-chemical properties resulting in differences in their toxicity profiles

serum (FBS, Gibco, Life Technologies, Belgium). Cells were incubated at 37 °C and 5% CO_2 and sub-cultured every third day. All cellular treatments were at 0, 2.5, 5, 7.5, 10, and 15 nM concentrations. All experiments were performed in triplicates.

Quantum dot nanoparticles

Both QDs used in this work were commercial products. CdSe/ZnS core/shell fluorescent NPs with NH_2 (Cytodiagnostics, Canada) and COOH (Invitrogen, UK) functional ligands were used. Details about the structure of the semiconductor part as well as the surface chemistry are not disclosed by the providers. These QDs had emission maxima of 664 nm (nominally 665 nm) and 585 nm (nominally 590 nm). These QDs have been previously thoroughly characterised (please see Additional file 1 for details) [18–20]. QD concentrations for exposure experiments were based on the concentrations of the QDs stocks as given by the suppliers. Cellular exposure stocks were prepared by diluting the QDs in sterile phosphate buffered saline (PBS). All concentration suspensions were

vortexed for 30 s prior to addition to the cell culture. Exposure to QDs were for 4 or 24 h.

QD uptake studies

Confocal microscopy and ICP-MS analyses were conducted to examine QD uptake into HFF-1 cells following 4 and 24 h exposure. Details can be found in Additional file 1.

Analysis of photo-stability of the QDs

The effect of the lowered pH levels in the intracellular environment on the photo-stability of the QDs was determined by examining the possible effects of altered pH levels. Experiments were performed as previously described [21]. More details on the methods used can be found in Additional file 1.

Cellular interaction with QDs

The consequence of cell QD interaction in terms of the generation of cytoplasmic and mitochondrial reactive oxygen species (ROS), the level of the lipidated LC3 protein (marker for autophagy), and cytoskeletal changes were investigated using high-content image analysis as detailed previously [14]. A detailed experimental section of these studies can be found in Additional file 1.

QD tracking studies

Single particle tracking (SPT) and confocal microscopy based analyses were conducted to track NH_2- or COOH-QDs in the intracellular environment, and to determine their colocalization with endosomes or lysosomes. Full details of the methodology can be found in Additional file 1.

Inductively coupled plasma mass spectrometry (ICP-MS)

Inductively coupled plasma mass spectrometry (ICP-MS) was conducted in order to determine the number of QDs excreted by the cells. For this end, cells were labelled with QDs at 2.5, 7.5, and 15 nM concentrations for 4 and 24 h. Cells were then washed three times with sterile PBS and supplemented with fresh culture media. Samples were collected from the culture supernatant at 0, 30, 120, 240, and 360 min. The amount of elemental cadmium and selenium in the samples was determined using ICP-MS (see Additional file 1: Section 4 for more details).

Gene expression studies

Two important human gene expression pathways, the human cytoskeletal regulatory and the cellular stress and toxicity pathways, were investigated using real time polymerase chain reaction (RT-PCR) arrays as described previously [14]. Briefly, 1.5×10^5 cells/mL were allowed to settle overnight, followed with incubation with 0, 2.5, 7.5, and 15 nM NH_2- or COOH-QDs for 24 h (see Additional file 1: Section 9 for more details).

Statistical analysis

All data are expressed as the mean ± standard deviation (SD), unless indicated otherwise. All experiments, except the PCR arrays, were analysed using the One Way Anova statistical method. Significance in the PCR arrays was determined based on twofold change from the control $\Delta\Delta Ct$ value.

Authors' contributions
The study was planned by BBM and SJS. Particle characterisation, data processing and the writing of part of Additional file 1 was done by WP and BP. ICP-MS was done by KK with guidance from WP. Single particle tracking and MATLAB data analysis were done by BBM and TFM with guidance and input from KB, SCD, JD, and UH. Toxicity and PCR data analysis and interpretation were done by BBM, SJS, SHD, and GJSJ. The manuscript was written by BBM with input of all co-authors. All authors read and approved the final manuscript.

Author details
[1] Biomedical NMR Unit/MoSAIC, KU Leuven Campus Gasthuisberg, Herestraat 49, 3000 Louvain, Belgium. [2] Institute of Life Science, Swansea University Medical School, Singleton Park, Swansea SA2 8PP, UK. [3] Faculty of Pharmaceutical Sciences, Ghent University, Harelbekestraat 72, 9000 Ghent, Belgium. [4] Center of Nano- and Biophotonics, Ghent University, Harelbekestraat 72, 9000 Ghent, Belgium. [5] Philipps University of Marburg, Renthof 7, 35032 Marburg, Germany. [6] CICBiomagune, San Sebastian, Spain.

Acknowledgements
We would like to thank Prof. Sebastian Munck and Miss Nicky Corthout from the VIB Centre for Biology of Disease, Belgium for the technical guidance in the use of the IN Cell Analyzer 2000 and IN Cell Developer Toolbox software.

Competing interests
The authors declare that they have no competing interests.

Funding
Part of this work was supported by the FWO Vlaanderen (KAN 1514716N to BBM and KAN 1505417N to SJS). Flemish agency for Innovation by Science and Technology (IWT SBO MIRIAD/130065 and NanoCoMIT/140061) and the KU Leuven program financing IMIR (PF 2010/017). Funding of the above listed agencies was applied to design, perform and analyze all cell-based experiments. The UGent consortium NB Photonics supported the single particle tracking experiments performed and the European commission (Grant FutureNanoNeeds to WJP) supported the nanomaterials characterisation part of this work.

References
1. Ko EY, Lee JI, Jeon JW, Lee IH, Shin YH, Han IK. Size tunability and optical properties of CdSe quantum dots for various growth conditions (vol 62, pg 121, 2013). J Korean Phys Soc. 2013;62:1358–1358.
2. Moghaddam MM, Baghbanzadeh M, Sadeghpour A, Glatter O, Kappe CO. Continuous-flow synthesis of CdSe quantum dots: a size-tunable and scalable approach. Chem Eur J. 2013;19:11629–36.
3. Howarth M, Liu W, Puthenveetil S, Zheng Y, Marshall LF, Schmidt MM, Wittrup KD, Bawendi MG, Ting AY. Monovalent, reduced-size quantum dots for imaging receptors on living cells. Nat Methods. 2008;5:397–9.
4. Itano MS, Neumann AK, Liu P, Zhang F, Gratton E, Parak WJ, Thompson NL, Jacobson K. DC-SIGN and influenza hemagglutinin dynamics in plasma membrane microdomains are markedly different. Biophys J. 2011;100:2662–70.
5. Baba K, Nishida K. Single-molecule tracking in living cells using single quantum dot applications. Theranostics. 2012;2:655–67.
6. Biermann B, Sokoll S, Klueva J, Missler M, Wiegert JS, Sibarita JB, Heine M. Imaging of molecular surface dynamics in brain slices using single-particle tracking. Nat Commun. 2014;5:3024.
7. Clarke S, Pinaud F, Beutel O, You C, Piehler J, Dahan M. Covalent monofunctionalization of peptide-coated quantum dots for single-molecule assays. Nano Lett. 2010;10:2147–54.
8. Green M. Semiconductor quantum dots as biological imaging agents. Angew Chem Int Ed. 2004;43:4129–31.

9. Maity AR, Stepensky D. Efficient subcellular targeting to the cell nucleus of quantum dots densely decorated with a nuclear localization sequence peptide. ACS Appl Mater Interfaces. 2016;8:2001–9.

10. Hsu CY, Chen CW, Yu HP, Lin YF, Lai PS. Bioluminescence resonance energy transfer using luciferase-immobilized quantum dots for self-illuminated photodynamic therapy. Biomaterials. 2013;34:1204–12.

11. Derfus AM, Chan WCW, Bhatia SN. Probing the cytotoxicity of semiconductor quantum dots. Nano Lett. 2004;4:11–8.

12. Kirchner C, Liedl T, Kudera S, Pellegrino T, Munoz Javier A, Gaub HE, Stolzle S, Fertig N, Parak WJ. Cytotoxicity of colloidal CdSe and CdSe/ZnS nanoparticles. Nano Lett. 2005;5:331–8.

13. Yong KT, Law WC, Hu R, Ye L, Liu LW, Swihart MT, Prasad PN. Nanotoxicity assessment of quantum dots: from cellular to primate studies. Chem Soc Rev. 2013;42:1236–50.

14. Manshian BB, Moyano DF, Corthout N, Munck S, Himmelreich U, Rotello VM, Soenen SJ. High-content imaging and gene expression analysis to study cell-nanomaterial interactions: the effect of surface hydrophobicity. Biomaterials. 2014;35:9941–50.

15. Ambrosone A, Roopin M, Pelaz B, Abdelmonem AM, Ackermann LM, Mattera L, Allocca M, Tino A, Klapper M, Parak WJ, et al. Dissecting common and divergent molecular pathways elicited by CdSe/ZnS quantum dots in freshwater and marine sentinel invertebrates. Nanotoxicology. 2017;11:289–303.

16. Del Pino P, Yang F, Pelaz B, Zhang Q, Kantner K, Hartmann R, Martinez de Baroja N, Gallego M, Moller M, Manshian BB, et al. Basic physicochemical properties of polyethylene glycol coated gold nanoparticles that determine their interaction with cells. Angew Chem Int Ed Engl. 2016;55:5483–7.

17. Nel AE, Brinker CJ, Parak WJ, Zink JI, Chan WCW, Pinkerton KE, Xia T, Baer DR, Hersam MC, Weiss PS. Where are we heading in nanotechnology environmental health and safety and materials characterization? ACS Nano. 2015;9:5627–30.

18. Joris F, Manshian BB, Peynshaert K, De Smedt SC, Braeckmans K, Soenen SJ. Assessing nanoparticle toxicity in cell-based assays: influence of cell culture parameters and optimized models for bridging the in vitro-in vivo gap. Chem Soc Rev. 2013;42:8339–59.

19. Tsoi KM, Dai Q, Alman BA, Chan WC. Are quantum dots toxic? Exploring the discrepancy between cell culture and animal studies. Acc Chem Res. 2013;46:662–71.

20. Manshian BB, Soenen SJ, Al-Ali A, Brown A, Hondow N, Wills J, Jenkins GJ, Doak SH. Cell type-dependent changes in CdSe/ZnS quantum dot uptake and toxic endpoints. Toxicol Sci. 2015;144:246–58.

21. Soenen SJ, Demeester J, De Smedt SC, Braeckmans K. The cytotoxic effects of polymer-coated quantum dots and restrictions for live cell applications. Biomaterials. 2012;33:4882–8.

22. Comeau JW, Costantino S, Wiseman PW. A guide to accurate fluorescence microscopy colocalization measurements. Biophys J. 2006;91:4611–22.

23. Zhang LW, Monteiro-Riviere NA. Mechanisms of quantum dot nanoparticle cellular uptake. Toxicol Sci. 2009;110:138–55.

24. Soenen SJ, Rivera-Gil P, Montenegro JM, Parak WJ, De Smedt SC, Braeckmans K. Cellular toxicity of inorganic nanoparticles: common aspects and guidelines for improved nanotoxicity evaluation. Nano Today. 2011;6:446–65.

25. Abdelmonem AM, Pelaz B, Kantner K, Bigall NC, Del Pino P, Parak WJ. Charge and agglomeration dependent in vitro uptake and cytotoxicity of zinc oxide nanoparticles. J Inorg Biochem. 2015;153:334–8.

26. Oh N, Park JH. Surface chemistry of gold nanoparticles mediates their exocytosis in macrophages. ACS Nano. 2014;8:6232–41.

27. Beddoes CM, Case CP, Briscoe WH. Understanding nanoparticle cellular entry: a physicochemical perspective. Adv Colloid Interface Sci. 2015;218C:48–68.

28. DiCiccio JE, Steinberg BE. Lysosomal pH and analysis of the counter ion pathways that support acidification. J Gen Physiol. 2011;137:385–90.

29. Soenen SJ, Parak WJ, Rejman J, Manshian B. (Intra)cellular stability of inorganic nanoparticles: effects on cytotoxicity, particle functionality, and biomedical applications. Chem Rev. 2015;115:2109–35.

30. Soenen SJ, Montenegro JM, Abdelmonem AM, Manshian BB, Doak SH, Parak WJ, De Smedt SC, Braeckmans K. The effect of nanoparticle degradation on amphiphilic polymer-coated quantum dot toxicity: the importance of particle functionality assessment in toxicology [corrected]. Acta Biomater. 2014;10:732–41.

31. Matovic V, Buha A, Bulat Z, Dukic-Cosic D. Cadmium toxicity revisited: focus on oxidative stress induction and interactions with zinc and magnesium. Arh Hig Rada Toksikol. 2011;62:65–76.

32. Zhang W, Sun X, Chen L, Lin KF, Dong QX, Huang CJ, Fu RB, Zhu J. Toxicological effect of joint cadmium selenium quantum dots and copper ion exposure on zebrafish. Environ Toxicol Chem. 2012;31:2117–23.

33. Schweiger C, Hartmann R, Zhang F, Parak WJ, Kissel TH, Rivera Gil P. Quantification of the internalization patterns of superparamagnetic iron oxide nanoparticles with opposite charge. J Nanobiotechnol. 2012;10:28.

34. Vercauteren D, Deschout H, Remaut K, Engbersen JF, Jones AT, Demeester J, De Smedt SC, Braeckmans K. Dynamic colocalization microscopy to characterize intracellular trafficking of nanomedicines. ACS Nano. 2011;5:7874–84.

35. Brandenberger C, Muhlfeld C, Ali Z, Lenz AG, Schmid O, Parak WJ, Gehr P, Rothen-Rutishauser B. Quantitative evaluation of cellular uptake and trafficking of plain and polyethylene glycol-coated gold nanoparticles. Small. 2010;6:1669–78.

36. Jiang X, Rocker C, Hafner M, Brandholt S, Dorlich RM, Nienhaus GU. Endo- and exocytosis of zwitterionic quantum dot nanoparticles by live HeLa cells. ACS Nano. 2010;4:6787–97.

37. Nold P, Hartmann R, Feliu N, Kantner K, Gamal M, Pelaz B, Huhn J, Sun X, Jungebluth P, Del Pino P, et al. Optimizing conditions for labeling of mesenchymal stromal cells (MSCs) with gold nanoparticles: a prerequisite for in vivo tracking of MSCs. J Nanobiotechnol. 2017;15:24.

38. Braun GB, Friman T, Pang HB, Pallaoro A, Hurtado de Mendoza T, Willmore AM, Kotamraju VR, Mann AP, She ZG, Sugahara KN, et al. Etchable plasmonic nanoparticle probes to image and quantify cellular internalization. Nat Mater. 2014;13:904–11.

39. Stern ST, Adiseshaiah PP, Crist RM. Autophagy and lysosomal dysfunction as emerging mechanisms of nanomaterial toxicity. Part Fibre Toxicol. 2012;9:20.

40. Peynshaert K, Manshian BB, Joris F, Braeckmans K, De Smedt SC, Demeester J, Soenen SJ. Exploiting intrinsic nanoparticle toxicity: the pros and cons of nanoparticle-induced autophagy in biomedical research. Chem Rev. 2014;114:7581–609.

41. Seleverstov O, Zabirnyk O, Zscharnack M, Bulavina L, Nowicki M, Heinrich JM, Yezhelyev M, Emmrich F, O'Regan R, Bader A. Quantum dots for human mesenchymal stem cells labeling. A size-dependent autophagy activation. Nano Lett. 2006;6:2826–32.

42. Li XM, Chen N, Su YY, He Y, Yin M, Wei M, Wang LH, Huang W, Fan CH, Huang Q. Autophagy-sensitized cytotoxicity of quantum dots in PC12 cells. Adv Healthc Mater. 2014;3:354–9.

43. Soenen SJ, Demeester J, De Smedt SC, Braeckmans K. Turning a frown upside down: exploiting nanoparticle toxicity for anticancer therapy. Nano Today. 2013;8:121–5.

44. Nel A, Xia T, Madler L, Li N. Toxic potential of materials at the nanolevel. Science. 2006;311:622–7.

45. Hoet PH, Nemery B, Napierska D. Intracellular oxidative stress caused by nanoparticles: what do we measure with the dichlorofluorescein assay? Nano Today. 2013;8:223–7.

46. George S, Xia TA, Rallo R, Zhao Y, Ji ZX, Lin SJ, Wang X, Zhang HY, France B, Schoenfeld D, et al. Use of a high-throughput screening approach coupled with in vivo zebrafish embryo screening to develop hazard ranking for engineered nanomaterials. ACS Nano. 2011;5:1805–17.

47. Siddiqui MA, Alhadlaq HA, Ahmad J, Al-Khedhairy AA, Musarrat J, Ahamed M. Copper oxide nanoparticles induced mitochondria mediated apoptosis in human hepatocarcinoma cells. PLoS ONE. 2013;8:e69534.

48. Wang XM, Hamza M, Wu TX, Dionne RA. Upregulation of IL-6, IL-8 and CCL2 gene expression after acute inflammation: correlation to clinical pain. Pain. 2009;142:275–83.

49. Prietl B, Meindl C, Roblegg E, Pieber TR, Lanzer G, Frohlich E. Nano-sized and micro-sized polystyrene particles affect phagocyte function. Cell Biol Toxicol. 2014;30:1–16.

50. Manshian BB, Pfeiffer C, Pelaz B, Himmelreich U, Parak WJ, Soenen SJ. High-content imaging and gene expression approaches to unravel the effect of surface functionality on cellular interactions of silver nanoparticles. ACS Nano. 2015;9:10431–44.

51. Wilson WR, Hay MP. Targeting hypoxia in cancer therapy. Nat Rev Cancer. 2011;11:393–410.

52. Poon E, Harris AL, Ashcroft M. Targeting the hypoxia-inducible factor (HIF) pathway in cancer. Exp Rev Mol Med. 2009;11:e26.

53. White CD, Khurana H, Gnatenko DV, Li Z, Odze RD, Sacks DB, Schmidt VA. IQGAP1 and IQGAP2 are reciprocally altered in hepatocellular carcinoma. BMC Gastroenterol. 2010;10:125.

54. Yamashiro S, Abe H, Mabuchi I. IQGAP2 is required for the cadherin-mediated cell-to-cell adhesion in *Xenopus laevis* embryos. Dev Biol. 2007;308:485–93.

55. Jin SH, Akiyama Y, Fukamachi H, Yanagihara K, Akashi T, Yuasa Y. IQGAP2 inactivation through aberrant promoter methylation and promotion of invasion in gastric cancer cells. Int J Cancer. 2008;122:1040–6.

56. Daniele T, Di Tullio G, Santoro M, Turacchio G, De Matteis MA. ARAP1 regulates EGF receptor trafficking and signalling. Traffic. 2008;9:2221–35.

57. Guo FJ, Liu Y, Huang JA, Li YH, Zhou GH, Wang D, Li YL, Wang JJ, Xie PL, Li GC. Identification of Rho GTPase activating protein 6 isoform 1 variant as a new molecular marker in human colorectal tumors. Pathol Oncol Res. 2010;16:319–26.

58. Tay CY, Cai P, Setyawati MI, Fang W, Tan LP, Hong CH, Chen X, Leong DT. Nanoparticles strengthen intracellular tension and retard cellular migration. Nano Lett. 2014;14:83–8.

59. Zaarur N, Meriin AB, Bejarano E, Xu X, Gabai VL, Cuervo AM, Sherman MY. Proteasome failure promotes positioning of lysosomes around the aggresome via local block of microtubule-dependent transport. Mol Cell Biol. 2014;34:1336–48.

9

Characteristics and properties of nano-LiCoO$_2$ synthesized by pre-organized single source precursors: Li-ion diffusivity, electrochemistry and biological assessment

Jean-Pierre Brog[1], Aurélien Crochet[2], Joël Seydoux[1], Martin J. D. Clift[3] (iD), Benoît Baichette[1], Sivarajakumar Maharajan[1], Hana Barosova[3], Pierre Brodard[4], Mariana Spodaryk[5] (iD), Andreas Züttel[5] (iD), Barbara Rothen-Rutishauser[3] (iD), Nam Hee Kwon[1*] (iD) and Katharina M. Fromm[1*] (iD)

Abstract

Background: LiCoO$_2$ is one of the most used cathode materials in Li-ion batteries. Its conventional synthesis requires high temperature (>800 °C) and long heating time (>24 h) to obtain the micronscale rhombohedral layered high-temperature phase of LiCoO$_2$ (HT-LCO). Nanoscale HT-LCO is of interest to improve the battery performance as the lithium (Li$^+$) ion pathway is expected to be shorter in nanoparticles as compared to micron sized ones. Since batteries typically get recycled, the exposure to nanoparticles during this process needs to be evaluated.

Results: Several new single source precursors containing lithium (Li$^+$) and cobalt (Co^{2+}) ions, based on alkoxides and aryloxides have been structurally characterized and were thermally transformed into nanoscale HT-LCO at 450 °C within few hours. The size of the nanoparticles depends on the precursor, determining the electrochemical performance. The Li-ion diffusion coefficients of our LiCoO$_2$ nanoparticles improved at least by a factor of 10 compared to commercial one, while showing good reversibility upon charging and discharging. The hazard of occupational exposure to nanoparticles during battery recycling was investigated with an in vitro multicellular lung model.

Conclusions: Our heterobimetallic single source precursors allow to dramatically reduce the production temperature and time for HT-LCO. The obtained nanoparticles of LiCoO$_2$ have faster kinetics for Li$^+$ insertion/extraction compared to microparticles. Overall, nano-sized LiCoO$_2$ particles indicate a lower cytotoxic and (*pro-*)inflammogenic potential in vitro compared to their micron-sized counterparts. However, nanoparticles aggregate in air and behave partially like microparticles.

Keywords: Single source precursors, Nano-LiCoO$_2$, Li$^+$ Diffusion coefficient, Li-ion batteries, Nanoparticle hazard

Background

Lithium cobalt oxide LiCoO$_2$ has been the most commonly used cathode material in rechargeable Li-ion batteries since Goodenough first introduced the reversible reaction of Li-ions in the structure [1]. The structures of Li$_{1-x}$CoO$_2$ have been extensively studied as a function of lithium de-intercalation, leading to several phase transformations from rhombohedral with 0.06 < x < 0.25 [2–5], via monoclinic with x = 0.5 [2, 3], to hexagonal for 0.66 < x < 0.83 [6, 7], and a second hexagonal phase, O1, for 0.88 < x < 1 [6–8].

The layered structure of lithiated LiCoO$_2$ exhibits two crystal structures depending on the temperature during synthesis and the preparation method. LiCoO$_2$ produced at low temperature (~400 °C) (LT-LCO) has a cubic spinel structure with the space group $Fd3\,m$ [9, 10] while the phase synthesized at high temperature (>850 °C,

*Correspondence: namhee.kwon@unifr.ch; katharina.fromm@unifr.ch
[1] Department of Chemistry, University of Fribourg, Chemin du Musée 9, 1700 Fribourg, Switzerland
Full list of author information is available at the end of the article

HT-LCO) has a rhombohedral layered structure [11]. LT-LCO shows a large hysteresis between the intercalation and de-intercalation of lithium ions [12–14], which is due to the mixing of Co^{3+} and Li^+ in the structure, preventing the formation of layered pathways for Li-ion diffusion. The material is therefore calcined at higher temperature to yield HT-LCO, which possesses alternating planes of Co^{3+} and Li^+ cations in the hexagonal ABCABC oxygen packing [15], providing superior electrochemical properties in Li-ion batteries [16].

Industrially, two starting materials, typically Li_2CO_3 and Co_3O_4, are heated in a two-step process to yield first at a temperature of <600 °C for 24 h under O_2 the LT-LCO. A second calcination step at 900 °C for >12 h under O_2 [17] yields the HT-LCO [18–20]. Such a prolonged calcination process at high temperature causes however coarsening of the particles and evaporation of lithium [21]. Various synthetic methods have thus been investigated to avoid the high temperature process, with the aim to obtain the rhombohedral layered structure of HT-LCO, e.g. sol–gel [22–25], hydrothermal [26], or precipitation [16]. However, low temperature syntheses formed mostly the cubic spinel $LT-LiCoO_2$, which is not favorable for Li^+ insertion/extraction. Thus, calcination at high temperature >800 °C was always required in a second step to use the so-produced material in Li-ion battery cathodes [16].

Another access to the layered structure of HT-LCO uses metal–organic single source precursors based on alkoxides or aryloxides, in which the metal ions are already preorganized. Indeed, the synthesis of heterobimetallic alkoxides and/or aryloxides can provide a facile route for obtaining soluble, volatile, and generally monomeric species, that can thus serve as valuable precursors for making metal oxides under rather mild conditions [27–36]. For example, Buzzeo published homoleptic cobalt phenolate compounds of the type $K_2[Co(OAr)_4]$ (OAr = $OC_6F_5^-$ or $3,5-OC_6H_3(CF_3)_2^-$), in which the effect of fluorination of phenoxide on $(K18C6)_2[Co(OAr)_4]$ is highlighted [37]. Boyle et al. published lithium cobalt double aryloxide compounds obtained from $LiN(SiMe_3)_2$, $Co(N(SiMe_3)_2)_2$ in THF and subsequent addition of an aryl alcohol. They obtained nanoparticles of $LiCoO_2$ by thin film formation [38], but did not characterize them electrochemically. Nanoparticles of HT-LCO have the advantage to offer shorter diffusion lengths for the Li-ions as compared to the commercial, micron-sized particles from which only ~50% of Li-ions can be used [26, 35]. On the other hand, since batteries are typically also shredded upon recycling, the use of nanomaterials in batteries might present a certain danger, which requires a risk management for new materials.

In this context, we present here new molecular precursors using simple ligands such as phenoxide and alkoxides with a low amount of carbon atoms that can produce nano-HT-LCO at quite low temperature. We have tested the new materials for their electrochemical properties in cathodes and their Li-ion diffusion coefficients were determined. In order to evaluate possible material hazards, the nanoparticles of HT-LCO were exposed directly at the air–liquid interface (ALI) using a well-established in vitro multicellular lung model [39]. The lung was chosen as an experimental tissue, since it can be considered by far the most important portal of entry for aerosolized nanoparticles into the human body [40–46]. Although various aspects of nanoparticles toxicity have already been described and studied in the recent literature, almost no studies were carried out in the domain of battery cathode nanoparticles.

Methods

Materials and reagents

Cobalt chloride ($CoCl_2$) (dry or hydrated with two H_2O), lithium phenoxide (LiOPh) in tetrahydrofuran (THF), lithium iso-propoxide (LiO^iPr) in THF, ethanol (technical grade and analytical grade), tetramethylethylenediamine (TMEDA), dioxane, dimethoxyethane (DME), pyridine (Py), heptane and micron-sized $HT-LiCoO_2$ were purchased from Sigma-Aldrich (Switzerland). Lithium tert-butoxide (LiO^tBu) in THF, lithium methoxide (LiOMe) in methanol, lithium ethoxide (LiOEt) in THF and THF (dry and over molecular sieves) were purchased from Acros Organics (Belgium). Deionized water was produced in house by double distillation.

Synthesis of bimetallic complexes [47]

All experiments were carried out under an inert argon atmosphere, using Schlenk techniques [48]. All solvents were bought dried and were stored over molecular sieve. The elemental analysis of the compounds turned out to be difficult to obtain due to the instability of most compounds in air, based on the loss of (coordinated) solvent.

The compounds $[Co(OPh)_4Li_2(THF)_4]$ (**1**), $[Co(OPh)_4Li_2(THF)_4]\cdot THF$ (**2**), $[Co(OPh)_4Li_2(THF)_2(H_2O)(THF)_2]_2$ (**3**), $[Co(OPh)_4Li_2(TMEDA)_2]$ (**4**), $[Co(OPh)_4Li_2(dioxane)_2]_n$ (**5**), $[Co(OPh)_4Li_2(DME)_2]$ (**6**), $[Co(OPh)_4Li_2(Py)_4]$ (**7**), $[Co_2(O^tBu)_6Li_4(THF)_2]$ (**8**), $[Co_2(O^tBu)_2(OPh)_4Li_2(THF)_4]$ (**9**), $[Co_2(O^iPr)_6Li_2(THF)_2]$ (**10**), $[Co_2(OEt)_{12}Li_8(THF)_{8-10}]$ (**11**), and $[Co_2(OMe)_6Li_2(THF)_2(MeOH)_2]$ (**12**) were synthesized using $CoCl_2$ as starting material and reacting it with the corresponding lithium aryloxide or alkoxide. In a typical reaction procedure, dried $CoCl_2$ is dissolved in dry THF under heating to reflux. After stirring for 30 min, aliquots of LiOR (R = Ph, ^tBu, ^iPr, Et, Me) are added. The mixture is heated to reflux, stirred during 30 min and then concentrated. Layering the concentrated solution with a non-solvent, respectively solvent exchange lead to single crystalline material for compounds **1–5** and **9**, while powders were obtained for **6–8** and **10–12**. Table 1 resumes the reaction conditions for all compounds, with detailed synthesis protocols and IR-analyses given in the Additional file 1: Text 1.

Calcination to LiCoO$_2$

Among the so obtained precursors, compounds **1, 8–12** were heated up to 450 °C for 1 h and 500 °C for 2 h at an average rate of 18 °C/min under an air flow of 8 l/min in a muffle furnace equipped with an evacuation smoke-stack for combustion gases. The black powder obtained was then cooled to room temperature within 5 min in air. The black/grey powder was next washed by centrifugation three times with water and two times with ethanol in order to remove LiCl. The clean and dry oxide nanopowder was finally annealed using an average ramp of 17 °C/min up to 600 °C for 80 min to remove low temperature oxide phase impurities. These materials were used for the biohazard tests. LiCoO$_2$ prepared with LiOMe and LiO^{t}Bu was calcined further until 700 °C for 30 min to measure charge/discharge capacities at different current densities.

Characterization
Single crystal X-ray structures
Single crystals of compounds **1–5** and **9** were mounted on a loop and all geometric and intensity data were taken from these crystals. Data collection using Mo-K$_{\alpha 1}$ radiation ($\lambda = 0.71073$ Å) was performed at 150 K on a STOE IPDS-II diffractometer equipped with an Oxford Cryosystem open flow cryostat [49]. Absorption correction was partially integrated in the data reduction procedure [50]. The structure was solved by SIR 2004 and refined using full-matrix least-squares on F^2 with the SHELX-97 package [51, 52]. All heavy atoms could be refined anisotropically. Hydrogen atoms were introduced as fixed contributors when a residual electronic density was observed near their expected positions. Diffraction data sets for compounds **1–5** are unfortunately incomplete due to decomposition of the single crystals, resulting in poor data sets and R-values for the compounds. However, the isotropic attribution of heavy atoms is unambiguous.

Crystallographic data (excluding structure factors) for the structures in this paper have been deposited with the Cambridge Crystallographic Data Center, 12 Union Road, Cambridge CB21EZ, UK. Copies of the data can be obtained on quoting the depositing numbers CCDC-1527018 (**1**), 1527022 (**2**), 1527023 (**3**), 1527020 (**4**), 1527019 (**5**), and 1527021 (**9**); E-mail: deposit@ccdc.cam.ac.uk). Important crystal data for these compounds are given in the Additional file 1: Table S1.

Other characterizations
For powder XRD measurements, a Stoe IPDS II theta, equipped with monochromated Mo-K$_{\alpha 1}$ radiation (0.71073 Å) was used in order to avoid X-ray fluorescence of the cobalt but also a Stoe STADIP, equipped with monochromated Cu-K$_{\alpha 1}$ radiation (1.540598 Å) and Mythen detector. TGA was recorded on a Mettler Toledo TGA/SDTA851e in closed aluminium crucibles with a pin hole. Specific surface area was measured on a Micromeritics Gemini V series BET with a pre-treatment

Table 1 The reactants, synthetic conditions and the yields of the compounds 1, 8-12

Compound	Formula	Reactants in synthesis	Yields (%)
1	[Co(OPh)$_4$Li$_2$(THF)$_4$]	CoCl$_2$ (0.1 g, 0.77 mmol) + 4 LiOPh 1 M in THF (3.1 ml, 3.1 mmol)	82
2	[Co(OPh)$_4$Li$_2$(THF)$_4$]·THF	Idem as **1**, but −24 °C under argon	56
3	[Co(OPh)$_4$Li$_2$(THF)$_2$(H$_2$O)(THF)$_2$]$_2$	Idem as **1**, but −24 °C in air	<10
4	[Co(OPh)$_4$Li$_2$(TMEDA)$_2$]	Idem as **1**, recrystallized from TMEDA	69
5	[Co(OPh)$_4$Li$_2$(dioxane)$_2$]$_n$	Idem as **1**, recrystallized from dioxane	95
6	[Co(OPh)$_4$Li$_2$(DME)$_2$]	Idem as **1**, recrystallized from DME	47
7	[Co(OPh)$_4$Li$_2$(Py)$_4$]	Idem as **1**, recrystallized from pyridine	39
8	[Co$_2$(O^{t}Bu)$_6$Li$_4$(THF)$_2$]	CoCl$_2$ (585 mg, 4.5 mmol) + 3 LiO^{t}Bu 1 M in THF 13.5 ml (13.5 mmol)	87
9	[Co$_2$(O^{t}Bu)$_2$(OPh)$_4$Li$_2$(THF)$_4$]	CoCl$_2$ (500 mg, 3.85 mmol) + LiO^{t}Bu (3.9 ml, 3.9 mmol) + LiOPh 1 M in THF (7.7 ml, 7.7 mmol)	85
10	[Co$_2$(O^{i}Pr)$_6$Li$_2$(THF)$_2$]	CoCl$_2$ (500 mg, 3.85 mmol) + 3 LiO^{i}Pr 2 M in THF (5.8 ml, 11.6 mmol)	92
11	[Co$_2$(OEt)$_{12}$Li$_8$ (THF)$_{8-10}$]	CoCl$_2$ (500 mg, 3.85 mmol) + 6 LiOEt 2 M in THF (11.6 ml, 23.2 mmol)	89
12	[Co$_2$(OMe)$_6$Li$_2$(THF)$_2$(MeOH)$_2$]	CoCl$_2$ (500 mg, 3.85 mmol) + 3 LiOMe 2 M in THF (5.3 ml, 11.7 mmol) and MeOH	90

under vacuum at 150 °C for one night. SEM images were recorded on Phenom Desktop SEM and a FEI XL 30 Sirion FEG with Secondary Electron and EDS Energy Dispersive Spectrometer detectors. SEM samples were prepared by spraying them on a carbon tape glued on a SEM holder to reproduce the spraying in the exposure chamber. All images were obtained without sputter coating pretreatment. TEM images were recorded on a FEI/Philips CM-100 Biotwin. Raman spectra were recorded with a confocal micro-Raman spectrometer, HORIBA LabRAM HR800, combined with an optical microscope Olympus BX41, using a red laser at 633 nm for excitation, attenuated with filters in order to avoid thermal degradation of the scotch tape used as sample holder. The Li^+ and $Co^{2+/3+}$ ion concentrations were determined by inductively coupled plasma optical emission spectroscopy (ICP-OES) using a Perkin Elmer Optima 7000DV.

The muffle furnace used for combustion and tempering is equipped with a eurotherm thermal controller (Tony Güller Naber Industrieofenbau, Zurich, Switzerland).

Metal ion release

A metal ion release test was conducted to assess the amount of potential metal ion dissolution from the tested compounds. 100 mg of each of the micro- and nanoparticles were immersed in 10 ml of deionised water at pH 7 and pH 4.5 for 24 h. The concentrations of the metal ions were then determined using ICP measurements (Additional file 1).

Statistical and data analysis

The microparticles of $LiCoO_2$ are represented in black and the nanoparticles in grey bars. Data are the mean ± the standard error of the mean (SEM) and are absolute values. Values were considered significantly different compared to the negative control with $p < 0.05$ using a one way Anova with a post hoc Tukey test (*nanoparticles, #microparticles).

Electrodes and electrochemical tests
Preparation of the electrodes

0.5 g of the nanoscale-$LiCoO_2$ and 10 wt% SFG graphite with respect to $LiCoO_2$ were ball milled in a horizontal set-up (Retch MM 400) for 15 min at a frequency of 30 Hz. The ball milling jar had a volume of 10 ml and contained two stainless steel balls of 10 mm in diameter. The electrode paste was prepared in a glass tube, starting with polyvinylidene fluoride (PVDF) (10 wt% with respect to $LiCoO_2$) and 0.5 ml of N-methyl-2-pyrrolidone (NMP), which were stirred by a mechanical stirrer for 30 min until PVDF was completely dissolved. 2 wt% of ABG graphite with respect to $LiCoO_2$ was then added and the mixture was stirred for 15 min. Then, the ball milled composite

powder (0.6 g) of graphite and $LiCoO_2$ were added to the PVDF/graphite/NMP mix and stirred for a half an hour. The so-obtained paste of PVDF/NMP/graphite/$LiCoO_2$ was spread onto an aluminum foil by the doctor-blade method and dried overnight at 120 °C. The overall weight ratio of the composite made of nano-$LiCoO_2$ (active material), carbon and binder was around 78:12:10.

Cell assembly

All compounds used were dried to avoid HF formation in the electrolyte and were assembled in a glove box under argon (MBraun, Germany) having <0.1 ppm of water and oxygen. Typically, the $LiCoO_2$ electrode was assembled in a coin cell using lithium metal as anode, a few drops of an ethyl carbonate (EC) and diethylene carbonate (DEC) mixture in a 1:1 volume ratio with 1 M $LiPF_6$ and 2 wt% of vinylene carbonate as electrolyte with respect to solvents and $LiPF_6$ as well as a Celgard separator.

Battery tests

A potentiostat, Princeton Applied Research 273A, and an Arbin battery test instrument (version 4.27) were used to examine the electrochemical properties of the carbon-nano-$LiCoO_2$ composite electrodes. Charge and discharge capacities of coin cells were measured by an Arbin 2000 battery test instrument at different current densities of C/20, C/10, C/5, C/2 and 1C. The voltage window was set between 2.6 and 4.4 V vs. Li^+/Li. The current densities between C/20 and 1C were based on the practical capacity of 140 mAh/g.

Li-ion diffusion coefficients were evaluated by cyclic voltammetry at a sweep rate of 1, 0.7, 0.5, 0.2 and 0.1 mV/s between 3.5 and 4.4 V vs. Li^+/Li.

The discharge kinetic of $LiCoO_2$ electrodes was investigated at various current densities between 20C and C/20. The $LiCoO_2$ coin cells were re-charged until 4.4 V vs. Li^+/Li at 20C current density and then rested for 3 min. The electrode was discharged at the same current density of 20C until 2.6 V. This procedure was repeated at various lower current densities until C/20 (so-called deep discharge). By this procedure, the capacity vs. the discharge current can be determined directly. The sum of all capacities, obtained at different discharge currents is the maximum discharge capacity of the battery:

$$C_{max} = I_1 \cdot t_1 + I_2 \cdot t_2 + \cdots + I_n \cdot t_n.$$

The equilibrium potentials of $LiCoO_2$ electrodes were measured with the pulsed cycle method (3 min with applied current, followed by 3 min rest) in the range of potentials between 2.6 and 4.2 V vs. Li^+/Li. The equilibrium charge/discharge current was C/10 (15 mA/g). These procedures were described in detail by Spodaryk et al. [53].

The exchange current densities were calculated from the Tafel plot, i.e. dependence of current vs. overpotential. Currents ($\pm i$), starting from the smallest to the highest, were alternatively applied and the potentials during the current flow were measured. From the overpotential (the difference between measured potential with the applied current and equilibrium potential, i.e. the potential which the electrode reaches during rest time), the exchange current densities were calculated. The detailed method is described by Chartouni et al. [54].

Electrochemical impedance spectroscopy (EIS) data were obtained using a potentiostat/galvanostat PGSTAT302N with FRA module (Metrohm Autolab). Impedance spectra of the Li-ion batteries were measured in the range of working frequencies from 10 mHz to 100 kHz. The range was built using a logarithmic distribution. The voltage modulation amplitude was 10 mV. The EIS spectra were analysed using fitting procedure in NOVA 1.4 software from Metrohm Autolab. The accuracy of the potentials measurements is ± 2 mV, of the current $\pm 2\%$ and of the capacity $\pm 2\%$.

The values of the elements from the equivalent circuit model (Additional file 1: Figure S10) were obtained by the following formulas:

$$Z_{Ri} = R_i$$

where R_i is contact resistance or charge transfer resistance, Ohm,Constant phase element (CPE), which models the behavior of an imperfect capacitor or of a double layer, calculated by:

$$Z_Q = \frac{1}{Y_0(j\omega)^n}$$

where Y_0 is admittance of an ideal capacitance, siemens S; n is an empirical constant, $0 < n < 1$ (n is frequency independent and in the case $n = 1$ formula describes an ideal capacitor, $n = 0$—resistor, $n = 0.5$—Warburg impedance); j is imaginary part of impedance; ω is angular frequency, rad/s, $\omega = 2\pi f$; f is frequency, Hz.

The Warburg impedance is provided by:

$$Z_W = \frac{1}{Y_0\sqrt{j\omega}}$$

Lung cell cultures

All in vitro exposure experiments in this study were conducted with a 3D triple cell co-culture model of the human epithelial tissue barrier cultured at the ALI. This system has previously been described in detail [39, 55]. Briefly, the model consists of a layer of human alveolar type II-like epithelial cells (A549, derived from the American Type Culture Collection) with human monocyte-derived macrophages (MDM) on the apical side (upper chamber) and monocyte-derived dendritic cells (MDDC)

on the basolateral side (lower chamber). A549 epithelial cells were cultured at a density of 0.5×10^6 cells/ml in cell culture medium RPMI 1640 (supplemented), on BD Falcon cell culture inserts (high pore density PET membranes, 4.2 cm^2 growth area, 3.0 μm pore size; *Beckton Dickinson AG, Switzerland*). The cell culture densities of MDM and MDDC were 5×10^4 and 25×10^4 cells/insert, respectively [56].

Human blood monocytes were isolated from different, individual buffy coats received from the Swiss blood donation service (Bern, Switzerland) (i.e. different donor for each exposure), using CD14$^+$ MicroBeads as described previously [57]. Due to this, variations in the background between different sets of cell cultures were expected to occur. Co-cultures were incubated for 24 h under suspension conditions in order to allow cell–cell habituation. Subsequently, cell culture medium was extracted from the apical layer to allow formation of the ALI over a period of 24 h in the incubator prior to particle exposures.

Air–liquid interface cell exposure system

The dry powder insufflator (DP-4, Penn Century, USA) was used to pulverise the LiCoO$_2$ particles. The particle exposure system consisted of a closed chamber ($15 \times 15 \times 35$ cm) coated with aluminium foil and equipped with a quartz crystal microbalance (QCM) for the in situ determination of the amount of material deposited. As the material settles onto the QCM, the frequency of the crystal changes (ΔF). The ΔF value (Hz) calculated from the recorded frequency values before and after deposition of material is converted to deposited mass per area (μg/cm^2) as described in [58].

To avoid electrostatic blocking of the needle, aggregation, asymmetric deposition and low deposition yield, a stainless steel needle without bevel of 2 mm Ø and 7 cm of length was used as pulverization means with a gas expulsion flow of ~120 ml/s of air in two pulse of ~0.5 s for each exposure.

Particles exposures

As described for the aerosolisation of dry volcanic ash particles [59] the pulverisation of the dry powder of nanoparticles produces a radial distribution of the particles at the bottom of the chamber. In order to obtain a regular and reproducible distribution of particles on the cells, the 6-well culture plates were placed in such a way that the inserts holding the triple cell co-cultures and the QCM balance were disposed equidistant from the centre in a cross-like pattern as drawn in the scheme below (Fig. 1).

Two inserts/wells were used for each of the three different concentrations of nanoparticles and microparticles. Experiments were repeated 3–4 times for each of the two

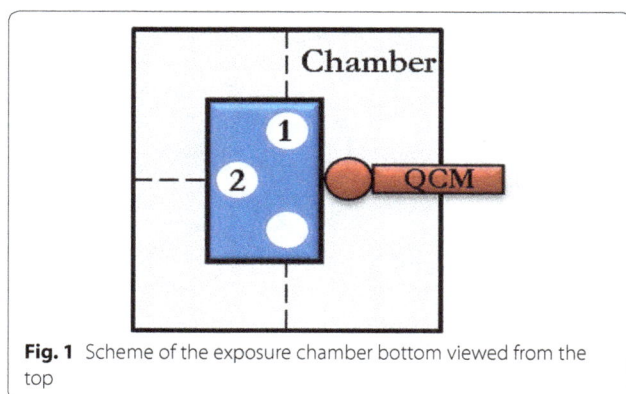

Fig. 1 Scheme of the exposure chamber bottom viewed from the top

$$4\ LiOPh + CoCl_2 \xrightarrow{\ solv\ } [Co(\mu\text{-}OPh)_4Li_2(solv)_x] + 2\ LiCl$$

solv = THF, TMEDA, pyridine, DME or dioxane.

Scheme 1 General reaction scheme for obtaining compounds **1–7**

particle sizes chosen (micronsize commercial particles and homemade nanoparticles). The pulverisation process took place over a period of about 1 month with each week a different blood donor source.

The samples (wells) were incubated overnight at 37 °C and 5% CO_2. The day after incubation, the supernatant was removed and replaced with 2 ml of culture medium.

Biological assays

Cytokine and chemokine quantification

The pro-inflammatory response of the triple co-culture after exposure to $LiCoO_2$ particles was quantified using the amount of the pro-inflammatory mediators which are tumor necrosis factor α (TNF-α) and interleukin-8 (IL-8) using commercial ELISA development kit and the related supplier protocol. The positive control for the pro-inflammatory proteins was treated with lipopolysaccharide 1 µg/ml (LPS) for 24 h.

Optical microscopy/LSM microscopy

After the exposure, cells were fixed and labelled as previously described by Lehmann et al. [56]. In short, samples were stained with a 250 µl mix of a 1:50 dilution of phalloidin-rhodamine for cell cytoskeleton and 1:100 dilution of 4′,6-diamidino-2-phenylindole (DAPI) for the cell nuclei. Coverslips were then mounted onto microscope slides using Glycergel and imaged by LSM.

Results

1-Solid states structures

Compounds **1–7** were obtained by reacting $CoCl_2$ with LiOPh in THF, followed by crystallization in THF under different conditions (temperature, presence of water or not, leading to compounds **1–3**) or by eliminating the THF solvent and replacing it with other mono- or bisdentate ligands, like TMEDA, dioxane, DME, or pyridine (**4–7**). A general reaction scheme (Scheme 1) resumes the family of compounds obtained. We describe here the

single crystal structures of compounds **1–5**, on which we base our structural discussion. For compounds **6** and **7**, the single crystal structures could not be determined as the single crystal quality was poor; yet, the chemical analyses confirm a chemical composition in analogy to the other five compounds.

Among the compounds, different structure types could be identified depending on the solvent present. For compounds **1–7**, the core of the structure is essentially based on one central cobalt ion which is tetrahedrally coordinated by four phenolate entities, bridging pairwise to two lithium ions. The coordination spheres of the lithium cations are completed by coordinating solvent molecules, leading either to molecular entities or a coordination polymer in case of **5**. Figure 2 shows as an example of such a core structure the one of compound **1**. In compound **3**, the terminal ligands of one of the two Li-ions have been formally replaced by two water molecules, which act as bridging ligands between two $[Li_2Co(OPh)_4]$ cores, leading thus to a dimer-type structure. Detailed structure descriptions for **1–5** with distances and angles are given in the Additional file 1: Text 2, while a resume is given in Table 6.

Compounds 8–12

For the compounds **8–12**, the aim was to test ligands other than aryloxides, such as alkoxides, and to also mix aryloxides and alkoxides as ligands. The synthesis used is similar to the one for compound **1** (Scheme 2), but replacing the LiOPh with alkoxides or using a mix of both.

Since the precursor compounds **8, 10, 11** and **12** did not afford single crystals, other methods were used to approach their structure. In possible analogy to compound **8**, the sodium compound $[Na_2Co_2(O^tBu)_6(thf)_2]$ was described in the literature [60]. Since the sodium ions are coordinated by four ligands, similar to the preferred coordination of Li^+, and since Co^{2+} tends to a tetrahedral coordination [61], we propose a similar structure for the lithium compound **8** (Fig. 3). The TGA and NMR measurements confirm that there are two THF molecules per three O^tBu ligands and the ICP measurement gives a ratio of one lithium for one cobalt ion.

The compounds **10–12** were also analyzed by TGA and NMR to determine the amount of ligand and solvent remaining in the solid state structure and the ratio between the ligand and the coordinating solvent

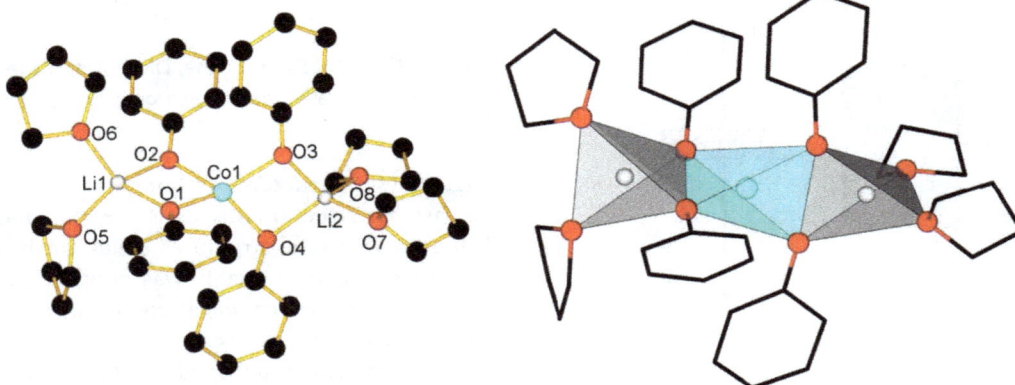

Fig. 2 Labelled view of the molecular structure of **1**, H-atoms are omitted for clarity (*left*); coordination polyhedra in **1** (*right*)

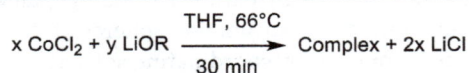

Scheme 2 General reaction equation for the synthesis of compound **8–12**

molecules. ICP measurements and argentometric titrations of chloride (Additional file 1: Table S3) were also performed to evaluate the ratio of lithium per cobalt ions and the amount of LiCl remaining in the material. The results are resumed in Table 2.

From the synthesis, we observed that three equivalents of ligand are required to form carbonate-free $LiCoO_2$ from this precursor **10**. The low amount of impurity of mainly Li_2CO_3 after combustion indicates that there is no excess of unreacted lithium precursor. We also found one Li^+ for one Co^{2+} ion in the complex as well as two THF molecules. From this data we propose that the O^iPr-compound possesses a structure similar to the O^tBu-precursor **8** (Fig. 4). Using the same method for the compound **12** and based on the findings shown in Table 2, we can propose a similar structure as for **8** (Fig. 4). The extra methanol molecules are difficult to assess since both methanol and THF have almost the same boiling point. Finally, NMR measurements are not

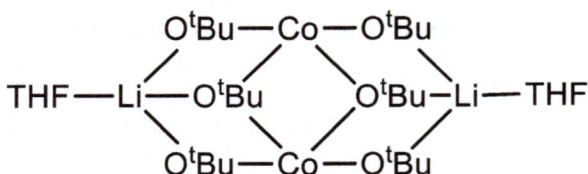

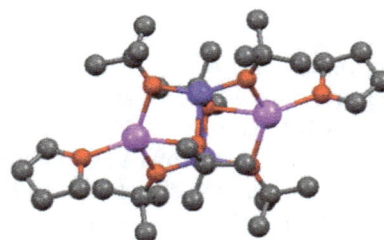

Fig. 3 Proposed structure for **8** (*left*) based on the [Na₂Co₂(O^{t}Bu)₆(thf)₂] compound (*right*, *dark blue* Co, *violet* Na, *red* O, *grey* C; H-atoms omitted described in [60]

Table 2 Combined results from TGA, NMR, ICP and argentometric titration for compounds 8–12

Compound no—reagent	Ligand eq. vs. Co eq.	Solvent molecules per complex	Free lithium (eq.)	LiCl (eq.)
8—LiO^{t}Bu	3	4 (residual THF)	1 Li per Co	2 Li per Co
9—LiO^{t}Bu + LiOPh	1 + 2 (3)	4	1 Li per Co	2 Li per Co
10—LiO^{i}Pr	3	2 THF	1 Li per Co	2 Li per Co
11—LiOEt	6	4–5 THF	4 Li per Co	2 Li per Co
12—LiOMe	3	2 THF/2 MeOH	1 Li per Co	2 Li per Co

helpful since the broadening of the signals (due to the paramagnetic influence of the cobalt ion) hides most of the possible peak shifts.

The compound **11** is the only one which does not follow this rule of three ligands per Co^{2+} and requires six ligands per Co^{2+} to form the desired oxide without impurities of Co_3O_4. An open double heterocubane structure is proposed, as it combines the minimum amount of ligands, the amount of free lithium for coordination, the amount of THF and the preferred coordination of lithium ions (4) and cobalt ions (4,6) as determined by TGA, NMR, ICP and argentometric titration (Fig. 4).

Compound **9** is an interesting mixed ligand compound as it forms molecules of $[(thf)_2Li(\mu\text{-}OPh)_2Co(\mu\text{-}O^tBu)]_2$ where the two O^tBu groups act as bridging ligands between two Co^{2+} ions. The OPh ligands bridge pairwise

between the cobalt and lithium ions, while two THF molecules complete the coordination of the lithium ions (Fig. 5). A detailed description with distances and angles is given in the Additional file 1: Table S1 and Text 2. The bond valence sums are >2 for both cobalt ions and >1 for both lithium ions, indicating sufficient good coordination of the metal ions by their ligands, as it is also the case for compounds **1–5** (Table 6).

Thermal decomposition to LiCoO₂

Among all compounds, **2** and **3** are difficult to handle as they lose their solvent molecules very quickly. The compounds **4–7** are not well suited for the formation of oxide at low temperature because of their relatively high boiling point, high carbon content and molecular weight. The following investigations for the formation of $LiCoO_2$

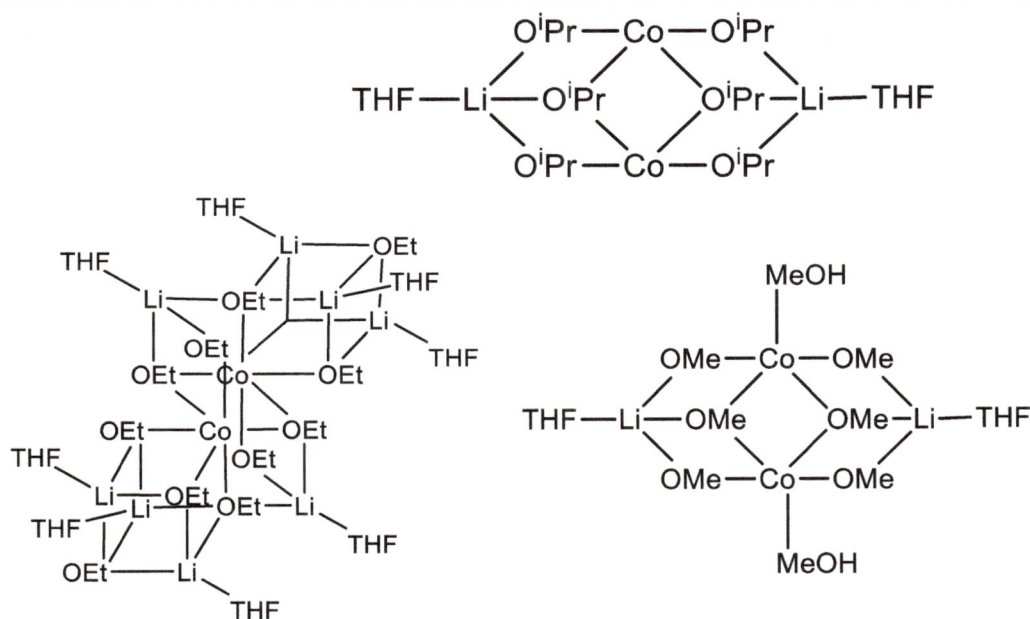

Fig. 4 Proposed structure of compound **10** (*top*), **11** (*left bottom*) and **12** (*right bottom*)

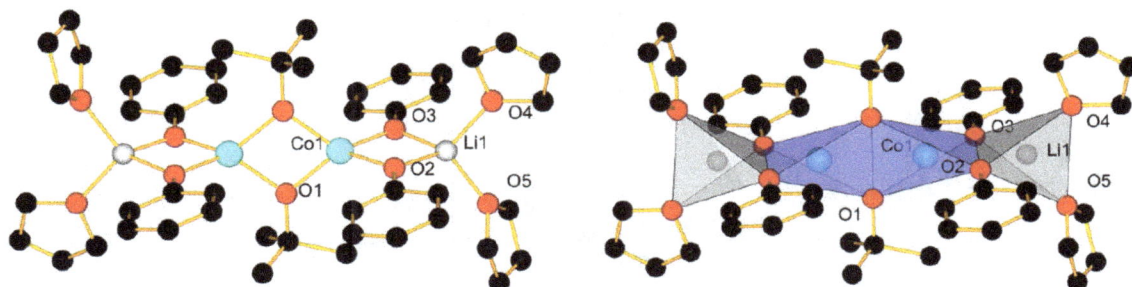

Fig. 5 Molecular view of compound **9** measured by XRD. H-atoms are omitted for clarity

were thus limited to compounds containing THF and the less carbon containing compounds, hence **1** and **8** to **12**.

In order to use these compounds as precursors for the manufacturing of LiCoO$_2$, TGA measurements under oxygen atmosphere with open crucible were performed on the chosen compounds (Fig. 6). The general decomposition process of these complexes begins with the loss of the coordinated and residual non-coordinated solvent molecules before 120 °C (THF B.P. 66 °C, MeOH 65 °C). At higher temperature, between ca. 100 and 400 °C depending on the precursor, the combustion process occurs: it consists of an oxidation of the Co^{2+} to Co^{3+} and of the ligand carbon backbone combustion. Above the temperature of 450 °C, the masses remain quasi constant (Fig. 6). The completed combustion temperature and the detail thermal measurement information are described in Additional file 1: Tables S4 and S5.

Based on the minimum temperature of decomposition of the complexes determined by TGA, combustion tests were performed at different temperatures. Heating to the minimal temperature of decomposition of the precursors of 300 °C for 1 h lead to the formation of the HT-LCO phase with some byproducts (Li$_2$CO$_3$) (Fig. 7a). Since Li$_2$CO$_3$ is highly soluble in water, it was removed after rinsing. We believe that the formation of HT-LCO at such a low temperature is possible due to the preorganization of metal ions within the heterobimetallic single source precursors. We decided nevertheless to increase the decomposition temperature by 50–100 °C compared to the decomposition temperature of the compounds in order to reduce the amount of byproducts, and for comparison purposes, the temperature was set to 450 °C for 1 h for all compounds.

After indexation of the powder diffractograms obtained after combustion at 450 °C, all of the tested precursors (**1**, **8–12**) afforded LiCoO$_2$ with low amounts of impurities that could not be detected by powder X-ray analysis after washing with water, hence less than 5% (Fig. 7). Heating to the minimal temperature of decomposition of the precursors of 300 °C for 1 h leads to the

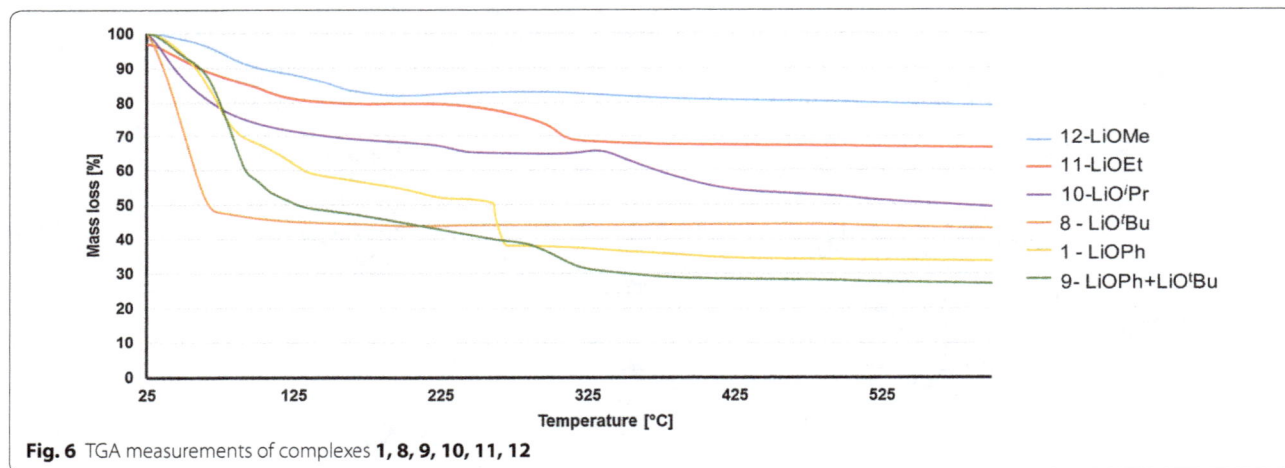

Fig. 6 TGA measurements of complexes **1, 8, 9, 10, 11, 12**

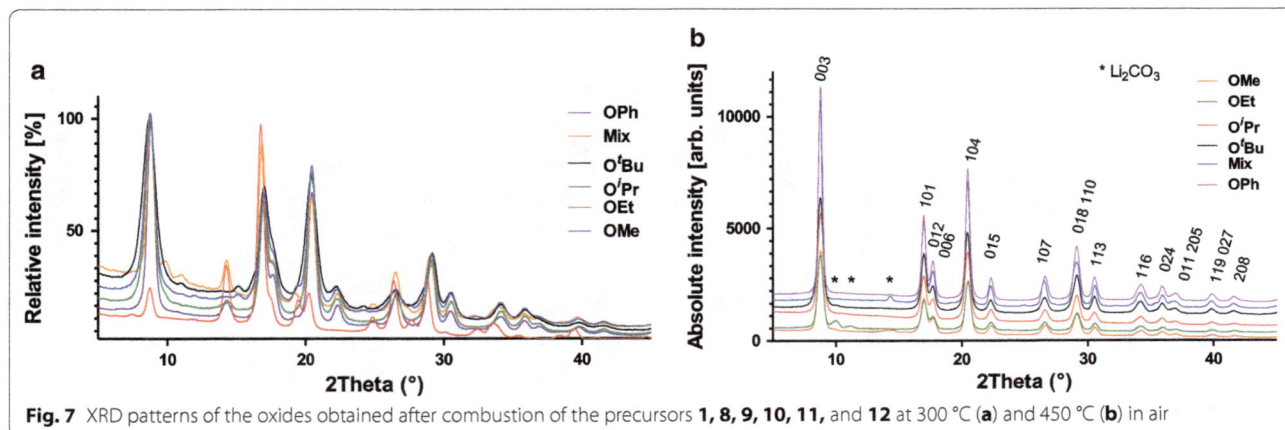

Fig. 7 XRD patterns of the oxides obtained after combustion of the precursors **1, 8, 9, 10, 11,** and **12** at 300 °C (**a**) and 450 °C (**b**) in air

formation of the HT-LCO phase with some byproducts (among which Li_2CO_3). A Rietveld refinement of the different diffractograms, taken on a Mo source, was performed to determine the exact phase of the oxide. The lattice cell parameters from the different precursors correspond to a slightly distorted HT-LCO, with the space group R-3 m. This small distortion of the unit cells arises from the fact that this material is composed of nanocrystallites which possess a more strain than standard micrometric crystallites. The c/a ratio gives also an indication on the general cation ordering of the oxide phase. If the c/a ratio is 4.899 or lower, it means that it is a cation-disordered rock salt structure, also called the LT-LCO with a spinel structure ($Fd3$ m). Since this ratio c/a is higher than this value in all cases, it indicates that the high temperature phase has been obtained for all precursors (Table 3).

Another method to identify LT and HT phases of $LiCoO_2$ is to verify the peaks at 2 theta = 65–67° ($\lambda = Cu-K_{\alpha 1}$). The HT-LCO has two split peaks of the (108) and (110) planes while the LT-LCO has one single peak of the (440) plane at 65° [13, 62]. As shown in Fig. 8 below, all the materials prepared with O^tBu, O^iPr, OMe and OPh show two split peaks corresponding to the HT-LCO phase.

After thermal treatment at 450 °C, the morphologies of the materials prepared with different precursors were analyzed using SEM (Fig. 9). All the materials show polyhedral shapes but the materials obtained from LiO^iPr and LiOPh precursors formed rhombohedral and triangle shapes.

Since the detection limit in powder X-ray diffraction is 3–5%, Raman spectroscopy was used to complete the analysis. The HT-LCO possesses only two Raman active modes: A_{1g} (Co–O stretching) υ_1 at 595 cm^{-1} and E_g (O–Co–O bending) υ_2 at 485 cm^{-1}, while LT-LCO has four Raman active modes (A_{1g}, E_g, 2 F_{2g}) which are respectively at υ = 590, 484, 605 and 449 cm^{-1} and are due to the mixing of cations in the structure [63].

The Raman spectrum of our non-annealed nano-LCO obtained from compound 8 shows a contamination of the

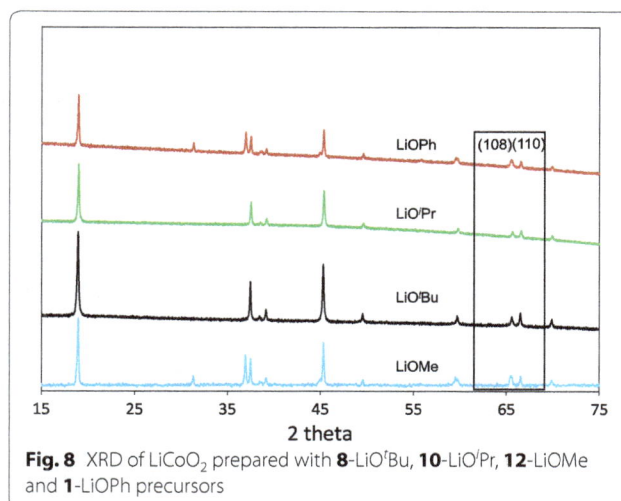

Fig. 8 XRD of $LiCoO_2$ prepared with **8**-LiO^{t}Bu, **10**-LiO^{i}Pr, **12**-LiOMe and **1**-LiOPh precursors

HT-LCO with the LT phase which can be easily removed by annealing at 600 °C for 1 h. No significant improvement can be observed for a 700 °C annealing (Fig. 10). In order to avoid particle growth due to coalescence and ripening, the duration and temperature of annealing have to be minimized, hence we used the 600 °C annealed nanoparticles for the biological assays described later.

ICP measurements on the nano-LCO obtained from **8** and on commercial micron-sized LCO were carried out and the ratio between Li$^+$ and Co^{3+} ions was calculated: we found 0.96 ± 0.02 Li$^+$ ions per Co^{3+} ion for the nano-LCO (Additional file 1: Table S4). Thus the stoichiometry is a little bit lower than the optimal 1:1 stoichiometry ratio. This can be explained at least partly by the washing steps during which part of the Li$^+$ can be washed away, the mechanical stress induced by ultrasounds and the shear stress of the centrifuge and the annealing in which the Li$^+$ and Co^{3+} ions can diffuse out of the oxide into the crucible. The ICP measurements of the micro-LCO give a Li$^+$ content of 1.01 ± 0.02 which is the optimal ratio for the HT-LCO.

Morphologies and determination of the particle and crystallite sizes

The crystallite and particle sizes were assessed via the Scherrer equation (X-ray) and the BET equation (gas adsorption), respectively. The details are described in the Additional file 1: Equation S1 – S5.

Table 4 gives the summary of specific surface area, different sizes of particles and crystallites obtained under identical combustion conditions (temperature, time, speed of heating/cooling and atmosphere composition) depending on the starting complexes.

Table 3 Cell parameters of the $LiCoO_2$ formed using different precursors and HT-$LiCoO_2$ Ref. [61]

Compound	a	c	c/a	Volume (Å^3)
HT-$LiCoO_2$ [61]	2.8156(6)	14.0542(6)	4.99	96.49(4)
1 (LiOPh)	2.8193(2)	13.930(3)	4.94	95.88 (3)
8 (LiO^{t}Bu)	2.8179(3)	13.949(3)	4.95	95.93(4)
9 (LiOPh + LiO^{t}Bu)	2.8139(3)	13.970(4)	4.96	95.79(4)
10 (LiO^{i}Pr)	2.8199(1)	13.936(2)	4.94	95.97(2)
11 (LiOEt)	2.8144(2)	13.942(2)	4.95	95.64(2)
12 (LiOMe)	2.8199(2)	13.956(3)	4.95	96.11(3)

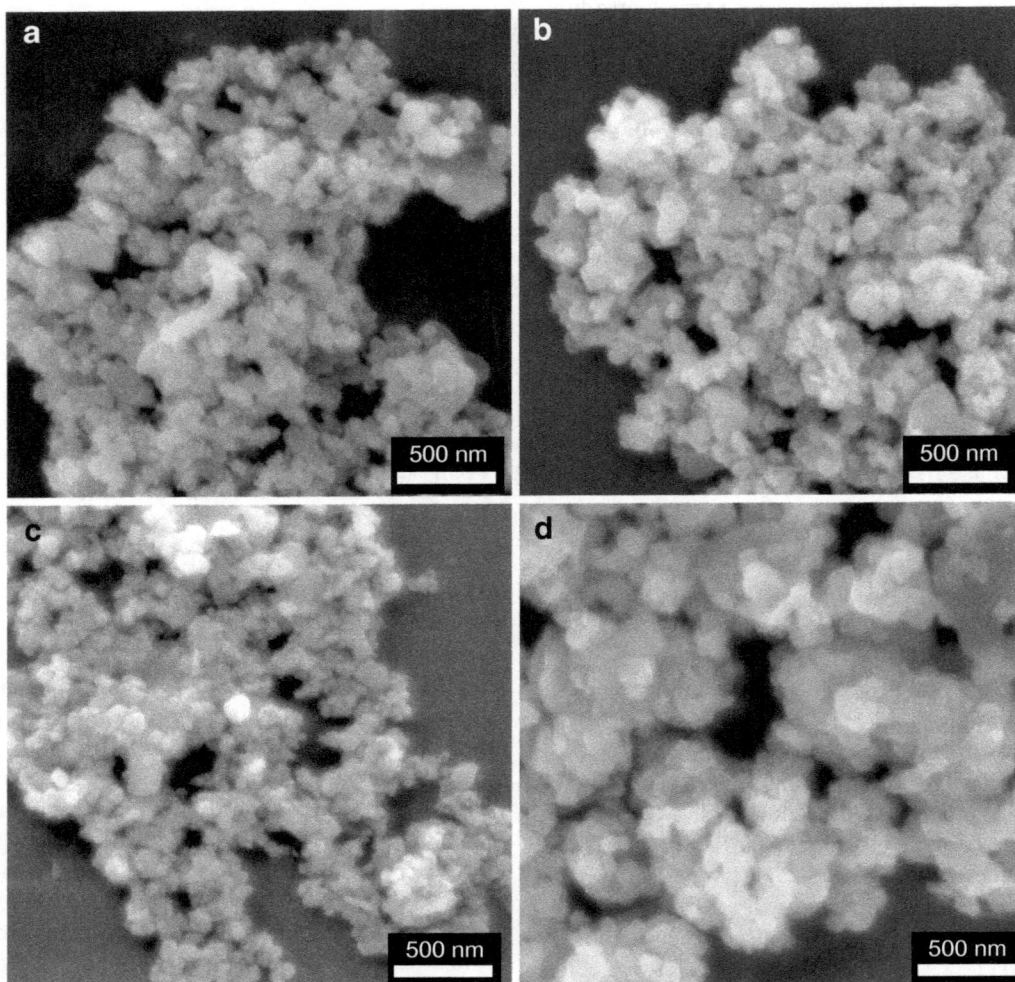

Fig. 9 SEM images of LiCoO$_2$ prepared with **8**-LiO^{t}Bu (**a**), **10**-LiO^{i}Pr (**b**), **12**-LiOMe (**c**), and **1**-LiOPh (**d**) at 450 °C for 1 h

The morphologies of the particles were investigated by SEM images (Fig. 11). The shapes of the particles obtained from the different precursors are similar and submicron. It is also noted that the material always tends to form large aggregates due to its high surface area.

Electrochemistry and Li-ion diffusion

Finally, in order to learn if the size of particles has a direct influence on the Li-ion diffusion, cyclic voltammetry of LiCoO$_2$ electrodes was performed on two different particles sizes: 40 and 15 nm coming from the precursors **8** and **12**, respectively after a prolonged ball milling of 1 h instead of 15 min.

Figure 12a shows the cyclic voltammograms of LiCoO$_2$ electrode prepared with LiO^{t}Bu precursor at different scan rates between 0.1 and 1 mV/s.

When Li$^+$ is extracted from LiCoO$_2$, Co^{3+} in LiCoO$_2$ is oxidized and electron is released (LiCo^{3+}O$_2$→Li$_{1-x}$Co$^{4+/3+}$O$_2$ + xe$^-$ + xLi$^+$). On the other hand, oxidized Li$_{1-x}$CoO$_2$ is reduced and electron is uptaken when Li$^+$ is re-inserted into Li$_{1-x}$CoO$_2$ (Li$_{1-x}$Co$^{4+/3+}$O$_2$ + xe$^-$ + xLi$^+$→LiCo^{3+}O$_2$). Therefore, the current increased where the redox reactions of Co^{3+}/Co^{4+} occurred above 3.9 V for anodic peaks and between 3.6 and 3.9 V vs. Li$^+$/Li for cathodic peaks. The CVs and the maximum current peaks of the compound **12** are shown in Additional file 1: Figure S9.

The Li-ion diffusion coefficient can be determined from these cyclic voltrammograms by using the Randle–Sevcik equation. The Randles–Sevcik equation [63]:

$$Ip = \left(2.69 \times 10^5\right) n^{3/2} A\, D_{Li}^{1/2} C\, v^{1/2} \tag{1}$$

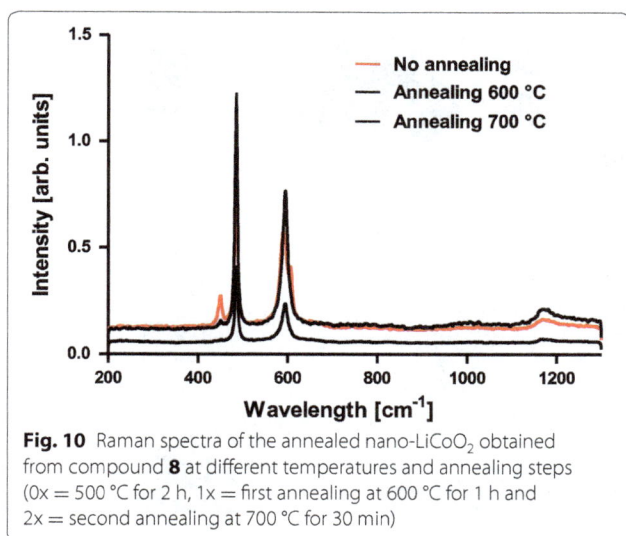

Fig. 10 Raman spectra of the annealed nano-LiCoO$_2$ obtained from compound **8** at different temperatures and annealing steps (0x = 500 °C for 2 h, 1x = first annealing at 600 °C for 1 h and 2x = second annealing at 700 °C for 30 min)

with Ip the peak current; n the number of transfer electrons; A the surface area of the electrode; C the concentration of reactants; and v the scan rate.

The plot of the square root of the scan rate vs. the anodic or cathodic peaks gives the slopes which represent the square root of the Li$^+$ ion diffusion coefficient value, D_{Li+} (Fig. 12b).

The Li$^+$ ion diffusion coefficients (D_{Li+}) of our nanoparticles were 2.3×10^{-5} and 4.5×10^{-6} cm^2 s^{-1} for **8**-LiO^{t}Bu and **12**-LiOMe, respectively while the one of commercial HT-LCO was 2×10^{-7} cm^2 s^{-1} (Table 5). The values obtained from nanoparticles are 20–100 higher than the standard value for HT-LCO [64]. Thus the kinetics with Li$^+$ ions are much faster in nanoscale LCO than

in micron-LCO. When we compare the values of diffusion coefficients of 15 and 40 nm of nano-LCO, the larger particle size of 40 nm has even higher diffusion coefficient. It will be explained in the discussion part later.

Electrochemical properties

After D_{Li+} was determined, the battery properties of our nanoscale LCO materials were investigated. The charge/discharge current is expressed as a C-rate to evaluate battery capacities at various current values. A C-rate is a measure of the rate at which a battery is discharged relative to its maximum capacity. The current density and C-rate are determined by the nominal specific capacity of 150 mAh/g. For example, the current densities are 150 and 7.5 mA/g at 1C (a battery is charged in 1 h) and C/20 (a battery is charged in 20 h), respectively. Figure 13 shows the discharge capacities of LiCoO$_2$ electrodes prepared by the precursors 1-LiOPh, 8-LiO^{t}Bu, 10-LiO^{i}Pr and 12-LiOMe. Depending on the precursor used in the synthesis, the specific capacity varies. 10-LiO^{i}Pr and 1-LiOPh derived LiCoO$_2$ electrodes obtained superior capacities to the ones obtained with 8-LiO^{t}Bu precursors. The mean specific capacity of LiCoO$_2$ derived from 1-LiOPh was 210 mAh/g at C/20, which is 77% of the theoretical capacity of 272 mAh/g, while LiCoO$_2$ from the LiO^{t}Bu precursor had 124 mAh/g (46% of the theoretical value) at the same rate.

After cycling of charge/discharge at different current densities, we disassembled the batteries for all four samples and rinsed the LiCoO$_2$ electrodes to verify their structures. XRD in Fig. 14 shows that all the cycled LiCoO$_2$ electrodes have two peaks at (108) and (110) corresponding to the HT-LCO phase, hence the structure is unchanged after cycling.

Table 4 The specific surface area, mean particle size and crystallite size of LiCoO$_2$ prepared with different precursors

SSA (m^2/g)

Annealed	1-LiOPh	8-LiO^{t}Bu	9-(LiOPh + LiO^{t}Bu)	10-LiO^{i}Pr	12-LiOMe	11-LiOEt
500 °C	9.46 (025)	16.50 (0.2)	9.62 (0.2)	11.50 (0.3)	19.70 (0.12)	not measured
600 °C	2.59 (0.015)	12.50 (0.14)	0.95 (0.03)	3.65 (0.05)	8.00 (0.07)	0.95 (0.01)
700 °C	0.50 (0.02)	6.10 (0.17)	0.78 (0.02)	3.04 (0.05)	5.50 (0.05)	0.34 (0.02)

Particle size (P)* and crystal size(C)** (nm)

	P(1)	C(1)	P(8)	C(8)	P(9)	C(9)	P(10)	C(10)	P(12)	C(12)	P(11)	C(11)
500 °C	126 (2)	50 (2)	72 (1)	60 (2)	124 (2)	75 (2)	103 (2)	40 (1)	60 (1)	50 (4)	Not measured	Not measured
600 °C	459 (2)	45 (2)	95 (1), 40***	45 (3)	1251 (26)	150 (3)	326 (3)	75 (1)	149 (1), 15***	45 (1)	1251 (9)	110 (3)
700 °C	2376 (60)	90 (2)	195 (4)	55 (2)	1529 (26)	185 (1)	391 (4)	295 (1)	216 (1)	170 (2)	3494 (130)	285 (1)

* The mean particle size was determined by the equation of d = K/(ρ×S$_{BET}$), where K is the shape factor, ρ is the density of the material (5.05 g/cc). and S$_{BET}$ is the specific surface area of the material

** Crystal size was determined using Scherrer equation d = Kλ/(B cosθ), where d is the mean crystallite size in volume-weight, λ is the wavelength of the X-rays, B is the width of a peak at a half maximum due to size effects assuming that there is no strain, K is a constant value of 0.89, and θ is the incident angle

*** Particle sizes were obtained after 1 h of ball milling

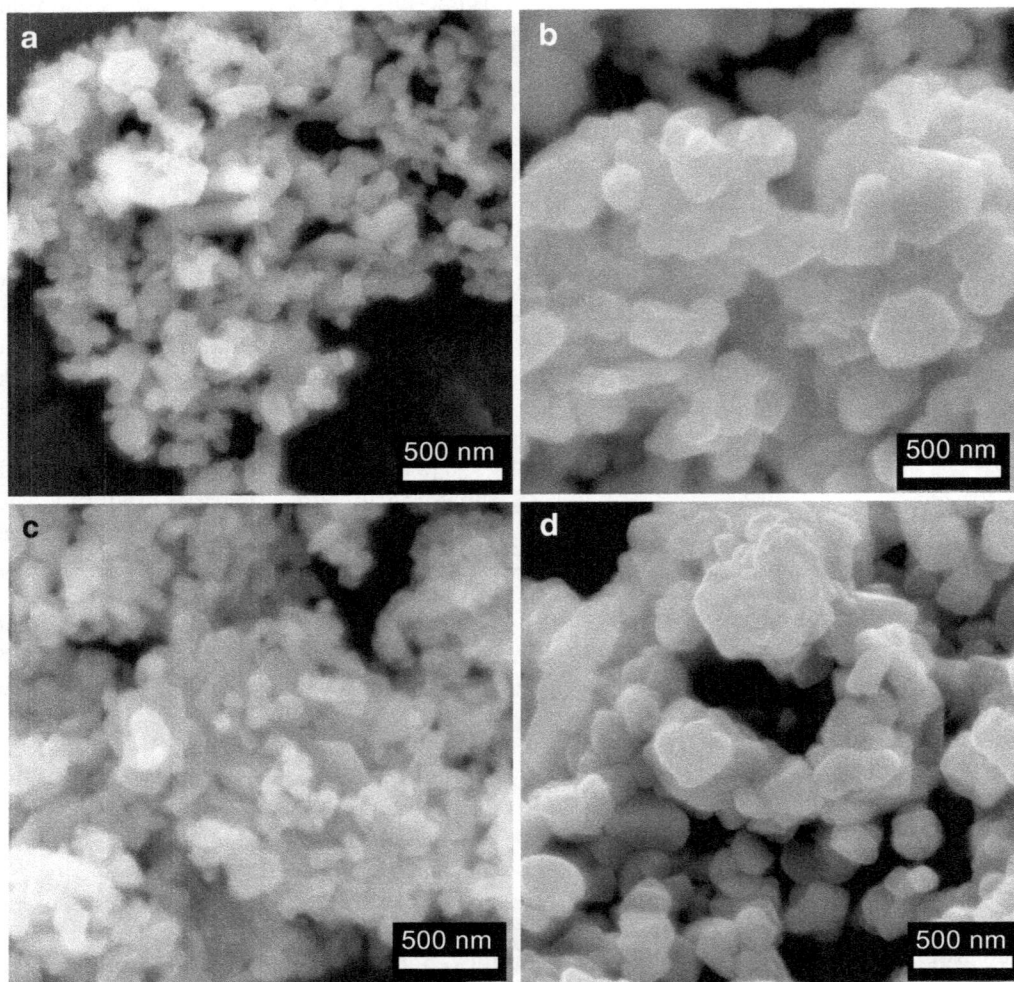

Fig. 11 SEM images of LiCoO₂ materials prepared with LiOᵗBu (**a**), LiOⁱPr (**b**), LiOMe (**c**), LiOPh (**d**) annealed at 600 °C

The equilibrium charge/discharge curves of the $LiCoO_2$ electrodes obtained from LiOPh, LiOᵗBu and LiOMe precursors were investigated as shown in Fig. 15. The markers are measured when the current is not applied to the battery while the dashed lines are recorded when the current is applied. They show the plateau of equilibrium charge curves at 3.9 V and discharge at 3.8 V vs. Li^+/Li. The coulombic efficiency of the $LiCoO_2$ electrodes from LiOPh reached >95% with relatively low polarization between charge and discharge process (Fig. 15a). In case of the $LiCoO_2$ electrode from LiOᵗBu (Fig. 15b), the coulombic efficiency reached also >95% but both charge and discharge processes result in half of the capacities compared to these of the electrodes from LiOPh. Moreover, the potentials during charging with the applied current (dashed lines on the graphs) are higher in Fig. 15b compared to these in Fig. 15a, c.

The deep discharge process was evaluated to estimate how fast the battery can reach the maximum discharge capacity of the different $LiCoO_2$ electrodes. Figure 16 exhibits that the $LiCoO_2$ electrode from LiOPh precursor, (a), can reach 99% of its maximum capacity (120 mAh/g) within 9 min (at 5.2 C) due to the fast kinetic reaction of Li^+ ion insertion/extraction. Of course, this maximum capacity remained at any lower current densities, showing the plateau on the right side in Fig. 16a. On the other hand, the electrode from LiOⁱPr precursor, (b), can be discharged to 90% of its maximum capacity (122 mAh/g) at much lower current density of 0.44 C (about 26 min) than (a). (b) can reach 85% (104 mAh/g) of its maximum discharge capacity within 6 min (at 7 C). Thus, this deep discharge measurement supports that the discharge capacities at higher current densities (>C/2) are lower in $LiCoO_2$ electrode with LiOPh than those in LCO with

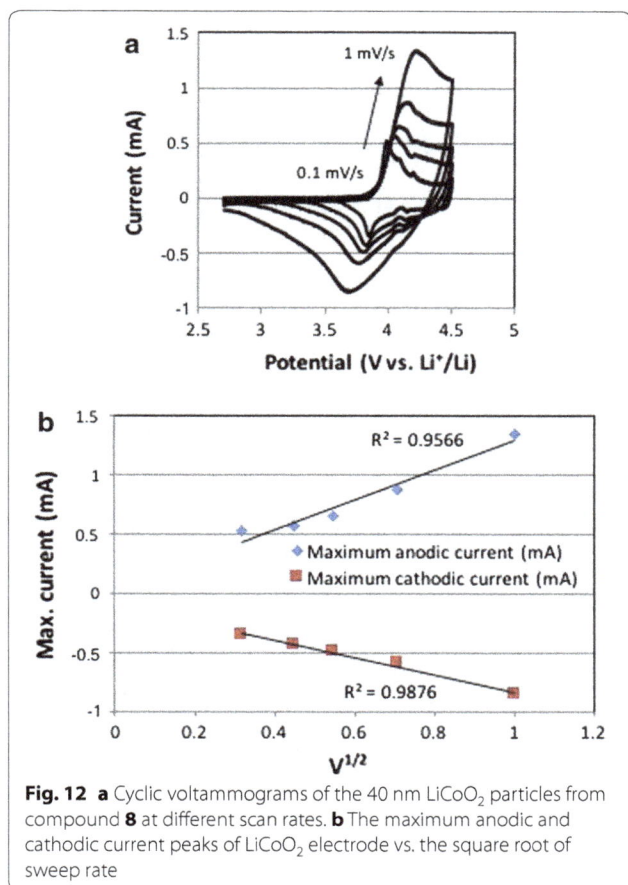

Fig. 12 a Cyclic voltammograms of the 40 nm $LiCoO_2$ particles from compound **8** at different scan rates. **b** The maximum anodic and cathodic current peaks of $LiCoO_2$ electrode vs. the square root of sweep rate

Table 5 Size and Li-ion diffusion coefficient comparison between two precursors, 8 and 12, and HT-LCO Ref. [64]

Compounds/precursors	Size	D_{Li} ($cm^2 s^{-1}$)
HT-LCO [64]	11 µm	2×10^{-7}
8—LiO^{t}Bu	40 nm	2.3×10^{-5}
12—LiOMe	15 nm	4.5×10^{-6}

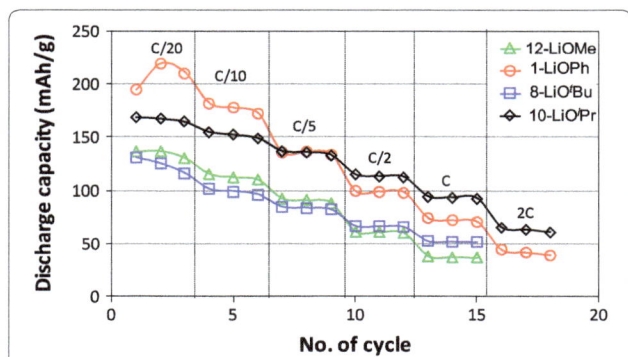

Fig. 13 Discharge capacities of $LiCoO_2$ electrode. $LiCoO_2$ materials were prepared by **8**-LiO^{t}Bu, **10**-LiO^{i}Pr, **12**-LiOMe and **1**-LiOPh

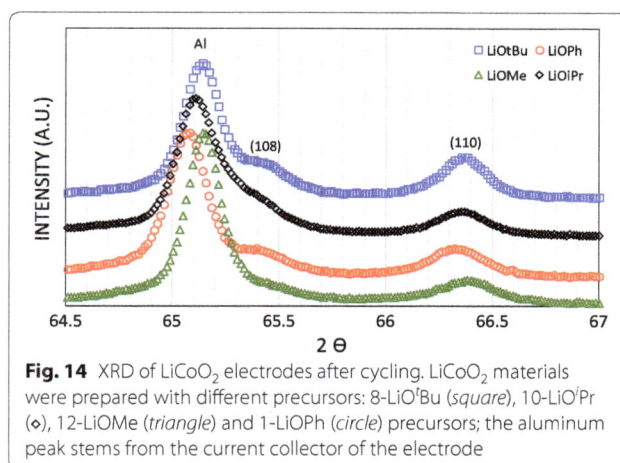

Fig. 14 XRD of $LiCoO_2$ electrodes after cycling. $LiCoO_2$ materials were prepared with different precursors: 8-LiO^{t}Bu (*square*), 10-LiO^{i}Pr (◊), 12-LiOMe (*triangle*) and 1-LiOPh (*circle*) precursors; the aluminum peak stems from the current collector of the electrode

LiO^{i}Pr (Fig. 16). Therefore, the kinetics of the electrode (a) obtained from LiO^{i}Pr is an order of magnitude faster than (b) (obtained from LiOPh) at high current densities.

The Nyquist plots presented for electrodes with different precursors were obtained in the frequency range of 100 kHz–0.01 Hz at 25 °C (Fig. 17). The EIS spectra of the electrodes with LiOPh and LiOMe precursors are similar in shape with one semicircle and Warburg branch, while the electrode obtained from LiO^{t}Bu precursor shows hodographs with two semicircles without Warburg impedance. After fitting the EIS data, the equivalent circuit models were proposed (Additional file 1: Figure S11).

The ion transfer resistance and total impedance of electrodes with different precursors increase in the following sequence: LiOPh < LiOMe < LiO^{t}Bu, which is in good agreement with the discharge capacities and equilibrium charge/discharge curves.

Hazard assessment of particles
Particle aerosolisation
Nanoparticles obtained from precursor **8**, which was annealed at 600 °C for 1 h, were compared to a commercially obtained, micron-sized $LiCoO_2$ sample. A dry powder insufflator was used to aerosolise both materials for direct deposition onto the surface of the multicellular epithelial tissue barrier model. Initially, following aerosolisation, the deposition of the two particle types was characterised in terms of their mass deposition, particle size, as well as their distribution and morphology.

The cell-delivered dose was monitored using an integrated quartz crystal microbalance (QCM) and showed a dose-dependent deposition of the both samples, i.e. 0.81 ± 0.2, 0.55 ± 0.14 and 0.16 ± 0.05 µg for nanoparticles, and 3.92 ± 0.78, 1.46 ± 0.63 and 0.51 ± 0.18 µg for microparticles. It was, however, not possible to achieve the same range of deposited concentrations for both

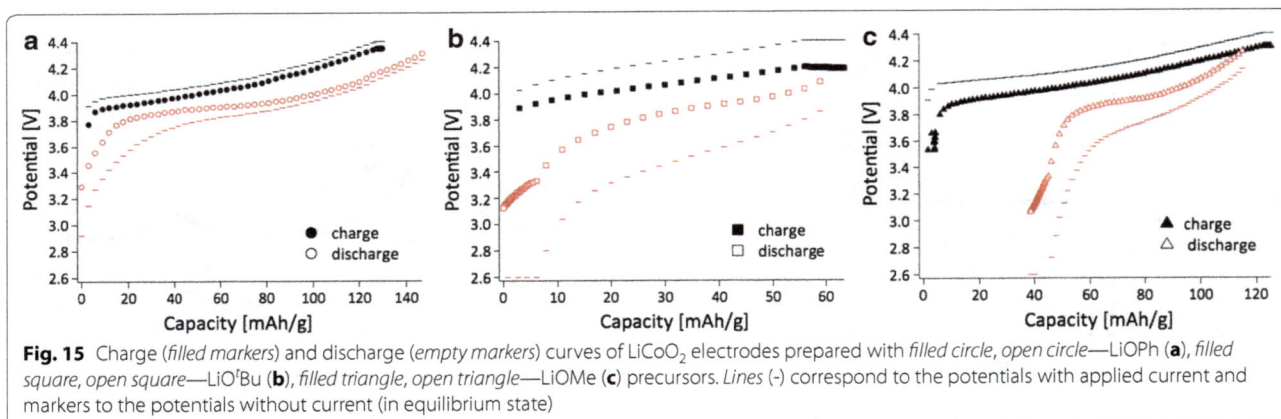

Fig. 15 Charge (*filled markers*) and discharge (*empty markers*) curves of LiCoO$_2$ electrodes prepared with *filled circle, open circle*—LiOPh (**a**), *filled square, open square*—LiO^{t}Bu (**b**), *filled triangle, open triangle*—LiOMe (**c**) precursors. *Lines* (-) correspond to the potentials with applied current and markers to the potentials without current (in equilibrium state)

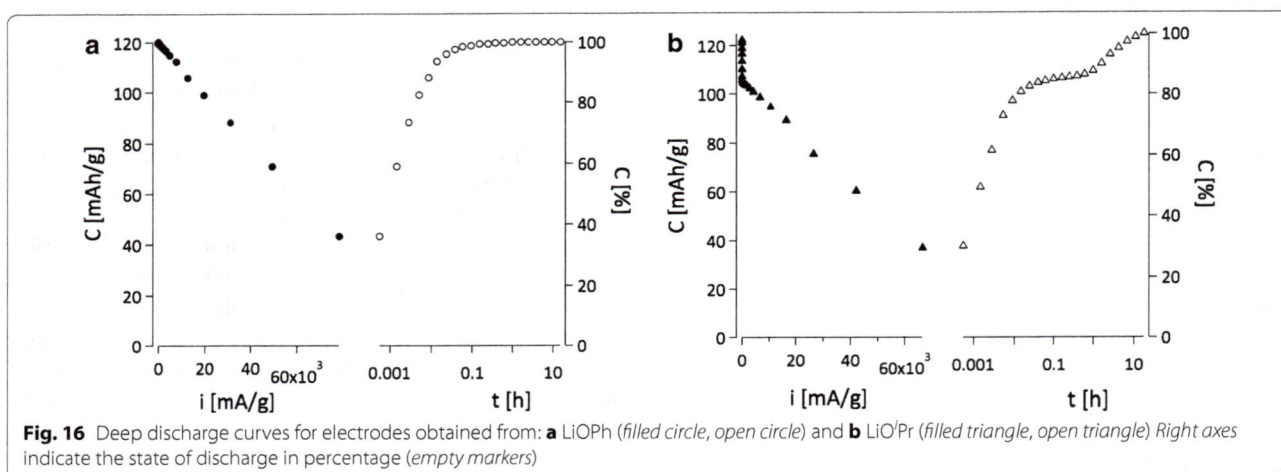

Fig. 16 Deep discharge curves for electrodes obtained from: **a** LiOPh (*filled circle, open circle*) and **b** LiO^{i}Pr (*filled triangle, open triangle*) *Right axes* indicate the state of discharge in percentage (*empty markers*)

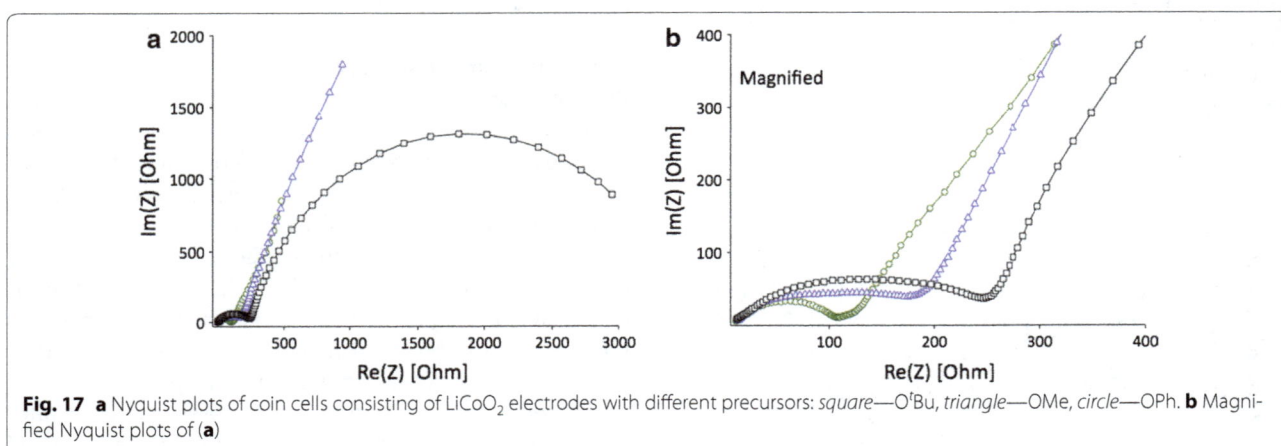

Fig. 17 **a** Nyquist plots of coin cells consisting of LiCoO$_2$ electrodes with different precursors: *square*—O^{t}Bu, *triangle*—OMe, *circle*—OPh. **b** Magnified Nyquist plots of (**a**)

nano- and micron-sized particles despite using the same initial feed concentration, as shown in Fig. 18a. Reason for this, apart from the different pulverisation methods, is that the microparticles can be considered to exhibit a higher density, and therefore greater tendency to agglomerate/aggregate leading to a higher surface density compared to the limited agglomeration/aggregation shown by the nanoparticles.

By using TEM it was observed that the pulverized nanoparticles of LCO formed agglomerates/aggregates ranging from nano-sized to micron-sized (ca. 0.05–50 µm). This could possibly be attributed to the low surface charge of the material (i.e.$\leq \pm 10$ mV). The average size of primary nanoparticles was estimated to be 64 ± 5 nm, as determined by the BET method, while the crystallite size was determined to be 60 ± 5 nm using the Scherrer equation. The micron-sized particles were noted to exhibit a size of 10–12 µm, as previously reported [16–21]. In terms of their morphology, nanoparticles were observed to show rhombohedral/tetrahedral shaped

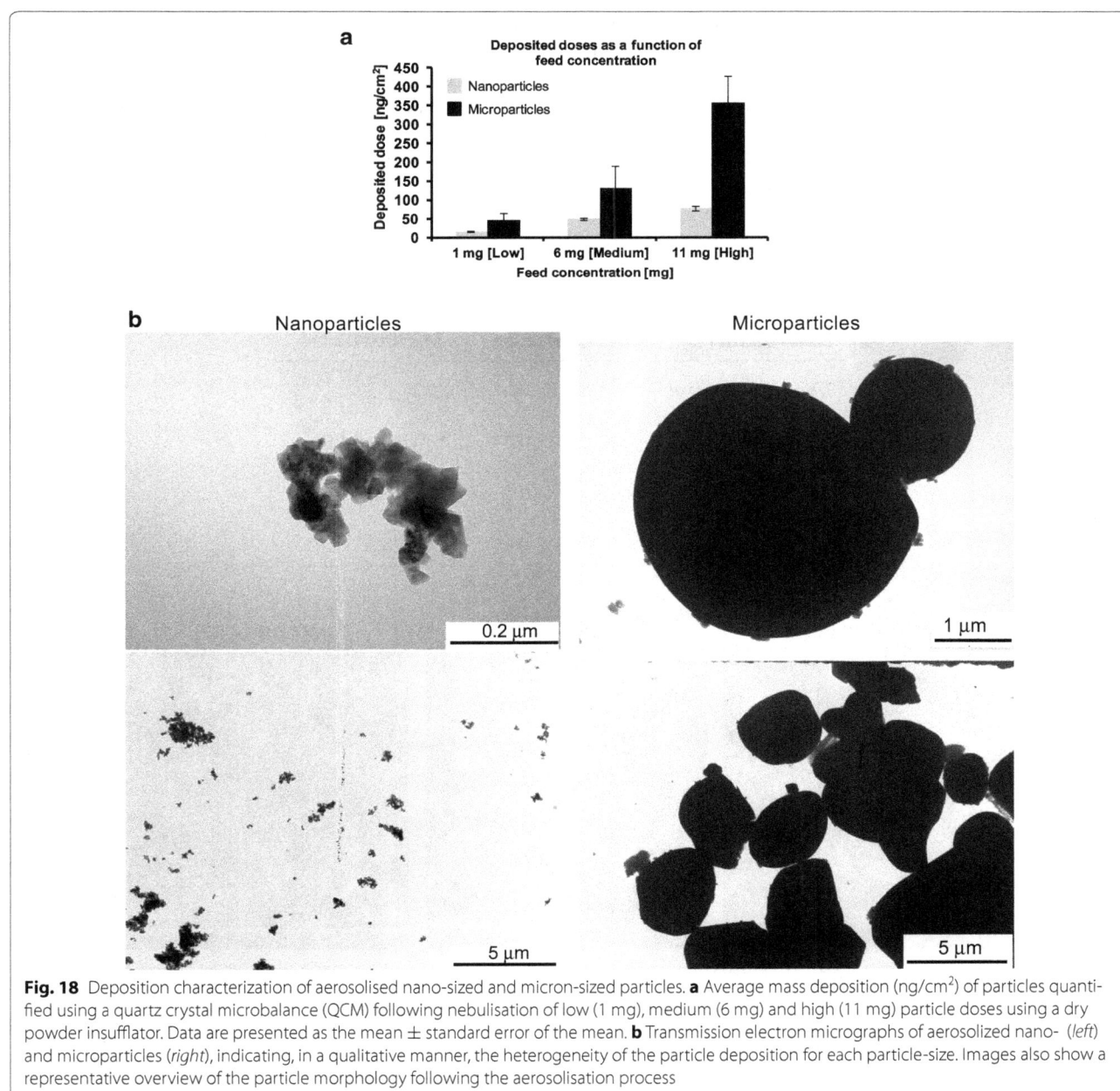

Fig. 18 Deposition characterization of aerosolised nano-sized and micron-sized particles. **a** Average mass deposition (ng/cm²) of particles quantified using a quartz crystal microbalance (QCM) following nebulisation of low (1 mg), medium (6 mg) and high (11 mg) particle doses using a dry powder insufflator. Data are presented as the mean ± standard error of the mean. **b** Transmission electron micrographs of aerosolized nano- (*left*) and microparticles (*right*), indicating, in a qualitative manner, the heterogeneity of the particle deposition for each particle-size. Images also show a representative overview of the particle morphology following the aerosolisation process

patterns, whereas the commercial microparticles were found to be irregular in shape, with most showing roundish shapes (Fig. 18b).

Cell death

After 24 h exposure, $LiCoO_2$ nanoparticles showed limited ability to cause cell death following their aerosolisation onto the in vitro multicellular epithelial tissue barrier model at each particle concentration tested (Fig. 19a). Both low and medium nanoparticle concentrations showed similar effects, whilst the highest concentration applied increased the level of cell death by 50% compared to the lower concentrations studied. This result can be attributed to an 'overload' scenario upon the cells at the highest concentration applied (Fig. 19b) [65]. It is important to note that although these values are significantly different from the negative control ($p > 0.05$) (i.e. cell culture media only), with the highest concentration applied showing a maximum of <15% cell death in the in vitro co-culture system, the findings indicate that the nanoparticles are not causing complete destruction of the cellular system but do induce a limited cytotoxic effect at these concentrations. Similar results were also evident following micron-sized $LiCoO_2$ particle exposures at each test concentration (Fig. 19a). In respect to these semi-quantitative results, it is also important to highlight that qualitative assessment, via confocal laser scanning microscopy, showed no morphological changes to the multicellular system following exposure to either particle type at the highest concentration applied for 24 h (Fig. 19b).

(Pro-)inflammatory response

No significant (pro-)inflammatory response (i.e. either TNF-α and IL-8 release) was observed following nanoparticle exposures across all concentrations tested (Fig. 20). Similar results were observed with the micronsized particles in terms of the TNF-α response from the multicellular system after 24 h exposure. However, microparticle exposures did show a significant increase ($p > 0.05$) in terms of the IL-8 response from the co-culture, in a concentration-dependent manner at this time point (Fig. 20).

a	Cell Death [%]	Nanoparticles	Microparticles
Negative control		0.71 ± 0.33	
Low		7.20 ± 0.45*	9.49 ± 0.09*
Medium		7.27 ± 0.25*	9.29 ± 0.34*
High		14.23 ± 0.09*	12.23 ± 0.50*

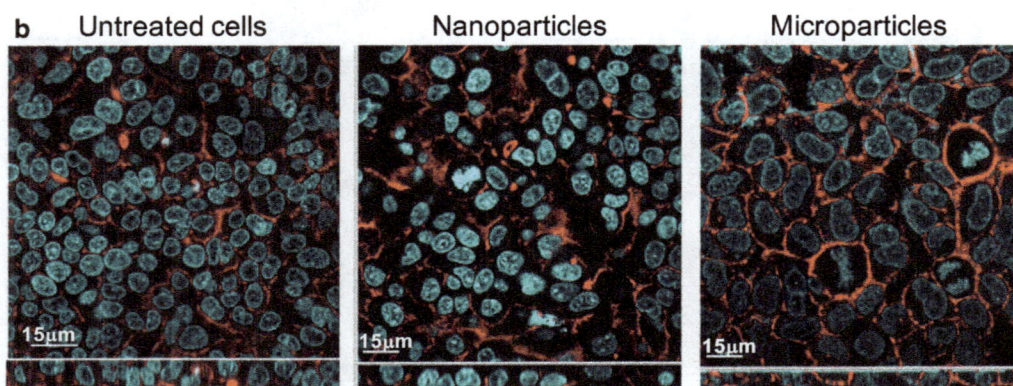

b Untreated cells   Nanoparticles   Microparticles

Fig. 19 Percentage (%) cell death levels and morphological analysis of the multicellular model of the epithelial tissue barrier following 24 h exposure to both $LiCoO_2$ nano-sized and micron-sized particles. **a** *Table* shows quantification of the average % cell death levels of propidium iodide stained cells at the three tested concentrations (low, medium and high), as analysed via one-colour flow cytometry analysis. *Asterisks* indicates a statistically significant increase in the % level of cell death within the multicellular in vitro system compared to the negative control (i.e. cell culture medium only) ($p > 0.05$) (n = 3). **b** Confocal laser scanning microscopy images show F-actin cytoskeleton (*red*) and the nuclei (*blue*) staining of the complete multicellular model following exposure to both particle sizes/types at the highest concentration tested after 24 h

Discussion

The general reaction of $CoCl_2$ and LiOPh for the generation of the precursors 1–7 is based on the LiCl-elimination and the formation of a mixed phenoxide with always the same metal ion ratio of 2:1 for Li:Co, as found in the core $[Li_2Co(OPh)_4]$ of the structures 1–7. The formation of this type of compound is in our hands independent of the amount of LiOPh added (between 1 and 6 equivalents). The core is always made of a central Co^{2+} ion which is surrounded in a (more or less distorted) tetrahedral way by four phenoxide ligands. Two by two, these O-donors act each as µ-bridging ligands to one Li^+ ion. The coordination sphere of the latter is then completed by either mono- or bidentate donor molecules stemming from the solvent. These coordinated solvent molecules influence the arrangement of the complexes with respect to each other. For instance, 0-dimensional compounds are obtained with monodentate terminal ligands like THF and pyridine or bidentate terminal ligands like DME and TMEDA, whereas bridging ligands such as dioxane lead to polymeric arrangements. In the $[Li_2Co(OPh)_4]$ cores (Fig. 21) of all compounds 1–5, for which the single crystal structures could be determined to satisfaction, the Co–O distances are between 1.938(4) and 1.978(4) Å long, while the angles O1–Co–O2 and O3–Co–O4 are very similar with 86°(±1°). The O2–Co–O3 and O1–Co–O4 angles are however more sensitive to the environment of the Li^+ cations (see Table 6), respectively packing effects, and vary between 112 and 127°.

The difference of composition between 1 and 2 originates from the crystallization technique. Indeed, 1 is prepared at room temperature with the addition of heptane for crystallization, while 2 is crystallized without any co-solvent at −24 °C. These two different methods give two different products: one thermodynamic compound 1 and one kinetic compound 2, which can be considered as solvates to each other [66].

In the structure of the compounds 1 to 7, an inherent stoichiometric ratio of two Li^+ for one Co^{2+} exists, hence excess of one equivalent Li^+ with respect to the desired $LiCoO_2$. During the firing, this excess of Li^+ in the precursor tends to form lithium carbonate either by reaction with the CO_2 in air or with the byproducts of the combustion. The carbonate can clearly be seen in powder X-ray diffractogram of the raw oxide. However, these impurities, as well as the main byproduct LiCl (formation of the precursors), can be easily washed away with water. Successful removal of LiCl was confirmed by powder X-ray diffraction as well as TEM/SEM.

For the compounds 8–12, except 11, the stoichiometric ratio is 1:1 for Li^+ to Co^{2+}, thus there is no excess Li^+ and hence almost no formation of lithium carbonate

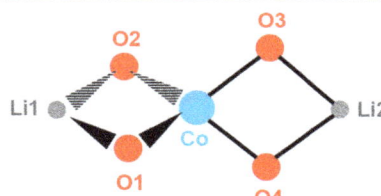

Fig. 21 Schematic representation with numbering of the $Li_2Co(OPh)_4$-core of compounds 1–7

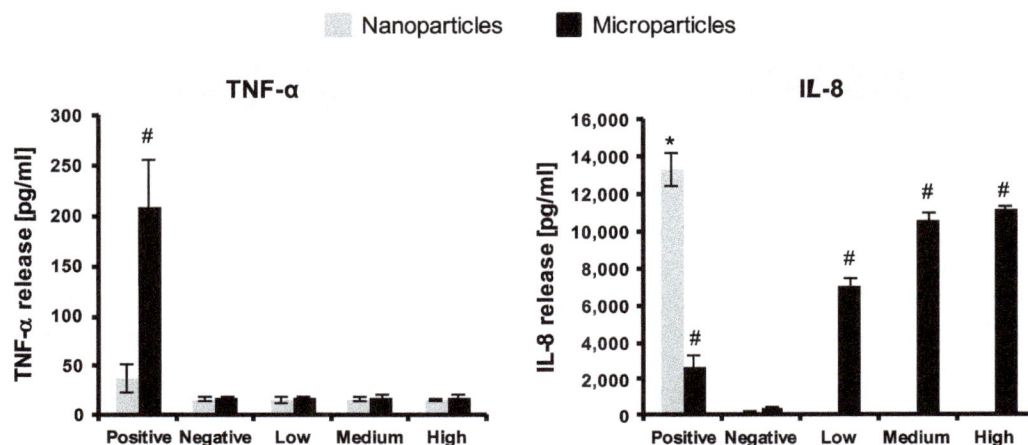

Fig. 20 (Pro-)inflammatory response of the multicellular epithelial tissue barrier following 24 h exposure to nano-sized and micron-sized nanoparticles at the three different test concentrations. *Graphs* show the results for the specific (pro-)inflammatory mediators chosen; tumor necrosis factor-α (TNF-α) and interleukin-8 (IL-8). Lipopolysaccharide ([100 µl of 1 µg/ml]) served as the positive assay control, whilst the negative control was cell culture medium only. Data is presented as the mean ± standard error of the mean. [#]indicates a statistically significant response ($p > 0.05$) compared to the negative control

(Additional file 1). While we produced our nanoscale materials in quite pure form by this washing step, the analysis of the commercial HT-LiCoO$_2$ shows that it contains some Li$_2$CO$_3$ impurities, which is one of the reactants of its synthesis.

The main physical/chemical differences in the final oxides obtained at 450 °C from **1** and **8–12** are the amount of impurities due to stoichiometric reasons and the size of the particles/crystallites obtained. Indeed, the LiOPh precursor **1** tends to form more impurities (carbonates, XRD in Additional file 1: Fig. S7) and a larger crystallite size. The amount of impurity is mainly due to the incorrect stoichiometric ratio in the starting structure of 2:1 for Li:Co, but also to a large amount of carbon atoms in the precursor. However, by decreasing the number of carbon atoms using alkoxide and by balancing the ratio between Co and Li to 1:1, better results in terms of size and smaller amounts of byproducts can be achieved. As shown in the Table 4, sizes as low as 60 nm of HT-LCO can be obtained.

We observed different LCO morphologies from the single source precursors. This could be related to the formation of LCO nuclei, which likely depend on the initial structure of the complex precursor. Not only the core structure, but also the arrangement of the molecules with respect to each other may play a role in the formation of different nuclei.

The redox potentials indeed confirm that the obtained nano-LiCoO$_2$ is in the HT-LCO phase. We also recognized that the oxidation of Co^{3+} to Co^{4+} (corresponding to Li$^+$ extraction from Li$_{1-x}$CoO$_2$) shows higher current than the reduction of Co^{4+} to Co^{3+} (Li$^+$ insertion into Li$_{1-x}$CoO$_2$). The cyclic voltammograms (CV) of both samples obtained from **12** and **8** show a HT-LCO CV profile with a low polarization and high potential, as expected from the X-ray diffraction pattern.

In terms of the Li$^+$ diffusivity, hence the kinetic with respect to Li$^+$ ions, we found it to be much faster in nanoscale LCO than in micron-LCO. In other words, the amount of Li$^+$ ions available for electrochemistry is larger in nanoscale LCO than that in micron-LCO due to the shorter path length of the Li$^+$ ion diffusion. The values obtained are >20 times higher than the standard value for HT-LCO [64]. In the best case measured in our hands, 77%, of all Li$^+$ ions were extracted from and re-inserted in the structure of nano-HT-LCO, while for the commercial material, only about 50% of Li$^+$ ions ($0 < x < 0.5$, Li$_{1-x}$CoO$_2$) can be used electrochemically in the rhombohedral layered structure of LiCoO$_2$. Further de-lithiation of commercial, micro-HT-LCO induces a phase transformation to the monoclinic system [17], resulting in irreversible capacity loss upon cycling. Therefore, the phase stability of LiCoO$_2$ is important during lithiation and de-lithiation in order to obtain high coulombic efficiency and longer cycleability of battery. This is what we could show for the nano-HT-LCO after battery cycling by analyzing the material by XRD. Hence, our LCO materials prepared by heterobimetallic single source precursors are stable upon cycling and provide fast electrochemical reactions with Li$^+$ ions due to nanosized particles.

Table 6 Comparison of compounds 1 to 5 and 9

	1	2	3	4		5	9
Tetrahedral volume of Co (Å^3)	3.332	3.384	3.344	3.364		3.339	3.229
Quadratic elongation	1.093	1.088	1.092	1.079		1.100	1.120
Angle variance (°^{2})	378.27	360.86	374.83	322.26		401.40	486.42
O1–Co (Å)	1.961 (6)	1.960 (7)	1.95 (1)	1.954 (3)	1.958 (4)	1.954 (4)	1.947 (3) 1.949 (3)
O2–Co (Å)	1.948 (5)	1.957 (6)	1.93 (1)	1.963 (4)	1.955 (3)	1.962 (4)	1.960 (3)
O3–Co (Å)	1.963 (5)	1.946 (6)	1.961 (8)	1.954 (3)	1.966 (3)	1.978 (4)	1.958 (4)
O4–Co (Å)	1.972 (6)	1.960 (6)	1.966 (7)	1.952 (4)	1.961 (4)	1.938 (4)	/
Mean O–Co (Å)	1.961	1.956	1.952	1.956	1.960	1.958	1.953
O1–Co–O2 (°)	84.9 (2)	86.5 (3)	85.7 (1)	86.8 (2)	86.6 (2)	85.6 (2)	80.2 (1)
O3–Co–O4 (°)	86.0 (2)	85.2 (3)	85.1 (4)	87.5 (1)	87.0 (2)	84.8 (2)	83.9 (2)
O1–Co–O4 (°)	122.4 (2)	121.0 (3)	125.6 (4)	118.6 (1)	122.7 (2)	126.5 (2)	124.4 (2)
O2–Co–O3 (°)	118.4 (2)	120.0 (3)	123.0 (4)	117.6 (2)	125.0 (1)	112.4 (2)	125.2 (1)
Mean O–Co–O (°)	85.45 120.4	85.85 120.5	85.4 124.3	86.98 120.98		85.2 119.45	82.0 124.8
BVS on Co	1.93	1.96	1.98	1.96	1.94	1.95	1.97
BVS on Li	1.17 1.14	1.13 1.14	1.13 1.07	1.19 1.17	1.13 1.14	1.19 1.25	1.18

The LCO materials prepared from various complexes showed different specific capacities. This difference may be related to several parameters such as the homogeneity of particle size, ball milling and the shape of LCO particles. Also, when a particle size distribution is broad, the specific capacity can be less good than the one from the narrower size distributed particles. The large size difference can lead to different Li^+ diffusion kinetics. However, the larger particles can be broken during ball milling and the size distribution becomes narrower, improving the kinetics of Li^+ diffusion and finally the specific capacity. The shape of LCO particle can also affect the diffusion of Li^+ because Li^+ diffuses in a specifically oriented layer of the structure.

The smaller particle size provides higher diffusion kinetics with Li^+ because the higher surface area of nano-$LiCoO_2$ provides more Li^+ ions to be released and uptaken into/from the electrolyte. In addition to the high surface area, there is another parameter governing the diffusion kinetics, which is the orientation of Li^+ diffusion path in the lattice structure of $LiCoO_2$. Li^+ is located in one layer of the $LiCoO_2$ lattice cell, diffusing in one preferred orientation. Thus, the length of Li^+ diffusion path in $LiCoO_2$ also affects the diffusion kinetics. We reported that the diffusion of Li^+ is not only related to the size of particle but also the shape of particle due to the preferred diffusion direction and its length in the lattice structure [67, 68]. In this regard, the higher diffusion coefficient of 40 nm (compound **8**) is probably coming from the shorter diffusion path of Li^+ in a single particle although the compound **12** has a smaller size of 15 nm.

We also found that $LiCoO_2$ produced from LiO^tBu has a larger overpotential and higher resistance than the one obtained from LiOPh. On the other hand, the $LiCoO_2$ electrode formed from LiOMe reached >120 mAh/g of charge capacity. However, the discharge capacity was 90 mAh/g with 70% of coulombic efficiency. These differences of equilibrum charge/discharge curves can be explained by different kinetics at equilibrium state.

The deep discharge measurement supports that the discharge capacities at higher current densities (>C/2) are lower in $LiCoO_2$ electrode with LiOPh than those in LCO with LiO^iPr (Fig. 16). Therefore, the kinetics of the electrode (a) obtained from LiO^iPr is an order of magnitude faster than (b) (obtained from LiOPh) at high current densities.

The electrochemical properties of batteries are influenced by not only the active material but also the composite, consisting of carbon and the active material [69]. The structural morphology and the physicochemical properties of composite affect the electron transfer and lithium ion diffusion in the electrode [64]. An ongoing follow-up study is hence the optimization of the electrode composites for each nanoscale HT-LCO material as a function of precursor.

In terms of the biological assessment, such studies had never been done on nanoscale LCO and are quite rare for battery materials in general. We found both nano- and micro-LCO to be relatively low toxic in the lung model which we used. The (pro-)inflammatory response upon exposure to nano-LCO was nil across all tested concentrations, while it was dose-dependent for micro-LCO. Neither nanoparticles nor micro-LCO induce a cytotoxic effect at the tested concentrations which leads to more than 15% cell death. In terms of the surface charges of nano and microparticles, we estimate it is low since both particles rather stick together [70].

Conclusions

A series of 12 new precursors containing lithium and cobalt ions in ratios of 2:1 or 1:1 with different aryl- and alkoxide ligands have been prepared and characterized. Their thermal decomposition leads to the formation of nanoscale HT-$LiCoO_2$ with the size of the so obtained nanoparticles depending on the precursor. Also, precursors with a 1:1 ratio of Li^+ to Co^{2+} lead to quite pure product, while the precursors with a 2:1 ratio gave Li_2CO_3 as byproduct. The use of our precursors allowed lowering the production temperature and time for the generation of HT-$LiCoO_2$ as a preorganisation of the metal ions takes place in the starting material. The nanomaterials of $LiCoO_2$ showed a superior Li-ion diffusivity by a factor of 20–100 compared to commercial $LiCoO_2$, depending on the precursor used to generate the cathode material. The electrochemical performance was varied depending on the precursors. $LiCoO_2$ with LiOPh and LiO^iPr provided higher specific capacities while $LiCoO_2$ with LiOMe and LiOtBu obtained lower specific capacities. Lithium ion diffusion coefficients of our nanoscale $LiCoO_2$ were >10 times higher than the one of microscale $LiCoO_2$ due to the shorter path length of lithium ion diffusion in nanomaterial of $LiCoO_2$. This means that high surface area of nanoscale $LiCoO_2$ can release and take Li^+ ions much more than micron $LiCoO_2$ material at the same condition.

To mimick conditions of recycling of batteries, nanopowders of $LiCoO_2$ were tested on a lung cell model. During the spraying of the powders, it was shown that the nanopowders tend to aggregate during the process due to a low zeta-potential. Nevertheless, they are slightly more toxic than the micron-scale material, while toxicity remained overall very low.

Additional file

Additional file1: Text 1. Synthesis of bimetallic compounds. **Table S1.** Crystal data. **Text 2.** Single crystal structure descriptions. **Text 3.** Argentometric titration. **Table S2.** Idealistic oxidation reactions of two types of compounds, precursors **1**, **5** with 2:1 and precursors **8**, **9** with 1:1 stoichiometric ratio between Li+ and Co2+. **Table S3.** Results of the argentometric titration of chloride and ICP-measurements for lithium. **Table S4.** ICP analysis for Li+ and Co3+ of LiCoO2 obtained from different precursors. **Figure S6.** XRD study of commercial LCO, and nano-LCO obtained from LiOtBu before annealing and after annealing at 600°C and 700°C. **Figure S7.** XRD of LiCoO2 from 9-LiOPh calcined at 450°C before washing. The red line corresponds to HT-LCO and the blue lines are Li2CO3. **Table S5.** The combustion temperature and the thermal measurement conditions of the compounds **1**, **8-12**. **Table S6.** TGA weight loss in percentage [%] with associated steps of compounds **1**, **8-12**. **Equation S1-S5.** Determination of the particle and crystallite sizes. **Figure S8.** Morphologies of LiCoO2 prepared with different precursors at 450°C. **Figure S9.** (a) Cyclic voltammograms of the 15 nm LCO prepared from the compound 12 at different sweep rates. (b) The maximum anodic and cathodic current peaks of LiCoO2 electrode versus the square root of sweep rate. **Table S7.** Li+ diffusion coefficients determined for HT-LCO obtained from different precursors. **Figure S10.** Nyquist plot for LiCoO2 electrodes from LiOtBu with fit: filled markers – experimental points, open markers – fit points with error bars a) and corresponding equivalent circuit model b) with fitting report c). **Figure S11.** Nyquist plot obtained for LiCoO2 electrodes from LiOPh with fit: filled markers – experimental points, open markers – fit points with error bars a) and corresponding equivalent circuit model b) with fitting report c).

Abbreviations
HT-LCO: high temperature LiCoO$_2$; LT-LCO: low temperature LiCoO$_2$; THF: tetrahydrofuran; ALI: air-liquid interface; LiOPh: lithium phenoxide; LiO^{t}Bu: lithium tert-butoxide; LiOMe: lithium methoxide; LiOEt: lithium ethoxide; LiO^{i}Pr: lithium iso-propoxide; TMEDA: tetramethylethylenediamine; DME: dimethoxyethane; Py: pyridine; IR: infra-red; NMR: nuclear magnetic resonance; XRD: X-ray diffraction; M: molecular mass; a, b, c: unit cell dimensions; β: monoclinic angle; V: unit cell volume; Z: number of independent asymmetric units per unit cell; ρ_{calcd}: density; T: temperature at which single crystals were measured; Θ: theta angle for single crystal x-ray measurements; $GOOF$: goodness of fit; $R1$, $wR2$: quality factors; TGA: thermogravimetric analysis; STDA: simultaneous differential thermal analysis; BET: Brunauer, Emmett and Teller; SEM: scanning electron miscroscope; FEI: field electron and ion; EDS: energy dispersive X-ray spectroscopy; ICP-OES: inductively coupled plasma optical emission spectrometry; PVDF: polyvinylidene fluoride; NMP: N-methyl-2-pyrrolidone; ABG and SFG: name of graphite provided from the manufacturer; EC: ethylene carbonate; DMC: dimethyl carbonate; DEC: diethylene carbonate; C/20, C/10, C/5, C/2, 1C and 20C: charging for 20 h, 10 h, 5 h, 2 h, 1 h and 3 mins; EIS: electrochemical impedance spectroscopy; FRA: frequency response analyser; RPMI: Roswell Park Memorial Institute; PET: polyethylene terephthalate; CH: cluster of differentiation; DP: dry powder insufflator; ELISA: enzyme-linked immunosorbent assay; BVS: bond-valence-sum.

Authors' contributions
JPB, AC and SM synthesized and characterized the complexes and LCO. BB and MS performed the electrochemistry. AZ proposed the deep discharge analysis. JPB and HB performed the biological assessment with the guidance of MJDC and BRR. NHK guided the characterization, electrochemistry and the manuscript. KMF supervised the project and the manuscript. All authors read and approved the final manuscript.

Author details
[1] Department of Chemistry, University of Fribourg, Chemin du Musée 9, 1700 Fribourg, Switzerland. [2] Fribourg Center for Nanomaterials FriMat, University of Fribourg, Chemin du Musée 9, 1700 Fribourg, Switzerland. [3] Adolphe Merkle Institute, University of Fribourg, 1700 Fribourg, Switzerland. [4] College of Engineering and Architecture of Fribourg, University of Applied Sciences of Western Switzerland, Boulevard de Pérolles 80, 1705 Fribourg, Switzerland. [5] Laboratory of Materials for Renewable Energy (LMER), ISIC-SB, École Polytechnique Fédérale de Lausanne (EPFL), Valais/Wallis Energypolis, Rue de l'Industrie 17, 1951 Sion, Switzerland.

Acknowledgements
This study was supported by the Swiss National Science Foundation (National Research Program 64), the Swiss Competence Center for Energy Research (SCCER) Heat and Electricity Storage, the FriMat (the Fribourg Center for Nanomaterials), the NCCR "Bioinspired Materials" and the University of Fribourg.

Competing interests
The authors declare that there are no competing interests nor commercial interests.

Funding
This work was supported by the National Research Program 64, Project Number 406440_141604 from the Swiss National Science Foundation.

References
1. Mizushima K, Jones PC, Wiseman PJ, Goodenough JB. Li$_x$CoO$_2$ (0 < x<−1): a new cathode material for batteries of high energy density. Mater Res Bull. 1980;15:783–9.
2. Reimers JN, Dahn JR. Electrochemical and in situ X-ray diffraction studies of lithium intercalation in Li$_x$CoO$_2$. J Electrochem Soc. 1992;139:2091–7.
3. Ohzuku T, Ueda A. Solid-state redox reactions of LiCoO$_2$ (R$\bar{3}$m) for 4 volt secondary lithium cells. J Electrochem Soc. 1994;141:2972–7.
4. Ménétrier M, Saadoune I, Levasseur S, Delmas C. The insulator-metal transition upon lithium deintercalation from LiCoO$_2$: electronic properties and 7Li NMR study. J Mater Chem. 1999;9:1135–40.
5. Molenda J, Stoklosa A, Bak T. Modification in the electronic structure of cobalt bronze Li$_x$CoO$_2$ and the resulting electrochemical properties. Solid State Ionics. 1989;36:53–8.
6. Van der Ven A, Aydinol MK, Ceder G. First principles evidence for stage ordering in Li$_x$CoO$_2$. J Electrochem Soc. 1998;145:2149–55.
7. Van der Ven A, Aydinol MK, Ceder G, Kresse G, Hafner J. First-principles investigation of phase stability in Li$_x$CoO$_2$. Phy Rev B. 1998;58:2975–87.
8. Amatucci GG, Tarascon JM, Klein LC. CoO$_2$, the end member of the Li$_x$CoO$_2$ solid solution. J Electrochem Soc. 1996;143:1114–23.
9. Li W, Reimers JN, Dahn JR. Lattice-gas-model approach to understanding the structures of lithium transition-metal oxides LiMO$_2$. Phy Rev B. 1994;49:826.
10. Huang W, Frech R. Vibrational spectroscopic and electrochemical studies of the low and high temperature phases of LiCo$_{1−x}$M$_x$O$_2$ (M = Ni or Ti). Solid State Ion. 1996;86–88:395–400.
11. Antolini E. LiCoO$_2$: formation, structure, lithium and oxygen nonstoichiometry, electrochemical behaviour and transport properties. Solid State Ion. 2004;170(3–4):159–71. doi:10.1016/j.ssi.2004.04.003.
12. Gummow RJ, Thackeray MM, Wif D, Hull S. Structure and electrochemistry of lithium cobalt oxide synthesised at 400°C. Mater Res Bull. 1992;27:327–37.
13. Garcia B, Farcy J, Pereira-Ramos JP, Baffier N. Electrochemical properties of low temperature crystallized LiCoO$_2$. J Electrochem Soc. 1997;144:1179–84.
14. Rossen E, Reimers JN, Dahn JR. Synthesis and electrochemistry of spinel LT-LiCoO$_2$. Solid State Ion. 1993;62:53–60.

15. Orman HJ, Wiseman PJ. Cobalt(III) lithium oxide, CoLiO$_2$: structure refinement by powder neutron diffraction. Acta Crystallogr C. 1984;40:12–4.

16. Garcia B, Farcy J, Pereira-Ramos JP, Perichon J, Baffler N. Low-temperature cobalt oxide as rechargeable cathodic material for lithium batteries. J Power Sources. 1995;54:373–7.

17. Shao-Horn Y, Croguennec L, Delmas C, Nelson EC, O'Keefe MA. Atomic resolution of lithium ions in LiCoO$_2$. Nat Mater. 2003;2(7):464–7.

18. Kim J, Fulmer P, Manthiram A, Kim J, Fulmer P, Manthiram A. Synthesis of LiCoO$_2$ cathodes by an oxidation reaction in solution and their electrochemical properties. Mater Res Bull. 1999;34(4):571–9.

19. Myung ST, Kumagai N, Komaba S, Chung HT. Preparation and electrochemical characterization of LiCoO$_2$ by the emulsion drying method. J Appl Electrochem. 2000;30(9):1081–5.

20. Yoshio M, Tanaka H, Tominaga K, Noguchi H. Synthesis of LiCoO$_2$ from cobalt—organic acid complexes and its electrode behaviour in a lithium secondary battery. J Power Sources. 1992;40:347–53.

21. Antolini E. Lithium loss from lithium cobalt oxide: hexagonal Li$_{0.5}$Co$_{0.5}$O to cubic Li$_{0.065}$Co$_{0.935}$O phase transition. Int J Inorg Mater. 2001;3:721–6.

22. Oh IH, Hong SA, Sun YK. Low-temperature preparation of ultrafine LiCoO$_2$ powders by the sol–gel method. J Mater Sci. 1997;32(12):3177–82.

23. Yoon WS, Kim KB. Synthesis of LiCoO$_2$ using acrylic acid and its electrochemical properties for Li secondary batteries. J Power Sources. 1999;81–82:517–23.

24. Peng ZS, Wan CR, Jiang CY. Synthesis by sol–gel process and characterization of LiCoO$_2$ cathode materials. J Power Sources. 1998;72(2):215–20.

25. Sun YK, Oh IH, Hong SA. Synthesis of ultrafine LiCoO$_2$ powders by the sol–gel method. J Mater Sci. 1996;31(14):3617–21.

26. Burukhin A, Brylev O, Hany P, Churaqulov BR. Hydrothermal synthesis of LiCoO$_2$ for lithium rechargeable batteries. Solid State Ion. 2002;151:259–63.

27. Fromm KM. Synthesis and crystal structure of Li[{Ca$_7$(u$_3$-OH)$_8$I$_6$(thf)$_{12}$}$_2$(u-I)]·3THF, a unique H-bound dimer of a Ca$_7$-cluster on the way to sol–gels. Chem Commun. 1999;17:1659–60.

28. Fromm KM, Gueneau ED, Goesmann H. Synthesis and crystal structure of [IBa(OBut)$_4${Li(thf)}$_4$(OH)]: a mixed ligand heterometallic cluster with an unusual low coordination number for barium. Chem Commun. 2000;22:2187–8. doi:10.1039/b005638n.

29. Fromm KM, Gueneau ED, Bernardinelli G, Goesmann H, Weber J, Mayor-López MJ, et al. Understanding the formation of new clusters of alkali and alkaline earth metals. J Am Chem Soc. 2003;125(12):3593–604.

30. Maudez W, Häussinger D, Fromm KM. A Comparative study of (Poly) ether adducts of alkaline earth metal iodides—an overview including new compounds. Z Anorg Allg Chem. 2006;632(14):2295–8.

31. Maudez W, Meuwly M, Fromm KM. Analogy of the coordination chemistry of alkaline earth metal and lanthanide Ln^{2+}Ions: the isostructural zoo of mixed metal cages [IM (OtBu)$_4${Li(thf)}$_4$(OH)] (M = Ca, Sr, Ba, Eu), [MM'$_6$ (OPh)$_8$(thf)$_6$] (M = Ca, Sr, Ba, Sm, Eu, M' = Li, Na), and their derivatives with 1,2- Dimethoxyethane. Chem Eur J. 2007;13:8302–16.

32. Gschwind F, Sereda O, Fromm KM. Multitopic ligand design: a concept for single-source precursors. Inorg Chem. 2009;48(22):10535–47. doi:10.1021/ic9009064.

33. Maudez W, Fromm KM. The heterometallic clusters of trivalent rare earth metals of [Ln(OPh)$_6${Li(dme)}$_3$], with Ln$_{1/4}$ Eu and Sm. Helv Chim Acta. 2009;92(11):2349–56.

34. Gschwind F, Crochet A, Maudez W, Fromm KM. From alkaline earth ion aggregates via transition metal coordination polymer networks towards heterometallic single source precursors for oxidic materials. Chimia. 2010;64(5):299–302. doi:10.2533/chimia.2010.299.

35. Kwon NH, Brog JP, Maharajan S, Crochet A, Fromm KM. Nanomaterials meet Li-ion batteries. Chimia. 2015;69(12):734–6. doi:10.2533/chimia.2015.734.

36. Crochet A, Brog J-P, Fromm KM. Mixed metal multinuclear Cr(III) cage compounds and coordination polymers based on unsubstituted phenolate: design, synthesis, mechanism, and properties. Cryst Growth Des. 2016;16(1):189–99. doi:10.1021/acs.cgd.5b01084.

37. Buzzeo MC, Iqbal AH, Long CM, Millar D, Patel S, Pellow MA, et al. Homoleptic cobalt and copper phenolate A$_2$[M(OAr)$_4$] compounds: the effect of phenoxide fluorination. Inorg Chem. 2004;43:7709–25.

38. Boyle TJ, Rodriguez MA, Ingersoll D, Headley TJ, Bunge SD, Pedrotty DM, et al. A novel family of structurally characterized lithium cobalt double

39. Rothen-Rutishauser BM, Kiama SG, Gehr P. A three-dimensional cellular model of the human respiratory tract to study the interaction with particles. Am J Respir Cell Mol Biol. 2005;32(4):281–9. doi:10.1165/rcmb.2004-0187OC.

40. Gwinn M, Vallyathan V. Nanoparticles: health effects-pros and cons. Environ Health Perspect. 2006;114:1818–25.

41. Maynard AD, Robert JA, Butz T, Colvin V, Donaldson K, Oberdörster G, et al. Safe handling of nanotechnology. Nature. 2006;444:267–9.

42. Timbrell J. Biomarkers in toxicology. Toxicology. 1998;129:1–12.

43. Donaldson K, Stone V, Tran C, Kreyling W, Borm P. Nanotoxicology. Occup Environ Med. 2004;61(9):727–8.

44. Li J, Muralikrishnan S, Ng C, Yung L, Bay B. Nanoparticle-induced pulmonary toxicity. Exp Biol Med (Maywood). 2010;235:1025–33.

45. Borm P, Klaessig F, Landry T, Moudgil B, Pauluhn J, Thomas K, et al. Research strategies for safety evaluation of nanomaterials, part V: role of dissolution in biological fate and effects of nanoscale particles. Toxicol Sci. 2006;90(1):23–32.

46. Oberdorster G, Stone V, Donaldson K. Toxicology of nanoparticles: a historical perspective. Nanotoxicology. 2007;1:2–25.

47. Brog JP, Crochet A, Fromm KM, inventors. Lithium metal aryloxide clusters as starting products for oxide materials patent WO 2012000123 A1; 2011.

48. Shriver DF. The manipulation of air-sensitive Compounds. New York: McGraw-Hill; 1969.

49. Cosier J, Glazer AM. A nitrogen-gas-stream cryostat for general X-ray diffraction studies. J Appl Crystallogr. 1986;19:105–7.

50. Blanc E, Schwarzenbach D, Flack HD. The evaluation of transmission factors and their first derivatives with respect to crystal shape parameters. J Appl Cryst. 1991;24:1035–41.

51. Burla MC, Caliandro R, Camalli M, Carrozzini B, Cascarano GL, Caro LD, et al. SIR2004: an improved tool for crystal structure determination and refinement. J Appl Cryst. 2005;38:381–8.

52. Sheldrick GM. A short history of SHELX. Acta Crystallogr A. 2008;64(Pt 1):112–22. doi:10.1107/S0108767307043930.

53. Spodaryk M, Shcherbakova L, Sameljuk A, Zakaznova-Herzog V, Braem B, Holzer M, et al. Effect of composition and particle morphology on the electrochemical properties of LaNi 5-based alloy electrodes. J Alloys Compd. 2014;607:32–8.

54. Chartouni D, Kuriyama N, Otto A, Güther V, Nützenadel C, Züttel A, et al. Influence of the alloy morphology on the kinetics of AB5-type metal hydride electrodes. J Alloys Compd. 1999;285:292–7.

55. Blank F, Rothen-Rutishauser BM, Schurch S, Gehr P. An optimized in vitro model of the respiratory tract wall to study particle cell interactions. J Aerosol Med Depos Clear Effects Lung. 2006;19(3):392–405. doi:10.1089/jam.2006.19.392.

56. Maguire T, Novik E. Methods in bioengineering: alternatives to animal testing. Norwood: Artech House; 2010. p. 239–60. ISBN: 9781608070114

57. Steiner S, Mueller L, Popovicheva OB, Raemy DO, Czerwinski J, Comte P, et al. Cerium dioxide nanoparticles can interfere with the associated cellular mechanistic response to diesel exhaust exposure. Toxicol Lett. 2012;214(2):218–25.

58. Lenz AG, Karg E, Lentner B, Dittrich V, Brandenberger C, Rothen-Rutishauser B, et al. A dose-controlled system for air-liquid interface cell exposure and application to zinc oxide nanoparticles. Part Fibre Toxicol. 2009;6:32–48. doi:10.1186/1743-8977-6-32.

59. Tomašek I, Horwell CJ, Damby DE, Barošová H, Geers C, Petri-Fink A, et al. Combined exposure of diesel exhaust particles and respirable Soufrière Hills volcanic ash causes a (pro-)inflammatory response in an in vitro multicellular epithelial tissue barrier model. Part Fibre Toxicol. 2016;13:67–80. doi:10.1186/s12989-016-0178-9.

60. Anson CE, Klopper W. Li J-S, L. Ponikiewski, Rothenberger A. A Close Look at Short C-CH3···Potassium Contacts: synthetic and Theoretical Investigations of [M$_2$Co$_2$(μ$_3$-OtBu)$_2$(μ$_2$-OtBu)$_4$(thf)$_n$] (M = Na, K, Rb, thf = tetrahydrofuran). Chem Eur J. 2006;12(7):2032–8. doi:10.1002/chem.200500603.

61. Akimoto J, Gotoh Y, Oosawa Y. Synthesis and structure refinement of LiCoO$_2$ single crystals. J Solid State Chem. 1998;141(1):298–302.

62. Maiyalagan T, Jarvis KA, Therese S, Ferreira PJ, Manthiram A. Spinel-type lithium cobalt oxide as a bifunctional electrocatalyst for the oxygen evolution and oxygen reduction reactions. Nat Commun. 2014;5:1–8. doi:10.1038/ncomms4949.

63. Bard AJ, Faulkner LR. Electrochemical methods: fundamentals and applications. New York: Wiley; 2001.

64. Kwon NH, Yin H, Brodard P, Sugnaux C, Fromm KM. Impact of composite structure and morphology on electronic and ionic conductivity of carbon contained $LiCoO_2$ cathode. Elecrochim Acta. 2014;134:215–21.

65. Borm P, Cassee FR, Oberdorster G. Lung particle overload: old school -new insights? Part Fibre Toxicol. 2015;12:10–4. doi:10.1186/s12989-015-0086-4.

66. Brog J-P, Chanez C-L, Crochet A, Fromm KM. Polymorphism, what it is and how to identify it: a systematic review. RSC Adv. 2013;3(38):16905–31. doi:10.1039/c3ra41559g.

67. Xiao X, Liu X, Wang L, Zhao H, Hu Z, He X, et al. $LiCoO_2$ nanoplates with exposed (001) planes and high rate capability for lithium-ion batteries. Nano Res. 2012;5(6):395–401. doi:10.1007/s12274-012-0220-7.

68. Kwon NH, Yin H, Vavrova T, Lim JHW, Steiner U, Grobéty B, et al. Nanoparticle shapes of $LiMnPO_4$, Li^+ diffusion orientation and diffusion coefficients for high volumetric energy Li^+ ion cathodes. J Power Sources. 2017;342:231–40. doi:10.1016/j.jpowsour.2016.11.111.

69. Kwon NH. The effect of carbon morphology on the $LiCoO_2$ cathode of lithium ion batteries. Solid State Sci. 2013;21:59–65.

70. Lia C-C, Leeb J-T, Loa C-Y, Wu M-S. Effects of PAA-NH_4 addition on the dispersion property of aqueous $LiCoO_2$ Slurries and the cell performance of as-prepared $LiCoO_2$ cathodes. Electrochem SolidState Lett. 2005;8(10):A509–12. doi:10.1149/1.2012287.

10

Toxicity of nano- and ionic silver to embryonic stem cells: a comparative toxicogenomic study

Xiugong Gao[*], Vanessa D. Topping, Zachary Keltner, Robert L. Sprando and Jeffrey J. Yourick

Abstract

Background: The widespread application of silver nanoparticles (AgNPs) and silver-containing products has raised public safety concerns about their adverse effects on human health and the environment. To date, in vitro toxic effects of AgNPs and ionic silver (Ag^+) on many somatic cell types are well established. However, no studies have been conducted hitherto to evaluate their effect on cellular transcriptome in embryonic stem cells (ESCs).

Results: The present study characterized transcriptomic changes induced by 5.0 µg/ml AgNPs during spontaneous differentiation of mouse ESCs, and compared them to those induced by Ag^+ under identical conditions. After 24 h exposure, 101 differentially expressed genes (DEGs) were identified in AgNP-treated cells, whereas 400 genes responded to Ag^+. Despite the large differences in the numbers of DEGs, functional annotation and pathway analysis of the regulated genes revealed overall similarities between AgNPs and Ag^+. In both cases, most of the functions and pathways impacted fell into two major categories, embryonic development and metabolism. Nevertheless, a number of canonical pathways related to cancer were found for Ag^+ but not for AgNPs. Conversely, it was noted that several members of the heat shock protein and the metallothionein families were upregulated by AgNPs but not Ag^+, suggesting specific oxidative stress effect of AgNPs in ESCs. The effects of AgNPs on oxidative stress and downstream apoptosis were subsequently confirmed by flow cytometry analysis.

Conclusions: Taken together, the results presented in the current study demonstrate that both AgNPs and Ag^+ caused transcriptomic changes that could potentially exert an adverse effect on development. Although transcriptomic responses to AgNPs and Ag^+ were substantially similar, AgNPs exerted specific effects on ESCs due to their nanosized particulate form.

Keywords: Silver nanoparticles, Silver ion, Embryonic stem cell, Developmental toxicity, Transcriptomics

Background

The use of engineered nanoscale materials in consumer products has increased dramatically over the past decade. Only 54 consumer products claimed to contain nanomaterials in 2005, but the number has surged to more than 1600 today [1]. It has been estimated that of all the nanomaterials manufactured, silver nanoparticles (AgNPs) have the highest degree of commercialization [2], owing largely to their broad spectrum of antimicrobial activities against bacteria, fungi, and viruses, including HIV and SARS [3, 4]. Currently, a large variety of consumer products contain AgNPs, including food packaging materials, dietary supplements, cosmetics, textiles, electronics, household appliances, medical devices, water disinfectants, and room sprays [5].

The widespread application of AgNPs has raised public safety concerns about their adverse effects on human health and the environment. Data from in vitro studies demonstrated that AgNPs induce cytotoxicity and

---
*Correspondence: Xiugong.Gao@fda.hhs.gov
Division of Applied Regulatory Toxicology, Office of Applied Research and Safety Assessment, Center for Food Safety and Applied Nutrition, U.S. Food and Drug Administration, 8301 Muirkirk Road, Laurel, MD 20708, USA

genotoxicity through the production of reactive oxygen species (ROS), DNA damage, cell cycle arrest, ultimately leading to inflammation, apoptosis, and cell death [6, 7]. In addition, in vivo studies [8–10] demonstrated that AgNPs enter the blood circulation and accumulate in organs including the brain, kidneys, lungs, spleen, testes and primarily the liver. The ability of AgNPs to enter the blood stream [10] and cross through the blood–brain barrier [11] points toward the potential of these nanoparticles to migrate into the uterus, placenta and embryo thus causing developmental toxicity [12].

Embryonic stem cells (ESCs) have been shown to faithfully recapitulate stages of early embryo development and are increasingly used as an in vitro model for developmental toxicity testing [13]. An in vitro test has been developed over two decades ago to evaluate embryotoxicity of chemical compounds using mouse embryonic stem cells (mESCs) [14]. The so-called embryonic stem cell test (EST) was latterly validated by the European Committee for the Validation of Alternative Methods (ECVAM) [15] and is currently used in the pharmaceutical industry [16]. In addition, implementing toxicogenomics into the EST improves its application domain and predictability [17], and has been shown as a promising alternative method for developmental toxicity testing [18]. Despite these progresses, only a few studies reported the toxicity of AgNPs in ESCs [19–21]. Moreover, little is known about the mechanisms of AgNP toxicity in ESCs at the molecular level.

At present, there is no consensus on the mechanisms of action for AgNP cytotoxicity and research findings are equivocal. For example, Bouwmeester et al. [22] found in an in vitro intestinal epithelium coculture model that treatment with AgNPs induced regulation of the same set of genes as with silver nitrate ($AgNO_3$), and 6–17% of the silver content in the AgNPs suspensions was found in the ionic form (Ag^+). The authors therefore speculated that the observed gene regulation was exerted by Ag^+ released from the AgNPs. On the other hand, other studies suggest that AgNP toxicity is related to Ag^+ and to the nanosize as well [23, 24]. Thus, AgNP cytotoxicity is a complex phenomenon and more research is needed to distinguish cellular effects triggered by the nanosized particle from those by Ag^+ released from the nanoparticles.

The present work aimed at unraveling the molecular mechanisms of AgNP toxicity in ESCs. Using microarrays, we characterized transcriptomic changes induced by citrate-coated 20 nm AgNPs during spontaneous differentiation of a C57BL/6-derived mESC cell line. This cell line has been demonstrated to detect global gene expression changes induced by a variety of developmental toxicants [18]. To our knowledge, no toxicogenomic study on AgNPs has been reported hitherto in

ESCs. We also compared the gene expression changes to those induced by Ag^+ (silver acetate). We evaluated the potential toxicity of AgNPs with emphasis on oxidative stress and apoptosis by correlating cellular responses to gene expression patterns, which provides a mechanistic understanding of the toxicity of AgNPs in ESCs.

Methods
Materials
BioPure 20 nm AgNP citrate solutions, at AgNP concentration of 1.0 mg/ml, were purchased from nanoComposix (San Diego, CA). The AgNPs were extensively washed with the suspending solvent to remove residual reactants from the manufacturing process, and were sterile filtered and tested for endotoxin contamination before delivery. The stock solution as obtained from the manufacturer was stored at 4 °C in the dark throughout the study, and was diluted to the designated concentrations using medium or water and vortexed briefly (2500 rpm, 5 s.) before the characterization or the exposure experiments. ReagentPlus grade silver acetate was purchased from Sigma-Aldrich (St. Louis, MO), and solutions of 1.00 mg/ml silver ion (Ag^+) equivalent to 1.53 mg/ml silver acetate were prepared fresh before exposure. All other chemicals used in this study were of molecular biology grade and were obtained from Sigma-Aldrich unless otherwise stated.

AgNP characterization
The AgNPs were characterized using dynamic light scattering (DLS) and transmission electron microscopy (TEM) to confirm the nanoparticle size distribution reported by the manufacturer. To measure the hydrodynamic diameters, samples of AgNPs were suspended at 50 µg/ml in deionized (DI) water or in culture medium, immediately placed in a disposable cuvette, and analyzed at 25 °C for 2 min per run on a Brookhaven 90Plus/BI-MAS Particle Size Analyzer (Holtsville, NY). Each sample was run in triplicate, with the polydispersity index (PDI) results presented as an average of the three measurements.

To characterize AgNP structure, shape and size uniformity, TEM was performed at an accelerating voltage of 80 kV on a JEOL JEM-1011 transmission electron microscope (Peabody, MA) equipped with a bottom-mounted Gatan Orius SC1000A camera and a Gatan Microscopy Suite software platform (Pleasanton, CA). Grids were prepared by placing a 10 µl drop of AgNP sample solution (50 µg/ml) onto a formvar/carbon-coated copper 100 mesh grid.

The stability of the AgNP suspension was monitored at several time points during a 24 h incubation period by dilution in culture media to 5.0 µg/ml followed by

Ultraviolet–visible (UV–vis) spectroscopy using a SpectraMax i3 spectrometer from Molecular Devices (Sunnyvale, CA).

Silver concentration assessment using ICP-MS

Mass concentration of silver was determined with inductively coupled plasma mass spectroscopy (ICP-MS) on a 7700 series ICP-MS from Agilent Technologies (Santa Clara, CA) equipped with on-line internal standard delivery. Total silver was analyzed using m/z 107 and Y and In as internal standards. Calibration standards were prepared by dilution from a 1000 ppm silver standard from Inorganic Ventures (Christiansburg, VA). A calibration curve was verified for each analysis using dilutions from a 1 ppm silver standard from SPEX CertiPrep (Metuchen, NJ). To assess silver concentration in the nanoparticle suspensions, tubes were sonicated while an aliquot for dilution was taken out and acidified with 800 μl of concentrated nitric acid. The samples were then diluted to 10 ml with a 4% HNO_3 0.5% HCl solution. For analysis of the supernatants, AgNP suspensions were subjected to centrifugation at $25,000 \times g$ for 90 min, using a WX Ultra Series centrifuge with a F50L-24 × 1.5 rotor (Thermo Scientific). Supernatants were carefully separated from pellets and silver concentration assessed.

Pluripotent mouse embryonic stem cell culture

Pluripotent ESGRO complete adapted C57BL/6 mESCs, which have been pre-adapted to serum-free and feeder-free culture condition, were obtained from EMD Millipore (Billerica, MA) at passage 12 (with 80% normal male mouse karyotype). The cells were seeded in cell culture flasks (Nunc, Roskilde, Denmark) coated with 0.1% gelatin solution (EMD Millipore), and maintained at 37 °C in a 5% CO_2 humidified incubator at standard densities (i.e., between $5 \times 10^4/cm^2$ and $5 \times 10^5/cm^2$) in ESGRO Complete Plus Clonal Grade Medium (EMD Millipore). The medium contains leukemia inhibitory factor (LIF), bone morphogenic protein 4 (BMP-4), and a glycogen synthase kinase-3b inhibitor (GSK3b-I) to help maintain pluripotency and self-renewal of the ESCs. Cells were passaged every 2–3 days (when reaching 60% confluence) with ESGRO Complete Accutase (EMD Millipore) at about 1:6 ratio. C57BL/6 mESCs maintain a stable karyotype under the above passaging condition. The cells used in the current study were at passage 18.

Cell differentiation through embryoid body formation

Induction of differentiation was achieved through embryoid body (EB) formation in hanging drop culture following a procedure adapted from De Smedt et al. [25]. In brief, stem cell suspensions were prepared on ice at a concentration of 3.75×10^4 cells/ml in ESGRO Complete Basal Medium (EMD Millipore), which does not contain LIP, BMP-4, or GSK3b-I. About 50 drops (each of 20 μl) of the cell suspension were placed onto the inner side of the lid of a 10-cm Petri dish filled with 5 ml phosphate buffered saline (PBS) (EMD Millipore) and incubated at 37 °C and 5% CO_2 in a humidified atmosphere. After 3 days, EBs formed in the hanging drops were subsequently transferred into 6-cm bacteriological Petri dishes (Becton–Dickinson Labware, Franklin Lakes, NJ) and were exposed to AgNPs or Ag^+. The EBs had an average diameter of 330–350 μm.

Cytotoxicity assay

Cytotoxicity was measured both in adherent (monolayer) culture and in EB culture by MTS assay using the CellTiter 96 AQueous One Solution Cell Proliferation Assay kit from Promega (Madison, WI), following instructions from the manufacturer. For adherent culture, pluripotent C57BL/6 mESC colonies cultured in ESGRO Complete Plus Clonal Grade Medium were dissociated with ESGRO Complete Accutase and a single-cell suspension at 1.0×10^5 cells/ml was prepared in ESGRO Complete Basal Medium. The cells were seeded in 96-well cell culture grade flat bottom plates (Nunc) coated with 0.1% gelatin (EMD Millipore) at 100 μl/well (1.0×10^4 cells/well) and allowed to adhere overnight at 37 °C with 5% CO_2. After 24 h, 100 μl medium containing 2× final concentrations of AgNPs or Ag^+ (0.1–50 μg/ml) was added to the test wells. In control wells, the same volume of medium was added as a vehicle control. The treatment was maintained for 24 h. At the end of the exposure, 20 μl of CellTiter 96 AQueous One Solution Cell Proliferation Assay reagent was added to each well that contained 100 μl medium. After 3 h incubation at 37 °C, the resultant absorbance was recorded at 490 nm using a SpectraMax i3 plate reader (Molecular Devices). Each concentration was tested in sextuplicate and repeated six times. To correct for interference of AgNPs or Ag^+ on MTS assay, a parallel control plate was set up with identical concentrations of AgNPs or Ag^+ but without seeded cells. The control plate was treated otherwise the same way as the test plate. The readings of the control plate were then subtracted from the corresponding wells of the test plate, and the resultant values were used in the dose–response plot.

For cytotoxicity assay in EB state, hanging drops were set up as described above. After 3 days, EBs were subsequently transferred into 6-cm bacteriological Petri dishes (Becton–Dickinson Labware) and treated with designated concentrations of AgNPs or Ag^+ for 24 h. Afterwards, 50 EBs were harvested per compound concentration. The EBs were allowed to precipitate and supernatant was removed, and were subsequently dissociated using ESGRO Complete Accutase (EMD Millipore). Cells were then resuspended in 700 µl ESGRO Complete Basal Medium (EMD Millipore), and 100 µl of the single cell suspension was pipetted into each well (in sextuplicate) of 96-well cell culture grade flat bottom plates (Nunc). Subsequent MTS assay was the same as for the adherent culture described above. The experiment was repeated independently three times.

AgNP/Ag+ exposure and RNA isolation

EB differentiation cultures at day 3 were exposed to the desired concentrations of AgNPs or Ag+ for 24 h. EBs were collected after exposure. Three biological replicates were used for each condition. Treatment with AgNPs or Ag+ at the concentrations used did not affect EB sizes (data not shown). EBs were lysed in RLT buffer (Qiagen; Valencia, CA) supplemented with β-mercaptoethanol, homogenized by QIAshredder (Qiagen), and kept in a −80 °C freezer until further processing. Total RNA was isolated on the EZ1 Advanced XL (Qiagen) automated RNA purification instrument using the EZ1 RNA Cell Mini Kit (Qiagen) following the manufacturer's protocol, including an on-column DNase digestion. RNA concentration and purity (260/280 ratio) were measured with a NanoDrop 2000 UV–Vis spectrophotometer (NanoDrop Products, Wilmington, DE). Integrity of RNA samples was assessed by an Agilent 2100 Bioanalyzer (Santa Clara, CA) with the RNA 6000 Nano Reagent Kit from the same manufacturer.

RNA processing and microarray experiment

The total RNA samples were preprocessed for hybridization to Mouse Gene 2.0 ST Array (Affymetrix, Santa Clara, CA) using the GeneChip WT PLUS Reagent Kit (Affymetrix) following the manufacturer's protocol. For each sample, 100 ng of total RNA was used. Subsequent hybridization, wash, and staining were carried out using the Affymetrix GeneChip Hybridization, Wash, and Stain Kit and the manufacturer's protocols were followed. The chips were then scanned on Affymetrix GeneChip Scanner 3000 7G and the image (DAT) files were preprocessed using the Affymetrix GeneChip Command Console (AGCC) software v.4.0 to generate cell intensity (CEL) files. Prior to data analysis, all arrays referred to in this study were assessed for data quality using the Affymetrix Expression Console software v.1.3 and all quality assessment metrics (including spike-in controls during target preparation and hybridization) were found within boundaries. The data set has been deposited in Gene Expression Omnibus (GEO; http://www.ncbi.nlm.nih.gov/geo/) of the National Center for Biotechnology Information with accession number GSE79766.

Microarray data processing and analysis

The values of individual probes belonging to one probe set in CEL files were summarized using the robust multiarray average (RMA) algorithm [26] embedded in the Expression Console software v.1.3 (Affymetrix), which comprises of convolution background correction, quantile normalization, and median polish summarization. Normalized data for all samples were then analyzed by unsupervised principal component analysis (PCA) [27] and hierarchical cluster analysis (HCA) [28], using the U.S. FDA's ArrayTrack software system [29, 30], to identify patterns in the dataset and highlight similarities and differences among the samples. Subsequently, differentially expressed genes (DEGs) were selected using one-way analysis of variance (ANOVA) using the Affymetrix Transcriptome Analysis Console (TAC) software v.2.0. Prior to making comparisons, a gene filtering procedure was applied to exclude probesets that appeared to be unexpressed in all sample groups. For each comparison between two experimental groups, the fold change (FC) of every annotated gene, together with their corresponding p value, was used for selection of DEGs with cutoff values indicated in the text.

Gene ontology and pathway analysis

The DEGs were subjected to gene ontology (GO) analysis using the Database for Annotation, Visualization, and Integrated Discovery (DAVID) [31, 32] to find overrepresentations of GO terms in the biological process (BP) category (GOTERM_BP_FAT). As background, the *Mus musculus* (mouse) whole genome was used. Statistical enrichment was determined using default settings in DAVID. The statistically enriched GO terms were classified using the web tool CateGOrizer [33] based on GO Slim. Pathway analysis was conducted with the Ingenuity Pathway Analysis (IPA) software using default settings to identify canonical pathways and pathway interaction networks associated with the DEGs.

Measurement of oxidative stress and apoptosis by flow cytometric analysis

EBs were exposed to 5.0 µg/ml of AgNPs or Ag+ the same way as in the gene expression study described above. As a positive control for both oxidative stress and apoptosis, EBs were also treated with 1 µM staurosporine. After treatment, EBs were collected and dissociated into single cells using the Embryoid Body Dissociation Kit from Miltenyi Biotec (San Diego, CA) following instruction from the manufacturer. Subsequent flow cytometry measurements of oxidative stress and apoptosis were conducted on a Guava easyCyte 8HT Flow Cytometer from EMD Millipore with a kit from the same manufacturer. The MitoStress Kit allows for the simultaneous measurement of mitochondrial superoxide generation, detected by membrane permeant dye MitoSox Red, and phosphatidyl serine expression on the cell surface of apoptotic cells as assessed by Annexin V binding. Data collection was performed using the InCyte and the CellCycle programs, both included in the guavaSoft software suite (ver. 3.1.1), and instructions from the manufacturer were followed.

Statistical analysis

Statistical analysis for data other than the microarray data was performed in KaleidaGraph v.4.03 of Synergy Software (Reading, PA) using one way ANOVA followed by Tukey HSD post hoc test.

Results

Silver nanoparticle characterization

The physicochemical properties of the 20 nm AgNPs are summarized in Table 1. The values in water agree well with those provided by the manufacturer. TEM analysis indicated that the AgNPs were spherical in shape and uniform in size with an average diameter of 20.4 ± 3.2 nm in water (Fig. 1a) and of 20.2 ± 4.1 nm in medium (Fig. 1b). The particle size distribution was narrow, with few particles >30 or <10 nm present. As is typical for nanomaterials, the hydrodynamic diameter of the AgNPs in water measured by DLS was found to be slightly larger (29.3 nm) than the physical diameter measured by TEM (20.4 nm). The hydrodynamic diameter increased substantially in the medium (78.6 nm) (Fig. 1c), probably due to slight agglomeration. Two small peaks appeared in the low diameter range on the DLS plot, probably resulted from protein species found in the media. The AgNPs had a low PDI (0.048) in water but much higher (0.349) in medium, indicating good stability in water but not in the culture medium. UV–Vis analysis in media (Fig. 1d) revealed colloidal homogeneity at the beginning (0 h) as reflected in the surface plasmon resonance with a characteristic peak ~400 nm. However, the peak intensity decreased rapidly within the first 2 h and then slowly between 2 and 8 h, with the peak broadened and shifted towards higher wavelengths, suggesting agglomeration of the particles in the medium. After

8 h, no further decrease in peak intensity was observed, although the peak position further shifted from 408 nm (8 h) to 424 nm (24 h).

To determine the amount of Ag^+ released from AgNPs, AgNPs were soaked in the medium at various concentrations (1.0, 2.0, 5.0 and 10.0 µg/ml) at 37 °C. After 24 h, AgNPs were removed from the medium by ultracentrifugation, and Ag^+ in the supernatant was measured by ICP-MS (Additional file 1: Figure S1). At 5.0 µg/ml, the initial silver ion (Ag^+) fraction of the AgNP solution was 1.82% (0.091 µg/ml) but increased to 21.46% (1.073 µg/ml) after 24 h incubation at 37 °C.

Cytotoxicity of silver nanoparticles and silver ion on mouse embryonic stem cells

Differentiating mESCs, either in adherent culture or in EB culture, were exposed to varying concentrations of AgNPs or Ag^+ for 24 h, and cell viability was measured by MTS assay (Fig. 2). In adherent culture, both Ag^+ and AgNPs exhibited a significant concentration-dependent cytotoxicity at concentrations >1.0 µg/ml (Fig. 2a). However, Ag^+ appeared much more potent than AgNPs, causing nearly complete cell death at as low as 5.0 µg/ml. In comparison, >40% cells survived after exposing to 50 µg/ml of AgNPs. Interestingly, low concentrations of Ag^+ (up to 0.5 µg/ml) increased cell viability almost 20% relative to the unexposed control, suggesting accelerated mESC proliferation and/or differentiation. Similar stimulating effects were also observed for AgNPs, albeit the effects were not statistically significant.

In EB state, differentiating mESCs appeared more resistant to AgNP and Ag^+ cytotoxicity. Ag^+ did not cause significant cytotoxicity at concentrations <2.0 µg/ml, nor did AgNPs at concentrations <5.0 µg/ml. Concentration-dependent decrease of cell viability at higher concentrations was also less severe than in adherent culture for both AgNPs and Ag^+. At 50 µg/ml, >70% cells remained alive after exposure to AgNPs, and ~35% cells survived Ag^+ exposure.

The morphological changes of the differentiating EBs (Additional file 2: Figure S2) corroborated the dose–response data shown above. After the 3 day EB formation

Table 1 Physicochemical properties of silver nanoparticles

Dispersant	Concentration (mg/ml)	TEM diameter (nm)	DLS diameter (nm)	PDI	ζ-potential (mV)
2 mM citrate[a]	1.06	20.6 ± 3.6	26.2	–	−39.8
Water	1.01	20.4 ± 3.2	29.3	0.048	–
Medium[b]	–	20.2 ± 4.1	78.6	0.349	–

TEM transmission electron microscopy, *DLS* dynamic light scattering; *PDI* polydispersity index

[a] Values provided by the manufacturer for the lot used in the current study

[b] ESGRO Complete Basal Medium (EMD Millipore) used for EB formation and exposure

– Data not available

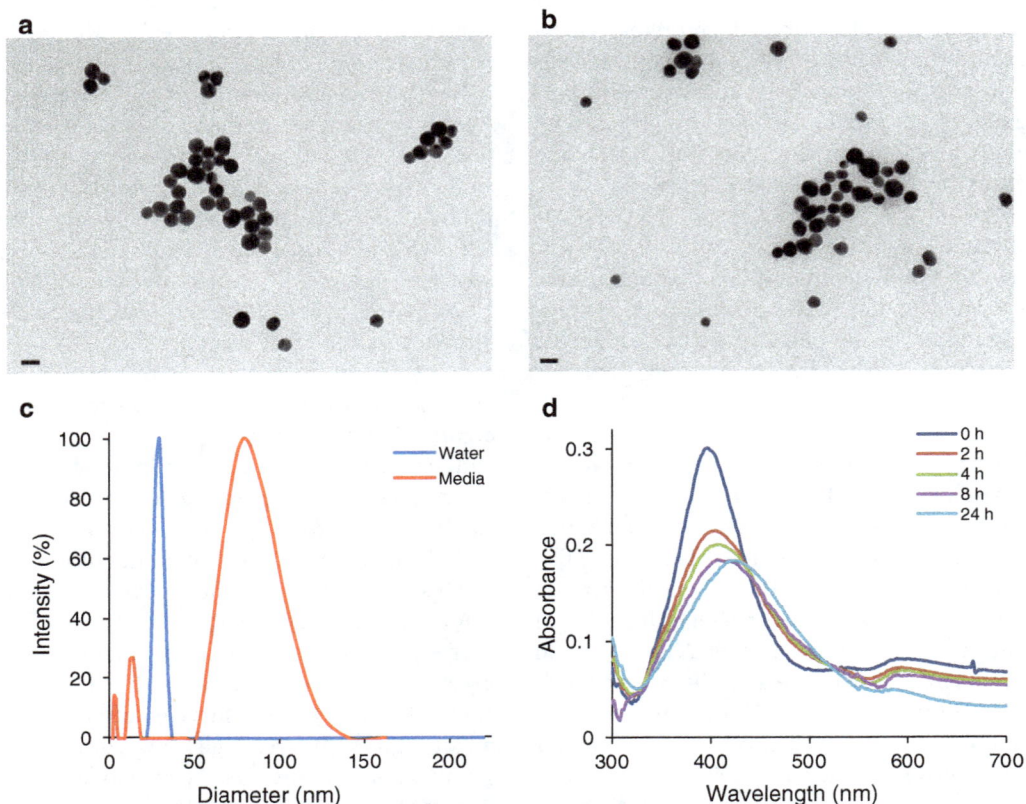

Fig. 1 Characterization of AgNPs. **a** Typical TEM images of AgNPs in water. **b** Typical TEM images of AgNPs in cell culture medium. The *size bar* at the *bottom left corner* of the images represents 20 nm. **c** Comparison of hydrodynamic size distribution of AgNPs in water and in cell culture medium. **d** Monitoring of AgNP stability by UV–vis spectra over 24 h incubation in cell culture medium

period and prior to exposure, the mESCs formed compact three-dimensional aggregates in spheroid shape, with a diameter of about 330–350 μm. During the following 24 h exposure period, the cells on the surface outgrew and formed a loose outer shell of 2–3 cells thick around the compact "core". In control samples and those of lower concentration exposure, the outer shell was intact with a smooth contour. However, when AgNPs exceeded 5.0 μg/ml, cells on the outer shell started to fall off. At 10 μg/ml and above, almost all cells on the outer shell fell off, whereas the core of the EBs was largely intact. In the case of Ag^+ exposure, cells stared to fall off the shell at 2.0 μg/ml. At 10 μg/ml and above, not only the outer shell completely disappeared, but the size of the core also decreased significantly.

The use of 5.0 μg/ml as the exposure concentration for AgNPs in the following microarray experiments was based on the cytotoxicity data described above. At this concentration, ~10% cell death (EC10) was observed for AgNPs in the EBs, as measured by MTS assay (Fig. 2b). Two concentrations of Ag^+ were used for comparison: 1.0 and 5.0 μg/ml. The lower concentration (1.0 μg/ml) corresponds to the concentration of Ag^+ released from 5.0 μg/ ml AgNPs into medium after 24 h incubation. The higher concentration (5.0 μg/ml) matches the total silver mass. Cell death in EBs after 24 h incubation at these concentrations of Ag^+ was 4 and 20%, respectively (Fig. 2b).

Global gene expression profiling

Cells in differentiating EBs were exposed for 24 h to 5.0 μg/ml of AgNPs, or to 1.0 or 5.0 μg/ml Ag^+ ions (Fig. 3a), and global gene expression changes were profiled using Affymetrix Mouse Gene 2.0 ST Array. Principal component analysis (PCA) of the microarray data showed that the biological triplicates of each treatment group (control, AgNP, and Ag^+) clustered roughly together and separated from one another except for the replicates of 1.0 μg/ml Ag^+ (Fig. 3b). Two of the replicates fell in the same area with the controls, but the remaining one fell close to the 5.0 μg/ml Ag^+ samples, far away from the other two. Due to the divergence between the replicates of the 1.0 μg/ml Ag^+ treatment group, they were excluded from further analysis. For the remainder of this report, treatment with 5.0 μg/ml Ag^+ will be simply referred to as treatment with Ag^+. Hierarchical clustering

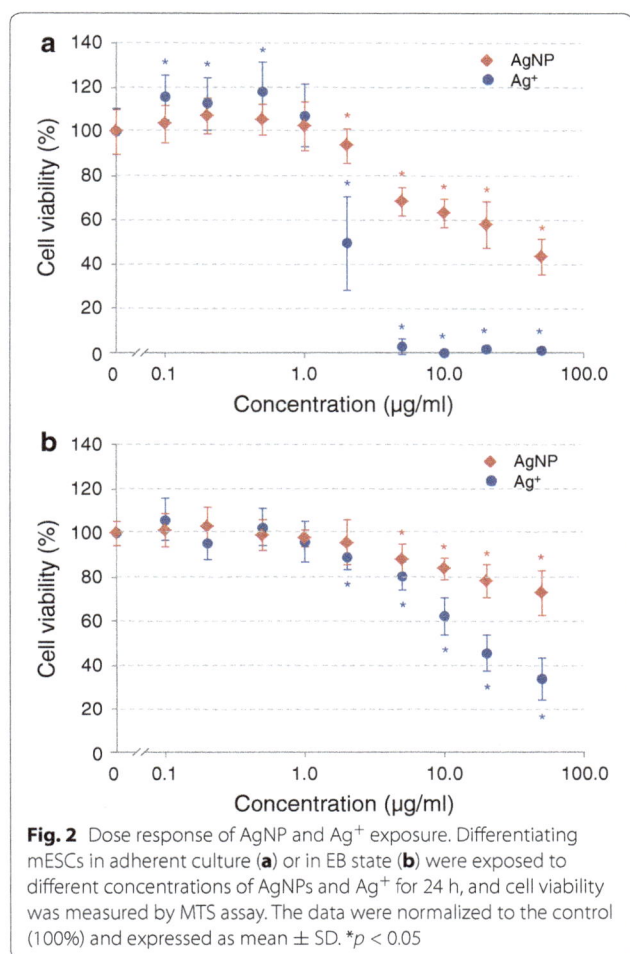

Fig. 2 Dose response of AgNP and Ag$^+$ exposure. Differentiating mESCs in adherent culture (**a**) or in EB state (**b**) were exposed to different concentrations of AgNPs and Ag$^+$ for 24 h, and cell viability was measured by MTS assay. The data were normalized to the control (100%) and expressed as mean ± SD. *$p < 0.05$

analysis (HCA) clustered the rest of the triplicates into respective treatment groups (Fig. 3c). In addition, the HCA indicated that the gene expression pattern of the samples treated with AgNPs was more related to that of the controls than to that of the Ag$^+$-treated samples.

The number of differentially expressed genes (DEGs) was considerably lower in cells exposed to AgNPs than in cells exposed to Ag$^+$ (Table 2). Using a fold change (FC) cutoff value of |FC| ≥ 1.5 and $p < 0.05$, 101 DEGs were identified in AgNP-treated cells (Additional file 3: Table S1), and 400 DEGs in Ag$^+$-treated cells (Additional file 4: Table S2). Among these genes, only 17 and 133 had |FC| ≥ 2.0 for AgNP- and Ag$^+$-treated cells respectively, indicating that in both cases the majority of the DEGs had a |FC| between 1.5 and 2.0. For both AgNP- and Ag$^+$-treated cells, the number of upregulated genes was smaller than that of downregulated genes. However, when directly comparing AgNP-treated cells to Ag$^+$-treated cells, the number of upregulated genes was higher than that of downregulated genes. There were 173 DEGs identified between AgNP-treated and Ag$^+$-treated cells (Additional file 5: Table S3).

The overlapping of DEGs between different treatment groups is plotted in the Venn diagrams shown in Fig. 4. Between the 43 genes upregulated by AgNP treatment and the 137 genes upregulated by Ag$^+$ treatment, only 18 genes (17 plus 1) were shared by both groups. Likewise, for the downregulated genes, 58 for AgNP-treated group and 263 for Ag$^+$-treated group, only 48 genes were common to both groups. These results suggest that there were substantial differences in the responses of cells to AgNPs and to Ag$^+$ albeit some similarities exit.

Functional annotation of differentially expressed genes

To unravel the cellular functions affected by exposure to AgNPs and to Ag$^+$, the DEGs were subjected to functional annotation using DAVID to find overrepresentations of gene ontology (GO) terms in the biological process (BP) category. The 101 genes regulated by AgNPs resulted in 120 GO terms (Additional file 6: Table S4). Using the CateGOrizer tool, these GO terms were grouped into 17 classes within the pre-defined set of parent/ancestor GO terms (Fig. 5). The 400 genes regulated by Ag$^+$ led to 322 GO terms (Additional file 7: Table S5), which were further grouped into 22 GO classes (Fig. 5). Despite the large differences in the numbers of DEGs between AgNPs and Ag$^+$, the functional classes enriched in these DEGs were strikingly similar between the two treatments. Fourteen classes were shared by AgNPs and Ag$^+$; on top of the list were *development, morphogenesis, metabolism, embryonic development,* and *cell differentiation.* Some other classes were only regulated by Ag$^+$ by not AgNPs, and vice versa. It is worthy to note that the class *response to stress* was enriched by AgNP treatment but not by Ag$^+$.

Pathways impacted by AgNPs and by Ag$^+$ were analyzed using IPA. Totally, 17 canonical pathways were affected by AgNPs and 27 by Ag$^+$ (Table 3). These pathways were broadly classified into four major categories. In both cases, the majority of the pathways identified fell into two categories, embryonic development and metabolism. Nevertheless, three canonical pathways related to cancer were found for Ag$^+$ but not for AgNPs. On the other hand, one pathway, *unfolded protein response*, was identified for AgNPs only.

It was noted that for the common pathways shared by AgNPs and Ag$^+$, there were always more genes involved in a particular pathway for Ag$^+$ than for AgNPs, hence lower p values for the Ag$^+$ pathways than their AgNP counterparts; and in almost all cases, the genes identified in AgNPs could be found in Ag$^+$, suggesting more potent effect of Ag$^+$ than AgNPs albeit of the same nature. This was also reflected by the pathway interaction networks shown in Fig. 6, where interactions among pathways impacted by Ag$^+$ were much more intense that those by AgNPs.

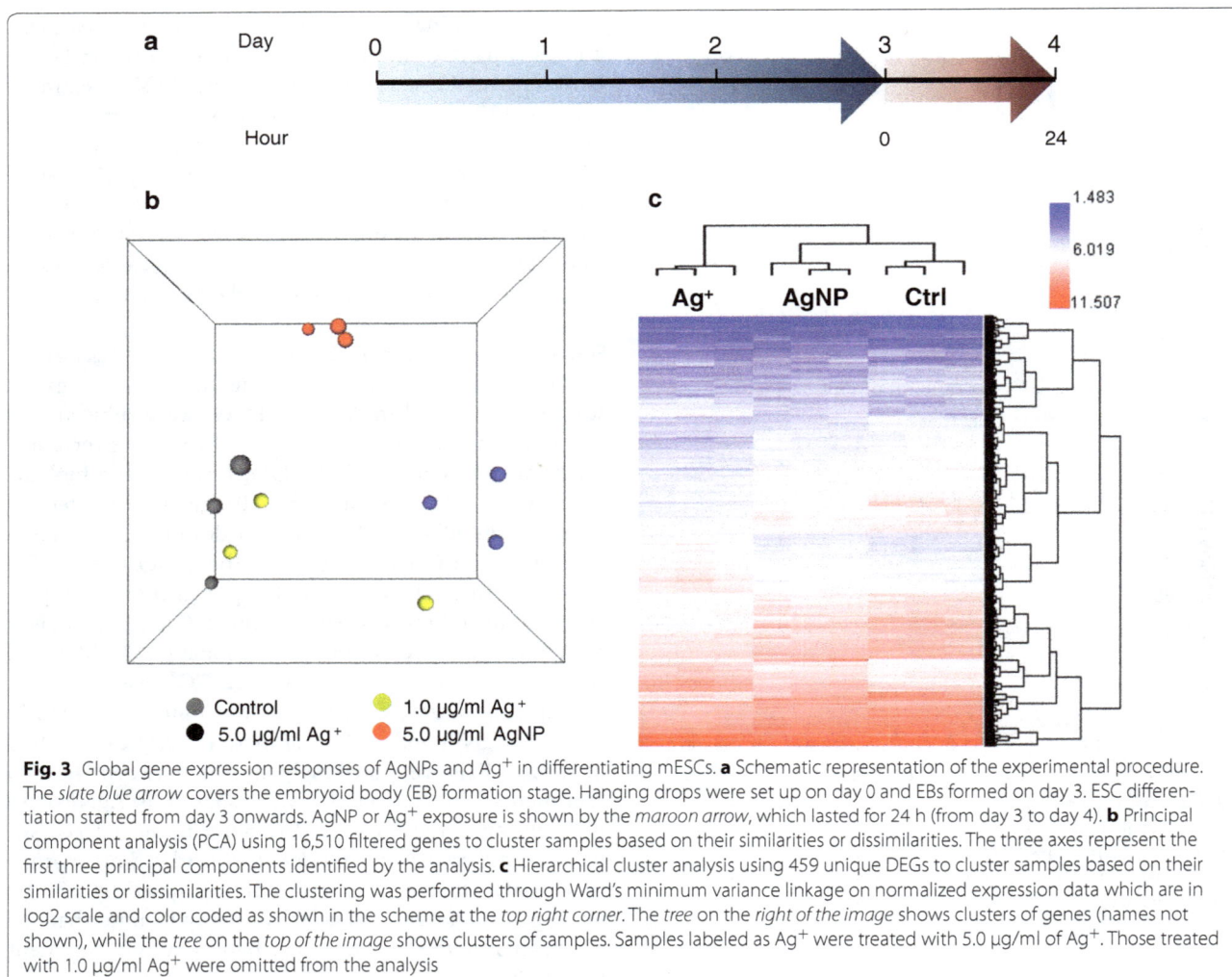

Fig. 3 Global gene expression responses of AgNPs and Ag$^+$ in differentiating mESCs. **a** Schematic representation of the experimental procedure. The *slate blue arrow* covers the embryoid body (EB) formation stage. Hanging drops were set up on day 0 and EBs formed on day 3. ESC differentiation started from day 3 onwards. AgNP or Ag$^+$ exposure is shown by the *maroon arrow*, which lasted for 24 h (from day 3 to day 4). **b** Principal component analysis (PCA) using 16,510 filtered genes to cluster samples based on their similarities or dissimilarities. The three axes represent the first three principal components identified by the analysis. **c** Hierarchical cluster analysis using 459 unique DEGs to cluster samples based on their similarities or dissimilarities. The clustering was performed through Ward's minimum variance linkage on normalized expression data which are in log2 scale and color coded as shown in the scheme at the *top right corner*. The *tree* on the *right of the image* shows clusters of genes (names not shown), while the *tree* on the *top of the image* shows clusters of samples. Samples labeled as Ag$^+$ were treated with 5.0 µg/ml of Ag$^+$. Those treated with 1.0 µg/ml Ag$^+$ were omitted from the analysis

Table 2 Number of differentially expressed genes ($|FC| \geq 1.5$, $p < 0.05$)

	All	Upregulated	Downregulated
AgNP vs Ctrl	101 (17)[a]	43 (6)	58 (11)
Ag$^+$ vs Ctrl	400 (133)	137 (41)	263 (92)
AgNP vs Ag$^+$	173 (39)	116 (32)	57 (7)

[a] Values in the parentheses indicate the number of genes that had $|FC| \geq 2.0$

Oxidative stress and apoptosis regulated by AgNPs

The *unfolded protein response* pathway identified for AgNPs (Table 3) prompted us to examine the expression of genes involved in oxidative stress, and several members of the heat shock protein and the metallothionein families were found significantly upregulated by AgNPs (Fig. 7). The increases in the expression of these genes in Ag$^+$-exposed cells were minimal and not significant except for Hspa1a, of which the expression in

Ag$^+$-exposed cells was significantly (with $p = 0.001086$) higher than the control, but with $|FC|$ marginally exceeded 1.5. These results suggest specific oxidative stress effect of AgNPs in ESCs.

To confirm these results, the effect of AgNPs on oxidative stress and downstream apoptosis were examined on the cellular level using flow cytometry (Fig. 8). Oxidative stress was assessed by mitochondrial superoxide generation which was detected by membrane permeant dye MitoSox Red. As shown in Fig. 8c, the percentage of cells positive for MitoSox Red staining (gated cells) increased significantly in AgNP treated cells (18.5%) as compared with control cells (1.4%). The mean intensity of yellow fluorescence (from MitoSox staining) of the entire cell population increased to 2.08-fold in cells treated with AgNPs in comparison with the control. As a comparison, treatment with 1 µM staurosporine, which has been reported to increases ROS production [34], resulted in 39.8% positive cells for MitoSox Red staining and

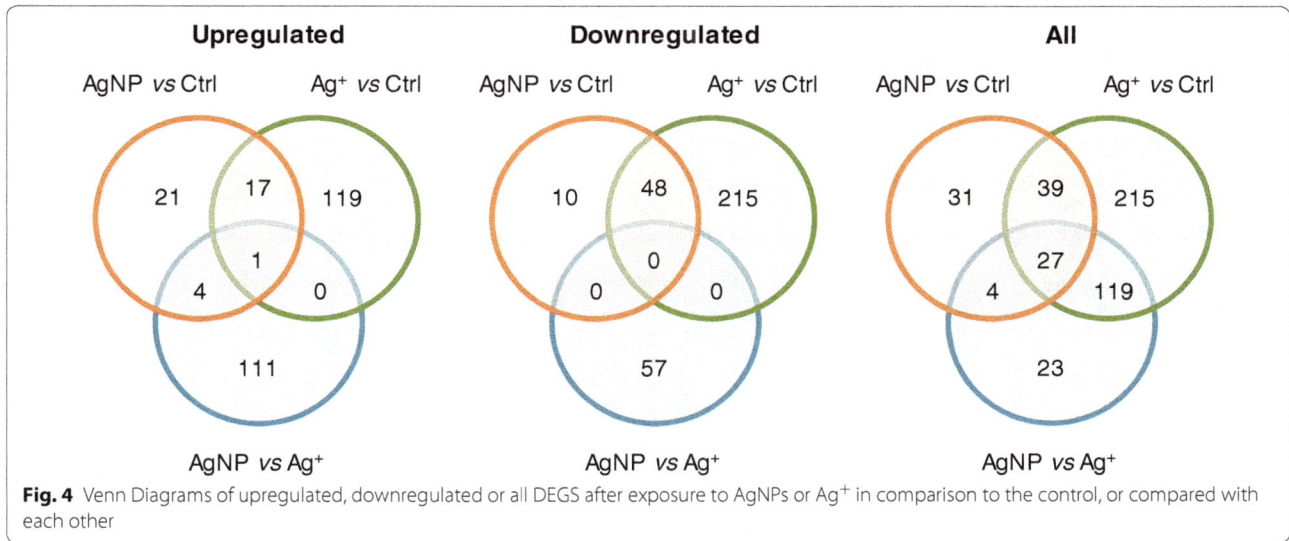

Fig. 4 Venn Diagrams of upregulated, downregulated or all DEGS after exposure to AgNPs or Ag+ in comparison to the control, or compared with each other

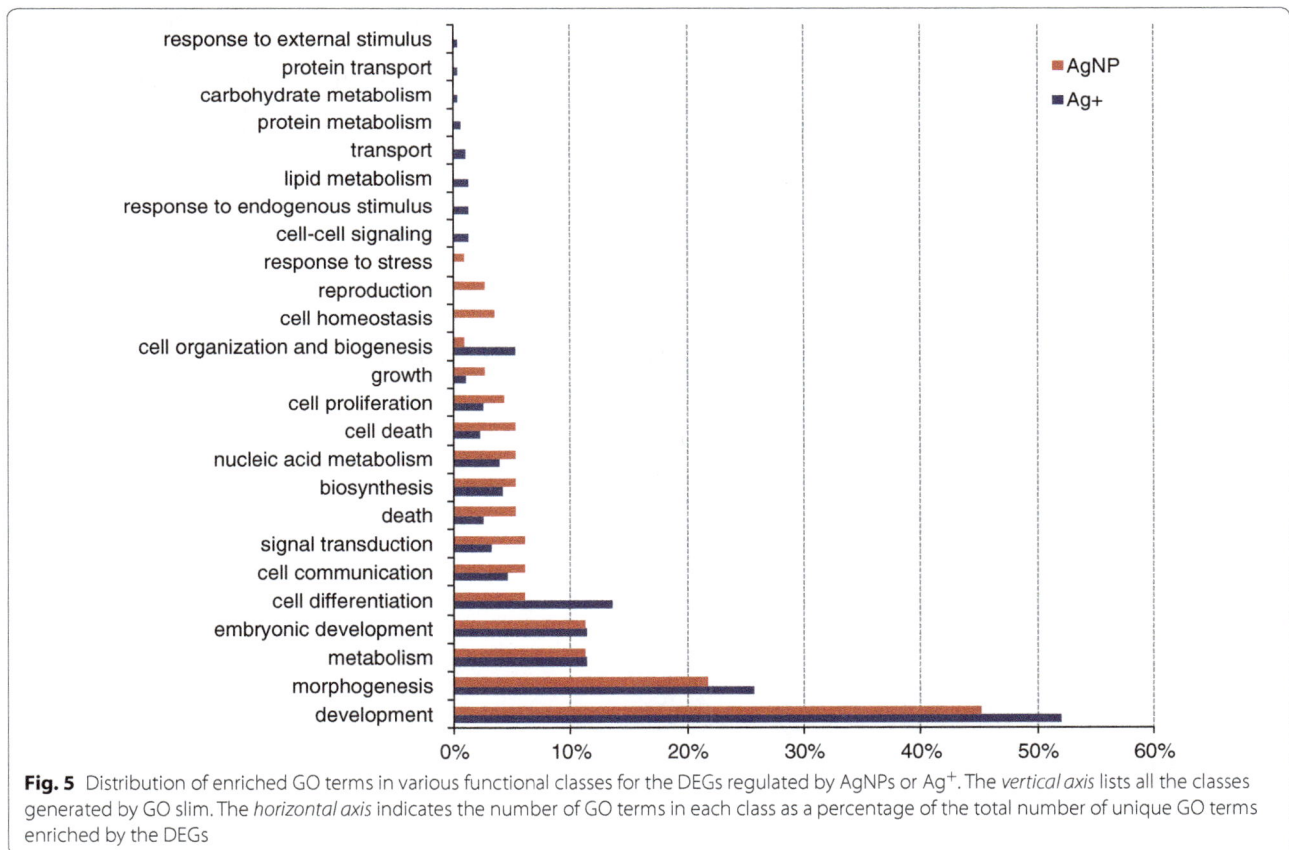

Fig. 5 Distribution of enriched GO terms in various functional classes for the DEGs regulated by AgNPs or Ag+. The *vertical axis* lists all the classes generated by GO slim. The *horizontal axis* indicates the number of GO terms in each class as a percentage of the total number of unique GO terms enriched by the DEGs

2.91-fold increase in mean fluorescence intensity compared with the control.

One of the hallmarks of apoptosis is the translocation of the membrane phospholipid phosphatidylserine to the external environment of the cells. Phosphatidylserine expression on the cell surface of apoptotic cells was assessed by binding of fluorescently labeled Annexin V in the current study. Similar to the results on oxidative stress as

Table 3 List of pathways impacted by AgNPs or Ag$^+$ and DEGs involved in the pathways

Ingenuity Canonical Pathways	AgNPs		Ag$^+$	
	p value	Molecules	p value	Molecules
Embryonic development				
Transcriptional regulatory network in embryonic stem cells	0.0005	HAND1, L1CAM, FOXC1	0.0000	HAND1, GATA6, L1CAM, EOMES, SKIL, HOXB1, FOXC1, GATA4
Human embryonic stem cell pluripotency	0.0178	SMAD7, SMAD6, BMP5	0.0074	BMP4, WNT3, BMP2, SMAD7, SMAD6, BMP5, FZD2
Mouse embryonic stem cell pluripotency			0.0002	LIFR, MYC, ID1, ID2, BMP4, T, ID3, FZD2
Role of NANOG in mammalian embryonic stem cell pluripotency			0.0001	LIFR, BMP4, T, WNT3, BMP2, GATA6, BMP5, FZD2, GATA4
Embryonic stem cell differentiation into cardiac lineages			0.0115	T, GATA4
Factors promoting cardiogenesis in vertebrates			0.0045	BMP4, WNT3, BMP2, BMP5, FZD2, GATA4
Cardiomyocyte differentiation via BMP receptors	0.0022	SMAD6, BMP5	0.0000	BMP4, BMP2, SMAD6, BMP5, GATA4
Regulation of the epithelial–mesenchymal transition pathway	0.0407	FOXC2, FGF14, PARD6B	0.0003	ETS1, FOXC2, ID2, FGF10, WNT3, FGF14, PARD6B, FGF3, NOTCH1, FZD2, FGF19
Hepatic fibrosis/hepatic stellate cell activation			0.0010	IGFBP4, COL4A1, FLT1, KLF6, SMAD7, LAMA1, BAMBI, IGFBP5, COL4A2, KDR
Axonal guidance signaling			0.0000	SLIT3, SHH, PAPPA, BMP4, WNT3, BMP2, UNC5B, L1CAM, HHIP, SLIT2, ROBO3, BMP5, NTN1, EFNB2, PRKAR2B, GLIS1, TUBB4A, FZD2, PTCH2, NRP1, UNC5C
BMP signaling pathway	0.0034	SMAD7, SMAD6, BMP5	0.0003	BMP4, PRKAR2B, BMP2, SMAD7, SMAD6, BMP5, CHRD
ERK5 signaling			0.0209	MYC, RPS6KA6, SGK1, CREB3L4
FGF signaling			0.0141	FGF10, FGF14, CREB3L4, FGF3, FGF19
Netrin signaling			0.0040	PRKAR2B, UNC5B, NTN1, UNC5C
Notch signaling			0.0004	DLL1, LFNG, DLL3, HES7, NOTCH1
RAR activation	0.0427	ALDH1A2, SMAD7, SMAD6	0.0148	PRKAR2B, CYP26A1, BMP2, ALDH1A2, SMAD7, SMAD6, ADCY8, RBP1
Sonic Hedgehog signaling			0.0001	SHH, PRKAR2B, GLIS1, HHIP, PTCH2
TGF-β signaling	0.0490	SMAD7, SMAD6		
eNOS signaling	0.0191	Hspa1b, HSPA1A/HSPA1B, KDR	0.0007	PRKAR2B, FLT1, HSPA1A/HSPA1B, PRKAA2, AQP8, ADCY8, KDR, LPAR3, NOSTRIN
VEGF family ligand-receptor interactions	0.0389	KDR, NRP1		
Metabolism				
Choline biosynthesis III	0.0490	HMOX1		
Corticotropin releasing hormone signaling			0.0389	SHH, PRKAR2B, CREB3L4, ADCY8, PTCH2
FXR/RXR activation			0.0200	TTR, APOB, VTN, VLDLR, MTTP, FGF19
Heme degradation	0.0166	HMOX1		
Histamine degradation	0.0490	ALDH1A2		
NAD biosynthesis II (from tryptophan)	0.0490	TDO2		
Retinoate biosynthesis I			0.0174	BMP2, ALDH1A2, RBP1
Serotonin degradation	0.0145	UGT2B28, ALDH1A2		
Sulfate activation for sulfonation			0.0331	PAPSS2
Tryptophan degradation to 2-amino-3-carboxymuconate semialdehyde	0.0288	TDO2		
Tyrosine biosynthesis IV			0.0490	PCBD1

Table 3 continued

Ingenuity Canonical Pathways	AgNPs		Ag$^+$	
	p value	Molecules	p value	Molecules
Vitamin-C transport			0.0251	SLC23A1, GLRX
Xenobiotic metabolism signaling	0.0178	UGT2B28, HMOX1, ALDH1A2, CES2		
Stress response				
Unfolded protein response	0.0209	Hspa1b, HSPA1A/HSPA1B		
Cancer				
Molecular mechanisms of cancer			0.0013	SHH, Naip1 (includes others), BMP4, WNT3, BMP2, SMAD7, SMAD6, BMP5, MYC, CCND2, PRKAR2B, ADCY8, FZD2, NOTCH1, PTCH2
Basal cell carcinoma signaling			0.0000	SHH, BMP4, WNT3, GLIS1, BMP2, HHIP, BMP5, FZD2, PTCH2
Bladder cancer signaling			0.0155	MYC, FGF10, FGF14, FGF3, FGF19

shown above, the percentage of cells positive for Annexin V binding (gated cells) increased significantly in AgNP treated cells (16.8%) as compared with control cells (1.9%), and the mean intensity of red2 fluorescence (from Annexin V binding) in cells treated with AgNPs was 2.04-fold as high as in control cells. For comparison, treatment with 1 μM staurosporine, a known apoptosis inducer [34], resulted in 38.9% positive cells for Annexin V binding and 2.91-fold increase in mean fluorescence intensity compared with the control.

Interestingly, cells treated with Ag$^+$ did not show significant changes in either the percentage of gated cells or the mean fluorescence intensity, for both MitoSox Red staining and Annexin V binding, in comparison with the control cells (Fig. 8c).

Discussion

The three-dimensional (3D) assembly of stem cells in the form of EB spheroids facilitates cellular interactions that promote morphogenesis, analogous to the multicellular, heterotypic tissue organization that accompanies embryogenesis [35]. The complex interactions between heterologous cell types result in the induction of differentiation of stem cells to derivatives of all three embryonic germ layers [36]. Therefore, ESC differentiation in EBs has been considered to faithfully recapitulate stages of early embryo development and is increasingly used as an in vitro model for developmental toxicity testing [13]. In addition, implementing toxicogenomics in ESC differentiation has been shown as a sensitive method for the detection of a variety of developmental toxicants [18]. In the present study, we characterized transcriptomic changes induced by citrate-coated 20 nm AgNPs during spontaneous differentiation of mESCs, and compared to those induced by Ag$^+$ (silver acetate) under otherwise identical conditions. Despite the large differences in the numbers of DEGs, functional annotation and pathway analysis of the regulated genes revealed overall similarities between AgNPs and Ag$^+$. Functions and pathways related to embryonic development and metabolism appeared on top of the lists for both treatments. Functional changes related to metabolism are thought to be necessary for cells to cope with the toxic insult of AgNPs and Ag$^+$ [37]. Those related to embryonic development suggest that both AgNPs and Ag$^+$ have the potential to cause developmental toxicities when in contact with the differentiating embryo. It has been reported that AgNPs induce distinct developmental defects in zebrafish embryos [38]. In a recent in vivo study, it was found that silver acetate exposure caused adverse effects on reproduction and postnatal development in rats [39].

Human exposure to nanoparticles may occur via inhalation, ingestion, dermal absorption or, in some cases, artificially induced via inoculation. Once systemically available, these nanoparticles appear capable of spreading to most organ systems and may even cross biological barriers. Several studies demonstrated that certain nanoparticles can penetrate the placenta barrier, reach the fetus, and evoke embryotoxic effects [40]. Since the application of AgNPs is expected to further increase in the future, long term exposure and potential accumulation of AgNPs in the human body may result. The results presented here point to the potential of AgNPs to cause developmental toxicity once these particles migrate into the uterus, cross the placenta and reach the embryo.

Proper physicochemical characterization of nanoparticles should be performed in the relevant dispersion medium prior to conducting toxicity studies [41, 42]. This is because complex interactions exists between a particle and its surrounding microenvironment, including attractive or repulsive forces between particles, and between particles and biological substances in the dispersion medium such as salts (ions) and proteins. These factors

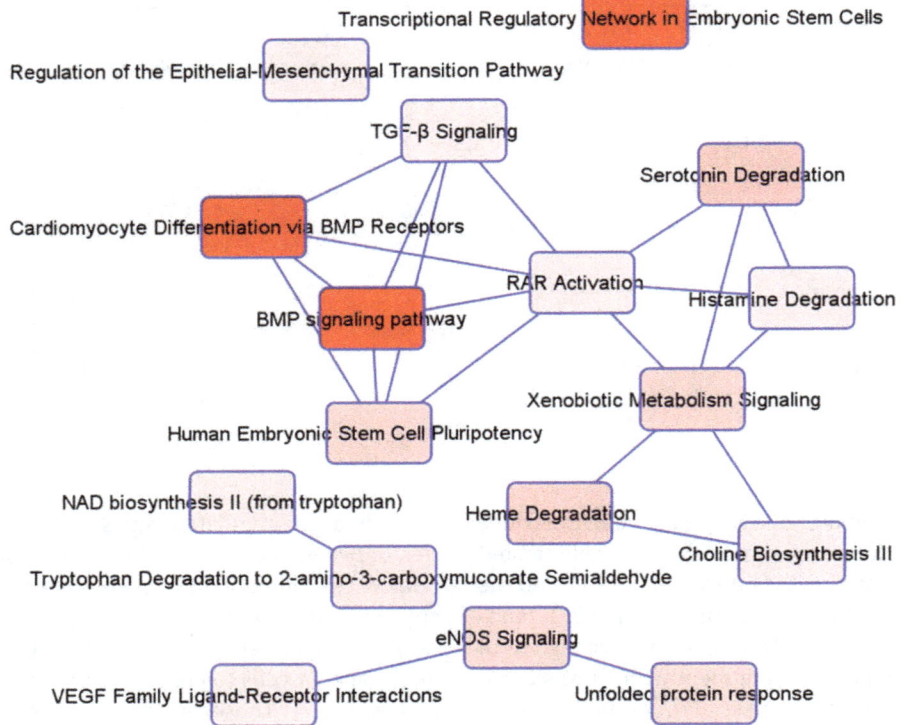

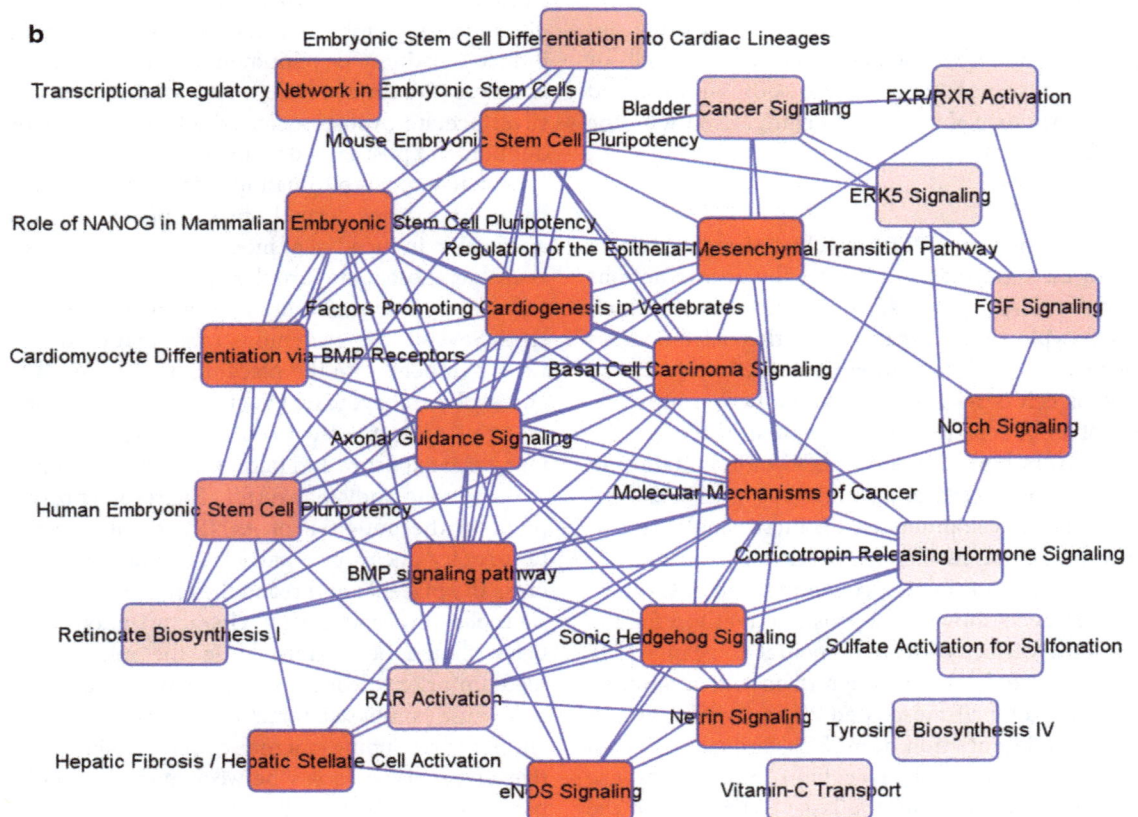

(See figure on previous page.)
Fig. 6 Pathway interaction networks for DEGs affected by **a** AgNPs and **b** Ag$^+$. Each *rectangular box* represents a pathway affected by the DEGs with the name indicated. The *darkness* of the *red color* of each *box* represents the p value for enrichment of each pathway—the *darker the color*, the *lower* the p value. A *line* linking two *boxes* represents an interaction between two pathways. The length of the line is arbitrary

affect both the hydrodynamic size and the surface charge of the nanoparticles, and can impact on their agglomeration status, which may in turn affect the extent of toxicity. It has been considered that larger agglomerates of nanoparticles are less toxic than monodispersed particles or smaller aggregates [41]. The DLS and UV–Vis data suggest there was a dynamic interplay between the AgNPs and the culture medium, starting immediately upon contact and lasting throughout the duration of the exposure, leading to increased agglomeration of the AgNPs. Aggregation of nanoparticles in cell culture media has been reported previously by others [19, 37, 40]. However, it has to be noted that there are significant limitations to the techniques used

for the characterization, which complicates the interpretation of the data. For both DLS and UV–Vis, characteristic peaks for AgNPs overlaps with those of the medium. Moreover, it is well known that DLS is a weight-averaged measurement biased towards larger particle sizes and especially towards agglomerates or aggregates. Thus, although the observed DLS curve showed an average hydrodynamic size of ~78 nm, the system may actually contain a very high number of smaller particles that are bioavailable and can interact with the ESCs. From another point of view, it should be pointed out that the agglomeration of AgNPs found here does not diminish, but rather adds to, the significance of the current study. Since the culture medium for ESCs used here was similar to body fluid in composition, it is likely that AgNPs entering the human body would agglomerate to some extent. In this sense, the results found in the current study are meaningful for real life situations. Since unagglomerated (monodispersed) form of nanoparticles would be more toxic than their agglomerated form, the results reported here signify the importance of nanoparticle regulation in consumer products.

It was interesting to note that in the EB state ESCs were in general more resistant to AgNP and Ag$^+$ cytotoxicity than in adherent culture (Fig. 2). At 5.0 µg/ml, Ag$^+$ caused nearly complete cell death (97%) in adherent culture, while >80% cells survived in EBs. For AgNPs, cell viability at 5.0 µg/ml was 68% in adherent culture and 87% in EB state. At 50 µg/ml, 73 and 34% cells remained alive after exposure to AgNPs and Ag$^+$ respectively in EB culture. In comparison, in adherent culture only 43% cells survived AgNP exposure, and <2% cells survived after exposing to Ag$^+$. These results could be explained by the fact that in adherent culture, cells formed a monolayer whereby AgNPs or Ag$^+$ diffused freely throughout the medium, and thereby reached equilibrium where all cells were equally exposed to the same concentrations of AgNPs or Ag$^+$. In contrast, in 3D aggregates of EBs, a concentration gradient of exogenous or endogenous factors is established between the surrounding culture environment and the interior of the spheroids [35]. Therefore, concentration of Ag$^+$ in the interior of the EBs would be lower than that of exterior environment. The concentration is inversely related to EB size, with decreasing concentration from the surface toward the center of aggregates [35]. Due to the high cell packing density (Additional file 2: Figure S2), cells in the center of the EBs may be completely shielded and not exposed to Ag$^+$ at

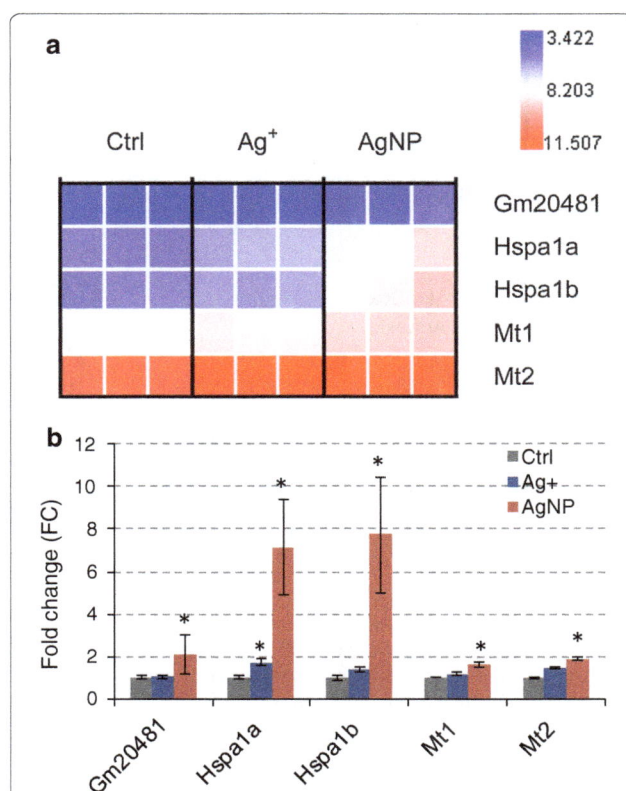

Fig. 7 Upregulation of several members of the heat shock protein and the metallothionein families by AgNPs, but not Ag$^+$. **a** Heat map showing the normalized expression intensity in the different treatment groups, which are in log2 scale and color coded as shown in the scheme at the *top right corner*. **b** *Bar graph* showing fold change (FC) of each member of the heat shock protein and the metallothionein families after exposure to AgNPs or to Ag$^+$. The FC values are relative to the controls (of which the mean value was set to 1)

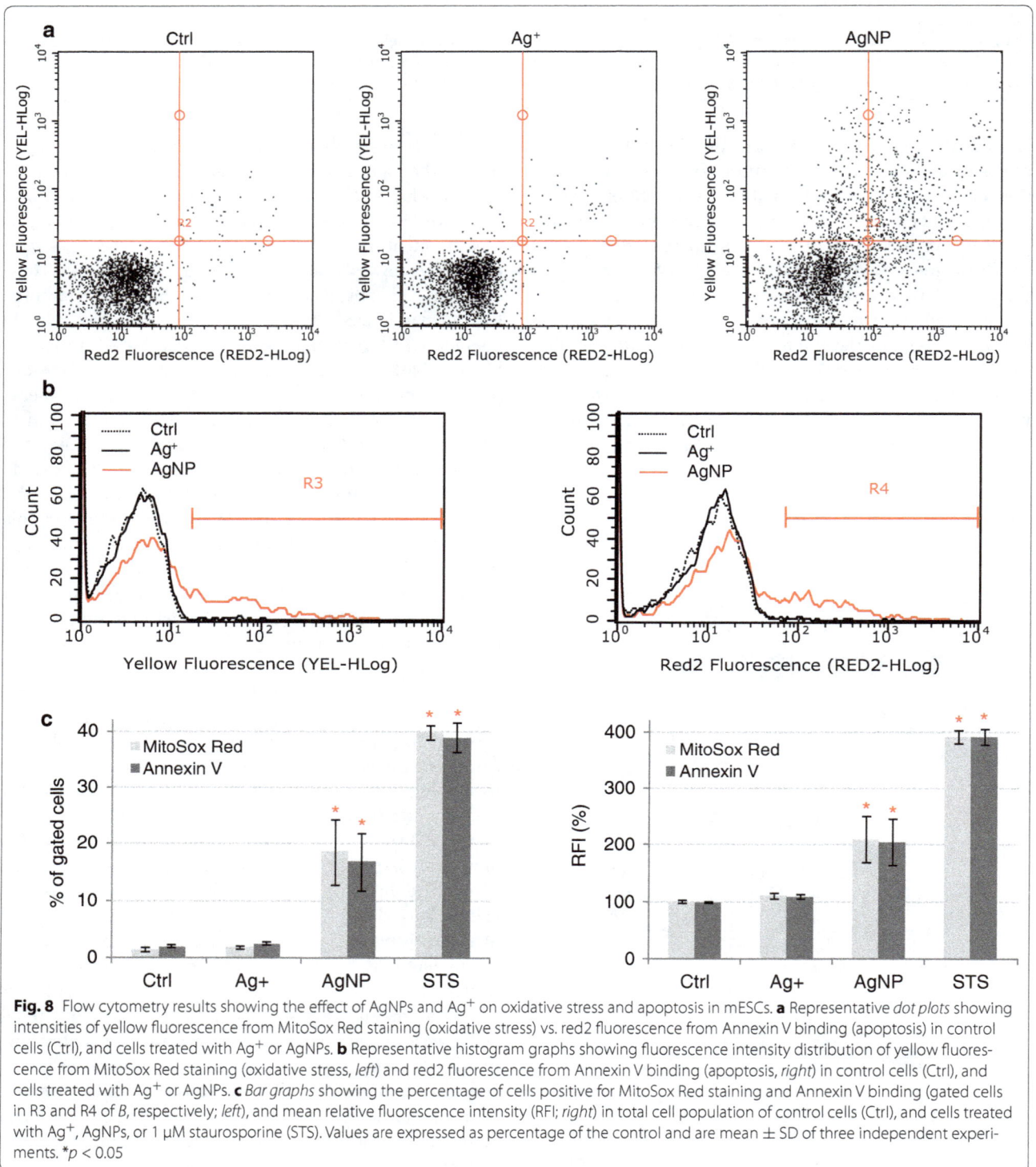

Fig. 8 Flow cytometry results showing the effect of AgNPs and Ag⁺ on oxidative stress and apoptosis in mESCs. **a** Representative *dot plots* showing intensities of yellow fluorescence from MitoSox Red staining (oxidative stress) vs. red2 fluorescence from Annexin V binding (apoptosis) in control cells (Ctrl), and cells treated with Ag⁺ or AgNPs. **b** Representative histogram graphs showing fluorescence intensity distribution of yellow fluorescence from MitoSox Red staining (oxidative stress, *left*) and red2 fluorescence from Annexin V binding (apoptosis, *right*) in control cells (Ctrl), and cells treated with Ag⁺ or AgNPs. **c** *Bar graphs* showing the percentage of cells positive for MitoSox Red staining and Annexin V binding (gated cells in R3 and R4 of *B*, respectively; *left*), and mean relative fluorescence intensity (RFI; *right*) in total cell population of control cells (Ctrl), and cells treated with Ag⁺, AgNPs, or 1 μM staurosporine (STS). Values are expressed as percentage of the control and are mean ± SD of three independent experiments. *$p < 0.05$

all. For AgNPs, the concentration in the interior of the EBs would be ever lower than Ag⁺ due to more severe steric hindrance imposed on the AgNPs as a result of their larger sizes, especially if agglomeration occurred.

It was also intriguing to note that in adherent cultures, low concentrations of Ag⁺ (up to 0.5 μg/ml) increased cell viability almost 20% relative to the unexposed control, suggesting accelerated mESC proliferation and/or differentiation

(Fig. 2a). Similar stimulating effects were also observed for AgNPs, albeit not statistically significant (Fig. 2a). Such an hormesis effect was also reported previously for AgNPs in HepG2 cells [24], and for silica nanoparticles in D3 mESCs [40], and could be explained as an adaptive response of cells to low levels of potentially toxic agents [40].

The toxicity of AgNPs has been demonstrated both in vitro and in vivo [43]. However, whether the observed toxicity is due to Ag^+ released from the AgNPs or related to the special properties of nanosized particles is not entirely clear, and is often a topic of rigorous debate. This is partly due to the fact that the release of Ag^+ from AgNPs is a dynamic process and is affected by many factors such as temperature, surface chemistry, and stabilizing agent [44]. In order to unravel the mechanism of toxicity, the effect of AgNPs on the gene expression in ESCs was studied here at a relatively low toxicity concentration (~EC10) to avoid cellular processes in necrotic or apoptotic cells overwhelming and leading to misinterpretation of the data. As comparison, two concentrations of Ag^+ were used, one approximates the maximum Ag^+ released from 5.0 μg/ml AgNPs after 24 h incubation in the medium (1.0 μg/ml; Additional file 1: Figure S1), the other was the equivalent mass concentration of silver in 5.0 μg/ml AgNPs (i.e., 5.0 μg/ml). At 1.0 μg/ml, Ag^+ only induced 47 DEGs (data not shown), compared to 101 DEGs regulated by 5.0 μg/ml AgNPs. This suggests that AgNP toxicity was not entirely due to Ag^+ released from the AgNPs. In the present study, we only measured Ag^+ release from AgNPs in the medium (without cells). AgNPs may also release Ag^+ inside the cell via a "Trojan-horse" mechanism, where the particles enter cells and are then ionized within the cell [45]. The purpose of the present study was to evaluate the effect of AgNPs and Ag^+ on the cellular transcriptome of ESCs, and therefore more systematic studies are needed in order to completely disentangle the effects of AgNPs from those of released Ag^+. In addition, a cellular uptake study of the AgNPs may help to confirm that the transcriptomic changes seen here were indeed caused by ingested AgNPs.

More importantly, microarray data analysis showed that several stress response proteins, including members of the heat shock proteins (HSPs) and the metallothionein (MT) families, were upregulated by AgNPs but not Ag^+. As the name suggests, HSPs are a group of proteins induced by heat shock, or hyperthermia. Their expression has also been found to increase when cells are exposed to an array of other stresses, including heavy metals and oxygen radicals [46], where they play a role in maintaining the correct folding of proteins by preventing protein aggregation or facilitating selective degradation of misfolded or denatured proteins [47]. MTs are a family of cysteine-rich proteins with several isoforms [48]. MTs have the capacity to bind heavy metals, both physiological and xenobiotic, through the thiol group of their cysteine residues [49]. It has been suggested that MTs not only are involved in the regulation of physiological metals and protect cells from metal toxicity, but also provide protection against oxidative stress [50]. The cysteines of MTs have been shown to bind oxidant radicals like superoxide and hydroxyl radicals [51], and MT expression has been implicated as a transient response to many forms of stress or injury [50]. Several previous studies [23, 24, 37, 52, 53] reported upregulation of HSPs and MTs in somatic cells following exposure to AgNPs. The upregulation of these proteins found in the current study suggests that AgNP exposure also induced cellular stresses and elicited cellular protective responses in ESCs. It was intriguing to note that the same concentration of Ag^+ did not induce such stress responses in the ESCs. The reason for this is not well understood but could be partly explained by the so-called Trojan horse theory. The plasma membrane functions to some degree as a natural barrier for metal ions, thereby protecting the cells from damage. However, nanoparticles circumvent this barrier when they are taken up by the cells via endocytic pathways, leading to the release of metal ions within the cells as a result of lysosome rupture, and subsequently generate free radicals within the cells [23]. Similar findings were reported previously [23, 24].

It has been reported that in several somatic cell types, augmented oxidative stress induced by AgNPs, through the generation of reactive oxygen species (ROS), may further lead to DNA damage and chromosomal aberrations [19, 53, 54]. Cells with damaged DNA will accumulate in the G2/M phase allowing the cells extra time to repair of damaged DNA, causing arrest in cell cycle progression [23, 54]. Cells with massive or irreversible DNA damage will not be able to repair the DNA effectively and undergo apoptosis at a later stage [19, 53, 54]. The results presented here indicate that this scenario may also hold true for ESCs. Flow cytometry results (Fig. 8) indicated that both oxidative stress and apoptosis increased significantly after treatment with AgNPs. Based on these results, the molecular mechanisms of ESC cellular responses against AgNPs are therefore speculated as following (Fig. 9): AgNPs enter the cell via endocytosis, release Ag^+ within the cell after lysosome rupture, and subsequently generate ROS. The increased oxidative stress further leads to DNA damage, causing cell cycle arrest at the G2/M phase in order to repair the damaged DNA. Cells unable to repair the DNA damages will eventually undergo apoptosis.

Kawata et al. [24] found a number of checkpoint related genes (BIRC5, BUB1B, CCNA2, CDC25B, CDC20, and CKS2) increased expression in HepG2 cells after exposure to AgNPs. However, despite the effect of AgNPs on

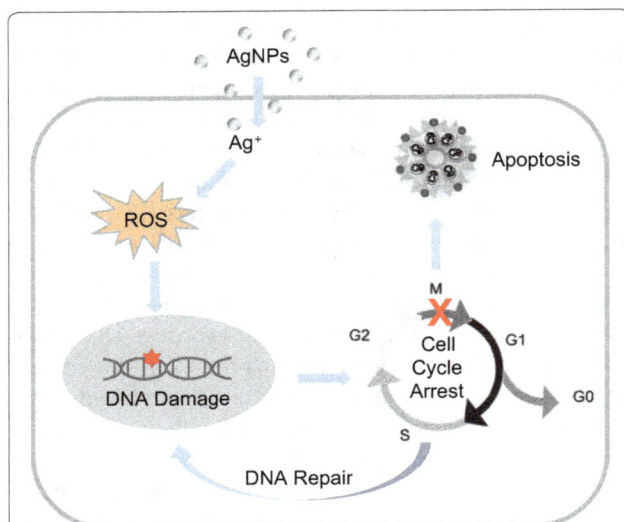

Fig. 9 Schematic representation of molecular mechanisms of AgNP toxicity in ESCs, which are similar to those previously reported for somatic cells. AgNPs enter ESCs via endocytosis, release Ag⁺ within the cell after lysosome rupture, and subsequently generate ROS. The elevated ROS may lead to DNA damage and cause the cell into cycle arrest in order to repair damaged DNA. Cells failed to do so will eventually undergo apoptosis

oxidative stress and apoptosis revealed at the protein level by flow cytometry, significant induction of genes associated with the cellular processes downstream of oxidative stress (DNA damage, cell cycle arrest, and apoptosis) was not observed at the mRNA level by the microarray study. This is probably because of the ephemeral nature of many gene expression changes. The correlation between mRNA and protein abundances in the cell has been reported to be notoriously poor [55]. One of the reasons for the poor correlation is that there is a temporal difference in cells responding to environment perturbations at the mRNA level and at the protein level [56]. There may have been detectable gene expression changes at the mRNA level; however, by the time when these changes are translated into the protein level, the altered mRNA expression of these genes may have already recovered to their normal level.

DNA damage by ROS production induced by AgNPs may further activate the p53 pathway [19], which plays important roles in carcinogenesis and tumor progressing. Therefore, AgNPs have the potential to cause carcinogenicity [24]. However, pathways related to cancer were not identified in cells exposed to AgNPs. On the contrary, three pathways were found in cells exposed to Ag⁺. The reason for this is not clear. However, it cannot be excluded that AgNPs would affect cancer pathways at higher concentrations than that used in the current study (5.0 μg/ml).

Conclusions

In this study, we characterized transcriptomic changes induced by AgNPs during spontaneous differentiation of mouse ESCs, and compared to those induced by Ag⁺. Overall, cellular responses to AgNPs and Ag⁺ in ESCs were substantially similar. In both cases, most of the functions and pathways impacted are related to embryonic development and metabolism, suggesting that both AgNPs and Ag⁺ have the potential to alter, reversibly or irreversibly, developmental pathways. However, specific effects on oxidative stress and apoptosis were observed for AgNPs. Taken together, our results indicate that the widespread application of AgNPs and silver-containing products may be a health concern. Long term exposure of humans to these products could potentially result in accumulation in the body and subsequently induce acute or chronic toxicity. Once systemically available, AgNPs has the potential to reach the embryo thus causing developmental toxicity. However, since information on whether these particles are able to transfer from the mother to the fetus across the placenta is currently lacking, the results of the current study have to be interpreted with caution until in vivo data becomes available.

Additional files

Additional file 1: Figure S1. Release of Ag⁺ from AgNPs into culture medium after 24 h incubation at 37 °C. (A) Ag⁺ released into the supernatant of the medium before (0 h) and after (24 h) incubation. (B) Total Ag⁺ in the AgNP suspension before (0 h) and after (24 h) incubation.

Additional file 2: Figure S2. Morphological changes of the differentiating EBs after exposure for 24 h to varying concentrations of AgNPs (A) or Ag⁺ (B). The concentrations of AgNPs or Ag⁺ used for the exposure (in μg/ml) are indicated by the numbers at the top left corner of each image.

Additional file 3: Table S1. List of 101 DEGs regulated by AgNPs. Gene expression profiles of AgNP-exposed samples were compared with the controls using one-way analysis of variance (ANOVA) based on the Welch t-test, and DEGs were selected by |FC| ≥ 1.5 and $p \leq 0.05$.

Additional file 4: Table S2. List of 400 DEGs regulated by Ag⁺. Gene expression profiles of Ag⁺-exposed samples were compared with the controls using one-way analysis of variance (ANOVA) based on the Welch t test, and DEGs were selected by |FC| ≥ 1.5 and $p \leq 0.05$.

Additional file 5: Table S3. List of 173 DEGs between AgNP- and Ag⁺-treated cells. Gene expression profiles of AgNP-exposed samples were compared with Ag⁺-exposed samples using one-way analysis of variance (ANOVA) based on the Welch t test, and DEGs were selected by |FC| ≥ 1.5 and $p \leq 0.05$.

Additional file 6: Table S4. Lists of 120 GO terms in the biological process (BP) category enriched from the 101 genes regulated by AgNPs. DAVID was used for the analysis. The *Mus musculus* (mouse) whole genome was used as background. Statistical enrichment was determined through a modified Fisher's exact test ($p < 0.1$) and count threshold >2 (default settings in DAVID).

Additional file 7: Table S5. Lists of 322 GO terms in the biological process (BP) category enriched from the 400 genes regulated by Ag⁺. DAVID was used for the analysis. The *Mus musculus* (mouse) whole genome was used as background. Statistical enrichment was determined through a modified Fisher's exact test ($p < 0.1$) and count threshold >2 (default settings in DAVID).

Abbreviations

3D: three-dimensional; AGCC: Affymetrix GeneChip Command Console; Ag^+: silver ion; $AgNO_3$: silver nitrate; AgNP: silver nanoparticle; ANOVA: analysis of variance; BMP-4: bone morphogenic protein 4; BP: biological process; DAVID: Database for Annotation, Visualization, and Integrated Discovery; DEG: differentially expressed gene; DI: deionized; DLS: dynamic light scattering; EB: embryoid body; ECVAM: European Committee for the Validation of Alternative Methods; ESC: embryonic stem cell; EST: embryonic stem cell test; FC: fold change; GEO: Gene Expression Omnibus; GO: gene ontology; GSK3b-I: glycogen synthase kinase-3b inhibitor; HCA: hierarchical cluster analysis; HSP: heat shock protein; ICP-MS: inductively coupled plasma mass spectroscopy; IPA: Ingenuity Pathway Analysis; LIF: leukemia inhibitory factor; mESC: mouse embryonic stem cell; MT: metallothionein; PBS: phosphate buffered saline; PCA: principal component analysis; PDI: polydispersity index; RMA: robust multi-array average; ROS: reactive oxygen species; STS: staurosporine; TAC: Transcriptome Analysis Console; TEM: transmission electron microscopy; UV-vis: ultraviolet-visible.

Authors' contributions

XG conceived and designed the study, carried the experimental work, analyzed the data, and wrote the manuscript. VDT performed AgNP characterization using DLS and TEM. ZK did the ICP-MS analysis for silver concentration assessment. RLS and JJY contributed to the conception and design of the study, intellectually accompanied the experimental work, critically reviewed the manuscript, and gave approval of the final version to be published. All authors read and approved the final manuscript.

Acknowledgements

We thank Dr. Marianna D. Solomotis for critically reading the manuscript. The findings and conclusions presented in this article are those of the authors and do not necessarily represent views, opinions, or policies of the U.S. Food and Drug Administration.

Competing interests

The authors declare that they have no competing interests.

Funding

This work was supported by internal funds of the U.S. Food and Drug Administration.

References

1. The Woodrow Wilson International Center. Consumer products inventory. http://www.nanotechproject.org/cpi/. Accessed Apr 2016.
2. Henig RM. Our silver-coated future. 2007. http://archive.onearth.org/article/our-silver-coated-future. Accessed Feb 2015.
3. Lara HH, Garza-Treviño EN, Ixtepan-Turrent L, Singh DK. Silver nanoparticles are broad-spectrum bactericidal and virucidal compounds. J Nanobiotechnol. 2011;9:30. doi:10.1186/1477-3155-9-30.
4. You C, Han C, Wang X, Zheng Y, Li Q, Hu X, Sun H. The progress of silver nanoparticles in the antibacterial mechanism, clinical application and cytotoxicity. Mol Biol Rep. 2012;39:9193–201. doi:10.1007/s11033-012-1792-8.
5. Scientific Committee on Emerging and Newly Identified Health Risks (SCENIHR). Opinion on nanosilver: safety, health and environmental effects and role in antimicrobial resistance, 2014. http://ec.europa.eu/health/scientific_committees/emerging/docs/scenihr_o_039.pdf. Accessed Apr 2016.
6. Bartłomiejczyk T, Lankoff A, Kruszewski M, Szumiel I. Silver nanoparticles—allies or adversaries? Ann Agric Environ Med. 2013;20:48–54.
7. Zhang T, Wang L, Chen Q, Chen C. Cytotoxic potential of silver nanoparticles. Yonsei Med J. 2014;55:283–91. doi:10.3349/ymj.2014.55.2.283.
8. Kim YS, Kim JS, Cho HS, Rha DS, Kim JM, Park JD, Choi BS, Lim R, Chang HK, Chung YH, Kwon IH, Jeong J, Han BS, Yu IJ. Twenty-eight-day oral toxicity, genotoxicity, and gender-related tissue distribution of silver nanoparticles in Sprague–Dawley rats. Inhal Toxicol. 2008;20:575–83. doi:10.1080/08958370701874663.
9. Kim YS, Song MY, Park JD, Song KS, Ryu HR, Chung YH, Chang HK, Lee JH, Oh KH, Kelman BJ, Hwang IK, Yu IJ. Subchronic oral toxicity of silver nanoparticles. Part Fibre Toxicol. 2010;7:20. doi:10.1186/1743-8977-7-20.
10. Tang J, Xiong L, Wang S, Wang J, Liu L, Li J, Yuan F, Xi T. Distribution, translocation and accumulation of silver nanoparticles in rats. J Nanosci Nanotechnol. 2009;9:4924–32.
11. Tang J, Xiong L, Zhou G, Wang S, Wang J, Liu L, Li J, Yuan F, Lu S, Wan Z, Chou L, Xi T. Silver nanoparticles crossing through and distribution in the blood-brain barrier in vitro. J Nanosci Nanotechnol. 2010;10:6313–7.
12. Doran KS, Banerjee A, Disson O, Lecuit M. Concepts and mechanisms: crossing host barriers. Cold Spring Harb Perspect Med. 2013;3:a010090. doi:10.1101/cshperspect.a010090.
13. Tandon S, Jyoti S. Embryonic stem cells: an alternative approach to developmental toxicity testing. J Pharm Bioallied Sci. 2012;4:96–100. doi:10.4103/0975-7406.94808.
14. Heuer J, Bremer S, Pohl I, Spielmann H. Development of an in vitro embryotoxicity test using murine embryonic stem cell cultures. Toxicol In Vitro. 1993;7:551–6.
15. Genschow E, Spielmann H, Scholz G, Pohl I, Seiler A, Clemann N, Bremer S, Becker K. Validation of the embryonic stem cell test in the international ECVAM validation study on three in vitro embryotoxicity tests. Altern Lab Anim. 2004;32:209–44.
16. Paquette JA, Kumpf SW, Streck RD, Thomson JJ, Chapin RE, Stedman DB. Assessment of the embryonic stem cell test and application and use in the pharmaceutical industry. Birth Defects Res B Dev Reprod Toxicol. 2008;83:104–11. doi:10.1002/bdrb.20148.
17. van Dartel DA, Piersma AH. The embryonic stem cell test combined with toxicogenomics as an alternative testing model for the assessment of developmental toxicity. Reprod Toxicol. 2011;32:235–44. doi:10.1016/j.reprotox.2011.04.008.
18. Gao X, Yourick JJ, Sprando RL. Transcriptomic characterization of C57BL/6 mouse embryonic stem cell differentiation and its modulation by developmental toxicants. PLoS ONE. 2014;9:e108510. doi:10.1371/journal.pone.0108510.
19. Ahamed M, Karns M, Goodson M, Rowe J, Hussain SM, Schlager JJ, Hong Y. DNA damage response to different surface chemistry of silver nanoparticles in mammalian cells. Toxicol Appl Pharmacol. 2008;233:404–10. doi:10.1016/j.taap.2008.09.015.
20. Park MV, Neigh AM, Vermeulen JP, de la Fonteyne LJ, Verharen HW, Briedé JJ, van Loveren H, de Jong WH. The effect of particle size on the cytotoxicity, inflammation, developmental toxicity and genotoxicity of silver nanoparticles. Biomaterials. 2011;32:9810–7. doi:10.1016/j.biomaterials.2011.08.085.
21. Samberg ME, Loboa EG, Oldenburg SJ, Monteiro-Riviere NA. Silver nanoparticles do not influence stem cell differentiation but cause minimal toxicity. Nanomedicine. 2012;7:1197–209. doi:10.2217/nnm.12.18.
22. Bouwmeester H, Poortman J, Peters RJ, Wijma E, Kramer E, Makama S, Puspitaninganindita K, Marvin HJ, Peijnenburg AA, Hendriksen PJ. Characterization of translocation of silver nanoparticles and effects on whole-genome gene expression using an in vitro intestinal epithelium coculture model. ACS Nano. 2011;5:4091–103. doi:10.1021/nn2007145.
23. Foldbjerg R, Irving ES, Hayashi Y, Sutherland DS, Thorsen K, Autrup H, Beer C. Global gene expression profiling of human lung epithelial cells after exposure to nanosilver. Toxicol Sci. 2012;130:145–57. doi:10.1093/toxsci/kfs225.
24. Kawata K, Osawa M, Okabe S. In vitro toxicity of silver nanoparticles at noncytotoxic doses to HepG2 human hepatoma cells. Environ Sci Technol. 2009;43:6046–51.
25. De Smedt A, Steemans M, De Boeck M, Peters AK, van der Leede BJ, Van Goethem F, Lampo A, Vanparys P. Optimisation of the cell cultivation

methods in the embryonic stem cell test results in an increased differentiation potential of the cells into strong beating myocard cells. Toxicol In Vitro. 2008;22:1789–96. doi:10.1016/j.tiv.2008.07.003.

26. Irizarry RA, Hobbs B, Collin F, Beazer-Barclay YD, Antonellis KJ, Scherf U, Speed TP. Exploration, normalization, and summaries of high density oligonucleotide array probe level data. Biostatistics. 2003;4:249–64.

27. Ringnér M. What is principal component analysis? Nat Biotechnol. 2008;26:303–4.

28. Eisen MB, Spellman PT, Brown PO, Botstein D. Cluster analysis and display of genome-wide expression patterns. Proc Natl Acad Sci USA. 1998;95:14863–8.

29. Tong W, Cao X, Harris S, Sun H, Fang H, Fuscoe J, Harris A, Hong H, Xie Q, Perkins R, Shi L, Casciano D. Arraytrack-supporting toxicogenomic research at the U.S. Food and Drug Administration National Center for Toxicological Research. Environ Health Perspect. 2003;111:1819–26.

30. Tong W, Harris S, Cao X, Fang H, Shi L, Sun H, Fuscoe J, Harris A, Hong H, Xie Q, Perkins R, Casciano D. Development of public toxicogenomics software for microarray data management and analysis. Mutat Res. 2004;549:241–53.

31. Dennis G Jr, Sherman BT, Hosack DA, Yang J, Gao W, Lane HC, Lempicki RA. DAVID: database for annotation, visualization, and integrated discovery. Genome Biol. 2003;4:P3.

32. da Huang W, Sherman BT, Lempicki RA. Systematic and integrative analysis of large gene lists using DAVID bioinformatics resources. Nat Protoc. 2009;4:44–57.

33. Hu ZL, Bao J, Reecy JM. CateGOrizer: a web-based program to batch analyze gene ontology classification categories. Online J Bioinform. 2008;9:108–12.

34. Pong K, Doctrow SR, Huffman K, Adinolfi CA, Baudry M. Attenuation of staurosporine-induced apoptosis, oxidative stress, and mitochondrial dysfunction by synthetic superoxide dismutase and catalase mimetics, in cultured cortical neurons. Exp Neurol. 2001;171:84–97.

35. Kinney MA, Hookway TA, Wang Y, McDevitt TC. Engineering three-dimensional stem cell morphogenesis for the development of tissue models and scalable regenerative therapeutics. Ann Biomed Eng. 2014;42:352–67. doi:10.1007/s10439-013-0953-9.

36. Martin GR, Wiley LM, Damjanov I. The development of cystic embryoid bodies in vitro from clonal teratocarcinoma stem cells. Dev Biol. 1977;61:230–44. doi:10.1016/0012-1606(77)90294-9.

37. Xu L, Takemura T, Xu M, Hanagata N. Toxicity of silver nanoparticles as assessed by global gene expression analysis. Mater Express. 2011;1:74–9. doi:10.1166/mex.2011.1010.

38. Asharani PV, Lian WuY, Gong Z, Valiyaveettil S. Toxicity of silver nanoparticles in zebrafish models. Nanotechnology. 2008;19:255102. doi:10.1088/0957-4484/19/25/255102.

39. Sprando RL, Black T, Keltner Z, Olejnik N, Ferguson M. Silver acetate exposure: effects on reproduction and post natal development. Food Chem Toxicol. 2016. doi:10.1016/j.fct.2016.06.022.

40. Park MV, Annema W, Salvati A, Lesniak A, Elsaesser A, Barnes C, McKerr G, Howard CV, Lynch I, Dawson KA, Piersma AH, de Jong WH. In vitro

developmental toxicity test detects inhibition of stem cell differentiation by silica nanoparticles. Toxicol Appl Pharmacol. 2009;240:108–16. doi:10.1016/j.taap.2009.07.019.

41. Boverhof DR, David RM. Nanomaterial characterization: considerations and needs for hazard assessment and safety evaluation. Anal Bioanal Chem. 2010;396:953–61. doi:10.1007/s00216-009-3103-3.

42. Warheit DB. How meaningful are the results of nanotoxicity studies in the absence of adequate material characterization? Toxicol Sci. 2008;101:183–5.

43. Ge L, Li Q, Wang M, Ouyang J, Li X, Xing MM. Nanosilver particles in medical applications: synthesis, performance, and toxicity. Int J Nanomed. 2014;9:2399–407. doi:10.2147/IJN.S55015.

44. Kittler S, Greulich C, Diendorf J, Koller M, Epple M. Toxicity of silver nanoparticles increases during storage because of slow dissolution under release of silver ions. Chem Mater. 2010;22:4548–54. doi:10.1021/cm100023p.

45. Park EJ, Yi J, Kim Y, Choi K, Park K. Silver nanoparticles induce cytotoxicity by a Trojan-horse type mechanism. Toxicol In Vitro. 2010;97:34–41. doi:10.1016/j.tiv.2009.12.001.

46. De Maio A. Heat shock proteins: facts, thoughts, and dreams. Shock. 1999;11:1–12.

47. Gupta SC, Sharma A, Mishra M, Mishra RK, Chowdhuri DK. Heat shock proteins in toxicology: how close and how far? Life Sci. 2010;86:377–84. doi:10.1016/j.lfs.2009.12.015.

48. Hunziker PE, Kägi JH. Isolation and characterization of six human hepatic isometallothioneins. Biochem J. 1985;231:375–82.

49. Davis SR, Cousins RJ. Metallothionein expression in animals: a physiological perspective on function. J Nutr. 2000;130:1085–8.

50. Theocharis SE, Margeli AP, Koutselinis A. Metallothionein: a multifunctional protein from toxicity to cancer. Int J Biol Markers. 2003;18:162–9.

51. Ruttkay-Nedecky B, Nejdl L, Gumulec J, Zitka O, Masarik M, Eckschlager T, Stiborova M, Adam V, Kizek R. The role of metallothionein in oxidative stress. Int J Mol Sci. 2013;14:6044–66. doi:10.3390/ijms14036044.

52. Sahu SC, Zheng J, Yourick JJ, Sprando RL, Gao X. Toxicogenomic responses of human liver HepG2 cells to silver nanoparticles. J Appl Toxicol. 2015;35:1160–8. doi:10.1002/jat.3170.

53. Xu L, Li X, Takemura T, Hanagata N, Wu G, Chou LL. Genotoxicity and molecular response of silver nanoparticle (NP)-based hydrogel. J Nanobiotechnol. 2012;10:16. doi:10.1186/1477-3155-10-16.

54. AshaRani PV, Low Kah Mun G, Hande MP, Valiyaveettil S. Cytotoxicity and genotoxicity of silver nanoparticles in human cells. ACS Nano. 2009;3:279–90. doi:10.1021/nn800596w.

55. Maier T, Güell M, Serrano L. Correlation of mRNA and protein in complex biological samples. FEBS Lett. 2009;583:3966–73. doi:10.1016/j.febslet.2009.10.036.

56. Chechik G, Koller D. Timing of gene expression responses to environmental changes. J Comput Biol. 2009;16:279–90. doi:10.1089/cmb.2008.13TT.

11

Synthesis and characterization of crosslinked polyisothiouronium methylstyrene nanoparticles of narrow size distribution for antibacterial and antibiofilm applications

Sarit Cohen[1†], Chen Gelber[1†], Michal Natan[2], Ehud Banin[2], Enav Corem-Salkmon[1] and Shlomo Margel[1*]

Abstract

Background: Isothiouronium salts are well known in their variety of antimicrobials activities. The use of polymeric biocides, polymers with antimicrobial activities, is expected to enhance the efficacy of some existing antimicrobial agents, thus minimizing the environmental problems accompanying conventional antimicrobials.

Methods: The current manuscript describes the synthesis and characterization of crosslinked polyisothiouronium methylstyrene (PITMS) nanoparticles (NPs) of narrow size distribution by dispersion co-polymerization of the monomer isothiouronium methylstyrene with the crosslinking monomer ethylene glycol dimetacrylate.

Results and discussion: The effect of total monomer, crosslinker and initiator concentrations on the size and size distribution of the formed NPs was also elucidated. The bactericidal activity of PITMS NPs of 67 ± 8 nm diameter was illustrated for 4 bacterial pathogens: *Listeria innocua*, *Escherichia coli*, *Pseudomonas aeruginosa* and *Staphylococcus aureus*. In order to demonstrate the potential of these PITMS NPs as inhibitor of biofilm formation, polyethylene terephthalate (PET) films were thin-coated with the PITMS NPs. The formed PET/PITMS films reduced the viability of the biofilm of *Listeria* by 2 orders of magnitude, making the coatings excellent candidates for further development of non-fouling surfaces. In addition, PITMS NP coatings were found to be non-toxic in HaCaT cells.

Conclusions: The high antibacterial activity and effective inhibition of bacterial adsorption indicate the potential of these nanoparticles for development of new types of antibacterial and antibiofilm additives.

Keywords: Isothioronium methylstyrene, Dispersion polymerization, Nanoparticles, Antibacterial, Biofilm

Background

Microbial infection remains one of the most serious complications in several areas including medical devices, healthcare products, water purification systems, hospitals, dental office equipment and food packaging [1–4]. Antimicrobial activity is related to compounds that locally kill microorganisms or inhibit their growth, without being toxic to surrounding tissue. Much attention has been directed to the development of antimicrobials from both academic research and industry [5, 6]. Polymers with antimicrobial activity, polymeric biocides, are known to have enhanced efficacy compared to existing antimicrobial molecules, minimizing the environmental problems accompanying conventional antimicrobials [7]. In general, it has been reported that polycations exhibit antibacterial properties, as they interact with and disrupt bacterial or fungi cell membranes. The adsorption of polycations, which possess very high positive charge density, onto the usually negatively charged pathogenic

*Correspondence: shlomo.margel@mail.biu.ac.il
†Sarit Cohen and Chen Gelber contributed equally to this work
[1] Department of Chemistry, The Institute of Nanotechnology and Advanced Materials, Bar-Ilan University, 52900 Ramat Gan, Israel
Full list of author information is available at the end of the article

microorganism cell surface, is enhanced as compared to monomeric cations [8]. Compared to low molecular weight biocides, polymeric biocides are non-volatile and chemically stable, along with their reduced toxicity and increased efficiency and selectivity. Furthermore, their lifetime is significantly longer and they have low penetration through the skin [9].

Thiourea, isothiouronium compounds and their derivatives constitute an important class of compounds which exhibit wide range of antimicrobial activities and play important roles in many chemical and biological processes [10–13]. The role of the terminal amino group, belonging to the isothioronium group, seems to consist mainly of favoring the binding with peptide terminating acyl-D-alanyl-D-alanine [14]. Trani et al. [15] reported that the enhanced activity of isothiouronium compounds, compared to thiourea compounds which lack positively charged N-terminus group, may be due to the enhanced acidity of the NH moieties, thereby functioning as a better binder to Ac-D-Ala-D-Ala, a bacterial cell-wall model (tenfold lower binding constant).

Adhesion and subsequent growth of bacteria on surfaces cause the formation of biofilm. There is a growing demand for reliable antibacterial surfaces that can effectively minimize bacterial colonization [16]. Quaternary ammonium salts are widely used as 'cationic disinfectants' or biocidal coating to minimize the problems of biofouling [17, 18].

Listeria monocytogenes, a Gram-positive facultative anaerobe, ubiquitous in nature and common in foods of both plant and animal origin has emerged into a highly problematic and fatal foodborne pathogen throughout the world. An effective method of preserving foods from the effect of microbial growth, such as adding antimicrobial agents, would reduce the likelihood of foodborne outbreaks of listeriosis, and decrease economic losses to the food industry [5, 19, 20].

Vinylic monomers are routinely used for preparing organic polymeric NPs since they may contain many possible functional side groups [21, 22]. The present manuscript describes the synthesis and evaluation of the antimicrobial activity of novel polyisothioronium methylstyrene (PITMS) nanoparticles (NPs), by dispersion co-polymerization of the monomer isothioronium methylstyrene (ITMS) and the crosslinking monomer ethylene glycol dimethacrylate (EGDMA) in an aqueous continuous phase [18, 19]. The antimicrobial activity of the obtained PITMS NPs was evaluated using 4 bacterial pathogens: *Listeria innocua, Escherichia coli* (*E. coli*), *Pseudomonas aeruginosa* (*P. aeruginosa*) and *Staphylococcus aureus* (*S. aureus*). The potential use of these PITMS NPs as an inhibitor of biofilm formation was demonstrated with *Listeria* as a model.

Results and discussion

Synthesis and characterisation of PITMS NPs

PITMS NPs of narrow size distribution were prepared by dispersion co-polymerization of ITMS and EGDMA according to the experimental part. The polymerization yield of the obtained PITMS NPs was calculated to be 75 %. Figure 1 presents a TEM image (Fig. 1a) and a typical hydrodynamic size histogram (Fig. 1b) of the obtained PITMS NPs. The dry diameter and size distribution of these PITMS particles, as shown by the TEM image, are 19 ± 2 nm, while the hydrodynamic diameter and size distribution of these particles dispersed in water, as shown by the size histogram, are 67 ± 8 nm. The hydrodynamic diameter is larger than the dry diameter probably since it also takes into account swollen and surface-adsorbed water molecules.

FTIR spectra of the ITMS monomer and the PITMS NPs are shown in Fig. 2a. The FTIR spectrum of the PITMS NPs is similar to that of the monomer, except for the additional absorption peak at about 916 and 987 cm^{-1} corresponding to the vinylic C–H bending band indicating the lack of residual monomer within the polymeric particles. Instead, the peaks that appears at 1110 and 1730 cm^{-1} corresponding to the C–O and C=O stretching band of EGDMA.

X-ray diffraction patterns of the ITMS monomer and the PITMS NPs are illustrated in Fig. 2b. The XRD pattern of the ITMS monomer displays clear sharp and narrow diffraction peaks typical for crystalline materials. These X-ray diffraction patterns indicate the crystalline nature of the monomer. In contrast, the X-ray diffraction pattern of the PITMS NPs, suggests the existence of a fully amorphous phase of the polymer, probably due to the loss of the crystalline structure of the monomer by the radical polymerization process.

The stability of the nanoparticle dispersion was evaluated by their ζ–potential, as shown in Fig. 3a. Since a positive particle surface charge will create repulsion between the particles and may prevent aggregation, the ζ–potential of their dispersion indicates their stability. Figure 3a illustrates a consistent sharp decrease in the ζ–potential of the nanoparticles by increasing the pH of the aqueous continuous phase from 37 mV at pH 4.0 to −6.0 mV at pH 10.5. At the isoelectric point (around pH 10.2, as shown in Fig. 3a), the particles are not stable, due to possible aggregation. Increasing the pH of the continuous phase above 11.5 probably causes hydrolysis of the isothioronium groups into deprotonated thiol groups, as reported in the literature [23].

The thermal stability is an important factor when incorporating an external substance as an additive to polymer matrices. The thermal stability of the PITMS NPs aqueous dispersion, after drying, was evaluated by TGA, as

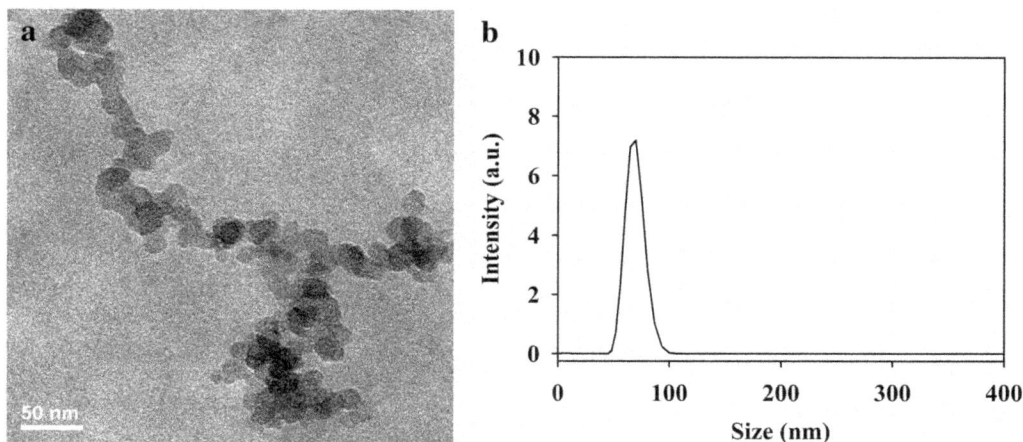

Fig. 1 TEM image (**a**) and hydrodynamic size histogram (**b**) of the PITMS NPs

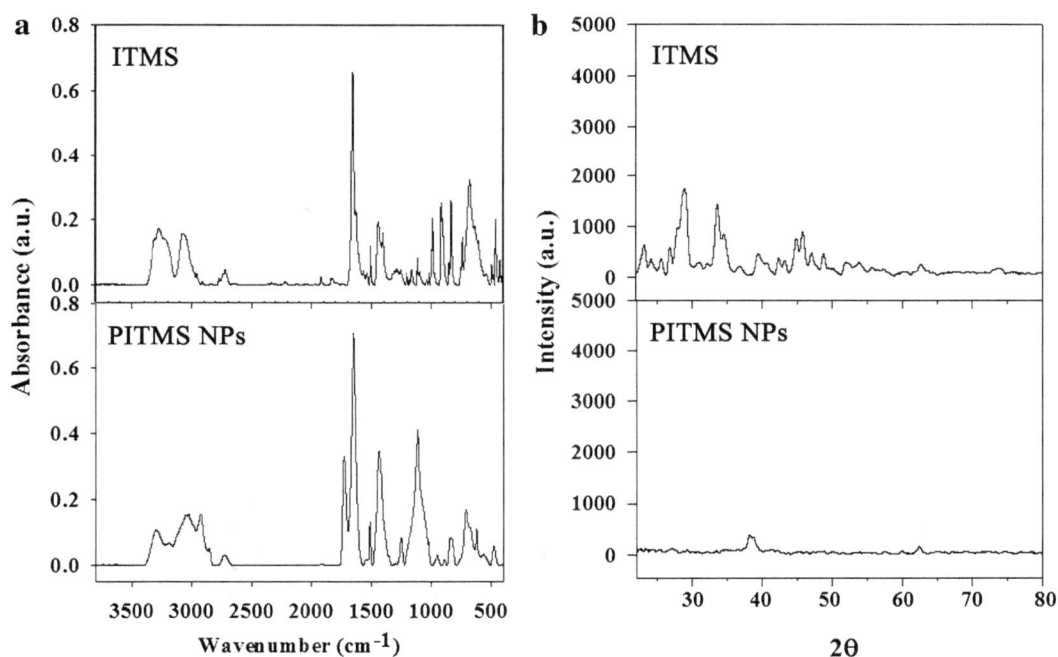

Fig. 2 FTIR spectra (**a**) and X-ray diffraction patterns (**b**) of the ITMS monomer and the PITMS NPs

illustrated in Fig. 3b. No mass loss was observes between room temperature to 160 °C indicating the thermal stability of the NPs in this temperature range. In the range of 160–270 and 270–450 °C, mass loss of 22 and 41 % was observed, attributed to the degradation of thiourea hydrochloride and the aromatic group from the PITMS NPs, respectively, as was indicated by the MS. In the range of 450–1000 °C, mass loss of 22 % was observed, probably attributed to the degradation of the polymer crosslinked carbon chain.

Effect of polymerization parameters on the size and size distribution of the PITMS NPs

Effect of the EGDMA concentration

The effect of the weight ratio [EGDMA]/[ITMS + EGDMA] on the hydrodynamic diameter and size distribution of the formed PITMS NPs was studied while retaining a constant total monomer ([ITMS] + [EGDMA]) concentration (0.5 g). Figure 4a shows that as the weight ratio of [EGDMA]/[total monomers] increases the diameter and the size distribution of

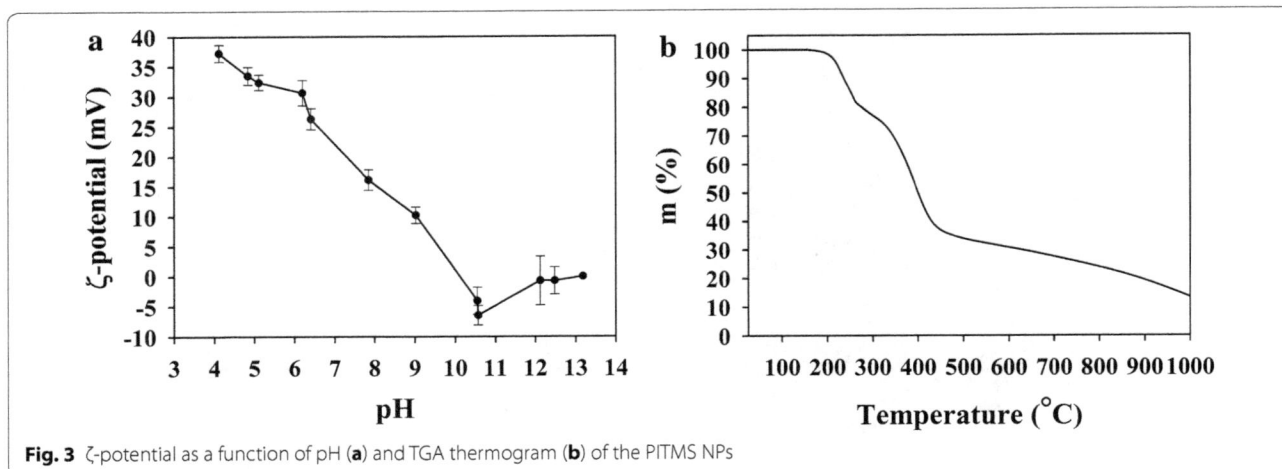

Fig. 3 ζ-potential as a function of pH (**a**) and TGA thermogram (**b**) of the PITMS NPs

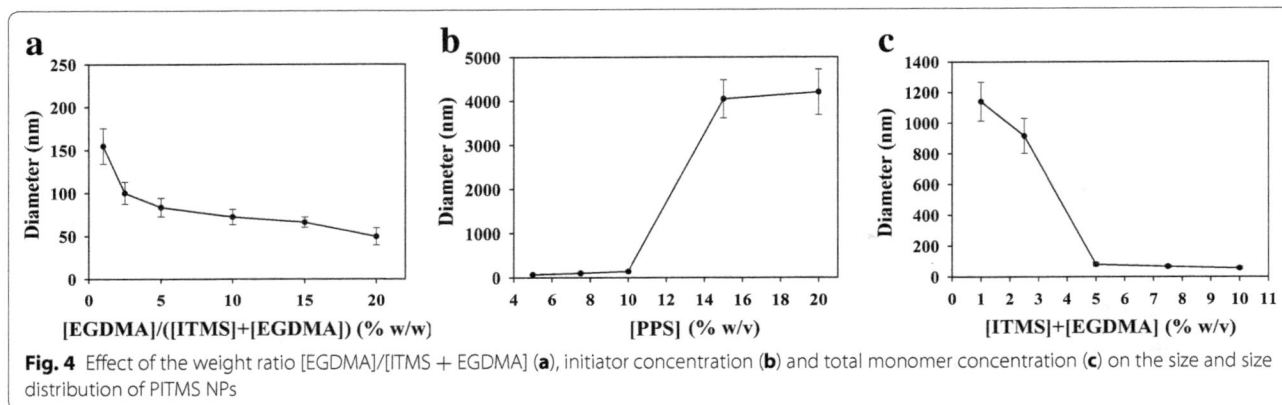

Fig. 4 Effect of the weight ratio [EGDMA]/[ITMS + EGDMA] (**a**), initiator concentration (**b**) and total monomer concentration (**c**) on the size and size distribution of PITMS NPs

the formed PITMS particles decreases. For example, raising the ratio from 1 to 2.5 and 5 % leads to a decrease in the average particle size from 155 ± 21 to 100 ± 13 and 83 ± 11 nm, respectively. This behavior may be explained by the fact that increasing the crosslinking decreases the ability of the growing nuclei to swell, resulting in smaller particles [24, 25].

Effect of the initiator concentration

The effect of the PPS concentration on the hydrodynamic diameter and size distribution of the PITMS particles was also elucidated. Increasing the PPS concentration leads to an increase in the diameter and size distribution of the formed PITMS particles as shown in Fig. 4b. For example, raising the weight % of PPS concentration from 5 to 7.5 and 10 % leads to an increase in the average particle size and size distribution from 70 ± 8 to 100 ± 12 and 139 ± 19 nm, respectively. A similar effect of influence of the initiator concentration on the size of the particles formed by dispersion polymerization was previously reported by Tseng, El-Aasser and others [26, 27].

According to their explanation, increasing the initiator concentration causes an increase in the oligomer radical concentration, and thus, in the concentration of precipitated oligomer chain. Due to this fact, along with the slow adsorption of the stabilizer, the aggregation process is enhanced, resulting in larger particles. The increase in the size distribution as the initiator concentration increases may be explained by the increase in the number of the oligomeric chains as the initiator concentration increases, thus favoring secondary nucleation during the particle growth stage, which increases the particle size distribution.

Effect of the total monomer concentration

The effect of total monomers concentration ([ITMS] + [EGDMA]) on the hydrodynamic diameter and size distribution of the PITMS particles showed that increasing the total monomer concentration leads to the formation of smaller PITMS particles with narrower size distributions as shown in Fig. 4c. For example, an increase in the total monomer concentration from 2.5 to

5 and 7.5 % leads to a decrease in the diameter and size distribution of the PITMS particles from 916 ± 113 to 81 ± 12 and 67 ± 9 nm, respectively.

The decrease in the average diameter as a function of the monomer concentration was previously reported for dispersion polymerization [28, 29] although many publications have reported the opposite behavior [26, 30]. This can be explained by the effect of the monomer concentration on the dispersion polymerization mechanism. Increasing the monomer concentration may affect the initial solvency of the reaction medium by decreasing or increasing (depending on the monomer type and the continuous phase) the solubility of the forming oligomers, so they may achieve shorter or longer chain lengths before precipitating. Earlier precipitation of the shorter oligomers eventually results in a larger number of smaller particles. In addition, increasing the monomer concentration can decrease or increase the solubility of the stabilizer, thus increasing or decreasing its adsorption on the growing particle. Both effects contribute to the change in the particle size.

Antibacterial activity of PITMS NPs

The antibacterial properties of PITMS NPs of 67 ± 8 nm were tested against *Listeria* bacteria, a common foodborne pathogen. As shown in Fig. 5, both 1 and 0.5 % PITMS NPs were able to kill all the tested bacteria following 24 h exposure, as opposed to the water-treated control, suggesting a potent antimicrobial activity of the PITMS particles. A 0.25 % particle concentration had only a partial bactericidal effect, implying that 0.5 % is the minimum inhibitory concentration (MIC) for PITMS

needed to inhibit growth under these experimental conditions. Similar results were obtained for three additional pathogenic bacteria: *E. coli*, *P. aeruginosa* and *S. aureus* as shown in the Additional file 1: Figure S1.

Antibiofilm activity of the PITMS NPs

In light of the antibacterial activity exerted by the PITMS, the inhibition of *Listeria* biofilm formation by the NPs was determined. As shown in Fig. 5b, a significant reduction in the biofilm formation of *Listeria* was detected; 2 logs for the PET/PITMS films, in comparison to a film containing the film former only or to a non-coated PET film.

Migration tests

Migration is the term used for the escape of additives from a polymeric host. Migration may limit the use of additives in plastic especially for food packing applications, pharmaceutical and other hygienic products. Crosslinked NPs that are compatible with the PET film and the film former may overcome this disadvantage, due to their large spatial structure, which reduces their migration while maintaining the activity. Hence, using appropriate NPs as an additive to plastic films will result in antibacterial properties and with decreased extractability and volatility. Indeed, no migration of the PITMS NPs to the continuous phase composed of 3 % aqueous acetic acid or 95 % ethanol were detected.

Optical properties of the PET/PITMS films

The optical properties of polymeric films are important for many applications, such as transparent food

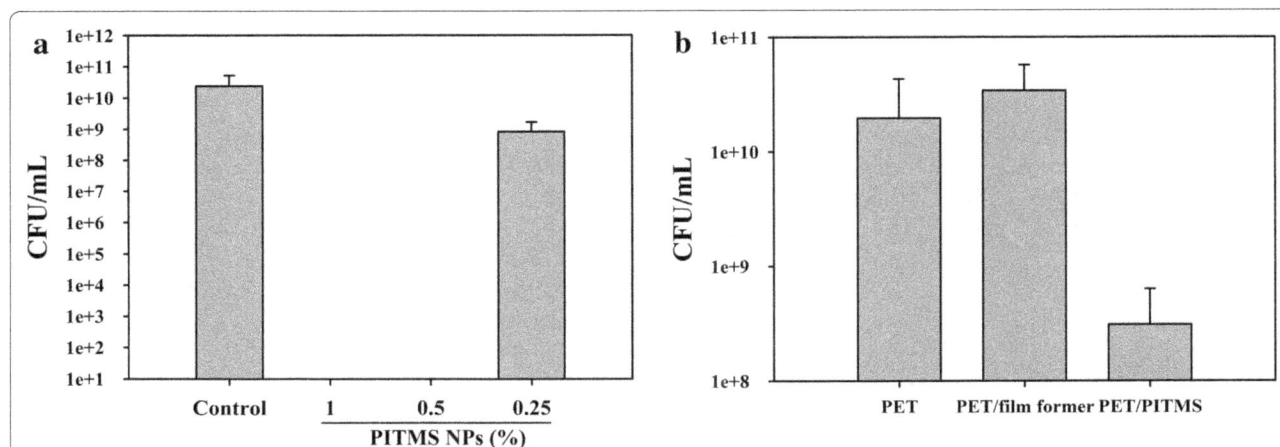

Fig. 5 Antibacterial activity of the PITMS NPs against *Listeria* (**a**). The bacteria were grown and treated with either PITMS NPs at the indicated concentrations or water (control). The results show the pattern observed in at least three independent experiments. Biofilm formation of *Listeria* on PET films (**b**). Data is presented as the mean ± SE. The representative calculations are an average of 3 experiments, with at least tetraplicates of each experiment group in each experiment. The results of the different experiment groups in each experiment differ significantly from each other (p < 0.05), e.g. PET/PITMS NPs differ significantly from PET or PET/film former)

packaging. The haze, clarity and transmittance of the PET/PITMS in comparison to a film containing the film former only or to a non-coated PET film are shown in Table 1. The results indicate the potential use of the PET/PITMS films as transparent films. There was no change in the optical properties of the various films after a year indicating that there is no migration during this time period.

Cellular cytotoxicity of PITMS NP coatings by LDH assay

In vitro cytotoxicity of the PITMS NP coatings was tested by LDH assay using human keratinocyte HaCaT cells [31]. HaCaT cell line is a spontaneously transformed human epithelial cell line from adult skin and the first permanent epithelial cell line that exhibits normal differentiation [32]. Lactate dehydrogenase, LDH, is an intracellular enzyme that catalyzes the reversible oxidation of lactate to pyruvate. Since LDH is mainly present in the cytosol, it is released into the supernatant only upon cell damage or lysis [33, 34]. When tested by the LDH quantitative assay, all the pre-incubated PET films had no cytotoxic effect on the HaCaT cell line (Fig. 6). PET/PITMS films are therefore suitable for food application, considering their non-toxicity.

Conclusions

The present study describes the synthesis and characterization of PITMS NPs of narrow size distribution. The effect of various polymerization parameters on the size and size distribution of the produced PITMS particles have been elucidated. This study also demonstrates that these NPs have excellent antibacterial activity against *Listeria*. This work integrates the advantages of polymer chemistry and technology with bacteriology, leading to possible developments in the formulation of new types of bactericides. The PET/PITMS films under the experimental conditions inhibited biofilm formation of *Listeria* by 2 orders of magnitude, making the coatings excellent candidates for further development of non-fouling surfaces. In future work we plan to check the antibiofilm activity of the NPs against additional types of bacteria and additional types of films in various coating thicknesses.

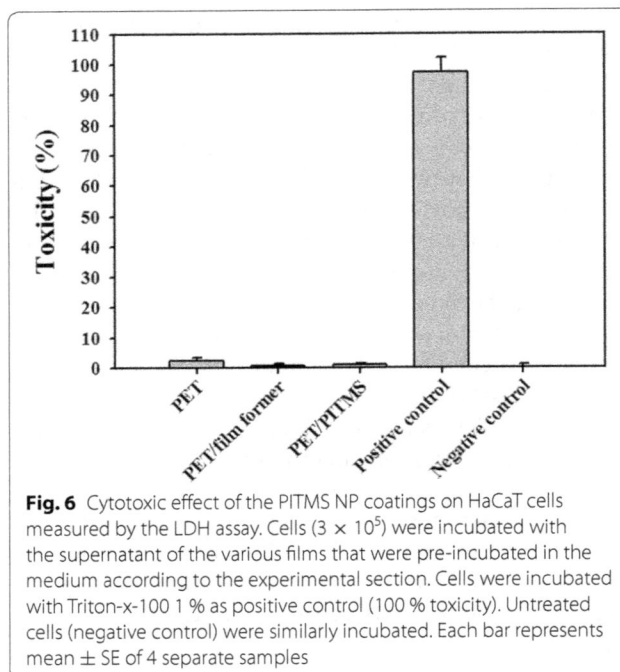

Fig. 6 Cytotoxic effect of the PITMS NP coatings on HaCaT cells measured by the LDH assay. Cells (3 × 10⁵) were incubated with the supernatant of the various films that were pre-incubated in the medium according to the experimental section. Cells were incubated with Triton-x-100 1 % as positive control (100 % toxicity). Untreated cells (negative control) were similarly incubated. Each bar represents mean ± SE of 4 separate samples

Methods

Materials

The following analytical-grade chemicals were purchased from Sigma-Aldrich (Israel) and used without further purification: thiourea, chloromethylstyrene (CMS; 97 %), potassium persulfate (PPS), ethylene glycol dimetacrylate (EGDMA), methanol, diethyl ether and tween 20. PET films of A4 size and 23 µm thick were obtained from Hanita Coatings RCA Ltd, Israel. Film former G-9/230 from ACTEGA Coating & Sealants, Wesel, Germany. Water was purified by passing deionized water through an Elgastat Spectrum reverse osmosis system (Elga Ltd, High Wycombe, UK). Dulbecco's minimum essential medium (DMEM) eagle, fetal bovine serum (FBS), 1 % glutamine, 1 % penicillin/streptomycin and mycoplasma detection kit from Biological Industries (Bet Haemek, Israel); cytotoxicity detection kit from BioVision, USA; the HaCaT cell line was a kind gift from Prof. Eli Shprecher, Molecular dermatology research laboratory, Tel-Aviv Sourasky medical center, Israel, and was cultured as previously described.

Synthesis of the isothioronium methylstyrene monomer

The monomer ITMS was synthesized according to the literature, as shown in Fig. 7. Briefly, thiourea (0.21 mol) was dissolved in methanol (60 mL), followed by the addition of *p*-chloromethylstyrene (CMS, 0.2 mol) to the solution. The reaction mixture was then stirred at room temperature for 24 h. Diethyl ether was then added to

Table 1 Haze, clarity and transmission of the PET, PET/film former and PET/PITMS films

	Haze (%)	Clarity (%)	Transmission (%)
PET	1.3 ± 0.1	99.2 ± 0.0	89.5 ± 0.1
PET/film former	2.5 ± 0.2	98.9 ± 0.04	90.6 ± 0.5
PET/PITMS	6.5 ± 0.4	94.5 ± 0.7	91.0 ± 0.1

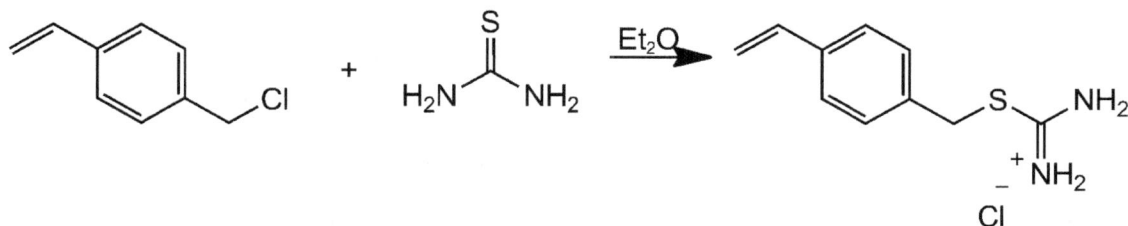

Fig. 7 Synthetic scheme for the ITMS monomer

precipitate the desired ITMS monomer. The filtered product was purified by dissolving it in ethanol and re-precipitation with ether (41.13 g, 0.18 mol, 90 %). The solid residue was analyzed by ^{1}H and ^{13}C NMR which showed the pure desired product.

Nuclear magnetic resonance (NMR) spectroscopy was performed on Bruker AC 400 MHz spectrometer. ^{1}H and ^{13}C NMR spectra were recorded in deuterated dimethyl sulfoxide.

^{1}H NMR (400 MHz, DMSO-d_6) δ_H in ppm: 4.53 (s, 2H, CH$_2$-S), 5.28 (d, 10.8 Hz, 1H, CH$_2$=CH [trans]), 5.85 (d, 17.6 Hz, 1H, CH$_2$=CH [cis]), 6.73 (dd, 10.8 Hz and 17.6 Hz, 1H, CH$_2$=CH [gem]), 7.41 (d, 8.2 Hz, Arom-CH), 7.48 (d, 8.2 Hz. Arom-CH), 9.32 (s, 4H, Isothiouronium).

^{13}C NMR (400 MHz, DMSO-d_6) δ_C in ppm: 34.8 (CH$_2$-S), 114.9 (**C**H$_2$=CH), 126.6 and 129.1 (Arom-CH), 133.6 and 135.8 (Arom C), 137.4 (CH$_2$=**C**H), 170.3 (C-Isothiouronium).

MS (CI+): 117 (CH$_2$CHArCH$_2$$^+$, 100 %), 193 (M+, 10.5 %).

Synthesis of the polyisothioronium methylstyrene NPs

In a typical experiment, PITMS NPs with dry diameter of 67 ± 8 nm were prepared by adding ITMS (425 mg), EGDMA (75 mg), PPS (25 mg) and Tween 20 (100 mg) to water (10 mL). The mixture was shaken at 73 °C for 15 h. The obtained NPs dispersed in water were then isolated from impurities by dialysis. The formed PITMS particles were washed with water at 60 °C in order to remove traces of the monomers and excess reagents. The effect of various polymerization parameters, e.g., total monomers, initiator and crosslinker concentrations, on the size and size distribution, and the polymerization yield of the ITMS to produce the particles was also elucidated.

Coating of the PET films with the PITMS NPs

PET films were coated with the NPs aqueous dispersion using a formulation with a ratio of 1:1 PITMS polymer to the film former (G-9/230). PITMS NPs of 67 ± 8 nm diameter dispersed in water (4 %) were first dispersed in

the G-9/230 film former 4 % aqueous solution (1:1 v/v). The obtained aqueous dispersion was then spread on the 23 μm thick PET films with a 6 μm (wet thickness) Mayer rod, followed by drying the PITMS coating on the PET films over night at room temperature.

Antibacterial assay

The antibacterial activity of the PITMS NPs of 67 ± 8 nm diameter was evaluated using the Gram-negative bacteria *E. coli* C600 and *P. aeruginosa* PAO1, and the Gram-positive bacteria *S. aureus* FRF1169 and *Listeria* ATCC 33,090, as the experimental models. All the bacterial strains used in this study were grown overnight in Luria Bertani (LB, Difco) media under shaking (250 rpm) at 37 °C. On the following day, the overnight cultures were each diluted into twofold concentrated LB medium to obtain a concentration of 2×10^5 colony-forming units (CFU/mL). The bacterial suspensions were incubated overnight with equivalent volumes of either PITMS (1, 0.5 and 0.25 %) or sterilized water (control). In the following day, tenfold serial dilutions were carried out and the bacterial cells were plated on LB agar plates, followed by their incubation at 37 °C for 20 h. Cell growth was monitored and determined by viable cell count and expressed as colony forming units (CFU/mL).

Static biofilm formation assay

The antibiofilm activity of the PET/PITMS films against *Listeria* bacteria was evaluated according to a protocol that was previously published [35] and compared to a control PET film and a PET film coated with the film former only using the Gram-positive bacteria *Listeria* ATCC 33,090 as the experimental model. Briefly, the bacteria were grown overnight in tryptic soy broth (TSB, DIFCO) growth medium. In the following day, bacterial cells were diluted in TSB to obtain a working solution with an OD$_{595}$ of 0.3 (approximately corresponds to 3×10^8 CFU/mL). 1 mL from the stock solution was taken into each well in a 24-well plate (DE-GROOT). Each of the different films was added to the well (1 cm diameter). The plates were then incubated at 25 °C under gentle agitation (100 rpm) for 20 h. In the day after, the films were rinsed 3 times with distilled water to remove the unattached bacteria (i.e. planktonic cells) and subsequently the attached cells were scraped from the films using 250 μl of Tris–HCl (0.1 M, pH 7.2) and cell scrapers (Greiner Bio-one). 200 μl out of the 250 μl, used for scrapping the cells, were transferred into the first line of a 96-well plate (Greiner Bio-One), while the rest of the lines were filled with 180 μl of Tris–HCl (0.1 M, pH 7.2). Serial dilutions were carried out and the cells spotted onto NB agar plates, which were then incubated at 37 °C for 20 h. Cell growth was monitored and determined by a viable cell count. The experiments were conducted at least three independent times, with internal duplicates.

Characterization of the PITMS NPs and the PET/PITMS films

Fourier transform infrared (FTIR) analysis was performed with a Bruker Platinum-FTIR QuickSnap TM sampling modules A220/D-01. The analysis was performed with 13 mm KBr pellets that contained 2 mg of the detected material (ITMS or PITMS) and 198 mg KBr. The pellets were scanned over 50 scans at a 4 cm^{-1} resolution.

Electrokinetic properties (ζ-potential) as a function of pH were determined with Zetasizer (Zetasizer 3000 HSa, Malvern Instruments, UK). ζ-potential measurements were performed at a constant ionic strength of 0.1 M.

Dried particle size and size distribution were measured with a transmission electron microscope (TEM). SEM pictures were obtained with a JEOL, JSM-840 Model, Japan. For this purpose, a drop of dilute particles dispersion in distilled water was spread on a glass surface, and then dried at room temperature. The dried sample was coated with carbon in vacuum before viewing under SEM. The average particle size and distribution were determined by the measurement of the diameter of more than 100 particles with image analysis software (Analysis Auto, Soft Imaging System GmbH, Germany).

Hydrodynamic diameter and size distribution of the particles dispersed in double distilled (DD) water were measured at room temperature with a particle analyzer; model NANOPHOX (SympatecGmbH, Germany).

The thermal behavior of the PITMS NPs was determined by thermo gravimetric analysis (TGA) with a TA TGA Q500 instrument combined with mass spectrometer (MS) from Thermo-star Pfeiffer Inc.

The weight % polymerization yield of the ITMS to form PITMS NPs was calculated by the following expression:

$$\text{Polymerization yield (weight \%)}$$
$$= [W(\text{PITMS})/W(\text{ITMS} + \text{EGDMA})] \times 100$$

where W(PITMS) is the weight of the dried PITMS NPs and W(ITMS + EGDMA) is the initial weight of the ITMS and EGDMA monomers.

Film thicknesses were measured on a Millitron 1204 IC (Mahr Feinmesstechnik GmbH). The optical parameters transmittance, haze, and clarity of the films were measured on a BYK Gardner haze-gard plus in accordance with ASTM D1003 "Standard Test Method for Haze and Luminous Transmittance of Transparent Plastics". The PET, PET/film former and PET/PITMS films are irradiated with visible light; the transmitted intensity is then integrated by the instrument. Haze and clarity are per definition components of scattered light under wide angle (>2.5°) and narrow angle (<2.5°), respectively. Mean values and standard deviations of transmittance, haze, and clarity were obtained by taking the average over several measurements (at least 4 measurements each).

Migration test

A specimen of 0.5 dm^2 of each film (PET, PET/film former and PET/PITMS films) was incubated with 50 mL of 3 % acetic acid in distilled water or 95 % ethanol for 2 h at 70 °C. The migration of the NPs from the PET/PITMS films into the continuous phase was accomplished by weighing the PET/PITMS films before and after the incubation and measuring the absorbance spectrum of the filtrate using Carry 100 UV–VIS spectrophotometer (Agilent Technologies Inc.) at a range of 200–600 nm. In addition, the haze of the PET/PITMS films during time was also measured.

Cytotoxicity of the PITMS NP coating

In vitro cytotoxicity of the PITMS NP coatings was tested by using HaCaT cell line. The cell line is adherent to the used culture dishes. HaCaT cells were grown in DMEM-eagle that was supplemented with 10 % heat-inactivated fetal bovine serum (FBS), 1 % glutamine and 1 % penicillin/streptomycin. Cytotoxicity was performed in two steps. First, the PET, PET/film former and PET/PITMS films were incubated within the medium at 37 °C for 24 h in a humidified 5 % CO$_2$ incubator. The next step is incubation of the supernatant with HaCaT cell line at 37 °C for 48 h and then measuring the release of cytoplasmic lactate dehydrogenase (LDH) into the cell culture supernatants.

Cell cytotoxicity was assessed by measuring the release of LDH into cell culture supernatants. LDH activity was assayed using the Cytotoxicity Detection Kit according to the manufacturer's instructions. Cells (3 × 10^5 cells per well) were seeded and grown to 75–80 % confluency in 96 well plates before treatment with the films supernatants. Cell cultures that were not exposed to the films supernatants were included in all assays as negative controls. Cell cultures that were treated with 1 % Triton-x-100 were used as positive controls. The cell cultures were further incubated at 37 °C in a humidified 5 % CO_2 incubator and then checked for cellular cytotoxicity after of 48 h. The percentage of cell cytotoxicity was calculated using the formula shown in the manufacturer's protocol [33]. All samples were tested in tetraplicates.

Abbreviations

CFU: colony-forming units; CMS: *p*-chloromethylstyrene; DD: double distilled; DMEM: Dulbecco's minimum essential medium; *E. coli*: *Escherichia coli*; EGDMA: ethylene glycol dimethacrylate; FBS: fetal bovine serum; FTIR: Fourier transform infrared; ITMS: isothiouronium methylstyrene; LB: Luria–Bertani; LDH: lactate dehydrogenase; MIC: minimum inhibitory concentration; MS: mass spectrometer; NMR: nuclear magnetic resonance; NPs: nanoparticles; *P. aeruginosa*: *Pseudomonas aeruginosa*; PET: polyethylene terephthalate; PITMS: polyisothiouronium methylstyrene; PPS: potassium persulfate; *S. aureus*: *Staphylococcus aureus*; TEM: transmission electron microscope; TGA: thermo gravimetric analysis; TSB: tryptic soy broth.

Authors' contributions

SC and CG carried out the synthesis and characterization of the nanoparticles. MN and EB carried out the antibacterial and antibiofilm study. EC carried out the toxicity experiment. SM supervised the study, and participated in its design and coordination. All authors read and approved the final manuscript.

Author details

[1] Department of Chemistry, The Institute of Nanotechnology and Advanced Materials, Bar-Ilan University, 52900 Ramat Gan, Israel. [2] The Mina and Everard Goodman Faculty of Life Sciences, The Institute for Advanced Materials and Nanotechnology, Bar-Ilan University, 52900 Ramat Gan, Israel.

Acknowledgements

This Study was supported by Magnet Program of the Israeli Ministry of Economy and the Israeli P^3 Consortium. The authors thank Dr. Moira Nir for her assistance with the optical properties and Gila Jacobi for her assistance with the biological experiments.

Competing interests

The authors declare that they have no competing interests.

References

1. Jahid IK, Ha SD. A review of microbial biofilms of produce: future challenge to food safety. Food Sci Biotechnol. 2012;21:299–316.
2. Quintavalla S, Vicini L. Antimicrobial food packaging in meat industry. Meat Sci. 2002;62:373–80.
3. Bhattacharya S, Saha I, Mukhopadhyay A, Chattopadhyay D, Chand U. Role of nanotechnology in water treatment and purification: potential applications and implications. Int J Chem Sci Technol. 2013;3:59–64.
4. Szymańska J, Sitkowska J. Bacterial hazards in a dental office: an update review. Afr J Microbiol Res. 2012;6:1642–50.
5. Appendini P, Hotchkiss JH. Review of antimicrobial food packaging. Innov Food Sci Emerg Technol. 2002;3:113–26.
6. Marambio-Jones C, Hoek EMV. A review of the antibacterial effects of silver nanomaterials and potential implications for human health and the environment. J. Nanoparticle Res. 2010;12:1531–51.
7. Kenawy ER, Worley SD, Broughton R. The chemistry and applications of antimicrobial polymers: a state-of-the-art review. Biomacromolecules. 2007;8:1359–84.
8. Ikeda T, Yamaguchi H, Tazuke S. New polymeric biocides: synthesis and antibacterial activities of polycations with pendant biguanide groups. Antimicrob Agents Chemother. 1984;26:139–44.
9. Lu G, Wu D, Fu R. Studies on the synthesis and antibacterial activities of polymeric quaternary ammonium salts from dimethylaminoethyl methacrylate. React Funct Polym. 2007;67:355–66.
10. Sheetz DP, Steiner EC. *S-(vinylbenzyl)isothiouronium salts*, US Patent 3,642,879. Michigan, USA; 1972.
11. Bella MD, Tait A, Parenti C. S-Aryl (tetramethy1) isothiouronium salts as possible antimicrobial agents, I. Arch Pharm. 1986;456:451–6.
12. Badawi AM, Azzam EMS, Morsy SMI. Surface and biocidal activity of some synthesized metallo azobenzene isothiouronium salts. Bioorg Med Chem. 2006;14:8661–5.
13. Shah DR, Lakum HP, Chikhalia KH. Synthesis and in vitro antimicrobial evaluation of piperazine substituted quinazoline-based thiourea/thiazolidinone/chalcone hybrids. Bioorg Khim. 2015;41:209–22.
14. Barna JC, Williams DH. The structure and mode of action of glycopeptide antibiotics of the vancomycin group. Annu Rev Microbiol. 1984;38:339–57.
15. Trani A, Ferrari P, Pallanza R, Ciabatti R. Thioureas and isothioronium salts of the aglycone of teicoplanin. J Antibiot. 1989;62:1268–75.
16. Tiller JC, Liao CJ, Lewis K, Klibanov AM. Designing surfaces that kill bacteria on contact. Proc Natl Acad Sci USA. 2001;98:5981–5.
17. Gottenbos B, Van Der Mei HC, Klatter F, Nieuwenhuis P, Busscher HJ. In vitro and in vivo antimicrobial activity of covalently coupled quaternary ammonium silane coatings on silicone rubber. Biomaterials. 2002;23:1417–23.
18. Naves AF, Palombo RR, Carrasco LDM, Carmona-Ribeiro AM. Antimicrobial particles from emulsion polymerization of methyl methacrylate in the presence of quaternary ammonium surfactants. Langmuir. 2013;29:9677–84.
19. Nair MKM, Vasudevan P, Venkitanarayanan K. Antibacterial effect of black seed oil on Listeria monocytogenes. Food Control. 2005;16:395–8.
20. Mbata TI, Debiao LU. Saikia, A Antibacterial activity of the crude extract of Chinese green tea (*Camellia sinensis*) on Listeria monocytogenes. African J Biotechnol. 2008;7:1571–3.
21. Gluz E, Rudnick-Glick S, Mizrahi DM, Chen R, Margel S. New biodegradable bisphosphonate vinylic monomers and near infrared fluorescent nanoparticles for biomedical applications. Polym Adv Technol. 2014;25:499–506.
22. Baruch-Sharon S, Margel S. Synthesis and characterization of polychloromethylstyrene nanoparticles of narrow size distribution by emulsion and miniemulsion polymerization processes. Colloid Polym Sci. 2010;288:869–77.
23. Koval I. V Synthesis, structure, and physicochemical characteristics of thiols. Russ J Org Chem. 2005;41:631–48.
24. Galperin A, Margel D, Margel S. Synthesis and characterization of uniform radiopaque polystyrene microspheres for X-ray imaging by a single-step swelling process. J Biomed Mater Res Part A. 2006;79A:544–51.
25. Goldshtein J, Margel S. Synthesis and characterization of polystyrene/2-(5-chloro-2H-benzotriazole-2-yl)-6-(1,1-dimethylethyl)-4-methyl-phenol composite microspheres of narrow size distribution for UV irradiation protection. Colloid Polym Sci. 2011;289:1863–74.
26. Tseng CM, Lu YY, El-Aasser MS, Vanderhoff JW. Uniform polymer particles by dispersion polymerization in alcohol. J Polym Sci Part A Polym Chem. 1986;24:2995–3007.
27. Shen S, Sudol E, El-Aasser M. Control of particle size in dispersion polymerization of methyl methacrylate. J Polym Sci Part A Polym Chem. 1993;31:1393–402.
28. Horák D, Chaykivskyy O. Poly(2-hydroxyethyl methacrylate-co-N, O-dimethacryloylhydroxylamine) particles by dispersion polymerization. J Polym Sci Part A Polym Chem. 2002;40:1625–32.

29. Galperin A, Margel S. Synthesis and characterization of new radiopaque microspheres by the dispersion polymerization of an iodinated acrylate monomer for X-ray imaging applications. J Polym Sci Part A Polym Chem. 2006;44:3859–68.

30. Paine AJ, Luymes W, McNulty J. Dispersion polymerization of styrene in polar solvents. 6. Influence of reaction parameters on particle size and molecular weight in poly(N-vinylpyrrolidone)-stabilized reactions. Macromolecules. 1990;23:3104–9.

31. Schug TT, Berry DC, Shaw NS, Travis SN, Noy N. Opposing effects of retinoic acid on cell growth result from alternate activation of two different nuclear receptors. Cell. 2007;129:723–33.

32. Boukamp P, Petrussevska RT, Breitkreutz D, Hornung J, Markham A, Norbert FE. Normal keratinization in a spontaneously immortalized aneuploid human keratinocyte cell line. J Cell Biol. 1988;106:761–71.

33. Decker T, Lohmann-Matthes ML. A quick and simple method for the quantitation of lactate dehydrogenase release in measurements of cellular cytotoxicity and tumor necrosis factor (TNF) activity. J Immunol Methods. 1988;15:61–9.

34. Kolitz-Domb M, Margel S. Engineered narrow size distribution high molecular weight proteinoids, proteinoid-poly(L-lactic acid) copolymers and nano/micro-hollow particles for biomedical applications. J Nanomed Nanotechnol. 2014;5:1–11.

35. Shemesh R, Krepker M, Natan M, Banin E, Kashi Y, Nitzan N, Vaxman A, Segal E. Novel LDPE/halloysite nanotube films with sustained carvacrol release for broad-spectrum antimicrobial activity. RSC Adv. 2015;5:87108–17.

12

Nanostructured biosensor using bioluminescence quenching technique for glucose detection

Longyan Chen[1], Longyi Chen[1], Michelle Dotzert[2], C. W. James Melling[2] and Jin Zhang[1]* ⓘ

Abstract

Background: Most methods for monitoring glucose level require an external energy source which may limit their application, particularly in vivo test. Bioluminescence technique offers an alternative way to provide emission light without external energy source by using bioluminescent proteins found from firefly or marine vertebrates and invertebrates. For quick and non-invasive detection of glucose, we herein developed a nanostructured biosensor by applying the bioluminescence technique.

Results: Luciferase bioluminescence protein (Rluc) is conjugated with β-cyclodextrin (β-CD). The bioluminescence intensity of Rluc can be quenched by 8 ± 3 nm gold nanoparticles (Au NPs) when Au NPs covalently bind to β-CD. In the presence of glucose, Au NPs are replaced and leave far from Rluc through a competitive reaction, which results in the restored bioluminescence intensity of Rluc. A linear relationship is observed between the restored bioluminescence intensity and the logarithmic glucose concentration in the range of 1–100 μM. In addition, the selectivity of this designed sensor has been evaluated. The performance of the senor for determination of the concentration of glucose in the blood of diabetic rats is studied for comparison with that of the concentration of glucose in aqueous.

Conclusions: This study demonstrates the design of a bioluminescence sensor for quickly detecting the concentration of glucose sensitively.

Keywords: Bioluminescence, Biosensor, Nanoparticles, Glucose

Background

Glucose is one of the most important carbohydrates to provide energy for cell metabolism. Studies also show that glycolytic flux is closely related to enzyme levels. Control of glucose level is the key to the regulation of signaling protein-involved pathways, and to manage chronic diseases, e.g. diabetics [1]. Different techniques from electrochemical methods to fluorescence detections have been developed for monitoring glucose level in vitro and in vivo [2–4]. Biosensors by using luminescence techniques have advantages in directly converting bioprocess to luminescence signal through a non-toxic and cost-effective process [5]. Resonance energy transfer (RET) techniques including Förster/fluorescence RET (FRET) and bioluminescence RET (BRET) are distance-dependent non-radiation energy transfer, which have been used for detecting protein–protein interactions in vitro and in vivo [6–8].

Generally, both FRET and BRET utilize the resonance energy transfer from a fluorophore donor to a fluorophore acceptor. The resonance energy transfer is dependent on the distance between donor and acceptor. Recent developments on FRET biosensor unveil a quenching mechanism due to the resonance energy transfer between metallic nanoparticles (NPs) and fluorophore donor. The quenching mechanism allows luminescence biosensor to obtain high signal-to-noise ratio, in which the fluorescence intensity of the donor can be quenched by the acceptor made of metallic nanoparticles, and the

*Correspondence: jzhang@eng.uwo.ca
[1] Department of Chemical and Biochemical Engineering, University of Western Ontario, 1151 Richmond St., London, ON N6A 5B9, Canada
Full list of author information is available at the end of the article

fluorescence intensity of the donor can be restored if the distance between the donor and metallic NPs increases [9, 10]. Metallic NPs, e.g. gold nanoparticles (Au NPs), can be used as an acceptor to quench the fluorescence of a donor in a FRET sensor due to their large extinction coefficients and broad energy absorption in the visible range. The quenching efficiency is reverse proportional to the distance between nanoparticle and the donor in a designed FRET biosensor [11–13]. The FRET assays by using quenching mechanism have been applied in different organic-based assays for nucleic acid detection [14, 15].

Bioluminescence is a unique phenomenon in nature which can be found from firefly to marine vertebrates and invertebrates. Currently, bioluminescence techniques have been used in studying gene expression. BRET technique shows an advantage over FRET technique because it does not require the external energy, e.g. light source, to excite the donor. Therefore, BRET technique has been used for real-time and non-invasive detection [16, 17]. Normally, bioluminescence reaction happens when luciferase, an enzyme, reacts with the substrate, e.g. coelenterazine (CTZ) [18, 19]. Luciferase protein can be used as a donner in BRET biosensor [20, 21]. Quite recently, luciferase protein conjugated onto Au NP has been demonstrated for proteases detection [16]. The bioluminescence intensity of luciferase can be eliminated/quenched by the conjugated Au NPs through the resonance energy transfer process. To our best knowledge, no such system has been applied in detection of analytes other than proteases.

Here, we developed a bioluminescence sensor by applying bioluminescent quenching principle for detecting glucose in a homogeneous assay format. The nanostructured biosensor is composed of a donor, i.e. Renilla luciferase conjugated with β-cyclodextrin (β-CD-Rluc), and a quenching element made of Au NPs modified with phenylboronic acid (PBA) as shown in Scheme 1. It is known that PBA can react with glucose by forming cyclic ester [22, 23]. PBA has been used as a recognition element for glucose sensing [24, 25]. Once Au NPs modified with PBA react with β-CD-Rluc through a reversely covalent bond as shown in Scheme 1, Au NPs can quench the bioluminescence intensity generated by Rluc reacting with coelenterazine (CTZ). In the presence of glucose, the bioluminescence intensity will be restored due to the stronger interaction between glucose and PBA, which leads to the release of β-CD-Rlu from PBA-Au NPs. In this study, we have investigated the effect of the ratio of the donor to the quenching on the optimal bioluminescence intensity with/without glucose. The study demonstrates the restored bioluminescence intensity as a function of the concentration of glucose.

Methods

Renilla reniformis luciferase gene (*Rluc*) was purchased from Promega Inc. The vector pET-32a (EMD Millipore Inc.) was used to subclone the *Rluc*. The *E. coli* strain BL21 (DE3) (invitrogen) was used as the bacterial strain for the expression of proteins. Restriction enzymes were purchase from New England Biolabs Inc. Unless otherwise indicated, all the chemicals were purchased from Sigma-Aldrich. Ultrahigh quality water with a resistance of 18.2 MΩ cm (at 25 °C) was obtained from a Nanopure™ water system (Thermofisher) fitted with a 0.22 μM filter.

Preparation of Au NP-based acceptor of the BRET sensor

Citrate-stabilized gold nanoparticles (Au NPs) were synthesized by using the reported method [26]. In brief, 240 mL of aqueous solution containing 0.21 mM of $HAuCl_4$ (0.08 mg/mL) and trisodium citrate (13.8 mg) was mixed at room temperature. The mixture was added 5 mL of ice-cold 0.1 M $NaBH_4$ solution under vigorous stirring. The solution was aged overnight in a dark place with vigorous stirring. Citrate-stabilized Au NPs were further modified by 11-mercaptoundecanoic acid (MUA). 20 mL of citrate-stabilized Au NPs (60 nM, pH pre-adjusted to 10.3 by 2 M NaOH) was mixed with 19 mL of ethanol. 1 mL of MUA (10 mM) ethanol solution was added in the gold solution and stirred for 19 h (hrs). The mixture was then filtered through 0.45 μm acetate filter and purified by Amicon ultralfilter (100 kDa MWCO). Finally, the MUA modified particles were resuspended in 20 mL water (~60 nM).

MUA-modifying Au NPs (MUA-Au NPs) was further conjugated with 3-aminophenylboronic acid (PBA) through carbodiimide reaction. Briefly, MUA-Au NPs (60 nM) was mixed with PBA (10 mg/mL), 1-ethyl-3-(3-dimethylaminopropyl)-carbodiimide (EDC, 10 mg/mL), and *n*-hydroxysuccinimide (NHS, 10 mg/mL) for 2 h at room temperature with gently shaking. The conjugated particles were further purified by Amicon ultralfilter (MWCO 100 kDa) and resuspended in phosphate-buffered saline buffer (PBS, 10 mM, pH 7.4).

Preparation of Rluc protein

The entire coding sequences of *Rluc* was cloned to the MCS site of pET 32-a plasmid under two restriction sites (BamH I and Xho I). Two primers were designed for the cloning (forward: 5′AAAGGATCCAGCGGTGGTGGTGGTAGCATGACTTCGAAAGTTTATGATCCAG; reverse: 5′ TGTGCTCGAGTTGTTCATTTTTGAGAACTCGCTC 3′). A trx region from pET 32-a coding for thioredoxin protein is kept to maintain high level of recombinant protein expression [27]. A six-amino acid linker (SGGGGS) highlighted by the underline in the

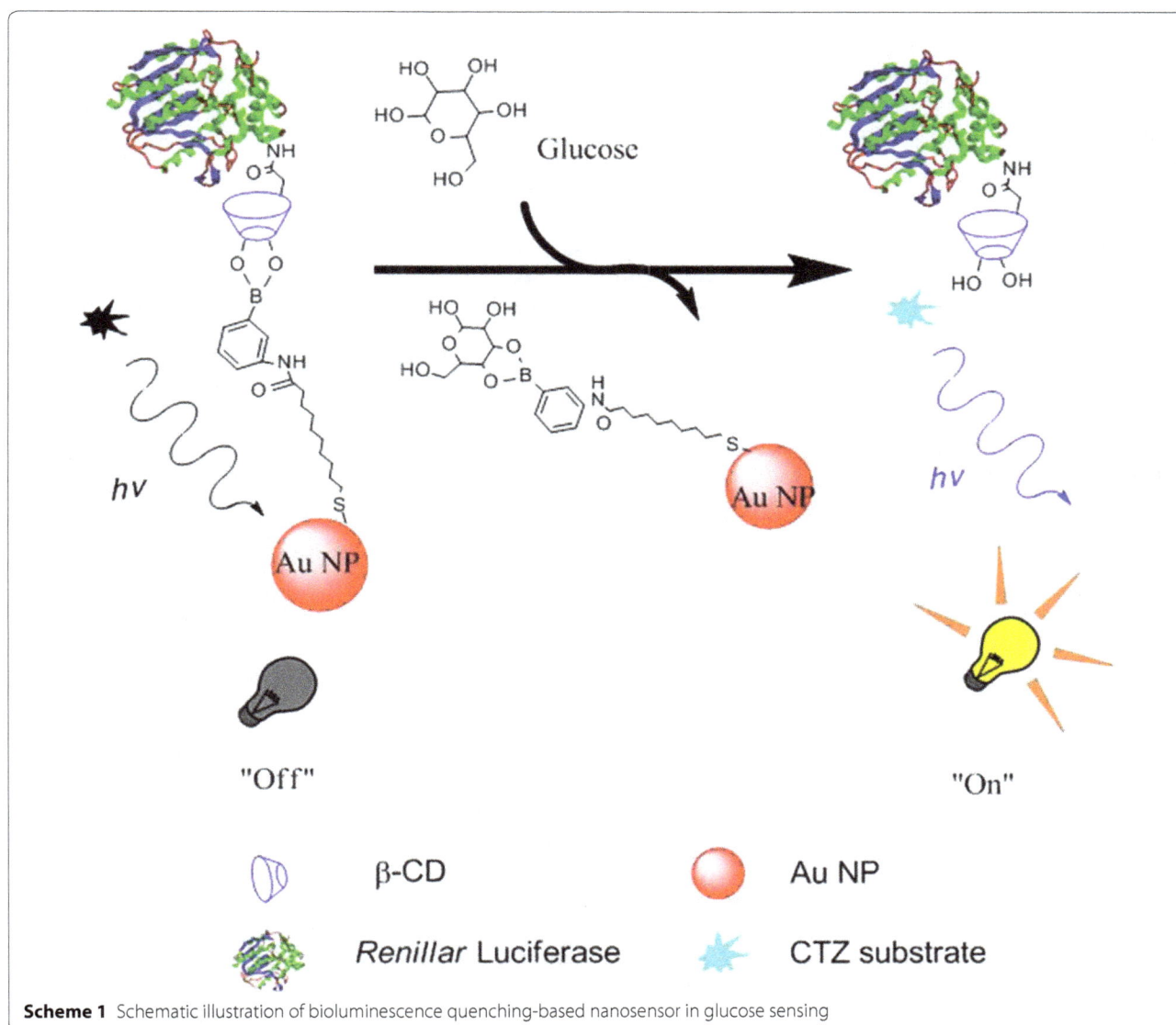

Scheme 1 Schematic illustration of bioluminescence quenching-based nanosensor in glucose sensing

forward primer was inserted after BamH I site to leave a flexible space for proper folding of Rluc protein. The successful construction of the plasmid was confirmed by DNA sequencing (Robarts Institute, Western University, London, ON).

For protein expression, the bacterial cells (*E. coli* k-12) with recombinant plasmid were grown from a single colony overnight at 37 °C in 5 mL of Luria–Bertani (LB) broth (containing 100 μg/mL ampicillin). After transferring to 500 mL of LB broth, the cells were allowed growing for 2 h at 37 °C. The cells were then cultured with isopropyl β-D-1-thiogalactopyranoside (IPTG) at a final concentration of 1 mM for 4 h at 20 °C. The cells were harvested by centrifugation at 12,000 rpm for 5 min

at 4 °C, and the pellet was resuspended in a lysis buffer [20 mM Tris/HCl buffer (pH 7.4), containing 500 mM NaCl, 5 mM imidazole, 0.2 mg/mL of lysozyme and 0.1% Triton-X-100] and sonicated with ice cooling for 15 s (s) followed by 30-s rest for 20 cycles by a Mandel Scientific Q500 sonicator. The suspension was centrifuged at 10,000 rpm at 4 °C for 30 min. The supernatant containing overexpression protein was purified via His-trap HP columns (GE lifescience, Inc.) by a syringe pump. After loading the samples, the expressed protein was eluted with an elution buffer [200 mM imidazole, 20 mM Tris/HCl buffer (pH 7.4), 500 mM NaCl and 10% glycerol] and dialyzed with PBS using Pur-A-Lyzer™ Mega Dialysis Kit (12KDa MWCO, Sigma-Aldrich). The purified protein was further concentrated by an Amicon Ultra centrifugal filter (ultra-15, MWCO 10 kDa, EMD Millipore Inc). The resultant Rluc protein solution was stored in aliquot at −80 °C. The concentration of the protein was determined by bicinchoninic acid (BCA) protein assay (Thermo scientific Inc).

Conjugation of β-cyclodextrin (β-CD) to Rluc

Dimethylformamide (DMF) solution containing 3.78 mg of succinyl-β-cyclodextrin (~2 μmol) was mixed with 250 μL of 10 mg/mL NHS and 400 μL of 16 mg/mL EDC. The mixture in 350 μL PBS was incubated for 2 h at room temperature under gently shaking. For conjugating reaction, 200 μL of above solution was mixed with 200 μL of 10 mg/mL Rluc solution in a PBS solution (final volume was maintained at 1 mL). The solution was further incubated overnight at 4 °C. The reaction was terminated by addition of 5 μL of ethanolamine. The β-CD labeled Rluc (β-CD-Rluc) was purified through a Nap-10 column (GE Healthcare) with PBS as an eluent. The labeled protein was collected by Amicon ultral filter (ultra-15) to desired concentration and stored at 4 °C for at least 4 weeks which can maintain over 90% activity. For long-term storage, the labeled protein was stored at −80 °C with complete retention of activity.

Carbohydrate assay

The amount of β-CD modified on a Rluc was determined by a sulfuric acid-phenol assay with slight modification [18]. The detailed results can be found from the supporting documents. In a typical test, a 30 μL of ice-colded glucose standard samples or β-CD-Rluc solution was mixed with 100 μL of concentrated sulfuric acid in microplate wells. 20 μL of aqueous 5% phenol solution was then added into the well. The plate was floated uncovered on near boiling (>90 °C) water bath for 5 min for color development, followed by cooling on ice for another 5 min. A microplate reader (Tecan M200 multimode microplate reader) was used to collect the absorbance of each well at 490 nm. The concentration of standard solution was plotted with the absorbance of each solution. The molar concentration of β-CD was thus calculated. The number of β-CD per protein was obtained by dividing the molar concentration of β-CD over the molar concentration of protein Rluc.

Evaluation of the bioluminescence quenching-based assay

3 μL of β-CD-Rluc (1 μM) was mixed with 100 μL of PBA-Au NPs (30 nM) in a PBS solution containing 0.1% bovine serum albumin (BSA). BSA was added to stabilize Rluc activity and reduce non-specific binding between protein and Au NPs. 10 μL of glucose solution (with different concentrations) was added in the above solution.

For the reactivity assay, 10 μL of all the tested substances with concentration of 50 μM was added into the above solution. The final incubation volume was 300 μL. The solution was then incubated at room temperature for 30 min. The bioluminescence was then collected by adding 1 μL of native coelenterazine (CTZ, 1 mg/mL in ethanol) in the above solution.

Male Sprague–Dawley rats, 8-weeks of age, were obtained from Charles River Laboratories. Rats were housed two per cage at constant temperature and humidity on a 12 h dark/light cycle, and had access to water and standard chow ad libitum. Ethics approval was obtained from the University of Western Ontario Research Ethics Board, in accordance with Canadian Council on Animal Care guidelines. The experimental protocol followed the Principles of Laboratory Animal Care (US NH publication No83-85, revised 1985). Diabetes was induced with multiple low-dose Streptozotocin (STZ) injections. STZ (20 mg/kg; Sigma Aldrich, Oakville, ON, Canada) in citrate buffer was injected into the intraperitoneal cavity for 5 consecutive days, and the blood samples of diabetic rats were measured by glucometer to verify the glucose concentration is larger than 20 mM. Diluted blood samples of diabetic rats with varied glucose concentrations, i.e. 100, 50, 20, 10, 5 and 1 μM, were prepared for evaluating the performance of the nanostructured sensor.

Materials characterization

Au nanoparticles with different surface modification were characterized by transmission electron microscope (TEM), UV–Vis spectrometer, and Fourier transform infrared spectroscopy (FTIR).

Transmission electron microscopy (TEM) image was obtained through a Philips CM-10 operating at 100 kV. Dynamic light scattering (DLS) measurements were performed on a Zetasizer Nano-ZS (Malvern Instruments Inc.) equipped with a He/Ne laser at 633 nm. UV–Vis spectroscopic measurements were obtained on a Cary 300 spectrometer. Fourier transform infrared (FT-IR) spectra were conducted within the 4000–500 cm^{-1} wavenumber range using a Bruker FTIR-IFS 55 spectrometry. Bioluminescence measurements were recorded by using a QuantaMaster 40 Spectrofluorometer (Photon Technology International Inc., London, ON). The software was set up at emission collection. The shutter at laser path was closed and the emission spectra were collected from 430 to 580 nm in a mode of 2 nm per step and integration time at 0.2 s.

Results

Carboxyl-modified β-CD was conjugated onto Rluc protein through the carbodiimide-mediated reaction, Native polyacrylamide gel electrophoresis (native-PAGE) was used to investigate the conjugation of carboxyl-modified β-CD to Rluc. It is noted that the β-CD conjugation could decrease protein positive charges because of the formation of primary amine group. Hence, the proteins with or without the conjugation of β-CD could be separated by native-PAGE on basis of protein surface charges [17]. In Fig. 1, the sample of β-CD-Rluc (lane 2) displays significant band shift as compared to the negative control, i.e. Rluc without conjugation (lane 1). Meanwhile, the conjugation of β-CD onto Rluc protein was evaluated by using sulfuric acid-phenol colorimetric assay for total carbohydrate after purification [18, 19]. In Additional file 1:

Figure S1, the typical peak at 490 nm to the sample of β-CD conjugating Rluc is observed, which confirms the successful conjugation of β-CD to the protein. The optimized molar ratio of β-CD to Rluc protein is estimated at 100:1, that is, 10 β-CD molecules on each Rluc molecule, which is determined through the standard curve (Additional file 1: Figure S1, ESI†). The precipitation of protein is observed when the molar ratio of β-CD to Rluc is larger than 100:1, which might be related to the high co-solvent (dimethylformamide) concentration in buffer [28]. TEM and DLS were carried out to measure the average particle size and size distribution of Au NPs. Figure 2a is the TEM micrograph of Au NPs. The spherical NPs have the average particle size around 8 ± 3 nm. While, the result of DLS indicates that average particle size is around 10 nm with large size distribution as shown in Fig. 2b. In addition, Fig. 3a shows the UV–Vis spectra of Au NPs with different modification. The citrate-stabilized Au NPs have typical surface plasmon resonance peak at 512 nm. The localized surface plasmon resonance (LSPR) of Au NP with the modification of MUA and PBA is centering at 525 and 550 nm, respectively. The broaden absorption band could be related to the changes in the local environment and chemical interface dampening [29]. The surface modifications on Au NPs was further investigated by FTIR spectra. In Fig. 3b, the peaks around 1344 cm^{-1} (to 1362 cm^{-1}) are attributed to B-O stretch from PBA [30]. Phenyl ring signature bands are observed at 1427 cm^{-1} (strong) and 1604 cm^{-1} (weak) to PBA-Au NPs [22]. The N–H bending band (red dash circle in Fig. 2b) is shown at 1581 cm^{-1}, which appears relative weak peak to the sample of PBA-Au NPs [25]. The amide I band of PBA-Au NPs associated with −C=O stretch of peptide bond spans

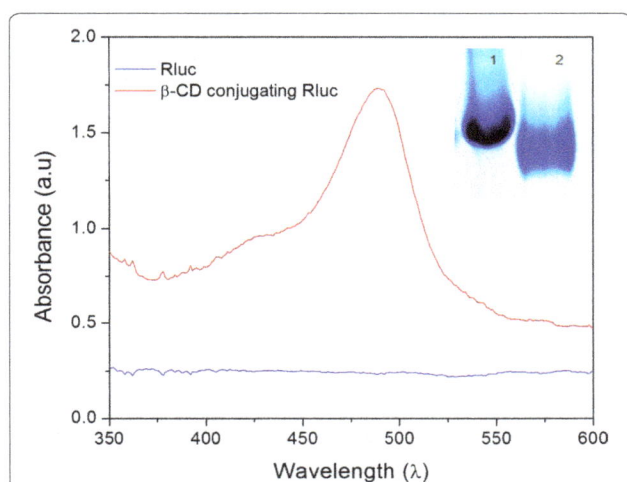

Fig. 1 UV–vis spectra of the Rluc with/without the conjugation of β-CD. The small inset image is the native-PAGE characterization, where *Lane 1* is Rluc without conjugation, and *Lane 2* refers β-CD conjugating-Rluc

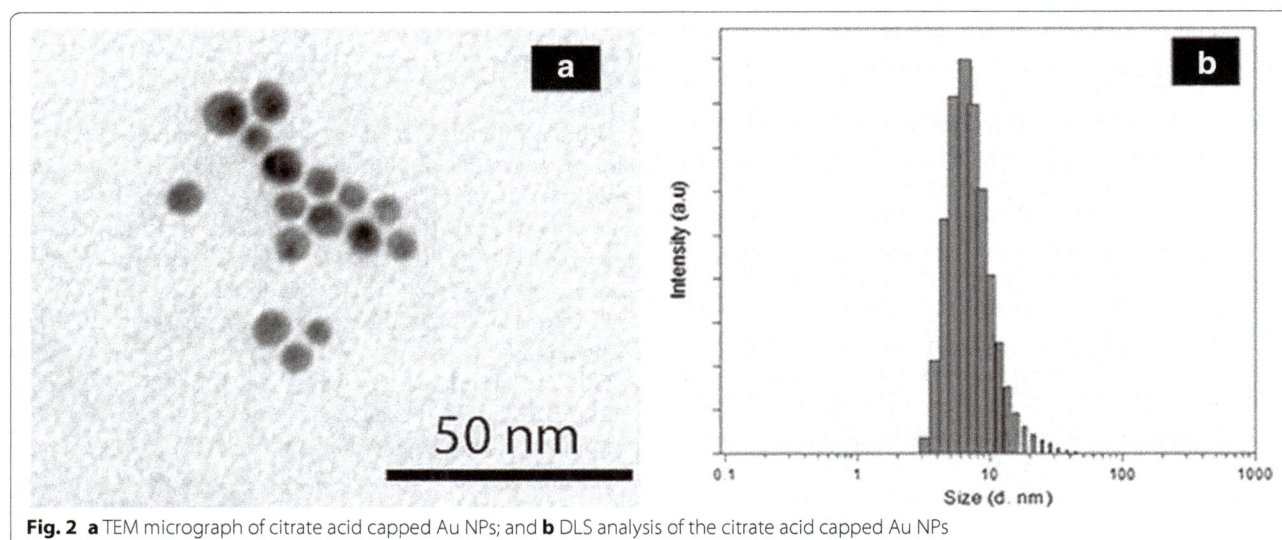

Fig. 2 a TEM micrograph of citrate acid capped Au NPs; and **b** DLS analysis of the citrate acid capped Au NPs

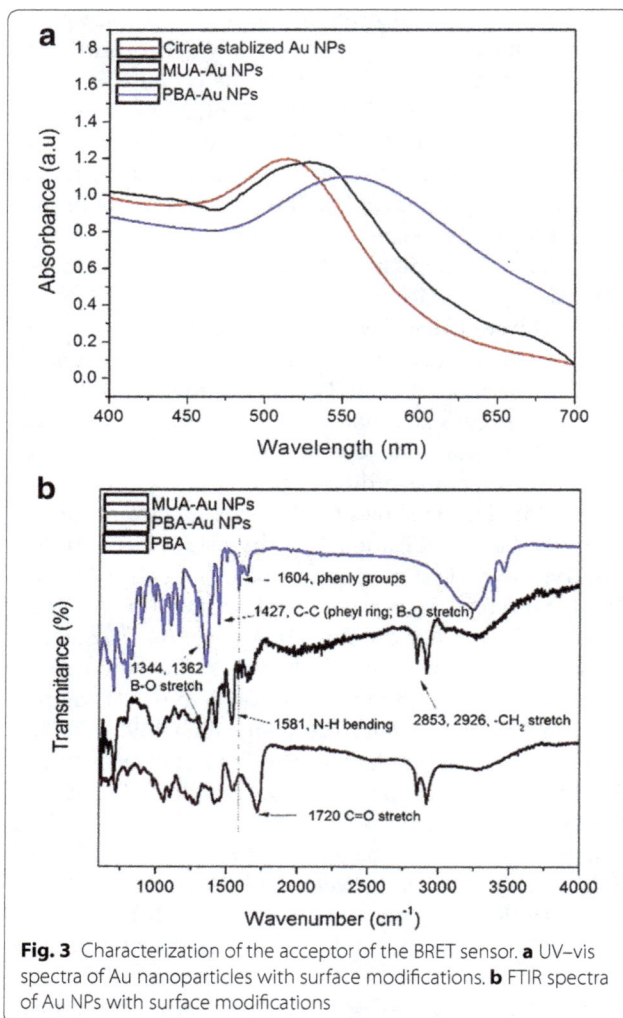

Fig. 3 Characterization of the acceptor of the BRET sensor. **a** UV–vis spectra of Au nanoparticles with surface modifications. **b** FTIR spectra of Au NPs with surface modifications

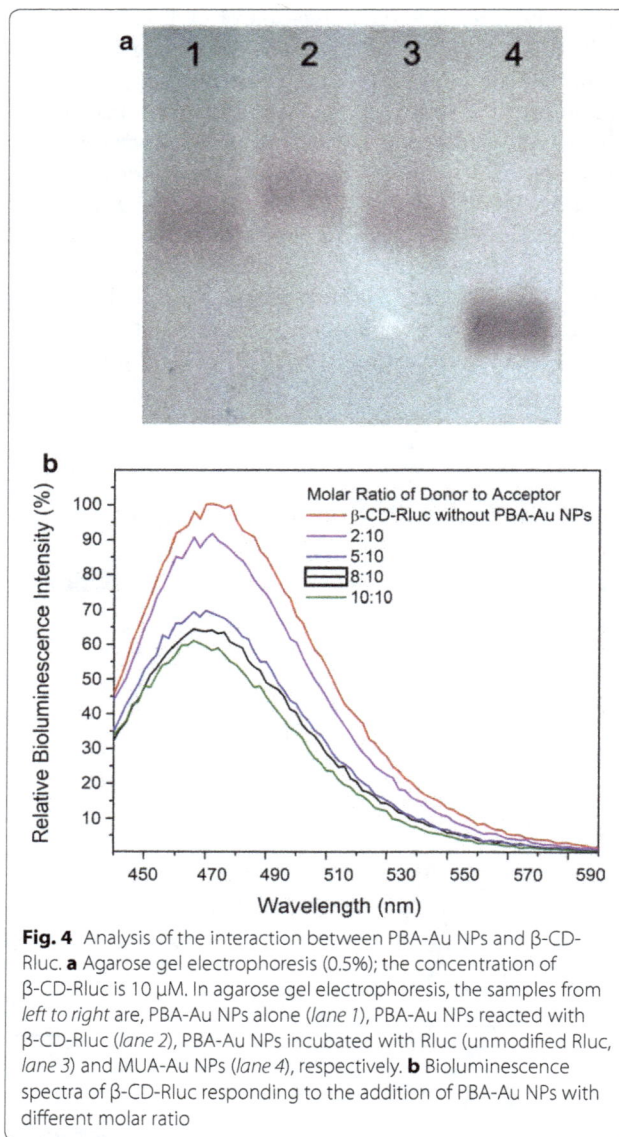

Fig. 4 Analysis of the interaction between PBA-Au NPs and β-CD-Rluc. **a** Agarose gel electrophoresis (0.5%); the concentration of β-CD-Rluc is 10 μM. In agarose gel electrophoresis, the samples from *left to right* are, PBA-Au NPs alone (*lane 1*), PBA-Au NPs reacted with β-CD-Rluc (*lane 2*), PBA-Au NPs incubated with Rluc (unmodified Rluc, *lane 3*) and MUA-Au NPs (*lane 4*), respectively. **b** Bioluminescence spectra of β-CD-Rluc responding to the addition of PBA-Au NPs with different molar ratio

from 1600 to 1700 cm^{-1} [31]. The results of FTIR indicates that successful modification of Au NPs with PBA.

PBA-Au NPs then react with β-CD-Rluc through the reversible boronic acid-diol interaction [22]. The interaction between PBA-Au NPs and β-CD-Rluc was investigated by agarose gel electrophoresis. In Fig. 4a, the mixture of PBA-Au NPs and β-CD-Rluc displays the slowest migrating band (lane 2) as compared to the negative controls, i.e. PBA-Au NPs alone (lane 1), PBA-Au NPs mixing with unmodified Rluc (lane 3), and MUA-Au NPs alone (lane 4). This band shift could be attributed to an increase of NPs in terms of particle size, due to the formation of β-CD-Rluc/PBA-Au NP complex. Without β-CD conjugation, Rluc/PBA-Au NP mixture shows similar band position to PBA-Au NP. Thereby, the interaction of conjugated protein with PBA-Au NP is specific due to the boronic acid-diol interaction. The bioluminescence

of Rluc is quenched by the conjugated Au NPs. Figure 4b shows that the bioluminescence intensity decreases gradually when adding more PBA-Au NPs in β-CD-Rluc by comparing to the signal from control group without NPs (red curve). The bioluminescence intensity can be quenched over ~58% when the molar ratio of PBA-Au NPs to β-CD-Rluc in the mixture is increased to 10/10 as shown in Fig. 4b.

Glucose intends to compete with β-CD-Rluc to react with PBA-Au NPs because the interaction between PBA and glucose is stronger than the reaction of β-CD with PBA, which, therefore, results in the restore of bioluminescence as shown in Scheme 1. The bioluminescence intensity of β-CD-Rluc as a function of the concentration of glucose was further investigated. In a typical assay, the nanostructured sensors (the molar ratio of NPs to protein = 10:10)were mixed with glucose at different concentrations, respectively.. The mixtures were incubated for 30 min. Figure 5 shows the bioluminescence intensity increases when the concentration of glucose in aqueous increases from 0 to 100 μM. It is noted that the positive control (PC) refers the β-CD-Rluc which has the same amount of β-CD-Rluc in the designed sensor.

The normalized bioluminescence intensity was calculated by using the Eq. 1;

Normalized Bioluminescence Intensity

$$= \frac{\text{Bioluminescence Intensity in the presence of glucose}}{\text{Bioluminescence Intensity of the positive control}}$$

(1)

In addition, we tested the sensor performance with blood sample of diabetic rats to evaluate the interference of other proteins in blood. Figure 6 indicates the correlation between the normalized bioluminescence intensity and the log form of the concentration of glucose in aqueous samples and in diluted blood samples. The normalized bioluminescence intensity increases linearly when the logarithmic glucose concentration increases in a range from 1 to 100 μM to both measurements in aqueous and in diluted blood samples as shown in Fig. 6. The detection limit for this assay is secured to 1 μM from three independent measurements. Meanwhile, the normalized bioluminescence intensity corresponding to the log form of glucose in blood samples is very close to the log form of glucose in aqueous sample. Consequently, this bioluminescence nanosensor can detect plasma glucose for diabetic rats.

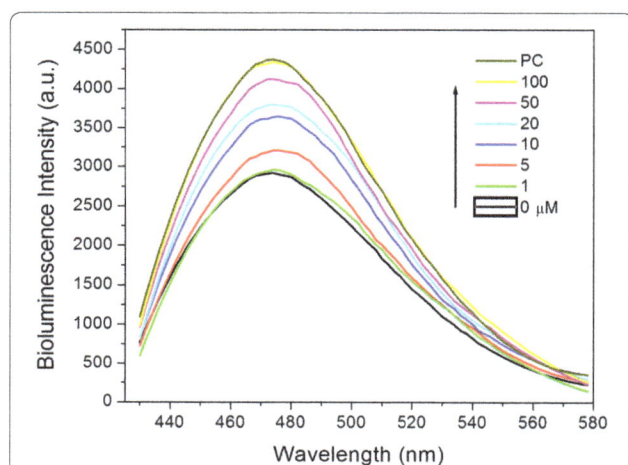

Fig. 5 Bioluminescence intensity with different glucose concentrations. PC refers the positive control, i.e. β-CD-Rluc alone without PBA-Au NPs

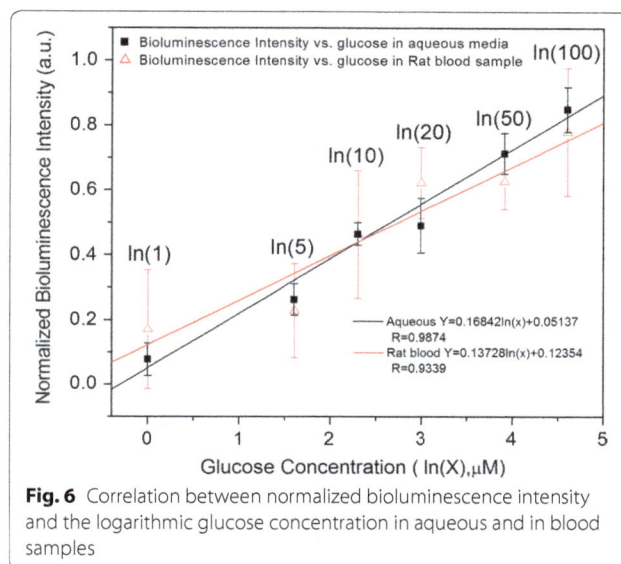

Fig. 6 Correlation between normalized bioluminescence intensity and the logarithmic glucose concentration in aqueous and in blood samples

Discussions

Bioluminescence reaction happens in nature when the enzyme, Rluc, reacts with a substrate, CTZ. We used the typical biological process to obtain pure and high amount of Rluc used as the donor in our designed BRET biosensor; first a plasmid containing Rluc gene was constructed, then the Rluc gene was cloned and expressed in bacterial system. High concentrated Rluc protein was obtained after the protein purification, and was conjugated to β-CD as shown in Fig. 1. Meanwhile, the acceptor was made by modified Au NPs which is able to quench the bioluminescence through the BRET mechanism. The particle size and size distribution were characterized by TEM and DLS as shown in Fig. 2. The discrepancy of the results in size distribution may result from the surface modification of Au NPs by PBS which cannot be shown up on the TEM micrographs as organic molecules have very low electron density. The surface modification of Au NPs were studied and verified by FTIR as shown in Fig. 3. In the presence of glucose, the reaction of glucose with PBA competes against the reaction of β-CD with PBA, which leads to the release of Au NPs from Rluc and results in the restoration of the bioluminescence of Rluc as shown in Fig. 4. The maximum number of β-CD conjugated onto one Rluc is estimated at 10 as shown in the supplementary materials. Figure 4b indicates that the bioluminescence intensity of Rluc can be quenched over ~58% when the molar ratio of PBA-Au NPs to β-CD-Rluc in the mixture is increased to 10:10. The restoration of bioluminescence has a linear relationship with the logarithmic concentration of glucose in a range from 1 to 100 μM as shown in Fig. 5.

To further study the sensitivity and selectivity of the designed sensor, the glucose-specific detection was evaluated by investigating the response of restored bioluminescence intensity to different molecules. The results indicate that the bioluminescent sensor does not react with amino acids, lipids, and common sugars, i.e. mannose, lactose and maltose as shown in Additional file 1: Figure S2. However, sucrose and fructose exhibited higher reactivity to the nanostructured biosensor than glucose. This could be due to higher affinity constants of fructose to PBA than glucose [32, 33]. Meantime, as sucrose is a disaccharide composed of glucose and fructose units linked via their anomeric carbons, it could be expected that one or both monosaccharides could react with PBA-Au NPs.

Though high concentration of fructose may interfere the glucose detection by using the bioluminescent sensor, it should be noted that the concentration of fructose in human blood (both healthy and diabetes subjects) is very low, almost one thousand-fold lower than glucose concentration in healthy subject [34]. Meantime, sucrose is not commonly present in human blood in healthy subjects, as it is rapidly absorbed and hydrolyzed into its monosaccharide units in the small intestine [35, 36]. In addition, Fig. 6 indicates that the results of the blood measurement for diabetic rats is very close to the results of the glucose measurement in aqueous. It is noted that the serum levels of albumin, triglycerides, total protein and glucose in rats is compatible with human's. STZ induced diabetic rats have been well accepted as an animal model for diagnosis and treatment of diabetes [37, 38]. Consequently, this nanostructured sensor by applying the bioluminescence quenching technique can be used to detect glucose with a high sensitive and high selective fashion.

Conclusions

In conclusion, we have developed a new nanostructured biosensor by applying bioluminescence quenching technique for detecting glucose. Bioluminescent protein, Rluc, covalently conjugated with β-CD, is used as the donor in the designed biosensor, while PBA-Au NPs act as the acceptor. The PBA-Au NPs interacts β-CD-Rluc via a reversible covalent bonding results in the quenched bioluminescence of Rluc. In the presence of glucose, it competes with β-CD to react with PBA, which releases Au NPs away from β-CD-Rluc, and results in the restoration of bioluminescence of Rluc. This paper also demonstrates the effect of the molar ratio of donor and acceptor on the bioluminescence quenching efficiency. The results indicate that normalized bioluminescence intensity increase linearly with increasing logarithmic glucose concentration in a range from 1 to 100 μM, with

detection limit of 1 μM. This nanostructured bioluminescence quenching-based sensor could be feasible for blood glucose level monitoring with sample pre-treatment (e.g. dilution).

Authors' contributions

The study was planned by LC and JZ. LYC participated the experiments on materials characterization and data analysis. LYC and JZ worked with MD and JM on testing glucose level in rats' tears. MD and JM provided the STZ-induced diabetic rats. LC and JZ wrote the manuscript with input of all other co-authors. All authors read and approved the final manuscript.

Author details

[1] Department of Chemical and Biochemical Engineering, University of Western Ontario, 1151 Richmond St., London, ON N6A 5B9, Canada. [2] School of Kinesiology, Faculty of Health Sciences, University of Western Ontario, London, ON N6A 5B9, Canada.

Acknowledgements

We are grateful for the financial support from the Natural Sciences and Engineering Research Council of Canada (NSERC).

Competing interests

The authors declare that they have no competing interests.

Funding

This work was supported by the Natural Sciences and Engineering Research Council of Canada (NSERC).

References

1. Guo X, Li H, Xu H, Woo S, Dong H, Lu F, Lange AJ, Wu C. Glycolysis in the control of blood glucose homeostasis. Acta Pharm Sin B. 2012;2:358–67.
2. Rahman MM, Ahammad AJS, Jin J-H, Ahn SJ, Lee J-J. a comprehensive review of glucose biosensors based on nanostructured metal-oxides. Sensors. 2010;10:4855.
3. Galant AL, Kaufman RC, Wilson JD. Glucose: detection and analysis. Food Chem. 2015;188:149–60.
4. Özcan L, Şahin Y, Türk H. Non-enzymatic glucose biosensor based on overoxidized polypyrrole nanofiber electrode modified with cobalt(II) phthalocyanine tetrasulfonate. Biosens Bioelectron. 2008;24:512–7.
5. Yuan L, Lin W, Zheng K, Zhu S. FRET-based small-molecule fluorescent probes: rational design and bioimaging applications. Acc Chem Res. 2013;46:1462–73.
6. Shrestha S, Deo S. Bioluminescence resonance energy transfer in bioanalysis. In: Daunert S, Deo SK, editors. Photoproteins in bioanalysis. Weinheim: Wiley; 2006. p. 95–111.
7. Deuschle K, Chaudhuri B, Okumoto S, Lager I, Lalonde S, Frommer WB. Rapid metabolism of glucose detected with FRET glucose nanosensors in epidermal cells and intact roots of Arabidopsis RNA-silencing mutants. Plant Cell. 2006;18:2314–25.
8. So M-K, Loening AM, Gambhir SS, Rao J. Creating self-illuminating quantum dot conjugates. Nat Protoc. 2006;1:1160–4.
9. Johansson MK. Choosing reporter-quencher pairs for efficient quenching through formation of intramolecular dimers. In: Didenko VV, editor. Fluorescent energy transfer nucleic acid probes: designs and protocols. Totowa: Humana Press; 2006. p. 17–29.

10. Hoebe RA, Van Oven CH, Gadella TWJ, Dhonukshe PB, Van Noorden CJF, Manders EMM. Controlled light-exposure microscopy reduces photobleaching and phototoxicity in fluorescence live-cell imaging. Nat Biotech. 2007;25:249–53.

11. Pons T, Medintz IL, Sapsford KE, Higashiya S, Grimes AF, English DS, Mattoussi H. On the quenching of semiconductor quantum dot photoluminescence by proximal gold nanoparticles. Nano Lett. 2007;7:3157–64.

12. Yun CS, Javier A, Jennings T, Fisher M, Hira S, Peterson S, Hopkins B, Reich NO, Strouse GF. Nanometal surface energy transfer in optical rulers, breaking the FRET barrier. J Am Chem Soc. 2005;127:3115–9.

13. Swierczewska M, Lee S, Chen X. The design and application of fluorophore-gold nanoparticle activatable probes. PCCP. 2011;13:9929–41.

14. Kim Y-P, Daniel WL, Xia Z, Xie H, Mirkin CA, Rao J. Bioluminescent nanosensors for protease detection based upon gold nanoparticle-luciferase conjugates. Chem Commun. 2010;46:76–8.

15. Chen L, Bao Y, Denstedt J, Zhang J. Nanostructured bioluminescent sensor for rapidly detecting thrombin. Biosens Bioelectron. 2016;77:83–9.

16. Fan F, Binkowski BF, Butler BL, Stecha PF, Lewis MK, Wood KV. Novel genetically encoded biosensors using firefly luciferase. ACS Chem Biol. 2008;3:346–51.

17. Emonet SF, Garidou L, McGavern DB, de la Torre JC. Generation of recombinant lymphocytic choriomeningitis viruses with trisegmented genomes stably expressing two additional genes of interest. Proc Natl Acad Sci. 2009;106:3473–8.

18. Cai Y, Bak RO, Krogh LB, Staunstrup NH, Moldt B, Corydon TJ, Schrøder LD, Mikkelsen JG. DNA transposition by protein transduction of the *piggyBac* transposase from lentiviral Gag precursors. Nucleic Acids Res. 2014;42:e28.

19. Dragulescu-Andrasi A, Chan CT, De A, Massoud TF, Gambhir SS. Bioluminescence resonance energy transfer (BRET) imaging of protein–protein interactions within deep tissues of living subjects. Proc Natl Acad Sci. 2011;108:12060–5.

20. Mo X-L, Fu H. BRET: NanoLuc-based bioluminescence resonance energy transfer platform to monitor protein-protein interactions in live cells. In: Janzen WP, editor. High throughput screening: methods and protocols. New York: Springer New York; 2016. p. 263–71.

21. Ferrier RJ, Prasad D. 1360. Boric acid derivatives as reagents in carbohydrate chemistry. Part VI. Phenylboronic acid as a protecting group in disaccharide synthesis. J Chem Soc (Resumed). 1965;0:7429–7432. doi:10.1039/JR9650007429.

22. James TD, Sandanayake KRAS, Shinkai S. Saccharide sensing with molecular receptors based on boronic acid. Angew Chem Int Ed Engl. 1996;35:1910–22.

23. Wu Q, Wang L, Yu H, Wang J, Chen Z. Organization of glucose-responsive systems and their properties. Chem Rev. 2011;111:7855–75.

24. Hall DG. Boronic acids: preparation, applications in organic synthesis and medicine. Weinheim: Wiley; 2006.

25. Hames BD. Gel electrophoresis of proteins: a practical approach. Oxford: OUP Oxford; 1998.

26. Fairbridge RA, Willis KJ, Booth RG. The direct colorimetric estimation of reducing sugars and other reducing substances with tetrazolium salts. Biochem J. 1951;49:423–7.

27. DuBois M, Gilles KA, Hamilton JK, Rebers PA, Smith F. Colorimetric method for determination of sugars and related substances. Anal Chem. 1956;28:350–6.

28. Brinkley M. A brief survey of methods for preparing protein conjugates with dyes, haptens and crosslinking reagents. Bioconjugate Chem. 1992;3:2–13.

29. Jain PK, Qian W, El-Sayed MA. Ultrafast cooling of photoexcited electrons in gold nanoparticle—thiolated DNA conjugates involves the dissociation of the gold—thiol bond. J Am Chem Soc. 2006;128:2426–33.

30. Brewer SH, Allen AM, Lappi SE, Chasse TL, Briggman KA, Gorman CB, Franzen S. Infrared detection of a phenylboronic acid terminated alkane thiol monolayer on gold surfaces. Langmuir. 2004;20:5512–20.

31. Timasheff SN, Fasman GD. Structure and stability of biological macromolecules. New York: Dekker; 1969.

32. Lorand JP, Edwards JO. Polyol complexes and structure of the benzeneboronate ion. J Org Chem. 1959;24:769–74.

33. Ayyub OB, Ibrahim MB, Briber RM, Kofinas P. Self-assembled block copolymer photonic crystal for selective fructose detection. Biosens Bioelectron. 2013;46:124–9.

34. Kawasaki T, Akanuma H, Yamanouchi T. Increased fructose concentrations in blood and urine in patients with diabetes. Diabetes Care. 2002;25:353–7.

35. Hewetson M, Cohen ND, Love S, Buddington RK, Holmes W, Innocent GT, Roussel AJ. Sucrose concentration in blood: a new method for assessment of gastric permeability in horses with gastric ulceration. J Vet Intern Med. 2006;20:388–94.

36. Sutherland LR, Verhoef M, Wallace JL, van Rosendaal G, Crutcher R, Meddings JB. A simple, non-invasive marker of gastric damage: sucrose permeability. Lancet. 1994;343:998–1000.

37. Chen L, Chen LY, Tse WH, Chen Y, McDonald MW, Melling J, Zhang J. Nanostructured biosensor for detecting glucose in tear by applying fluorescence resonance energy transfer quenching mechanism. Biosens Bioelectron. 2017;2017(91):393–9.

38. Zhang DL, Luo G, Ding XX, Lu C. Preclinical experimental models of drug metabolism and disposition in drug discovery and development. Acta Pharmaceutica Sinica B. 2012;2(6):549–61.

13

A *retro-inverso* cell-penetrating peptide for siRNA delivery

Anaïs Vaissière[1†], Gudrun Aldrian[2†], Karidia Konate[1†], Mattias F. Lindberg[1], Carole Jourdan[1], Anthony Telmar[1], Quentin Seisel[1], Frédéric Fernandez[3], Véronique Viguier[3], Coralie Genevois[4], Franck Couillaud[4], Prisca Boisguerin[1] and Sébastien Deshayes[1*]

Abstract

Background: Small interfering RNAs (siRNAs) are powerful tools to control gene expression. However, due to their poor cellular permeability and stability, their therapeutic development requires a specific delivery system. Among them, cell-penetrating peptides (CPP) have been shown to transfer efficiently siRNA inside the cells. Recently we developed amphipathic peptides able to self-assemble with siRNAs as peptide-based nanoparticles and to transfect them into cells. However, despite the great potential of these drug delivery systems, most of them display a low resistance to proteases.

Results: Here, we report the development and characterization of a new CPP named RICK corresponding to the *retro-inverso* form of the CADY-K peptide. We show that RICK conserves the main biophysical features of its L-parental homologue and keeps the ability to associate with siRNA in stable peptide-based nanoparticles. Moreover the RICK:siRNA self-assembly prevents siRNA degradation and induces inhibition of gene expression.

Conclusions: This new approach consists in a promising strategy for future in vivo application, especially for targeted anticancer treatment (e.g. knock-down of cell cycle proteins).

Keywords: Enantiomer, D-Amino acids, Retro-inverso, siRNA delivery, Cell penetrating peptides, Nanoparticle, Gene knock-down, Cancer

Background

Small interfering RNAs (siRNAs) have been widely considered as powerful tools to specifically control protein activation and/or gene expression post-transcriptionally [1], particularly for complex genotypic alterations occurring in cancer [2]. siRNAs can enter the RNA-induced silencing complex (RISC), which induces enzyme-catalyzed degradation of their complementary messenger RNAs (mRNAs) in cells, thus disrupting specific molecular pathways in various diseases [3]. The design versatility and the highly specific nature of these oligonucleotides highlight their potential as drugs of the future. However, biological barriers remain the main obstacle for the siRNAs delivery due

to their large molecular weight ($\sim$14 kDa) and their highly anionic ($\sim$40 negative charges) character [4, 5].

Currently, most delivery strategies including lipids, polycationic polymers, nanoparticles and peptide-based formulations are based on formulations protecting siRNAs from degradation by hydrolytic enzymes and at the same time mediate their cellular delivery. Several technologies have been proposed to tackle these problems [6–8]. Cell penetrating peptides (CPPs) are one of the most promising non-viral strategies to improve intracellular routing of large molecules including siRNA [6, 9, 10]. CPPs are usually short (up to 30 amino acids) peptides that originate from a wide variety of sources (e.g., humans, mice, viruses or purely synthetic) [6]. Based on their structural characteristics, CPPs can be divided into two classes: arginine-rich CPPs and amphipathic CPPs [11].

Amphipathic CPPs contain both hydrophilic and hydrophobic domains necessary for cellular internalization and interaction with the cargo. In primary

*Correspondence: Sebastien.Deshayes@crbm.cnrs.fr
†Anaïs Vaissière, Gudrun Aldrian and Karidia Konate contributed equally to this work
[1] Centre de Recherche de Biologie cellulaire de Montpellier, UMR 5237 CNRS, 1919 Route de Mende, 34293 Montpellier, France
Full list of author information is available at the end of the article

amphipathic CPPs, these domains are distributed according to their position along the peptide chain as shown for pVEC [12], MPG and Pep-1 [13]. Secondary amphipathic CPPs are another large class of peptides in which the separation between hydrophilic and hydrophobic domains occurs due to the secondary structure formation, such as α helices or β sheets [13]. Many of the most commonly used CPPs are members of this class such as penetratin [14], transportan [15], CADY [16] or C6M1 [17].

Strategies based on the conjugation of siRNA moieties to the CPP lacked efficacy and activated innate immunity in vivo [18]. Furthermore, the incorporation of the siRNA guide strand into the deep-pocket of RISC, essential for the silencing mechanism, is inhibited by covalent linkage between the CPP and the siRNA [4]. Therefore, non-covalent approaches based on electrostatic and hydrophobic interactions between the CPP and the cargo resulting in the nanoparticle formation are currently in the focus of drug delivery strategies [17, 19–22].

Previous investigations revealed that the conformational state of non-covalent CPPs plays an important role in the interaction with the cargo as well as in the self-assembly process leading to efficient peptide-based nanoparticles (PBN) [23]. In addition, the secondary structure of CPPs seems to control membrane interactions, peptide and siRNA entry and to influence the final cellular internalization route [24]. In this context, our recent comparative study revealed that siRNA-loaded nanoparticles formed by the CADY-K CPP (derived from CADY [16]) displayed a twofold higher luciferase knock-down efficiency than the parental peptide or other analogues [25]. CADY-K is thus an ideal candidate for further application especially with regards to ex vivo or in vivo siRNA delivery.

However, the in vivo development of CPPs could be compromised by degradation phenomena resulting from extracellular and/or intracellular proteases, probably partly explaining the low success of CPP application in clinical trials. Therefore, we decided to further improve our applied strategy by using nanoparticles formulated with a *retro-inverso* CPP. In 1995, Goodman and Chorev first evaluated the advantages of *retro-inverso*

peptides—peptides consisting in D-amino acids in the reverse sequence of the naturally occurring L-isoforms [26]. Subsequently, *retro-inverso* transformation has commonly been employed as a strategy for the synthesis of proteolytically stable peptide analogues while maintaining the structural features [27, 28]. Herein, we adopted a similar strategy to characterize the new CPP RICK (*Retro-Inverso CADY-K*). Bearing a high degree of the topochemical equivalence of L-isomer CPP, the *retro-inverso* version would exhibit similar bioactivity as the corresponding L-peptide with the advantage of a reduced endogenous protease degradation.

Here, we present a more in-depth analysis of the structural conformation of the RICK CPP and of the nanoparticle formulation/characterization in the presence of siRNA. To better understand the structural feature and the new highly potent peptide-based nanoparticle (PBN) for siRNA delivery, homologues were also analyzed such as the L-isoform (CADY-K) and the D-isoform (D-cady-k). siRNA-loaded RICK nanoparticles were evaluated in vitro in terms of stability and protease-resistance as well as in vitro by evaluating the gene knock-down (luciferase and cyclin B1) in U87 human glioblastoma cells. Overall, our results provide a comprehensive molecular basis of siRNA-loaded RICK nanoparticles for the further development of PBNs based on *retro-inverso* CPPs.

Methods
Materials
Dioleylphosphatidylglycerol (DOPG), dioleylphosphatidylcholine (DOPC) and 1-palmitoyl,2-oleoylphosphatidylcholine (POPC) phospholipids, cholesterol (Chol), sphingomyelin (SM) and labelled phosphocholine (TopFluorPC) [29] were purchased from Avanti Polar Lipids. RICK, CADY-K, and D-cady-k were purchased from LifeTein (sequences in Table 1). Atto633-labeling of the RICK peptide was performed as described in Additional file 1. Unlabeled and Cy3-labeled siRNA were obtained from Eurogentec and siRNA-Cy3b (labelled on 3′-end) from BioSynthesis. The different sequences are for anti-firefly luciferase (siFLuc): 5′-CUU-ACG-CUG-AGU-ACU-UCG-AdTdT (sense strand) and a scrambled version

Table 1 RICK, CADY-K and D-cady-k characterization by DLS

Peptides			Mean size (nm) at 0 h			Mean size (nm) at 72 h		
ID	Sequences	Properties	I (%)	Nb (%)	Pdl	I (%)	Nb (%)	Pdl
RICK	kwllrwlsrllrwlarwlg	retro-inverso	92 ± 15	32 ± 4	0.24 ± 0.01	91 ± 14	37 ± 9	0.25 ± 0.02
CADY-K	GLWRALWRLLRSLWRLLWK	L-AA	116 ± 22	36 ± 3	0.30 ± 0.05	124 ± 26	28 ± 1	0.33 ± 0.07
D-Cady-k	glwralwrllrslwrllwk	D-aa	90 ± 16	28 ± 7	0.30 ± 0.02	82 ± 3	32 ± 2	0.29 ± 0.02

All CPP:siRNA complexes were formed at R = 20 with using a siRNA concentration of 500 nM in an aqueous solution of 5% glucose for mean size acquisition of the anti-luciferase (siSCR): 5′-CAU-CAU-CCC-UGC-CUC-UAC-UdTdT-3′ (sense strand) used as control. The sequence for the siRNA anti-cyclinB1 (siCycB1) is 5′-GAA-AUG-UAC-CCU-CCA-GAA-AdTdT-3′ (sense strand). The *siRNA stock solutions* were prepared in RNase-free water. *Peptide stock solutions* as well as CPP:siRNA complexes were prepared as published recently [25]. *Large Unilamellar Vesicles (LUVs)* were prepared by the extrusion method from a lipid mixture of DOPC/SM/Chol (2:2:1) as previously reported [25]. *Giant Unilamellar Vesicles (GUVs)* were formed by using the hydration method [30] with several modification as mentioned in Additional file 1.

Structural evaluation
Circular dichroism (CD) measurements
CD spectra were recorded on a Jasco 810 (Japan) dichrograph in quartz suprasil cells (Hellma) with an optical path of 1 mm for peptide in solution or in the presence of liposomes vesicles. Same concentrations of peptide (40 µM) were used for each condition. Spectra were obtained from 3 accumulations between 190 and 260 nm with a data pitch of 0.5 nm, a bandwidth of 1 nm and a standard sensitivity.

Fluorescence spectrometry
Fluorescence experiments were performed on a PTI spectrofluorimeter at 25 °C in 5% glucose. Intrinsic Tryptophan fluorescence of RICK, CADY-K and D-cady-k at 5 µM was excited at 290 nm and emission spectrum was recorded between 320 and 390 nm, with a spectral bandpass of 2 and 6 nm for excitation and emission, respectively. Effect of siRNA was investigated through addition of siRNA to peptide solution in order to form complexes at a CPP:siRNA molar ratio of R = 20. All spectra are normalized to maximum of fluorescence of free peptides and plotted in relative fluorescence (%).

Nanoparticle formation
Agarose gel shift assay
CPPs:siRNA complexes were formed at different ratio in 5% glucose and pre-incubated for 30 min at room temperature. Each sample was analyzed by agarose gel (1% w/v) electrophoresis stained with GelRed (Interchim) for UV detection as descripted previously [25].

Dynamic light scattering (DLS) and zeta potential (ZP)
CPPs:siRNA nanoparticles were evaluated with a Zetasizer NanoZS (Malvern) in terms of mean size (Z-average) of the particle distribution and of homogeneity (PdI). Zeta potential was determined in 5% glucose supplemented with 5 mM NaCl, OptiMEM or DMEM. All results were obtained from three independent

measurements (three runs for each measurement at 25 °C).

Environmental scanning electron microscopy (ESEM) and transmission electron microscopy (TEM)
For environmental scanning electron microscopy experiments, RICK:siRNA complexes were especially formed in ultra-pure water in order to avoid interference due to 5% glucose. Complexes were prepared at a CPP:siRNA molar ratio of R = 20 and for a final peptide concentration of 40 µM. Drops of 10 µl were spotted onto a copper support. Particles were examined using an FEI Quanta 200 FEG Scanning Electron Microscope operated at 10.00 kV, with a pressure of 400 Pa and a magnification of 15,000 × (det Gaseous Scanning Electron Detector, Working Distance = 9.4 mm). For transmission electron microscopy (TEM), a drop of 5 µl of suspension is deposited on a carbon coated 300 mesh grid for 1 min, blotted dry by touching filter paper and then placed on a 2% uranyl acetate solution drop. After 1 min the excess stain is removed by touching the edge to a filter paper, the grid is dried at room temperature for few minutes and examined using a Jeol 1200EX2 Transmission Electron Microscope operating at 100 kV accelerating voltage. Data were collected with a SIS Olympus Quemesa CCD camera.

Evaluation of the CPP stability
HPLC analysis
100 µl of each peptide (12 µM in water) were incubated with 20 µl of porcine trypsin (105 µM) at 37 °C. Evolution of the peptide in the presence of the protease was followed by RP-HPLC (Waters) at 220 nm on a C8 Aquapore RP-300 7 µ, 100 × 2.1 mm column (Perkin Elmer) at different time points. The mobile phase consisted of: A: demineralized water (Milli-Q quality; Millipore) and B: acetonitrile (HPLC gradient, Carlo Erba, Peypin), containing 0.1 and 0.08% trifluoroacetic acid (Sigma Aldrich), respectively. The elution profile was: from 0 to 60% B in 55 min with a flow rate of 1 ml/min. Injection volume was 20 µl.

Fluorescence spectrometry
CPP:siRNA-Cy3b nanoparticles (R = 20, with 20 nM of siRNA) formulated in 100 mM Tris–HCl (pH 7.5) were incubated on a non-binding black 96-well plate (Greiner Bio-One) with different volumes of 0.05% trypsin (Life Technologies). After 24 h incubation the fluorescence was recorded (Ex = 544 nm and Em = 590 nm).

siRNA extraction
CPP:siRNA-Cy3 nanoparticles (R = 20, with 200 pmol of siRNA) were incubated with fetal bovine serum (FBS) (PAA) 24 h at 37 °C. Thereafter the siRNA-Cy3 was

extracted using the mirPremier™ microRNA isolation kit (Sigma-Aldrich) by following the instruction of the manufacturer. The siRNA-Cy3 was isolated on a denaturing urea (8 M) acrylamide gel (15%) and visualized using the pre-defined Cy3 filter on the Amersham Imager 600.

Cellular activity evaluation of CPP:siRNA nanoparticles
Culture conditions
The efficiencies of siRNA-mediated gene silencing were investigated in U87 cell lines (U87 MG, human glioblastoma) stably transfected with firefly and *Renilla* luciferase (FLuc-RLuc) encoding plasmid (details of cell line generation are given in the Additional file 1). Cells were grown in a complete medium: DMEM (+GlutaMAX™ supplement, Life Technologies), with 100 units/ml penicillin (Life Technologies), 100 mg/ml streptomycin (Life Technologies), 10% heat-inactivated fetal bovine serum (FBS) (PAA), non-essential amino acids NEAA 1X (LifeTechnologies) and selection antibiotics hygromycin B (Invitrogen) (50 µg/ml) and Blasticidine (Gibco) (2 µg/ml). All the cells were maintained in a humidified incubator with 5% CO_2 at 37 °C.

Transfection experiments
For Luciferase assay, 5000 cells were seeded 24 h before experiment into 96-well. The next day, nanoparticles were formed by mixing siRNA and CPPs (equal volumes, "siRNA on CPPs") in 5% glucose water, followed by an incubation of 30 min at 37 °C. In the meantime, the growth medium covering the cells was replaced by 70 µl of fresh pre-warmed serum-free DMEM. 30 µl of the nanoparticle solutions were added directly to the cells and after 1 h 15 of incubation, 100 µl DMEM + 20% FBS was added to each well without withdrawing the transfection reagents, and cells were then incubated for another 36 h. The experimental procedure was designed to test CPP:siRNA nanoparticles at a peptide:siRNA molar ratio of R = 20, containing siRNA concentrations of 5, 10 and 20 nM in the final volume of 200 µl. For trypsin pre-incubation assay, CADY-K:siFLuc and RICK:siFLuc nanoparticles were incubated in the presence of trypsin (5 µg/ml and 10 µg/ml for nanoparticles loaded with 10 nM and 20 nM siRNA, R = 20) for 24 h at 37 °C. Before adding the nanoparticles solution to the cells, a trypsin inhibitor was added. Cells were incubated for 1.5 h with the nanoparticles in serum-free DMEM. After the addition of DMEM supplemented with 20% FBS (final FBS concentration = 10%), cells were further incubated for 48 h and finally lysed for the luciferase detection.

For Western blot assay, 75,000 cells were seeded 24 h before experiment into 24-well plates. For standard incubation, the cells were incubated with 175 µl of fresh pre-warmed serum-free DMEM + 75 µl of the nanoparticle solutions. After 1.5 h of incubation, 250 µl DMEM + 20% FBS was added to each well without withdrawing the transfection reagents, and cells were then incubated for another 24 h. The experimental procedure was designed to test CPP:siRNA nanoparticles at a peptide:siRNA molar ratio of R = 20, containing siRNA concentrations of 5, 10 and 20 nM in the final volume of 500 µl. For transfection in the presence of serum, cells were incubated with nanoparticles in DMEM + 10% FBS. For serum pre-incubation assay, nanoparticles were pre-incubated for 2 h at 37 °C in 5% glucose supplemented with 10% FBS. Then cells were incubated for 1.5 h with the nanoparticles in serum-free DMEM. After the addition of DMEM + 20% FBS (10% FBS final), cells were further incubated for 24 h and finally lysed for CyclinB1 western blotting detection.

For microscopy assay (spinning Disk), 400,000 cells were seeded 24 h before imaging into a Fluoro Dish from World Precision Instruments (tissue culture dish with cover glass bottom, dish diameter = 35 mm/glass diameter = 23 mm/glass thickness 0.17 mm). Before microscopy imaging, cells were washed and covered with 1600 µl of complete medium. 400 µl of nanoparticles (Atto633-RICK:siRNA-Cy3b; R = 20), formed as previously described in 5% glucose water, were directly added on the cells at the very beginning of imaging.

Measurement of cell cytotoxicity
Evaluation of cytotoxicity induced by the nanoparticles was performed using Cytotoxicity Detection Kit[Plus] (LDH, Roche Diagnostics) on 50 µl of supernatant, by following the manufacturer instructions.

Luciferase reporter gene silencing assay
The evaluation of siRNA delivery using the different vectors was carried out by measuring the remaining luciferase firefly (FLuc) and luciferase *Renilla* (RLuc) activity in cell lysates. Briefly, after 48 h, the medium covering the cells was carefully removed and replaced by 50 µl of 0.5× Passive Lysis Buffer (PLB; Promega). After 30 min of shaking at 4 °C, plates containing the cells were centrifuged (10 min, 1800 rpm, 4 °C) and 5 µl of each cell lysate supernatant were finally transferred into a white 96-well plate. FLuc and RLuc activities were quantified using a plate-reading luminometer (POLARstar Omega, BMG Labtech), measuring light emission over a 2 s reaction period immediately after injections of 100 µl of half-diluted Dual Luciferase Assay Reagents (Promega) per well. The results were expressed as percentage of relative light units (RLU) in non-treated cells (%FLuc and %RLuc), then normalized on %RLuc to obtain the Relative Luc Activity (%FLuc/%RLuc).

Western blotting

Transfected cells washed in PBS, and lysed in buffer 150 mM sodium chloride, 1.0% Triton X-100, 0.1% SDS (sodium dodecyl sulphate), 50 mM Tris pH 8.0 including protease inhibitors (SigmaFAST). Cells were incubated for 5 min on ice with 130 µl/24-well lysis buffer. There-after, cells were scrapped and transferred in a 1.5 ml tube. After 5 min on ice, the cell lysates were centrifuged (10 min, 16,100*g*, 4 °C), supernatants were collected and protein concentrations were determined using the Pierce BCA Protein Assay (ThermoFisher). Cell extracts (0.25–0.375 µg/µl) were separated by 4–20% Mini-PROTEAN® TGX™ Precast Gel (Bio-Rad). After electrophoresis, samples were transferred onto Trans-Blot® Turbo™ Mini PVDF Transfer membrane (Bio-Rad). As antibodies (all from Cell Signaling), we used anti-cyclin B1 mouse mAb V152, anti-Vinculin rabbit mAb E1E9V, anti-mouse IgG HRP and anti-rabbit IgG HRP. Blots were revealed with the Pierce ECL plus Western blotting substrate (ThermoFisher) on an Amersham imager 600 (GE Healthcare Life Science). The signal intensities of the blots were quantified by Image J.

Membrane interaction and internalization of CPP:siRNA nanoparticle

GUV imaging by confocal microscopy

For the microscopy experiments, the GUVs were placed into a glass bottom cell culture dish (Greiner bio-one) and incubated with 400 nM Atto633-RICK:20 nM siRNA-Cy3b. Confocal images of GUVs were immediately obtained with an inverted LSM780 multi-photon microscope (Zeiss). The obtained confocal images were projected and treated with the software ImageJ.

Liposome leakage assay

Large unilamellar vesicles (LUV) reflecting the plasma membrane were prepared as described in detail in Additional file 1. Leakage was measured as an increase in fluorescence intensity upon addition of the CPP or CPP nanoparticle (500 nM final CPP concentration) to 2 ml of LUVs (100 µM) in buffer (20 mM HEPES, 145 mM NaCl, pH 7.4). 100% fluorescence was achieved by solubilizing the membranes with 0.1% (v/v) Triton X-100 resulting in the completely unquenched probe.

Spinning disk confocal microscopy

To study the entrance of the NPs inside living cells, we used an inverted microscope (Nikon Ti Eclipse) coupled to a spinning disk (ANDOR) system. An excitation laser was used to illuminate a 60× (1.4 numerical aperture) oil immersion objective. The scanning system microscope was equipped with a motorized stage ASI allowing sample scanning in x, y and z direction. The spinning disk

head was a Yokogawa CSU-X1 with 3 dichroic: mono, dual or quad-band filter. To visualize the fluorescence, we used as emission filter a 4 band pass filter (QUAD dichroic) center around 405, 488, 561 and 640 nm. Cells were maintained at 37 °C by a cage incubator (Okolab) all along the 2 h of measurements. Details concerning the acquisition parameters were provided in Additional file 1. No emission bleed through was observed between the different channels observed with these acquisition parameters. Emission photons were collected on an EM-CCD camera (iXon Ultra) and images were recorded every 2 for 120 min and projected with the software Andor IQ3. These acquisition parameters provide an image of 512 × 512 pixels with a pixel size of 0.15 µm. The image treatment was performed with the Imaris software for a 3D reconstruction.

Results and discussion

Structural characterization

As already described for the CADY peptide and its analogues, the conformational state has a strong importance for its ability to interact on one hand with the siRNA cargo and on the other hand with phospholipids constituting the main component of biological membranes [25, 31, 32]. Thus, the structural state of RICK was investigated by circular dichroism (CD) and compared to those obtained for its enantiomers CADY-K and D-cady-k. Secondary structure was determined when peptide is free in solution, mixed with siRNA at a peptide:siRNA molar ratio of R = 20, or when liposomes are added to the peptide:siRNA mix at a lipid/peptide ratio of r = 5 (Fig. 1a, b).

As previously identified, CD profile of free CADY-K has a minimum at 203 nm and a weak shoulder at 220 nm (Fig. 1a), suggesting a mainly random coil conformation with few helical contribution [25]. Similarly, in the presence of siRNA CADY-K adopts a right-handed α-helical structure (maximum at 191 nm, minima at 210 and 220 nm) which is not impacted by addition of zwitterionic large unilamellar vesicles (LUVs) made of DOPC/SM/Chol (2:2:1). As expected, CD spectra of D-cady-k are clearly mirror images of CADY-K profiles (Fig. 1a). The spectra of free D-cady-k, associated with siRNA and in the presence of liposomes have bands centered at the same wavelength of CADY-K signals. The conversion of L- to D-amino acids only induces an inversion of bands intensity while maintaining similar absolute amplitude of the L-form, resulting in a perfect symmetry as already observed for some antimicrobial peptides [33–35]. This indicates that the enantiomers CADY-K and D-cady-k have exactly the same conformational behavior: while free peptides are mainly disordered in solution, CADY-K and D-cady-k adopt right-handed and left-handed helical structure, respectively, in the presence of siRNA.

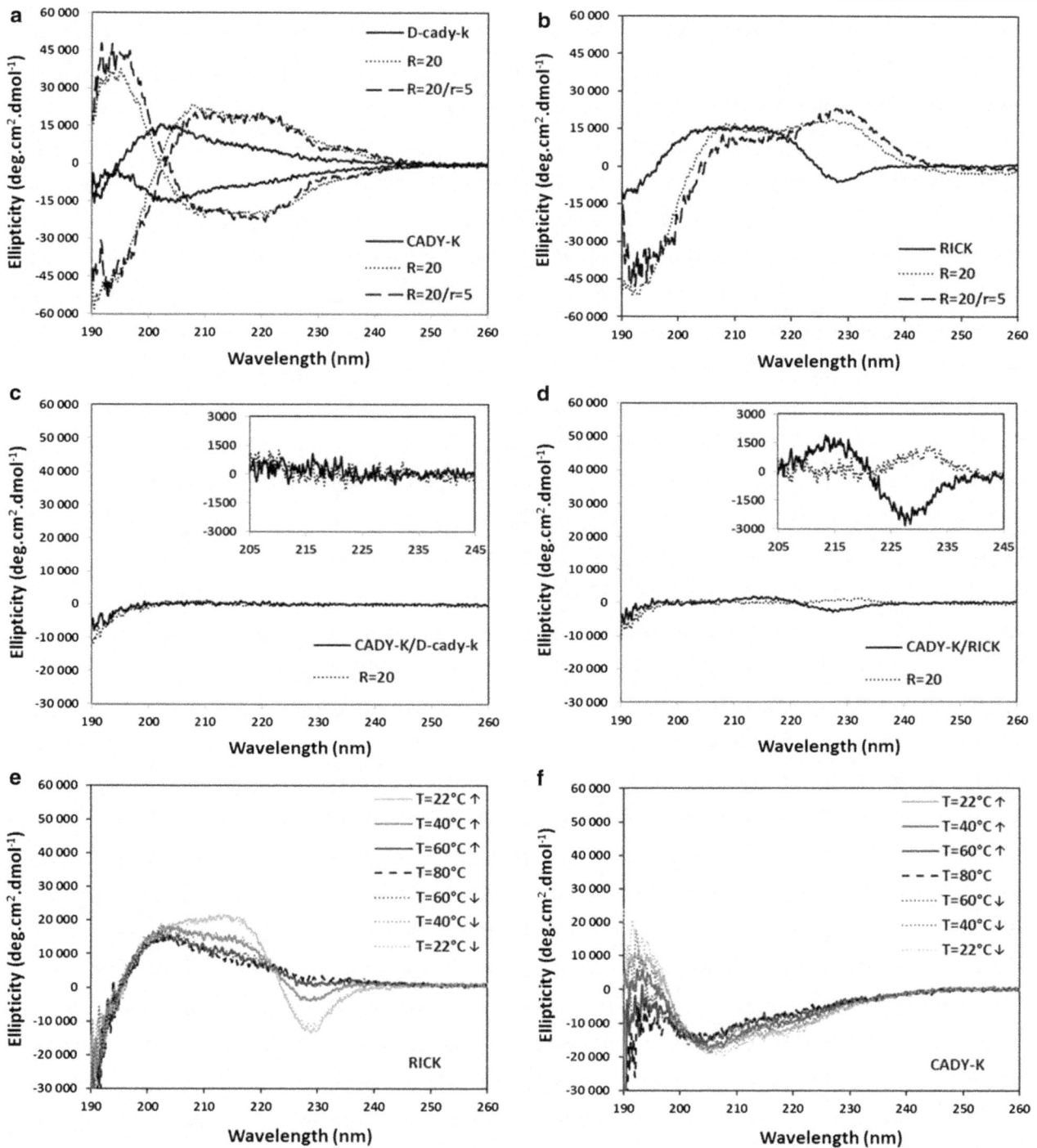

Fig. 1 Structural characterization of RICK and isoforms. **a, b** Circular dichroism (CD) profiles of CADY-K, D-cady-k and RICK in solution, associated to siRNA anti-FLuc at a 20/1 molar ratio (R = 20) and in the presence of LUVs composed of DOPC/SM/Chol (2:2:1) and at a final lipid/peptide ratio of 5 (r = 5). **c, d** CD spectra of CADY-K/D-cady-k and CADY-K/RICK mixtures in solution at an equimolar ratio and in the presence of siRNA at a final peptide:siRNA ratio R = 20. **e, f** Effects of heating/cooling cycle from 22 to 80 °C on CD spectra of RICK and CADY-K. CD signals are expressed in Ellipticity

Surprisingly CD spectra of RICK show two main differences (Fig. 1b). Firstly, the CD spectrum of free peptide consists in less defined extrema. A decrease of the signal around 190 nm does not reflect a net minimum and the positive band between 202 and 220 nm might be considered as a wide maximum suggesting a mix of several structures hardly identifiable/assignable. Secondly, contributions of close tryptophan/tryptophan interactions are clearly observed through the band at 228 nm which is negative for D-amino acids [36, 37]. These contributions suggest a cluster of tryptophans within the peptide sequence, as observed for tryptophan zipper peptides [38]. The addition of siRNA induces a net minimum at 192 nm with a significant maximum at 208 nm which suggest a main left-handed helical structure, although the critical region at 222 nm seems to be hidden by a pronounced positive band at 228 nm. The negative tryptophan band turns to a positive one, indicating a conformational change with a reorganization of the tryptophan residues.

In order to further understand structural differences observed for RICK, we studied the CD spectra of equimolar mixtures of enantiomers CADY-K/D-cady-k and CADY-K/RICK. Indeed, considering that all peptides might not interfere each other in their way to interact with siRNA and LUVs, investigations of these mixtures should underline structural variations between isoforms compared to the parent peptide CADY-K. Analyses of CADY-K/D-cady-k mixtures, with or without siRNA, result in the absence of any CD signal (Fig. 1c). The mean molar ellipticity does not vary and remains at zero level (Fig. 1c, insert; Additional file 1: Figures S1), as expected for a mix of the L- and the D-enantiomer (compensation of each contribution) [39]. These observations clearly confirm the similar structure of CADY-K and D-cady-k and furthermore underline the absence of specific interactions between both isomers. In contrast, analyses of CADY-K/RICK mixtures reveal a slight inflection centered at 228 nm, the tryptophan region, which varies in the presence of siRNA (Fig. 1d, inset; Additional file 1: Figure S1). Although weak, this effect suggests that the *retro-inverso* peptide contains different tryptophan/tryptophan interactions upon its complexation to siRNA.

This structural particularity of RICK was then analyzed through thermal unfolding to evaluate the influence of the temperature on these hydrophobic interactions. Applying heating/cooling cycle from 22 to 80 °C, we could detect several variations in the CD profiles of RICK (Fig. 1e), whereas no significant change was observed for CADY-K (Fig. 1f). An increase of temperature above 40 °C induced a denaturation of RICK mainly characterized by a loss of amplitude at 215 and 228 nm. These modifications form an isodichroic point at 220 nm

suggesting an equilibrium between two structural states of RICK: folded at low temperature versus denatured at high temperature. In addition the CD profile of heated RICK (60/80 °C) is superimposable with the CD spectrum of D-cady-k at 25 °C, suggesting the same random coil conformation (Additional file 1: Figure S2). In conclusion, these analyses reveal that RICK adopts a structure which is sensitive to thermal unfolding. Similar thermal unfolding, associated to a loss of 228 nm contribution, was described for tryptophan zipper β-hairpins (Trpzip) based on the unfolding of the zipper [40, 41]. Thus, in a similar manner, our data suggest that tryptophan residues of RICK establish tryptophan/tryptophan interactions forming a kind of tryptophan zipper. Finally, these interactions change in the presence of siRNA, leading to a conformational change in a left-handed helix with peptide/siRNA interactions.

Nanoparticle characterization

Peptide-based nanoparticles (PBN) involve formation of stable particles through interactions between carrier peptide and the cargo molecule [17, 19, 21, 23]. The conformational changes detected by CD suggested that RICK interacts with siRNA. To assess complexation of siRNA, we investigated the formation of RICK:siRNA complexes by agarose shift assay (Fig. 2a). siRNA alone migrated into the agarose gel but when complexed with CPPs, the peptides prevented oligonucleotide migration in a molar ratio-dependent manner. For better quantification, the fluorescence signals of the CPP:siRNA complexes are represented relative to the signal intensity of siRNA alone (=100%). Although a slight difference could be noticed with CADY-K at a molar ratio R = 10, both peptides (RICK and CADY-K) were clearly able to complex siRNA in a similar manner with optimal complexation at peptide:siRNA molar ratios of R = 20 and R = 40 (Fig. 2a).

We then monitored the intrinsic tryptophan fluorescence of RICK and CADY-K with or without siRNA. Fluorescence emission spectra of free peptides revealed that CADY-K was characterized by a maximum of fluorescence intensity centered at 348 nm, corresponding to tryptophan usually exposed to the aqueous solution [42] (Fig. 2b). In contrast RICK had a maximum of fluorescence at 345 nm, which was slightly shifted to lower wavelength (blue shift) compared to CADY-K. This suggested a more hydrophobic environment for the tryptophan of RICK, such as the tryptophan/tryptophan cluster observed by CD. In the presence of siRNA, at a final peptide:siRNA molar ratio of R = 20, a similar change was detected in the spectra of both peptides. A net decrease of fluorescence intensity and a shift of the maximum of fluorescence to 340 nm were observed

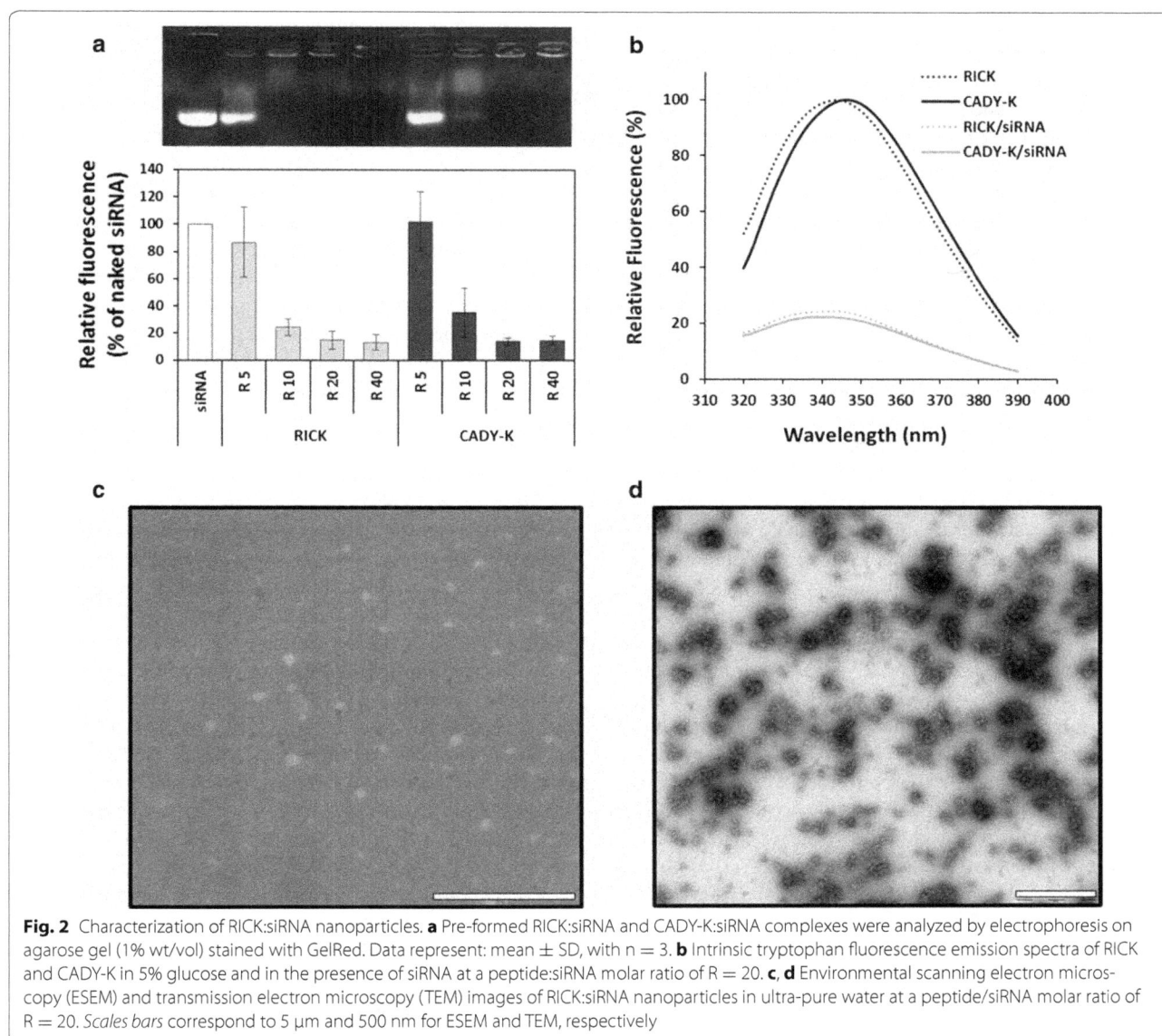

Fig. 2 Characterization of RICK:siRNA nanoparticles. **a** Pre-formed RICK:siRNA and CADY-K:siRNA complexes were analyzed by electrophoresis on agarose gel (1% wt/vol) stained with GelRed. Data represent: mean ± SD, with n = 3. **b** Intrinsic tryptophan fluorescence emission spectra of RICK and CADY-K in 5% glucose and in the presence of siRNA at a peptide:siRNA molar ratio of R = 20. **c, d** Environmental scanning electron micros-copy (ESEM) and transmission electron microscopy (TEM) images of RICK:siRNA nanoparticles in ultra-pure water at a peptide/siRNA molar ratio of R = 20. *Scales bars* correspond to 5 μm and 500 nm for ESEM and TEM, respectively

(Fig. 2b). Previous investigations indicated that the intrinsic tryptophan fluorescence of peptides might undergo a strong quenching in the presence of nucleic acids based on aromatic stacking effects [16, 43]. This aromatic stacking process might be also associated to the blue shift, from 348 to 340 nm for CADY-K and from 345 to 340 nm for RICK, which suggests a more hydropho-bic environment of the tryptophan. Our results were in agreement with the fact that short peptides are able to undergo tryptophan fluorescence quenching when inter-acting with single-stranded nucleic acids as well as DNA duplex [44–46]. For example, the strong binding of the KWGK peptide to a 21-mer duplex involves intercala-tion and stacking interactions of the tryptophan with the oligonucleotide GC regions. In this context, the quench-ing of tryptophan fluorescence was due to an electron transfer from indole of the tryptophan side chain (in the excited state) to purine and pyrimidine bases [46]. Taken together, these results suggested that the siRNA was able to destabilize the RICK tryptophan-zipper-like structure (loss of Trp/Trp interactions) in favor of the RNA bases/ tryptophan interactions.

Colloidal features of RICK:siRNA complexes were char-acterized and compared to CADY-K and D-cady-k. Size and homogeneity were determined for each complex (5% glucose, R = 20). Dynamic light scattering (DLS) investi-gations enable to express particle size in intensity, volume or number [47]. Intensity measurements (%) revealed that

RICK, CADY-K and D-cady-k formed peptide-based nanoparticles with diameter of ~100 nm with polydispersity index (PDI) of ~0.28 (Table 1; Additional file 1: Figure S3). Because it is well known that the diffused light intensity (%) is dependent of the measured particle size, we decided to evaluate also the size distribution based on number (%). Indeed, the percentage of intensity resulting from light diffused by bigger particles is one million fold stronger than those from small ones. Thus, signals of small particles are under evaluated because they are not visualized in size distribution based on intensity recording. Size distribution based on number (%) obtained using algorithms could give us a more accurate information about size distribution. This nanoparticle size representation suggests the presence of smaller nanoparticles in addition to those found by the intensity-based size distribution. These smaller particles have diameters of ~30 nm for RICK:siRNA and its isoforms (Table 1; Additional file 1: Figure S3). More importantly, both measured size values did not change significantly even after 72 h storage at 4 °C, which is in agreement with previous experiments with CADY or CADY-K [25, 48].

Charge surface of nanoparticles was also evaluated by zeta potential (ZP) measurements revealing similar values of ~40 mV for all three CPP:siRNA complexes in 5% glucose (Table 2). Same values were obtained in an aqueous solution containing 5 mM NaCl or in 5% glucose for the parental peptide CADY:siRNA [25, 48]. However, it could be interesting to evaluate ZP in culture media used during transfection (OptiMEM and serum free DMEM). Table 2 showed the ZP changes to more neutral values; however, the overall charge remained in the positive range, which is required for the cellular translocation. Only the addition of 50% serum induced a negative ZP of −12 mV [25] but this condition is not relevant to in vitro transfection conditions. Yet, we should note here that Lindberg et al. measured a negative ZP for CADY:siRNA. This difference with our findings could be due to many parameters such as formulation, siRNA length, molar ratio, concentration etc. [49].

To obtain more details regarding their size but also their shape, RICK:siRNA nanoparticles were analyzed

by environmental scanning electron microscopy (ESEM) and transmission electron microscopy (TEM). Several samples were prepared in ultra-pure water as the use of 5% glucose might interfere with the imaging. Analyses of ESEM samples revealed globular nanoparticles as already observed for CADY peptide [48]. Measurements of diameters indicated a mean size centered at 300 nm and ranging from 150 to 550 nm (Fig. 2c). These sizes were consistent with the different hydrodynamic diameters identified by intensity (%) measurements, suggesting a similar Gaussian size distribution.

To gain more details on the nanoparticle shape, we performed TEM measurements which allow a better resolution to measure small nanoparticles. TEM investigations revealed nanoparticles with a mean size of ~120 nm and also smaller nanoparticles with a mean diameter of ~21 nm (Fig. 2d; Additional file 1: Figure S4). Even if the TEM sample preparation could induce self-assembly phenomena, the obtained images are in good agreements with DLS results in size distribution based on number (%) and intensity (%) which highlighted the presence of small (~30 nm) and bigger (~100 nm) nanoparticles in solution (Table 1; Additional file 1: Figure S3).

These observations were in agreement with the nanoparticle formation of CADY:siRNA determined by molecular modeling or by atom force microscopy (AFM), as previously demonstrated [48]. We could hypothesize that RICK:siRNA nanoparticles were formed in the same manner: first the formation of ~20 nm spheres composed of RICK:siRNA (R = 20) which self-associate in a "beads necklace"-like manner to nanoparticles of ~100 nm. Such assembly could be also compared to the "large branching agglomerates" described for the parental peptide CADY in the presence of miRNA [50].

Proteolysis resistance

In order to prevent proteolysis, peptides composed of L-amino acids are usually synthesized using their corresponding unnatural D-isomer. Because this replacement can induce a loss of biological activity due to a different side chain orientation, *retro-inverso* peptides—the D-amino acids in the reverse sequence of the naturally occurring L-peptides—can be used as a peptidomimetic approach.

In order to investigate the stability of RICK compared to CADY-K, we analyzed the siRNA release of formulated nanoparticles (CPP:siRNA-Cy3b, R = 20) by fluorescence spectroscopy after 24 h of incubation with increasing trypsin concentrations (Fig. 3a). At the beginning the mean fluorescence of the siRNA-Cy3b alone (start) is around 240,000. As expected, the fluorescence was quenched in the same way when RICK or CADY-K was added (mean fluorescence of ~50,000 equal to a 4.6-fold

Table 2 Zeta potential (ZP) of RICK, CADY-K and D-cady-k

ID	ZP (mV)		
	5% glucose[a]	OptiMEM	DMEM
RICK	40 ± 2	14 ± 3	14 ± 1
CADY-K	38 ± 1	13 ± 1	12 ± 3
D-Cady-k	40 ± 1	n.d.	n.d.

All CPP:siRNA complexes were formed at R = 20 using a siRNA concentration of 500 nM in an aqueous solution of 5% glucose, OptiMEM or DMEM

[a] For zeta potential (ZP) assessment the solution was supplemented with 5 mM NaCl (OptiMEM and DMEM contained NaCl)

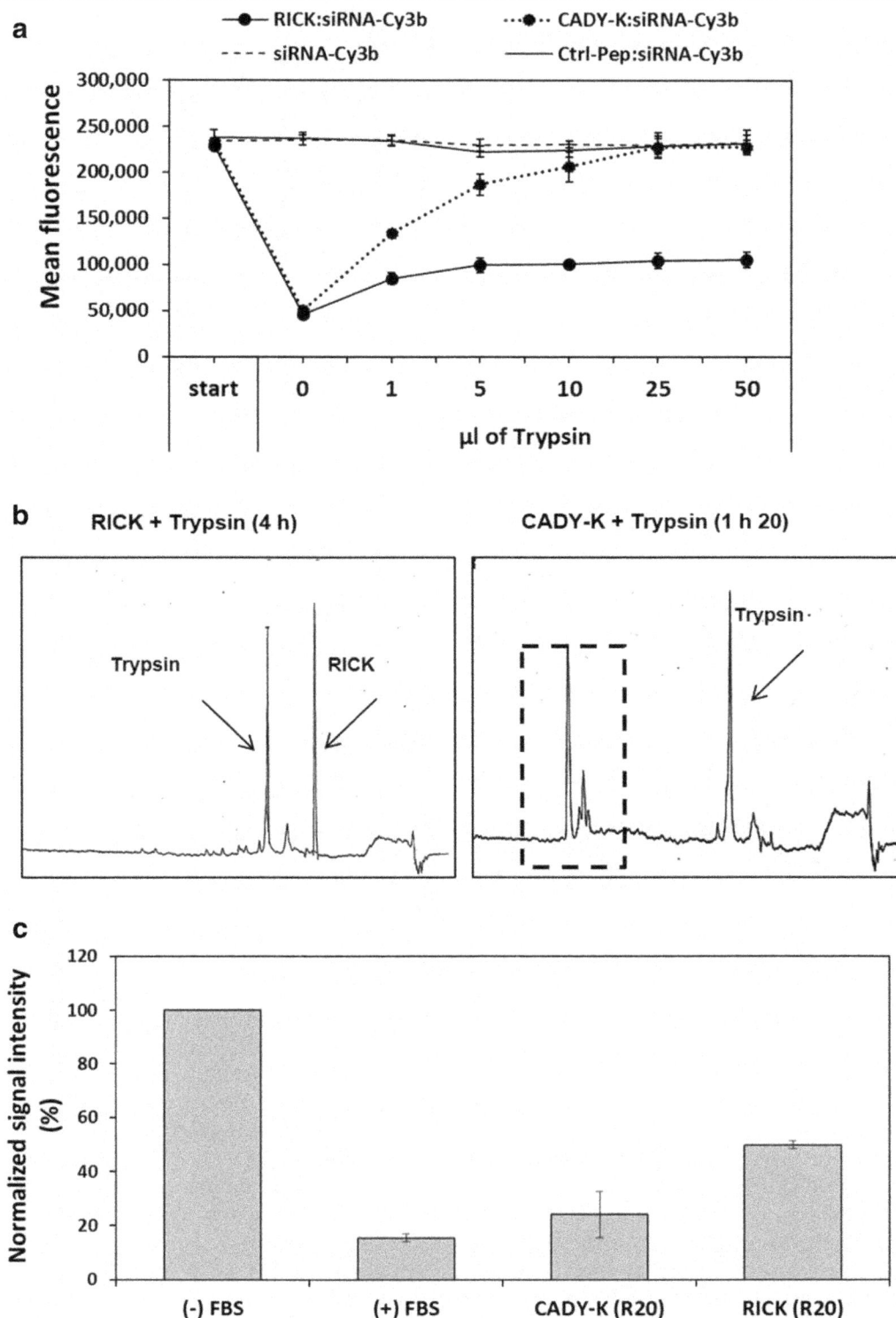

Fig. 3 Analysis of the proteolysis resistance of the RICK peptide and the consequence on the PBN stability. **a** Fluorescence emission of siRNA-Cy3b alone or in the presence of RICK, CADY-K and a control peptide (Ctrl-pep) at a peptide molar ratio of R = 20 in the presence of increasing trypsin amounts. **b** HPLC chromatograms of RICK and CADY-K in the presence of Trypsin. The *dotted square* represent the degraded CADY-K. Degradation products are given in Additional file 1: Table S1. **c** siRNA recovery after serum incubation. Cy3-labeled siRNA alone or complexed to CADY-K or RICK (R = 20) were incubated with serum. siRNAs were extracted from the sample and visualized in an acrylamide gel. Resulting fluorescence intensities were normalized to untreated siRNA (−FBS)

decrease). Control experiments with siRNA-Cy3b alone or in the presence of a peptide not able to form nanoparticles (Ctrl-Pep:siRNA) revealed no change in the fluorescence intensity during the whole assay.

In contrast, by adding an increasing amount of trypsin, we observed a progressive release of the siRNA of the previously formulated CADY-K:siRNA nanoparticles. At the highest trypsin concentration the initial siRNA fluorescence was completely restored, probably due to the total CADY-K degradation. Applying the same conditions to the RICK:siRNA nanoparticles, nearly no change in the mean fluorescence intensity was observed, confirming the stability of the complex in the presence of trypsin.

HPLC analyses were performed to evaluate the impact of trypsin on RICK and CADY-K peptide sequence. At a time point zero (t = 0), we observed two peaks on the HPLC spectra corresponding to the peptide and the trypsin. As expected, the HPLC chromatograms show that CADY-K was completely degraded within 80 min of trypsin incubation, whereas RICK is stable over time (4 h and more) (Fig. 3b). To confirm that the degradation products of CADY-K were due to trypsinization, we collected the corresponding HPLC fractions for MS/MS analysis (Additional file 1: Table S1).

Finally, we evaluated the efficient siRNA-Cy3 protection by RICK or CADY-K nanoparticle formation in the presence of serum. After a 24 h incubation at 37 °C, the siRNA was extracted from the sample preparation and quantified by Cy3 detection after acrylamide gel migration. The measured signal intensities were normalized to the conditions without serum (−FBS) (Fig. 3c). When the siRNA was directly incubated with serum (+FBS), 84% of the siRNA signal is lost, revealing its rapid degradation. In contrast, in the presence of CADY-K 24% of the siRNA was rescued, and more importantly, 50% when the siRNA was protected by RICK. Due to the *retro-inverso* properties, we obtained a twofold higher resistance to proteases for RICK compared to CADY-K.

In sum, we demonstrated through three approaches that RICK is more stable against proteolysis phenomena (trypsin or serum) compared to CADY-K. To our knowledge, only Seelig and Coll are working on *retro-inverso* CPP in a non-covalent strategy (riDOM:plasmid) [51, 52]. However, they did not look at trypsin or serum resistance of their peptide. Furthermore, only few publications comparing different enantiomers in terms of protease stability are available. For example, the SAP CPP enantiomers exhibit the same behavior in terms of cellular delivery but D-isomer showed better stability to proteases than the L-form [53], confirming our results.

Membrane interaction and internalization of RICK:siRNA nanoparticles

Once we demonstrated the proteolysis resistance of RICK peptide, we looked at RICK behavior in the presence of lipid bilayers. Giant unilamellar vesicles (GUVs) are often used to mimic the cell plasma membrane in a simplified environment [54]. With size ranging from 1 to 100 µm, they provide good models to investigate interactions between lipid bilayers and non-lipid molecules such as CPPs or nanoparticles. The ability of nanoparticles to target, label, and/or penetrate lipid bilayers has resulted in great interest in correlating these phenomena with nanoparticle penetration into cells or even with nanoparticle-related cytotoxicity. To evaluate the interaction of RICK nanoparticles with lipid membranes, we decided to generate GUVs of simple composition using POPC in order to minimize phase separation phenomena during GUV formation. First, we incubated the GUVs with siRNA-Cy3b or with Atto633-RICK alone. As shown in Fig. 4a, the siRNA was homogeneously distributed in the surrounding solution without association to the lipid bilayer, whereas the labeled peptide immediately stuck to the GUV's membrane. RICK peptide (positively charged) was naturally attracted by lipids (negatively charged), whereas, the siRNA (negatively charged) was rather repulsed.

When the GUVs were incubated with the nanoparticles Atto633-RICK:siRNA-Cy3b, we noticed that both molecules stuck predominantly to the membrane (Fig. 4a). Furthermore, when the GUVs were first incubated with the siRNA-Cy3b and thereafter the non-labelled RICK peptide was added to this mixture (to avoid Cy3b quenching), addition of the peptide led to a fluorescent profile change. Indeed, we could observe a reorganization of siRNA-Cy3b that progressively associates in a dotted pattern to the GUV's membranes (Additional file 1: Figure S5). The irregular dotted pattern was probably due to an unregular distribution of unlabeled RICK peptide and RICK:siRNA-Cy3b nanoparticles or to a different nanoparticles formulation in this assay. The obtained results demonstrated that the RICK peptide was able to recruit siRNA molecules to the GUV's membrane by its dual interaction with siRNA and lipid membranes. Attraction of RICK for lipid membranes was further confirmed by Langmuir adsorption at an air–water interface (CMC ~100 nM, Πsat ~35 mN/m) (Additional file 1: Figure S6A) as well as by the enhancement of intrinsic tryptophan fluorescence emission associated to a 10 nm blue-shift in the presence of an excess of liposomes (Additional file 1: Figure S6B).

In order to evaluate the lipid membrane interaction and/or transduction of the nanoparticles, we performed leakage assays using LUVs with a complex lipid

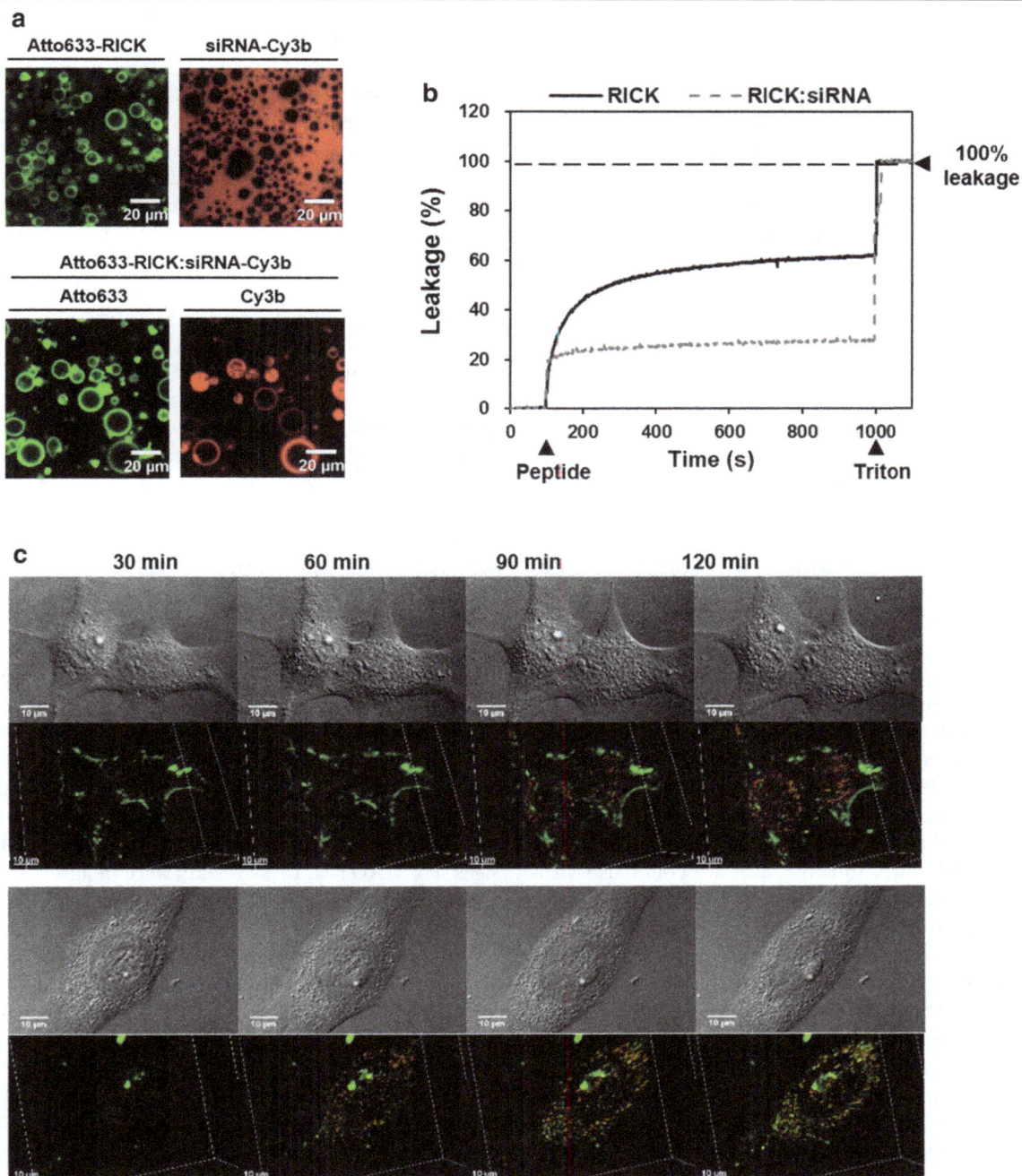

Fig. 4 Membrane interaction and internalization of RICK:siRNA nanoparticles on GUVs, LUVs and living cells. **a** Representative fluorescence microscopy images of GUVs incubated with 400 nM Atto633-RICK or 20 nM siRNA-Cy3b alone or with the formulated nanoparticles (400 nM Atto633-RICK:20 nM siRNA-Cy3b). *Bars* represent 20 μm. **b** Comparison of the leakage properties of 500 nM RICK alone and 500 nM RICK:25 nM siRNA nanoparticles on LUVs [DOPC/SM/Chol (2:2:1)]. Peptides/nanoparticles were injected at 100 s and the Triton (positive control) at 1000 s (n ≥ 2). **c** Representative 3D confocal microscopy images of the Atto633-RICK:siRNA-Cy3b internalization in living U87 cell lines. Images were taken every 30 min during 2 h, to follow fluorescence entrance and repartition inside the cell. For the 3D representation, the images were treated with Imaris software. *Bars* represent 10 μm

composition [DOPC/SM/Chol (2:2:1)] reflecting the plasma membrane. In the absence of peptides (or nanoparticles), no leakage was observed based on the low permeability of the phospholipid vesicle membrane to the fluorophore/quencher mix (Fig. 4b). Addition of peptide alone on the LUVs induced a significant increase in fluorescence revealing an important leakage. At the endpoint (after 15 min incubation), we obtained a leakage of 60% compared to the Triton (positive control). In contrast, the RICK:siRNA nanoparticle leakage was twofold weaker (28%) than the one observed for the peptide alone (60%). This is in agreement with the fact that in the RICK:siRNA complexes one part of the peptide interacts with the siRNA and the other part with the lipids.

Finally, we looked at the cellular internalization of the nanoparticles in living U87 human glioblastoma cells using spinning disk confocal microscopy (Fig. 4c; movie in Additional file 2). One of the advantage of using this technique was the possibility of a 3D reconstruction of the cell and a better understanding of nanoparticles distribution inside the cell. Furthermore, fluorescence quantification of siRNA-Cy3b and RICK-Atto633 allowed us to determine nanoparticles behavior once they reached the cell and crossed the plasma membrane. Based on this quantification, in the first 30 min, we could observe the progressive internalization of both RICK and the siRNA. After 90 min, the fluorescence patterns of both molecules were uniform in the cytoplasm reaching a maximum at 120 min. The fluorescence profile was exclusively cytoplasmic, no nanoparticles were found inside the nucleus, but it seems that the siRNA accumulated around the nucleus (at ~90 min). In some case, we could also visualize the accumulation of Atto633-RICK at the plasma membrane over the whole incubation period (2 h), which indicated the strong affinity of RICK peptide for lipids, while siRNA was never found at the plasma membrane. Furthermore, longer incubation (>24 h) of the nanoparticles did not give rise to higher cytosolic accumulation of Atto633-RICK or siRNA-Cy3b (data not shown).

In conclusion, we could state that nanoparticles had no lytic properties (no GUV or LUV destruction) and that the amphipathic peptide did not induce any membrane perturbation (no inner vesicle or tube formation) as reported by Ayala-Sanmartin and coll. for R9 and RW9 [55]. Finally, the RICK:siRNA internalization occurred immediately after applying the nanoparticles on cells. A minimal incubation time of 90 min should be sufficient to deliver the siRNA and then to induce a biological activity. Internalization mechanism of our RICK-based PBN is estimated to be through direct translocation based on the positive ZP value in transfection medium (Table 2) as well as on previous microscopic experiments with the

parental CADY peptide showing no co-localization with endosome markers [56].

Cellular evaluation OF RICK:siRNA nanoparticles

A luciferase gene silencing assay, in combination with an LDH cytotoxicity assay, was used to compare the efficiency of siRNA-loaded nanoparticles formulated with RICK, CADY-K and D-cady-k. For this purpose, we used an U87 cell line stably transfected with two plasmids containing firefly (FLuc) and *Renilla* (RLuc) luciferase reporter genes under transcriptional control of their own constitutive Cytomegalovirus (CMV) promoter. This cell line allowed us to knock-down FLuc (siFLuc) and to use RLuc as an internal control for the normalization (see scheme in Additional file 1: Figure S7). Based on our previous work [25] and on current structural characterization, peptide:siRNA nanoparticles have been formulated in 5% glucose, R = 20, with a siRNA targeting FLuc (siFLuc) at three different concentrations (5, 10 and 20 nM). In addition, a scrambled siRNA version (siSCR) was used as negative control. All the FLuc activities were normalized to the RLuc levels and to non-treated cells and then plotted as Relative Luc Activity (%).

As shown in Fig. 5a, the relative LUC activity (%) recorded for RICK:siRNA, CADY-K:siRNA and D-cady-k:siRNA nanoparticles revealed a specific and similar dose-dependent knock-down of FLuc reaching ~75% of inhibition at the highest siRNA concentration (20 nM). These knock-downs were specific for the used nanoparticles, because the same nanoparticles formulated with the siSCR show a Luc activity similar to the non-treated cells. For CADY-K, these results were in agreement with those reported previously on Neuro-2a-Luc[+] and B16-F10-Luc[+] cells [25].

With regard to cytotoxicity, all values were close to those measured for untreated cells (0%) and not higher than 20% (Fig. 5a). In this context, cytotoxicity evaluation was an important feature to avoid false-positive results because toxic effect would have an impact on both luciferase activities. Under the applied conditions (5000 cells per well), we observed at siRNA concentrations higher than 60 nM (=1.2 µM RICK concentration) that the normalized results were no longer coherent based on the unspecific knock-down of RLuc (data not shown).

In parallel, stability of the PBNs in the course of storage was investigated over the time with regard to future clinical applications. Although the DLS investigations have already indicated that nanoparticles were stable and conserved the same colloidal features after 72 h at 4 °C, we decided to control the FLuc knock-down efficiency of siRNA-loaded RICK PBN over the time. A stock solution of RICK:siRNA and CADY-K:siRNA complexes (5%

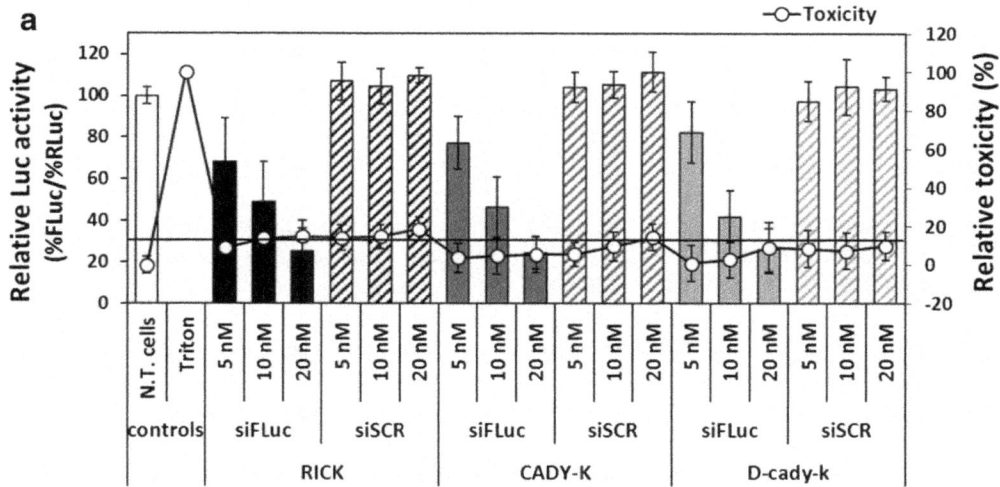

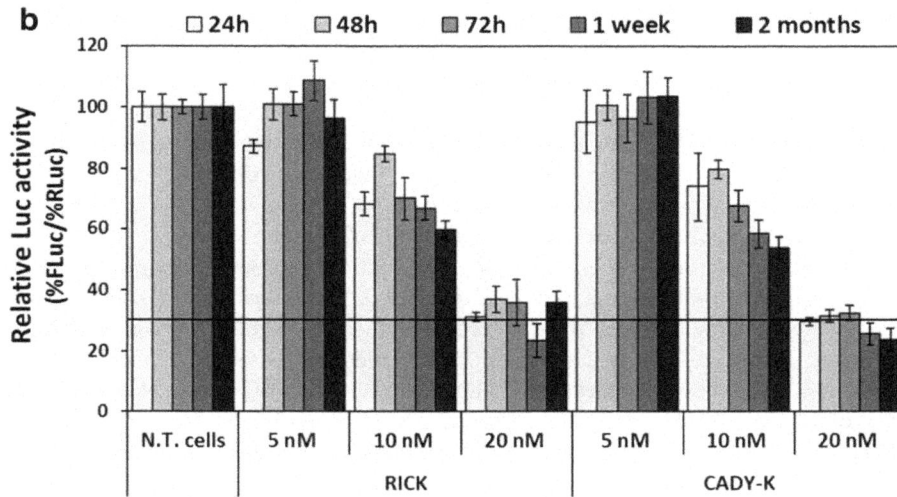

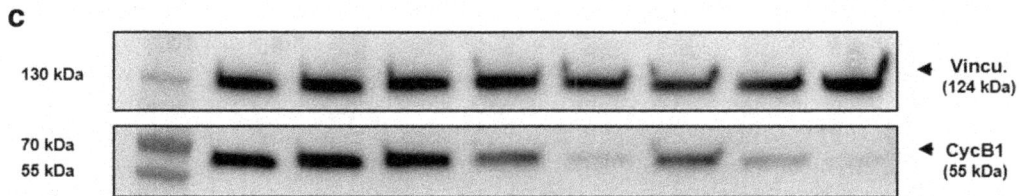

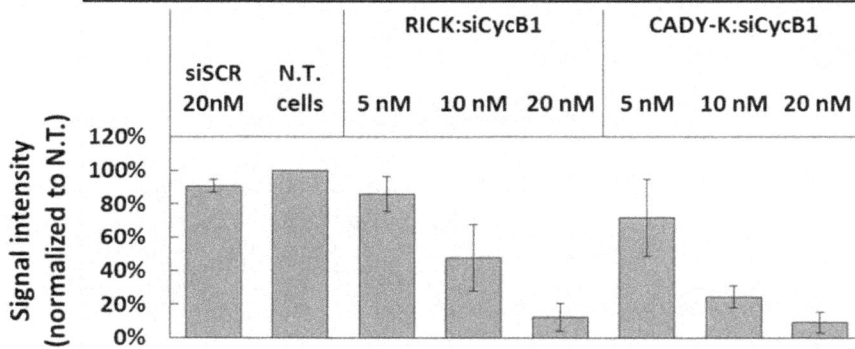

(See figure on previous page.)
Fig. 5 Cellular evaluation of RICK:siRNA nanoparticles. **a** Relative Luc activity (%FLuc/%RLuc) and relative toxicity (LDH quantification) after transfection with RICK-, CADY-K- and D-cady-k-based complexes in U87-FLuc-RLuc cells. siFLuc and siSCR correspond to siRNA anti-firefly Luciferase and its scrambled version, respectively. **b** Evaluation of RICK:siLuc and CADY-K:siLuc inhibiting efficiency after different PBN storage duration. Relative Luc activity (%FLuc/ %RLuc) was measured after PBN storage of 24, 48, 72 h, 1 week and 2 months. **c** Gene silencing on the endogenic protein cyclin B1. siSCR scrambled siRNA, N.T. non-treated cells, siCycB1 siRNA anti-Cyclin B1, Vincu. Vinculine

glucose, R = 20) were formulated, stored at 4 °C and aliquots were used for luciferase screening after 24, 48, 72 h, 1 week and 2 months (Fig. 5b). Results revealed no significant difference between the dose-dependent knock-down of FLuc, whatever the moment of the luciferase assay. As internal control, "freshly" prepared RICK:siRNA and CADY-K:siRNA nanoparticles were used to validate the obtained results. In all cases (fresh versus stored solution), the knock-down efficiency seemed to be slightly lower (~70% for 20 nM, ~35% for 10 nM and ~5% for 5 nM siRNA) than previously observed (~75% for 20 nM, ~55% for 10 nM and ~22% for 5 nM siRNA) (Fig. 5b and data not shown). Other samples of RICK:siRNA and CADY-K:siRNA nanoparticles evaluated after a long storage period (6 months and 1 year at 4 °C) confirmed that these nanoparticles were stable over the time and induced the same knock-down activity (data not shown).

To confirm the efficiency of the RICK:siRNA nanoparticles, gene silencing was also undertaken on the endogenic protein cyclin B1 which is known to be deregulated in most common cancers [57–60] instead of an overexpressed system (luciferase). After an optimization step (cell number per well and siRNA concentration), the knock-down efficiency of RICK:siRNA and CADY-K:siRNA nanoparticles (5% glucose, R = 20) were evaluated in a dose-dependent manner (5, 10 and 20 nM siRNA). The resulting protein expression (blot signal intensities) were first normalized to the loading control vinculin and thereafter to the non-treated condition (N.T cells) (Fig. 5c). As expected, we clearly observed a specific knock-down of Cyclin B1 (no effect using siSCR) for both types of nanoparticles. The only difference consisted in the different systems we used: in the overexpressed system, 20 nM siRNA were required for an 80% reduction of FLuc expression in 5000 cells whereas only 10 nM were necessary for the same percentage of Cyclin B1 knock-down in 75,000 cells. That discrepancy was directly due to the CMV-dependent overexpression of FLuc which is probably more than sevenfold higher than the endogenic expression of the Cyclin B1.

In a similar glioma cell line, Youn and coll. found ~40% reduction in luciferase expression using a myristic acid conjugated-CPP (transportan; TP) equipped with a transferrin receptor-targeting peptide (myr-TP-Tf) complexed with siRNA-FLuc, suggesting a high activity of RICK-based nanoparticles [61].

Finally, gene silencing abilities of RICK- and CADY-K-based nanoparticles were investigated after treatment with trypsin and through incubation with serum in order to evaluate biological activity for both peptides in proteolytic conditions. An overnight pre-incubation of nanoparticles in trypsin induced a net reduction of the FLuc knock-down efficiency for CADY-K compared to RICK (Fig. 6a). Only a ~30% decrease of FLuc activity was observed for CADY-K at 20 nM siRNA, compared to the 75% inhibition observed without trypsin treatment. In contrast, the level of inhibition observed for RICK remained similar to those obtained without trypsin treatment, i.e. ~60 and ~80% of FLuc silencing for 10 nM and 20 nM siRNA, respectively (Figs. 5a, 6a). These results clearly emphasized the resistance of RICK to proteases as shown previously by HPLC/MS (Fig. 3b; Additional file 1: Table S1).

In a similar manner, we addressed the effect of serum on nanoparticles through the monitoring of CyclinB1 knock-down. The presence of 10% FBS during the transfection induced a slight decrease of silencing efficiency, characterized by the necessity of using a twofold higher siRNA concentration (10, 20 and 40 nM instead of 5, 10 and 20 nM) to reach the same percentage of CyclinB1 inhibition in serum-free transfection media (Fig. 6b, compared to Fig. 5c). This is probably due to the interaction of the serum proteins with the nanoparticles (as shown by ZP changes) resulting in a reduced internalization which is translate by a lower knock-down activity. However, the CyclinB1 knock-down was the same for RICK-and CADY-K-based nanoparticles (inhibition of ~40, ~60 and ~90% of the protein expression for 10, 20 and 40 nM, respectively), suggesting the absence of any difference between both peptides in these conditions. We hypothesized that this phenomenon could be due to the fact that both nanoparticles were rapidly internalized (within minutes), thus strongly limiting degradation by proteases.

To simulate an in vivo administration (blood circulation before internalization), we performed a pre-incubation with 10% FBS (2 h at 37 °C before transfection). Under this condition, compared to RICK, a significant decrease of inhibition efficiency was observed for CADY-K (Fig. 6c). Comparing the nanoparticles loaded with 40 nM siRNA, the induced CyclinB1 knock-down by RICK:siRNA was 2.4-fold higher than the one obtained with CADY-K:siRNA (~25 and 60% for RICK:siCyCB1 and

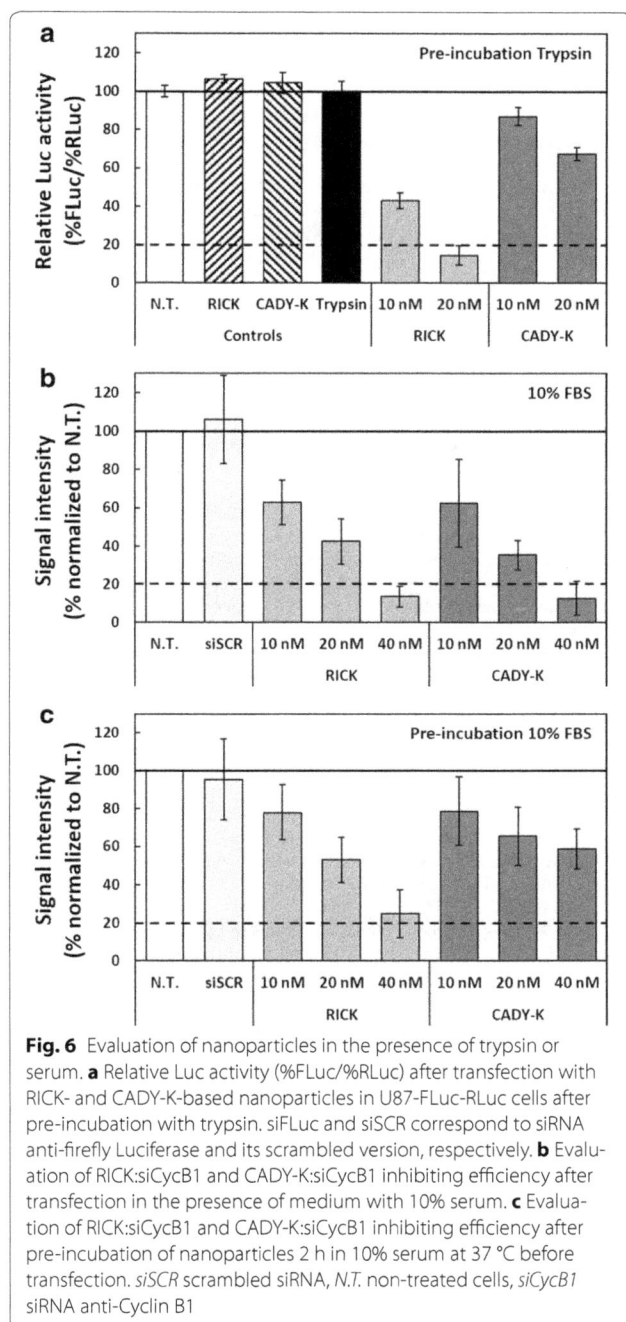

Fig. 6 Evaluation of nanoparticles in the presence of trypsin or serum. **a** Relative Luc activity (%FLuc/%RLuc) after transfection with RICK- and CADY-K-based nanoparticles in U87-FLuc-RLuc cells after pre-incubation with trypsin. siFLuc and siSCR correspond to siRNA anti-firefly Luciferase and its scrambled version, respectively. **b** Evaluation of RICK:siCycB1 and CADY-K:siCycB1 inhibiting efficiency after transfection in the presence of medium with 10% serum. **c** Evaluation of RICK:siCycB1 and CADY-K:siCycB1 inhibiting efficiency after pre-incubation of nanoparticles 2 h in 10% serum at 37 °C before transfection. *siSCR* scrambled siRNA, *N.T.* non-treated cells, *siCycB1* siRNA anti-Cyclin B1

CADY-K:siCycB1, respectively). This latter result confirmed the more pronounced stability for RICK in serum, as already observed for other D-isoform peptides [53].

All this data confirmed the high potential of RICK-based nanoparticles as a promising tool for siRNA delivery in vitro and in vivo.

Conclusions

In this report, we present the development and the characterization of a new peptide-based nanoparticle for siRNA cellular delivery. Based on previously reported secondary amphipathic peptide CADY-K [25], we designed the RICK peptide by combining the use of D-amino acids to the inversion of the peptide sequence. Our results revealed that this *retro-inverso* peptide RICK retained the main unique properties of CADY-K such as conformational versatility, formation of RICK:siRNA nanoparticles and remarkable efficacy in siRNA cellular delivery. Interestingly, in contrast to other *retro-inverso* peptides [51, 62], we observed a structural difference concerning the tryptophan cluster between CADY-K (L-isomer), D-cady-k (D-(isomer) and RICK, highlighting the important role of this amino acid. Moreover, because tryptophan has emerged as an important amino acid for membrane translocation [63, 64], a tryptophan clustering is a potential feature that should be taken into account for future PBN designs.

The main advantage of RICK-based nanoparticles lays in resistance to enzymatic degradation and the resulting protection of the siRNA. RICK-based nanoparticles are able to rapidly internalize siRNA into cells and to induce the inhibition of gene expression at a low dose (~75% knock-down of overexpressed luciferase and ~80% of endogenous CyclinB1 with RICK:20 nM siRNA) without any significant cytotoxicity. In the presence of serum, the activity is maintained by doubling the siRNA concentration (80% of CyclinB1 knock-down for RICK:40 nM siRNA). Compared to other CPP-based siRNA carriers, the measured knock-down effect is remarkable because in many cases higher siRNA concentrations are required for a less important knock-down efficiency [20, 65–67]. More importantly, RICK nanoparticles preserve the activity in proteolytic conditions (trypsin or serum pre-incubation) compared to CADY-K.

In summary, our results underline that RICK-based nanoparticles have the same biophysical properties (structure, size, ZP etc.) and the same biological in vitro efficiency (siRNA-induced knock-down) in serum-free or serum containing medium. However, RICK-based nanoparticles have the outstanding property to avoid siRNA degradation based on its own proteolytic stability as demonstrated by pre-incubation with trypsin and serum. This is an important prerequisite for its in vivo application since longer blood circulation is required. In parallel, we already consider the PEGylation of RICK to further improve RICK-based nanoparticle application in vivo [68].

Additional files

Additional file 1. Supplementary information about circular dichroism analyses of RICK and isomers mixtures, characterization of peptide-based nanoparticles by DLS and TEM, mass spectra analyses of CADY-K, characterization of membrane interactions by fluorescence microscopy and principle of the Dual luciferase evaluation.

Additional file 2. Cellular internalization of siRNA-Cy3b (red dots)/RICK-Atto633 (green dots) nanoparticles over 2 hours by confocal microscopy.

Abbreviations

CPP: cell-penetrating peptide; PBN: peptide-based nanoparticle; siRNA: short interfering RNA; LUV: large unilamellar vesicles; GUV: giant unilamellar vesicles; FBS: fetal bovine serum; CD: circular dichroism; DLS: dynamic light scattering; PDI: polydispersity index; ZP: zeta potential; ESEM: environmental scanning electron microscopy; TEM: transmission electron microscopy.

Authors' contributions

All authors contributed to experiments and data analyzing (SD, KK: CD and DLS measurements; AV and MFL: Luciferase/LDH assay; FF and VV: TEM and ESEM measurements; QS: Liposome leakage assay; GA: RICK labeling; GA, KK and CJ: Peptide stability experiments; SD and GA: design of the peptides; SD and AT: Western blots; AV: confocal microscopy; CG and FC: Genetically modified cell generation). PB and SD designed/supervised the experiments and wrote the manuscript. AV, GA, KK, FC, MFL helped to improve the manuscript. All authors read and approved the final manuscript.

Author details

[1] Centre de Recherche de Biologie cellulaire de Montpellier, UMR 5237 CNRS, 1919 Route de Mende, 34293 Montpellier, France. [2] Sys2Diag, UMR 9005-CNRS/ALCEDIAG, 1682 Rue de la Valsiere, 34184 Montpellier, France. [3] Microscopie Électronique et Analytique, Université de Montpellier, Place Eugène Bataillon, 34095 Montpellier, France. [4] EA 7435 IMOTION (Imagerie moléculaire et thérapies innovantes en oncologie), Université de Bordeaux, 146 rue Leo Saignat, 33076 Bordeaux, France.

Acknowledgements

We thank Dr. Eric Vivès for many helpful suggestions/comments and for critically reading the manuscript and Dr Cyril Favard as well as Naresh Yandrapalli for their critical advice on GUVs experiments. We thank Sylvain De Rossi (Montpellier RIO imaging microscopy platform) for technical advices as well as Guillaume Cazals (technical platform of the Institute of Biomolecules Max Mousseron, Montpellier, France) for LC–MS and LC–MS/MS experiments.

Competing interests

The authors declare that they have no competing interests.

Funding

This work was supported by the Fondation de la Recherche Médicale (DBS 20140930769), the Labex TRAIL TARGLIN (ANR-10-LABX-57) and the Centre National de la Recherche Scientifique.

References

1. Fire A, Xu S, Montgomery MK, Kostas SA, Driver SE, Mello CC. Potent and specific genetic interference by double-stranded RNA in Caenorhabditis elegans. Nature. 1998;391:806–11.
2. Devi GR. siRNA-based approaches in cancer therapy. Cancer Gene Ther. 2006;13:819–29.
3. Dykxhoorn DM, Novina CD, Sharp PA. Killing the messenger: short RNAs that silence gene expression. Nat Rev Mol Cell Biol. 2003;4:457–67.
4. de Fougerolles A, Vornlocher H-P, Maraganore J, Lieberman J. Interfering with disease: a progress report on siRNA-based therapeutics. Nat Rev Drug Discov. 2007;6:443–53.
5. Whitehead KA, Langer R, Anderson DG. Knocking down barriers: advances in siRNA delivery. Nat Rev Drug Discov. 2009;8:129–38.
6. Heitz F, Morris MC, Divita G. Twenty years of cell-penetrating peptides: from molecular mechanisms to therapeutics. Br J Pharmacol. 2009;157:195–206.
7. Gooding M, Browne LP, Quinteiro FM, Selwood DL. siRNA delivery: from lipids to cell-penetrating peptides and their mimics. Chem Biol Drug Des. 2012;80:787–809.
8. Lepeltier E, Bourgaux C, Couvreur P. Nanoprecipitation and the "Ouzo effect": application to drug delivery devices. Adv Drug Deliv Rev. 2014;71:86–97.
9. Dietz GPH, Bähr M. Delivery of bioactive molecules into the cell: the Trojan horse approach. Mol Cell Neurosci. 2004;27:85–131.
10. Järver P, Mäger I, Langel Ü. In vivo biodistribution and efficacy of peptide mediated delivery. Trends Pharmacol Sci. 2010;31:528–35.
11. Patel LN, Zaro JL, Shen W-C. Cell penetrating peptides: intracellular pathways and pharmaceutical perspectives. Pharm Res. 2007;24:1977–92.
12. Elmquist A, Lindgren M, Bartfai T, Langel Ü. VE-cadherin-derived cell-penetrating peptide, pVEC, with carrier functions. Exp Cell Res. 2001;269:237–44.
13. Deshayes S, Morris MC, Divita G, Heitz F. Cell-penetrating peptides: tools for intracellular delivery of therapeutics. Cell Mol Life Sci. 2005;62:1839–49.
14. Derossi D, Joliot AH, Chassaing G, Prochiantz A. The third helix of the Antennapedia homeodomain translocates through biological membranes. J Biol Chem. 1994;269:10444–50.
15. Pooga M, Hällbrink M, Zorko M, Langel UL. Cell penetration by transportan. FASEB J. 1998;12:67–77.
16. Crombez L, Aldrian-Herrada G, Konate K, Nguyen QN, McMaster GK, Brasseur R, et al. A new potent secondary amphipathic cell–penetrating peptide for siRNA delivery into mammalian cells. Mol Ther J Am Soc Gene Ther. 2009;17:95–103.
17. Jafari M, Xu W, Pan R, Sweeting CM, Karunaratne DN, Chen P. Serum Stability and physicochemical characterization of a novel amphipathic peptide C6M1 for siRNA delivery. PLoS ONE. 2014;9:e99797.
18. Moschos SA, Jones SW, Perry MM, Williams AE, Erjefalt JS, Turner JJ, et al. Lung delivery studies using siRNA conjugated to TAT(48–60) and penetratin reveal peptide induced reduction in gene expression and induction of innate immunity. Bioconjug Chem. 2007;18:1450–9.
19. Crombez L, Morris MC, Deshayes S, Heitz F, Divita G. Peptide-based nanoparticle for ex vivo and in vivo drug delivery. Curr Pharm Des. 2008;14:3656–65.
20. van Asbeck AH, Beyerle A, McNeill H, Bovee-Geurts PHM, Lindberg S, Verdurmen WPR, et al. Molecular parameters of siRNA–cell penetrating peptide nanocomplexes for efficient cellular delivery. ACS Nano. 2013;7:3797–807.
21. Regberg J, Srimanee A, Erlandsson M, Sillard R, Dobchev DA, Karelson M, et al. Rational design of a series of novel amphipathic cell-penetrating peptides. Int J Pharm. 2014;464:111–6.
22. Boisguérin P, Deshayes S, Gait MJ, O'Donovan L, Godfrey C, Betts CA, et al. Delivery of therapeutic oligonucleotides with cell penetrating peptides. Adv Drug Deliv Rev. 2015;87:52–67.
23. Deshayes S, Konate K, Aldrian G, Crombez L, Heitz F, Divita G. Structural polymorphism of non-covalent peptide-based delivery systems: highway to cellular uptake. Biochim Biophys Acta BBA Biomembr. 2010;1798:2304–14.
24. Eiríksdóttir E, Konate K, Langel Ü, Divita G, Deshayes S. Secondary structure of cell-penetrating peptides controls membrane interaction and insertion. Biochim Biophys Acta BBA Biomembr. 2010;1798:1119–28.
25. Konate K, Lindberg M, Vaissiere A, Jourdan C, Aldrian G, Margeat E, et al. Optimisation of vectorisation property: a comparative study for a secondary amphipathic peptide. Int J Pharm. 2016;509:71.

26. Chorev M, Goodman M. Recent developments in retro peptides and proteins—an ongoing topochemical exploration. Trends Biotechnol. 1995;13:438–45.

27. Aldrian-Herrada G, Desarménien MG, Orcel H, Boissin-Agasse L, Méry J, Brugidou J, et al. A peptide nucleic acid (PNA) is more rapidly internalized in cultured neurons when coupled to a retro-inverso delivery peptide. The antisense activity depresses the target mRNA and protein in magnocellular oxytocin neurons. Nucleic Acids Res. 1998;26:4910–6.

28. Brugidou J, Legrand C, Méry J, Rabié A. The retro-inverso form of a homeobox-derived short peptide is rapidly internalised by cultured neurones: a new basis for an efficient intracellular delivery system. Biochem Biophys Res Commun. 1995;214:685–93.

29. Kay JG, Koivusalo M, Ma X, Wohland T, Grinstein S. Phosphatidylserine dynamics in cellular membranes. Mol Biol Cell. 2012;23:2198–212.

30. Weinberger A, Tsai F-C, Koenderink GH, Schmidt TF, Itri R, Meier W, et al. Gel-assisted formation of giant unilamellar vesicles. Biophys J. 2013;105:154–64.

31. Konate K, Crombez L, Deshayes S, Decaffmeyer M, Thomas A, Brasseur R, et al. Insight into the cellular uptake mechanism of a secondary amphipathic cell-penetrating peptide for siRNA delivery. Biochemistry. 2010;49:3393–402.

32. Konate K, Rydstrom A, Divita G, Deshayes S. Everything you always wanted to know about CADY-mediated siRNA delivery* (*but afraid to ask). Curr Pharm Des. 2013;19:2869–77.

33. Dean SN, Bishop BM, van Hoek ML. Susceptibility of *Pseudomonas aeruginosa* biofilm to alpha-helical peptides: D-enantiomer of LL-37. Front Microbiol. 2011;2.

34. Chen Y, Vasil AI, Rehaume L, Mant CT, Burns JL, Vasil ML, et al. Comparison of biophysical and biologic properties of α-helical enantiomeric antimicrobial peptides. Chem Biol Drug Des. 2006;67:162–73.

35. Wade D, Boman A, Wåhlin B, Drain CM, Andreu D, Boman HG, et al. All-D amino acid-containing channel-forming antibiotic peptides. Proc Natl Acad Sci USA. 1990;87:4761–5.

36. Grishina IB, Woody RW. Contributions of tryptophan side chains to the circular dichroism of globular proteins: exciton couplets and coupled oscillators. Faraday Discuss. 1994;99:245–62.

37. Woody RW. Contributions of tryptophan side chains to the far-ultraviolet circular dichroism of proteins. Eur Biophys J. 1994;23:253–62.

38. Roy A, Bour P, Keiderling TA. TD-DFT modeling of the circular dichroism for a tryptophan zipper peptide with coupled aromatic residues. Chirality. 2009;21(Suppl 1):E163–71.

39. Noda M, Matoba Y, Kumagai T, Sugiyama M. A novel assay method for an amino acid racemase reaction based on circular dichroism. Biochem J. 2005;389:491–6.

40. Takekiyo T, Wu L, Yoshimura Y, Shimizu A, Keiderling TA. Relationship between hydrophobic interactions and secondary structure stability for Trpzip beta-hairpin peptides. Biochemistry. 2009;48:1543–52.

41. Chetal P, Chauhan VS, Sahal D. A Meccano set approach of joining trpzip a water soluble beta-hairpin peptide with a didehydrophenylalanine containing hydrophobic helical peptide. J Pept Res. 2005;65:475–84.

42. Vivian JT, Callis PR. Mechanisms of tryptophan fluorescence shifts in proteins. Biophys J. 2001;80:2093–109.

43. Deshayes S, Gerbal-Chaloin S, Morris MC, Aldrian-Herrada G, Charnet P, Divita G, et al. On the mechanism of non-endosomial peptide-mediated cellular delivery of nucleic acids. Biochim Biophys Acta BBA Biomembr. 2004;1667:141–7.

44. Helene C, Dimicoli JL. Interaction of oligopeptides containing aromatic amino acids with nucleic acids. Fluorescence and proton magnetic resonance studies. FEBS Lett. 1972;26:6–10.

45. Toulmé JJ, Hélène C. Specific recognition of single-stranded nucleic acids. Interaction of tryptophan-containing peptides with native, denatured, and ultraviolet-irradiated DNA. J Biol Chem. 1977;252:244–9.

46. Jain AA, Rajeswari MR. Binding studies on peptide-oligonucleotide complex: intercalation of tryptophan in GC-rich region of c-myc gene. Biochim Biophys Acta. 2003;1622:73–81.

47. Bhattacharjee S. DLS and zeta potential—what they are and what they are not? J Control Release. 2016;235:337–51.

48. Deshayes S, Konate K, Rydström A, Crombez L, Godefroy C, Milhiet P-E, et al. Self-assembling peptide-based nanoparticles for siRNA delivery in primary cell lines. Small. 2012;8:2184–8.

49. Lindberg S, Regberg J, Eriksson J, Helmfors H, Muñoz-Alarcón A, Srimanee A, et al. A convergent uptake route for peptide- and polymer-based nucleotide delivery systems. J Controll Release. 2015;206:58–66.

50. Urgard E, Lorents A, Klaas M, Padari K, Viil J, Runnel T, et al. Pre-administration of PepFect6-microRNA-146a nanocomplexes inhibits inflammatory responses in keratinocytes and in a mouse model of irritant contact dermatitis. J Control Release. 2016;235:195–204.

51. Québatte G, Kitas E, Seelig J. riDOM, a cell-penetrating peptide. Interaction with DNA and heparan sulfate. J Phys Chem B. 2013;117:10807–17.

52. Québatte G, Kitas E, Seelig J. riDOM, a cell penetrating peptide Interaction with phospholipid bilayers. Biochim Biophys Acta BBA Biomembr. 2014;1838:968–77.

53. Pujals S, Fernández-Carneado J, Ludevid MD, Giralt E. D-SAP: a new, noncytotoxic, and fully protease resistant cell-penetrating peptide. ChemMedChem. 2008;3:296–301.

54. Morales-Penningston NF, Wu J, Farkas ER, Goh SL, Konyakhina TM, Zheng JY, et al. GUV preparation and imaging: minimizing artifacts. Biochim Biophys Acta. 2010;1798:1324–32.

55. Lamazière A, Burlina F, Wolf C, Chassaing G, Trugnan G, Ayala-Sanmartin J. Non-metabolic membrane tubulation and permeability induced by bioactive peptides. PLoS ONE. 2007;2:e201.

56. Rydström A, Deshayes S, Konate K, Crombez L, Padari K, Boukhaddaoui H, et al. Direct translocation as major cellular uptake for CADY self-assembling peptide-based nanoparticles. PLoS ONE. 2011;6:e25924.

57. Malumbres M, Barbacid M. Cell cycle, CDKs and cancer: a changing paradigm. Nat Rev Cancer. 2009;9:153–66.

58. Aaltonen K, Amini R-M, Heikkilä P, Aittomäki K, Tamminen A, Nevanlinna H, et al. High cyclin B1 expression is associated with poor survival in breast cancer. Br J Cancer. 2009;100:1055–60.

59. Holtkamp N, Afanasieva A, Elstner A, van Landeghem FKH, Könneker M, Kuhn SA, et al. Brain slice invasion model reveals genes differentially regulated in glioma invasion. Biochem Biophys Res Commun. 2005;336:1227–33.

60. Soria J-C, Jang SJ, Khuri FR, Hassan K, Liu D, Hong WK, et al. Overexpression of cyclin B1 in early-stage non-small cell lung cancer and its clinical implication. Cancer Res. 2000;60:4000–4.

61. Youn P, Chen Y, Furgeson DY. A myristoylated cell-penetrating peptide bearing a transferrin receptor-targeting sequence for neuro-targeted siRNA delivery. Mol Pharm. 2014;11:486–95.

62. Petit MC, Benkirane N, Guichard G, Du AP, Marraud M, Cung MT, et al. Solution structure of a retro-inverso peptide analogue mimicking the foot-and-mouth disease virus major antigenic site. Structural basis for its antigenic cross-reactivity with the parent peptide. J Biol Chem. 1999;274:3686–92.

63. Bechara C, Pallerla M, Zaltsman Y, Burlina F, Alves ID, Lequin O, et al. Tryptophan within basic peptide sequences triggers glycosaminoglycan-dependent endocytosis. FASEB J. 2013;27:738–49.

64. Jobin M-L, Blanchet M, Henry S, Chaignepain S, Manigand C, Castano S, et al. The role of tryptophans on the cellular uptake and membrane interaction of arginine-rich cell penetrating peptides. Biochim Biophys Acta BBA Biomembr. 2015;1848:593–602.

65. Simeoni F, Morris MC, Heitz F, Divita G. Insight into the mechanism of the peptide-based gene delivery system MPG: implications for delivery of siRNA into mammalian cells. Nucleic Acids Res. 2003;31:2717–24.

66. Veldhoen S, Laufer SD, Trampe A, Restle T. Cellular delivery of small interfering RNA by a non-covalently attached cell-penetrating peptide: quantitative analysis of uptake and biological effect. Nucleic Acids Res. 2006;34:6561–73.

67. Lundberg P, El-Andaloussi S, Sütlü T, Johansson H, Langel U. Delivery of short interfering RNA using endosomolytic cell-penetrating peptides. FASEB J. 2007;21:2664–71.

68. Aldrian G, Vaissière, Konate K, Seisel Q, Vivès E, Fernandez F, et al. PEGylation rate influences peptide-based nanoparticles mediated siRNA delivery in vitro and in vivo. J Control Release. 2017. doi:10.1016/j.jconrel.2017.04.012.

14

Photoinduced effects of m-tetrahydroxyphenylchlorin loaded lipid nanoemulsions on multicellular tumor spheroids

Doris Hinger[1]* [ID], Fabrice Navarro[2,3], Andres Käch[4], Jean-Sébastien Thomann[2,3], Frédérique Mittler[2,3], Anne-Claude Couffin[2,3] and Caroline Maake[1]

Abstract

Background: Photosensitizers are used in photodynamic therapy (PDT) to destruct tumor cells, however, their limited solubility and specificity hampers routine use, which may be overcome by encapsulation. Several promising novel nanoparticulate drug carriers including liposomes, polymeric nanoparticles, metallic nanoparticles and lipid nanocomposites have been developed. However, many of them contain components that would not meet safety standards of regulatory bodies and due to difficulties of the manufacturing processes, reproducibility and scale up procedures these drugs may eventually not reach the clinics. Recently, we have designed a novel lipid nanostructured carrier, namely Lipidots, consisting of nontoxic and FDA approved ingredients as promising vehicle for the approved photosensitizer m-tetrahydroxyphenylchlorin (mTHPC).

Results: In this study we tested Lipidots of two different sizes (50 and 120 nm) and assessed their photodynamic potential in 3-dimensional multicellular cancer spheroids. Microscopically, the intracellular accumulation kinetics of mTHPC were retarded after encapsulation. However, after activation mTHPC entrapped into 50 nm particles destroyed cancer spheroids as efficiently as the free drug. Cell death and gene expression studies provide evidence that encapsulation may lead to different cell killing modes in PDT.

Conclusions: Since ATP viability assays showed that the carriers were nontoxic and that encapsulation reduced dark toxicity of mTHPC we conclude that our 50 nm photosensitizer carriers may be beneficial for clinical PDT applications.

Keywords: Nanoemulsion, Biocompatibility, mTHPC, Photodynamic therapy, Spheroids, Lipid nanoparticles

Background

A wealth of publications report on the development of promising novel nanoparticulate drug carriers including liposomes [1], polymeric nanoparticles [2], metallic nanoparticles [3] and lipid nanocomposites [4]. However, many of them contain components that would not meet safety standards of regulatory bodies such as the European Medicines Agency (EMA) or the US food and drug administration (FDA) [5]. Furthermore, due to difficulties

of the manufacturing processes, reproducibility and scale up procedures these drugs may eventually not translate into the clinics.

Recently, we have designed a novel lipid nanostructured carrier, namely Lipidots, consisting of nontoxic and FDA approved ingredients: wax and soybean oil serve as core components and lecithin as membranous hull with a polyethylene glycol (PEG) coating [6]. Containing only natural compounds, they are likely to be broken down and removed or recycled by the body [7]. Lipidots may be utilized and adapted for many different applications such as fluorescent imaging probes, contrast agent carriers, or targeted drug delivery [8]. They offer the possibility

*Correspondence: doris.hinger@uzh.ch
[1] Institute of Anatomy, University of Zurich, Winterthurerstrasse 190, 8057 Zurich, Switzerland
Full list of author information is available at the end of the article

to tune the viscosity of their lipid core, thereby adapting the release of an encapsulated compound to the desired profile [9]. Moreover, Lipidots can be manufactured with high colloidal stability at laboratory and industrial scales using ultrasonics or high pressure homogenization [6].

An interesting future application of Lipidots may be in the context of photodynamic therapy (PDT), a modality which is currently receiving increasing clinical attention as a promising anti-cancer treatment [10]. PDT principles rely on the activation of a light-sensitive drug (the photosensitizer, PS), which, through oxidative reaction cascades of type I and type II leads to the generation of cytotoxic reactive oxygen species (ROS) and strictly localized cell death. Remarkably, PDT has the potential to overcome disadvantages of standard oncologic regimes such as surgery, chemo- or radiotherapy because it is minimal invasive, bears little risk for the development of resistance and lacks severe side effects [11]. However, the efficiency of PDT critically depends on a high local accumulation of the PS at the tumor site. But since many potent PSs are hydrophobic, they tend to aggregate in aqueous environments (e.g. after intravenous injection), with negative consequences for their biodistribution and photoactivity, which can eventually lead to unsatisfactory therapeutic effects [12]. With the aim to improve PDT applications, various PSs have been entrapped into nano-carriers, including e.g. Photophrin, hypocrellin A, chlorin e6, tetraarylporphyrin, the near infrared dye indocyanine green [13] or the powerful FDA approved second generation PS m-tetrahydroxyphenylchlorin (mTHPC) [14].

In a previous study we have reported on the successful and reproducible encapsulation of mTHPC (generic name: Temoporfin) into Lipidots and their extensive characterization [15]. While our physico-chemical and photophysical data indicate that these particles may be well suited for PDT applications, results about their biological activity are only very preliminary yet [15]. In the present study we have thus set out to investigate PDT effects of mTHPC-loaded Lipidots for the first time in an advanced in vitro 3-dimensional (3D) head and neck cancer cell model. To estimate their potential for clinical PDT use, we produced Lipidots with two sizes (50 and 120 nm) and, after mTHPC encapsulation, compared their in vitro effects to free mTHPC in terms of light-induced toxicity, penetration properties, dispersion behaviour, PDT effects, cell death mechanisms and gene expression patterns.

Methods
Chemicals
MTHPC was obtained from Biolitec, Jena, Germany as powder. A stock solution of 1.47 mM (1 mg/mL) in 100 % ethanol was prepared and stored at 4 °C in the dark. 1,1′-dioctadecyl-3,3,3′,3′-tetramethylindodicarbocyanine

perchlorate (DiD) was purchased from Life Technologies (Carlsbad, USA). If not otherwise indicated, chemicals were purchased from Sigma-Aldrich, Buchs, Switzerland.

Nanoparticle preparation
Lipidots were prepared according to Delmas et al. [9] and Navarro et al. [15]. Briefly, the manufacturing process consists of mixing an aqueous phase and a lipid phase which are separately prepared, including on the one hand MyrjS40 surfactant dissolved into 1X phosphate buffered saline (PBS) (154 mM NaCl, 0.1 M Na_2HPO_4, pH 7.4) and on the other hand soybean oil and wax (Suppocire NB) under melted state. The ultrasonication step is performed using a VCX750 ultrasonic processor during 20 min (power input 190 W, 3-mm probe diameter, Sonics). MTHPC was incorporated into the lipid mixture as a concentrated solution in ethyl acetate and after vacuum elimination of organic solvent, the oily phase was added to the aqueous phase and emulsification was performed as previously described [15]. For 50 nm Lipidots, the dispersion is composed of 37.5 % (w/w) of lipid phase (with a lecithin/PEG surfactant weight ratio of 0.19 and a surfactant/core weight ratio of 1.20) whereas for 120 nm Lipidots, the dispersion is composed of 43.0 % (w/w) of lipid phase (with a lecithin/PEG surfactant weight ratio of 0.21 and a surfactant/core weight ratio of 3.0). The Lipidots were loaded with mTHPC (thereafter called M-Lipidots) at two different ratios of numbers of PS per nanoparticle for 50 and 120 nm-sized Lipidots, respectively (920 and 4600 molecules of mTHPC/particle, respectively). The mTHPC concentrations were determined by high-performance liquid chromatography (HPLC) analysis. HPLC of prepared samples was carried out on a Sunfire C18 column (250 mm × 4.6 mm, i.d. 5 μm) at 30 °C. The mTHPC compound was eluted at 2.10 min using a isocratic mobile phase of acetonitrile/H_2O trifluoroaceticacid, 0.1 %: 9/1 at 1 mL/min flow rate after injection of 30 μL. The UV detection is operated at 425 nm. The mTHPC concentrations were assessed using a calibration curve in the range of 1–12 μg/mL. For comparisons at constant PS content, all working solutions were diluted using PBS to obtain equivalent mTHPC amounts in solution to be added in cell culture media for PDT treatment (3.67, 7.34, 14.69 μM mTHPC content). For in vitro additional fluorescence imaging and flow cytometry purposes, dye-doped nanoparticles, thereafter called D-Lipidots, were prepared as previously described [16] by incorporating DiD lipophilic indocyanine into the oily core of 50 nm Lipidots.

Monolayer cell culture
CAL-33 tongue squamous cell carcinoma cells (DSMZ, Braunschweig, Germany), were grown in RPMI without

phenol red, 10 % FCS, 2 mM Glutamax (Life Technologies), and 1 % Penicillin/Streptomycin (LifeTechnologies). Cells were kept in 75 cm^2 cell culture flasks at 5 % CO_2 and 37 °C. Cell counting was performed with a Neubauer chamber (Laboroptik Ltd., Lancing, UK) on an aliquot of cells after staining with 0.1 % (w/v) nigrosin in PBS.

Spheroid cell culture

The bottoms of 96 well plates were coated with 65 µL 1.5 % (w/v) agarose (Life Technologies) in cell culture medium without supplements. 3D cell culture spheroids were prepared by putting 96 drops of 5000 CAL-33 cells in 10 µL complete cell culture medium on the inner side of the lid of a 96 well plate. Then the lids with the hanging drops were put back on the plates and incubated for 24 h. Thereafter, 190 µL of complete cell culture medium was added to the wells and the drops were spun down shortly in a centrifuge (Virion, Zürich, Switzerland) and incubated for another 72 h. By that time the spheroids had reached an average diameter of 200 µm and were immediately used for the experiments [17].

Light microscopy
Monolayer cells

CAL-33 cells were seeded on 12 mm glass cover slips (Karl Hecht, Sondheim, Germany) and incubated with 7.34 µM mTHPC or M-Lipidots or 1 µM D-Lipidots in cell culture medium for up to 28 h in the dark. The cover slips were washed twice with PBS and subsequently fixed for 20 min with 4 % (w/v) formaldehyde (FA)/PBS. After washing they were mounted on microscopic slides (Menzel, Braunschweig, Germany) with Glycergel (Dako, Glostrup, Denmark) and analyzed with a confocal laser scanning microscope (Leica SP5, Heerbrugg, Switzerland). MTHPC was excited at 488 nm and fluorescence was detected between 590–660 nm. Images were analyzed with the imaging software Imaris (Bitplane, Belfast, UK).

Spheroids

Spheroids were incubated with 7.34 µM of mTHPC or M-Lipidots in 100 µL cell culture medium for up to 28 h in 96 well plates in the dark. Spheroids were picked with a 1 mL pipette and transferred to microcentrifuge tubes. After washing twice with PBS spheroids were fixed in 4 % (w/v) FA/PBS for 1 h, washed in PBS and analyzed in 18-well µ-slides (IBIDI) by widefield fluorescence microscopy (Leica DMI 6000) or confocal laser scanning microscopy (Leica, SP5). Per time point, 3–5 images were acquired using differential interference contrast (DIC) and epifluorescence and mean fluorescence was calculated from regions of interest (ROIs) which were drawn around the cell assemblies in the DIC channel with Leica AS lite software. Confocal laser scanning microscopy (Leica SP5) was performed on 3–5 fixed spheroids per condition with a 20× objective (HC Plan APO). After spheroid integrity was confirmed by DIC imaging, optical sectioning was performed with an argon laser at 488 nm for excitation of mTHPC. Pictures from the center of the spheroids were taken and processed with the imaging software Imaris (Bitplane, Belfast, UK).

Cytotoxicity assessment

Spheroids were incubated with 3.67, 7.34 and 14.69 µM of mTHPC or M-Lipidots for 24 h in 96 well plates in the dark. Substance-mediated damage (i.e. dark toxicity) was assessed by either measuring spheroid areas as ROIs with widefield microscopy and the Leica AS imaging software or by means of an ATP luciferase viability assay (Promega, Fitchburg, USA). For the ATP luciferase viability assay 100 µL of Cell Viability Assay solution was added to each well after drug incubation, the contents were mixed by pipetting and the plate was transferred for 20 min to a shaker. Subsequently bioluminescence was measured in a microplate reader (Biotek, Vermont, USA).

Phototoxicity assessment

Spheroids were incubated with 3.67, 7.34 and 14.69 µM of mTHPC or M-Lipidots for 24 h in 96 well plates in the dark. Subsequently the plates were subjected to PDT by illuminating with white light from 2.5 cm above (3440 lx; fluorescent tube SYLVA-NIA standard F15 W/154, daylight) for 20 min. To ensure an even illumination, the outer rim of the well plates was never used for experimentation and the sequence of samples within the plate was changed between repetitions. Spheroid areas were microscopically determined as described above and cell survival was determined by ATP luciferase viability assay 5 h after irradiation as described above.

Apoptosis assay

Spheroids were incubated with 7.34 µM mTHPC or M-Lipidots for 24 h. After illuminating for 1 min (conditions as described above) spheroids were incubated for another 1.5 h with 100 µL 15 µM Hoechst 33342 and 30× Flica reagent (FAM Flica Poly Caspase kit, ImmunoChemistry Technologies, Enzo Life Sciences, Lausen, Switzerland). The spheroids were subsequently harvested with a 1 mL pipette and transferred to microcentrifuge tubes. After washing twice with wash buffer (FAM Flica Poly Caspase Kit) they were fixed for 1 h in fixing solution (FAM Flica Poly Caspase Kit) and analyzed in 18 well µ-slides (IBIDI) with a confocal laser scanning microscope (Leica SP5, Heerbrugg, Switzerland) within 24 h.

Electron microscopy

Spheroids were incubated for 24 h with 3.67 μM mTHPC or 50 nm M-Lipidots and irradiated for 1 min as described above. One hour after light treatment they were washed and fixed and sequentially treated with OsO_4 and uranylacetate. After dehydration they were embedded in Epon/Araldite and sections were contrasted with uranyl acetate and lead citrate. They were examined with a CM100 transmission electron microscope (FEI, Eindhoven, The Netherlands) or with an Auriga 40 scanning electron microscope (Zeiss, Oberkochen, Germany). For a more detailed description see Additional file 1.

Quantitative reverse transcription polymerase chain reaction (qRT-PCR)

A total of 120 spheroids were incubated with 3.67 μM mTHPC or 50 nm M-Lipidots for 24 h. After illuminating for 1 min as described spheroids were incubated for another 2 h, subsequently harvested with a 1 mL pipette and transferred to microcentrifuge tubes. They were washed twice with PBS and resuspended in 600 μL lysis buffer (Qiagen, Venlo, The Netherlands), vortexed vigorously and passed 30 times through a 1 mL syringe with a 20 gauge needle. Total RNA was extracted with the RNeasy Micro Kit (Qiagen) as described per manufacturer's instructions, processed with a cNDA synthesis kit (Qiagen) and the obtained cDNA used for a quantitative PCR array (Human Cancer Drug Targets RT^2 Profiler PCR Array, Qiagen). For further details please refer to the Additional file 1.

Flow cytometry

Flow cytometry analysis of the interaction of fluorescent D-Lipidots with cells was performed using a 9 colors FACS BD LSR2 equipped with lasers emitting at 488 and 633 nm (BD, Franklin Lakes, USA). CAL-33 cells were seeded at a density of 10^5 cells per well in 12 well plates and incubated for 24 h. D-Lipidots with a diameter of 50 nm were incubated at the corresponding concentration of 1 μM DiD in presence of cell monolayers for 2, 3 or 6 h in complete cell culture medium. Thereafter, cells were rinsed with PBS (×2), harvested by the addition of trypsin followed by a centrifugation, and then fixed with 2 % FA before flow cytometry analysis. 10,000 to 20,000 events were recorded. The data from fluorescence measurements at an emission wavelength of 660 nm for DiD were analyzed using DIVA v8.1 software (BD) by using the overlay option.

Statistical evaluation and graphical modelling

Two-way ANOVA of cell toxicity and phototoxicity data was analyzed from at least two independent experiments and five replicates per condition. Means

are plotted ± standard deviations. Statistics and graphical plots were established and analyzed with GraphPad Prism software (Graphpad Software, La Jolla, USA).

Results

Nanoparticle preparation

To investigate the effect of particle size and PS payload on transport and delivery, two series of nanoparticles were prepared with two different payloads. For the 50 nm nanoparticles, mTHPC was incorporated with a content of 920 molecules/particle whereas for 120 nm particles, the amount of mTHPC was estimated at 4600 molecules/particle. Therefore one 120 nm nanoparticle contains fivefold more molecules of mTHPC than one 50 nm nanoparticle. Expressed in equivalent mTHPC concentration (3.67, 7.34 and 14.69 μM) the solution of 50 nm nanoparticles contains fivefold more particles than the solution of 120 nm particles. As observed in our preliminary study [15], mTHPC was efficiently encapsulated into lipid nanoparticles without affecting neither the colloidal properties of the carrier nor photophysical properties of the loaded PS. Indeed, an aggregation of mTHPC inside the lipid core of nanoparticles can be observed only for 50 nm particle at high payload (>4 %w/w total lipid, data not published). Estimated from the whole excipients initially incorporated in the Lipidot formulation, mTHPC was loaded in our study at 2.8 and 1.0 % w/w for 50 and 120 nm particles, respectively (Table 1).

Particle size and size distribution of lipid nanoparticles

Dynamic light scattering (DLS) technique was used to determine the particle hydrodynamic diameter (in nm), particle size distribution (expressed by polydispersity index PDI) using Zetasizer Nano ZS (Malvern Instruments, France). At least three different nanoparticle batches (lipid dispersed phase weight fraction: 10 %) are measured per condition. Data were expressed as mean ± standard deviation of three independent measurements performed at 25 °C (Table 1).

Lipidot size drives uptake kinetics in CAL-33 cells

Using confocal laser scanning microscopy and CAL-33 monolayers and spheroids, uptake of 50 and 120 nm M-Lipidots was investigated over time and compared to free mTHPC (Fig. 1). In CAL-33 monolayer cultures, fluorescence of free mTHPC could be readily detected after 2 h of incubation as a diffuse signal throughout the cytoplasm, sparing the nucleus. In contrast, no fluorescence from our nanoparticle formulations was apparent at this time point. Only after 6 h both sizes of M-Lipidots were visible with the same distribution pattern as free mTHPC, however, the fluorescence was markedly weaker with 120 nm M-Lipidots compared

Table 1 Physicochemical characterization of Lipidots

	Lipid (mg/mL)	Number of particles/mL	MTHPC molecules/particle	mTHPC (μg/mL)	Drug loading[a]	Hydrodynamic diameter (nm)	Poly-dispersity index
50 nm M-Lipidot	50	7.27565×10^{14}	~920	722	2.8 %	47.7 ± 1.1	0.153 ± 0.01
120 nm M-Lipidot	50	5.26306×10^{13}	~4600	262	1.0 %	111.2 ± 2.2	0.103 ± 0.01
50 nm Lipidot[b]	50	7.27565×10^{14}	–	–	–	49.5 ± 1.5	0.170 ± 0.07
120 nm Lipidot[b]	50	5.26306×10^{13}	–	–	–	95.4 ± 3.4	0.120 ± 0.05

Data with standard deviation

[a] Expressed w/w of total lipids (included in nanoparticle formulation)

[b] Empty Lipidots

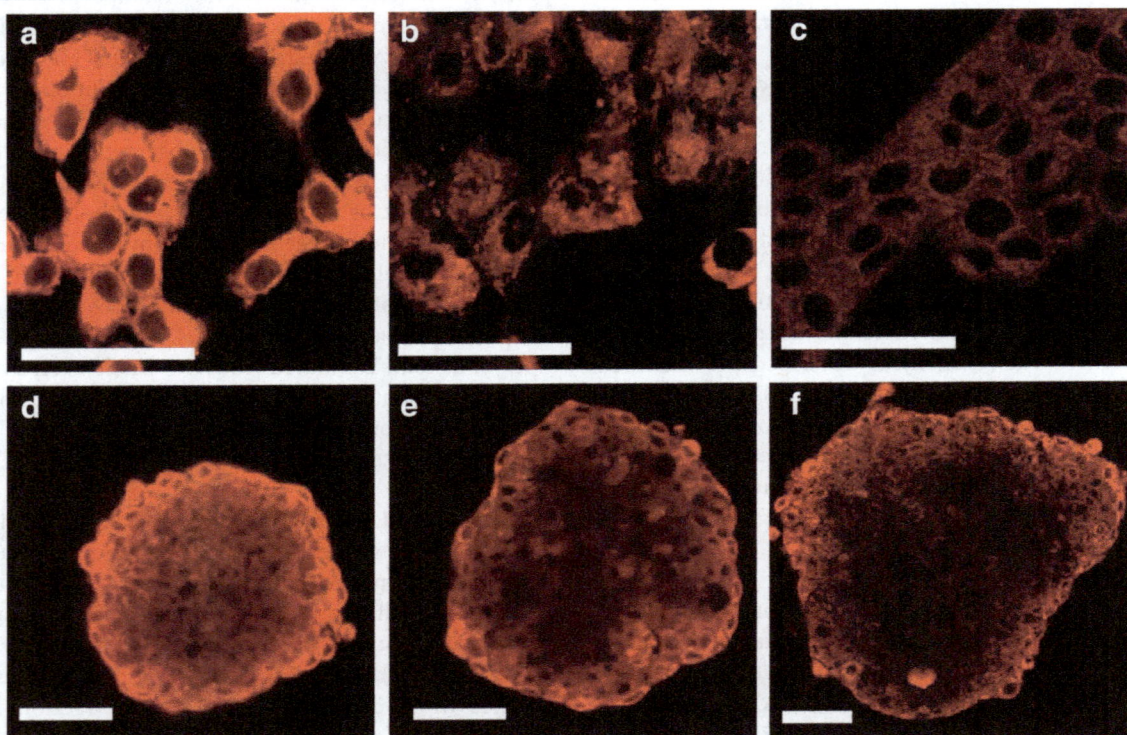

Fig. 1 Confocal laser scanning microscopy images of CAL-33 cells incubated for 28 h with free mTHPC (**a**, **d**), 50 nm M-Lipidots (**b**, **e**) and 120 nm M-Lipidots (**c**, **f**) in monolayers (**a–c**) and spheroids (**d–f**). Concentration for all treatments: 7.34 μM mTHPC. *Scale bar* 50 μm

to 50 nm M-Lipidots. The intracellular distribution pattern stayed similar until 28 h but fluorescence accumulated over time for all formulations (Fig. 1a–c).

To obtain further information with regard to uptake kinetics, flow cytometry was used to measure in CAL-33 the fluorescence of 50 nm D-Lipidots over time (Fig. 2). These 50 nm D-Lipidots show the same accumulation behavior as 50 nm M-Lipidots (Fig. 2a), but are better suited for flow cytometry applications. Data confirmed microscopic observations in CAL-33 cells, showing an increase of fluorescence intensity after 6 h of incubation as compared to earlier time points (Fig. 2b).

To better predict the in vivo behavior, uptake was then investigated in CAL-33 spheroids (Fig. 1d–f). In this 3D model of an avascular mini tumor, free mTHPC accumulated in the outer cell layer at about the same time as in monolayer cells (2 h), however, it took up to 6 h until the PS was penetrating further into the spheroid. Eventually it reached the spheroid core at 24 h with a modest overall fluorescence increase until 28 h. At these late time points, fluorescence signals showed a homogeneous distribution within the spheroid. The weaker fluorescent signals of 50 nm M-Lipidots were apparent in the outer cell layers after 4 h, and continued to penetrate slowly deeper into

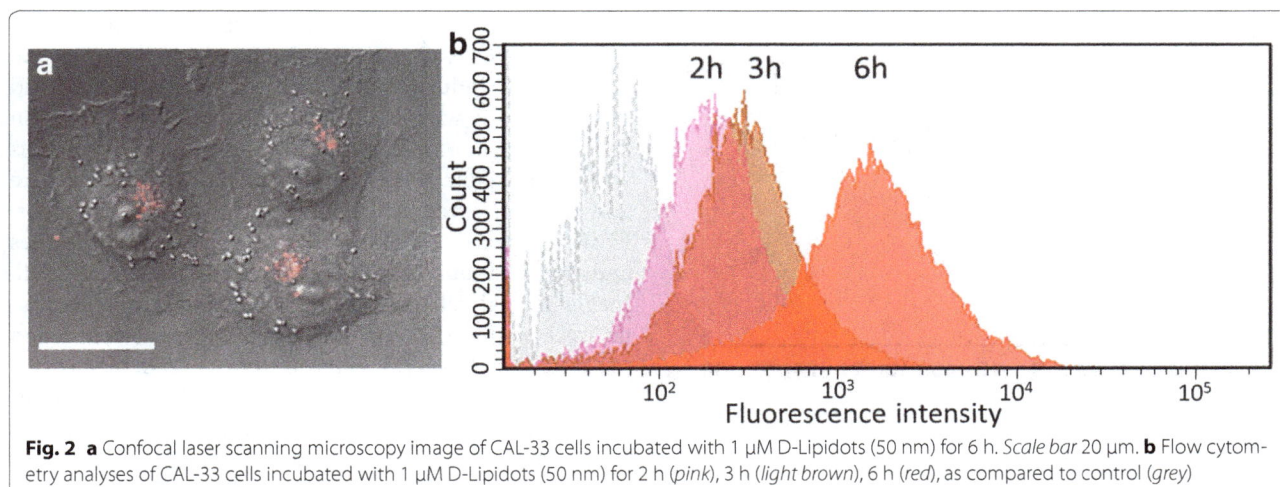

Fig. 2 **a** Confocal laser scanning microscopy image of CAL-33 cells incubated with 1 μM D-Lipidots (50 nm) for 6 h. *Scale bar* 20 μm. **b** Flow cytometry analyses of CAL-33 cells incubated with 1 μM D-Lipidots (50 nm) for 2 h (*pink*), 3 h (*light brown*), 6 h (*red*), as compared to control (*grey*)

the spheroid center. At 28 h the core was fluorescent, but the signal displayed a more punctuate and less homogeneous pattern. Compared to 50 nm M-Lipidots, penetration of 120 nm M-Lipidots was retarded, most of which did not reach the center even at 28 h as evidenced by a less fluorescent spheroid core.

Semiquantitative analyses of microscopy data confirmed that time dependent uptake curves were different between free mTHPC and M-Lipidots in the spheroid model (Fig. 3). Free mTHPC was taken up in a nonlinear, asymptotical way with high initial uptake rates and quickly decreasing rates over time whereas 120 nm M-Lipidots were taken up by the spheroid in an almost linear fashion during the whole time of the experiment at a very low initial uptake rate. The uptake curve of the 50 nm M-Lipidots presents an uptake in a nonlinear way but at a lower initial uptake rate as free mTHPC. Based on uptake studies, further studies were therefore performed after a 24 h exposure to the compounds.

The nanoformulations are less cytotoxic than the free substance at high drug concentrations

To obtain information about a possible cytotoxicity of our nanocarriers, we first tested empty Lipidots by means of an ATP luciferase viability assay that measures cell viability in CAL-33 spheroids (Fig. 4a). A comparison revealed that both 50 and 120 nm Lipidots are well tolerated for concentrations of particles corresponding to the equivalent mTHPC concentration from 0 to 14.69 μM ($\hat{=}$69.3–692.9 μg/mL lipid [50 nm]; 190.7 μg/mL–1.90 mg/mL lipid [120 nm]), with the smaller particles being slightly superior (p < 0.01). While the 50 nm particles did not exhibit any toxicity at the tested concentrations the 120 nm particles reduced viability by 10 %. As a next step, cytotoxic effects of PS-loaded M-Lipidots

were compared to free mTHPC in CAL-33 spheroids (Fig. 4b). While free mTHPC showed a clear toxicity (68 % viability) in the dark at the highest concentration tested (14.69 μM), encapsulation of mTHPC into Lipidots resulted in a significantly reduced dark toxic effect (78 % viability with the 50 nm Lipidots; 86 % viability with the 120 nm Lipidots, p < 0.001).

The 50 nm M-Lipidots show high photodynamic potency similar to free mTHPC

The PDT effects mediated by M-Lipidots or free mTHPC were investigated in CAL-33 spheroids (Figs. 5, 6). Our microscopic analyses showed that PDT with both free mTHPC and 50 nm M-Lipidots induced a pronounced and comparable destruction of the spheroids (Fig. 5). Although the size reduction was difficult to microscopically measure under conditions of high destruction, the results correlated with the respective ATP luciferase viability assays (Fig. 6b). The 50 nm Lipidots as well as free mTHPC reduced spheroid sizes by 100 % at higher concentrations (p < 0.001). However, after PDT with 120 nm M-Lipidots, even at the highest concentration (14.69 μM), only mild phototoxic effects were visible with size reductions by only 34 % (Figs. 5, 6a, p < 0.001). These limited PDT effects of 120 nm M-Lipidots could also be confirmed by ATP luciferase viability assays (Fig. 6b). Viability after PDT with the highest concentration (14.69 μM) was 1.8 % with mTHPC, 6.6 % with the 50 nm particles and 66.2 % with the 120 nm particles (p < 0.001).

Free mTHPC causes apoptosis and necrosis while 50 nm M-Lipidots cause mostly apoptosis

By "FLICA" apoptosis assays high pan-caspase activity was detected in CAL-33 spheroids after PDT with 50 nm M-Lipidots (Fig. 7c) and, to a lesser extent, after

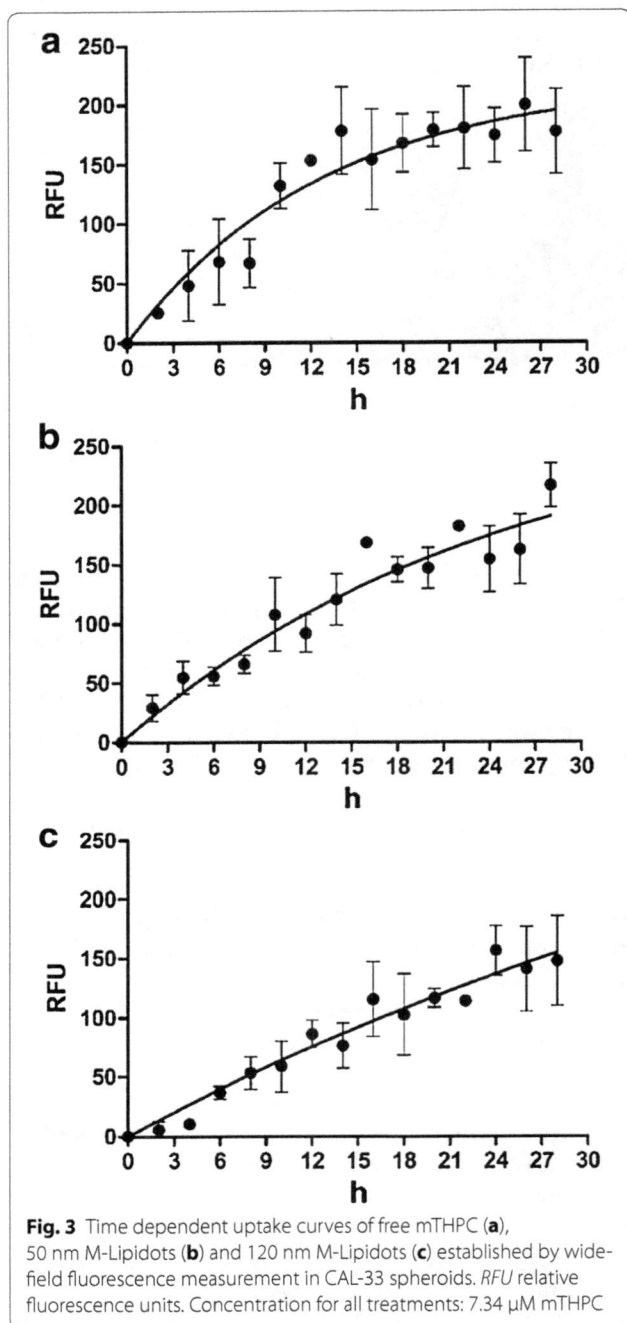

Fig. 3 Time dependent uptake curves of free mTHPC (**a**), 50 nm M-Lipidots (**b**) and 120 nm M-Lipidots (**c**) established by wide-field fluorescence measurement in CAL-33 spheroids. *RFU* relative fluorescence units. Concentration for all treatments: 7.34 µM mTHPC

controls showed intact spheroid structures and most cells displayed well preserved cell organelles (Fig. 8a, d). MTHPC-induced PDT seemed to disrupt spheroid structure as a whole, causing cells to die either in an apoptotic or in a necrotic manner (Fig. 8b, e). Apoptosis was recognizable by the condensed chromatin structure and well preserved cell membranes of some dying cells. However, necrotic features like destroyed cell organelles and membranous cellular debris were present as well. Inside several cells inclusion bodies with grainy deposits were visible that may be aggregated and contrasted mTHPC (Fig. 8g). PDT with 50 nm M-Lipidots was primarily damaging the spheroid center leaving an outer rim of cells intact under these conditions (Fig. 8c). In the spheroid center cells were primarily showing features of apoptotic cell death, as described above (Fig. 8f). Additionally, in the outer cell layer, close to the cytoplasmic membrane, vesicles with enclosed sphere-like structures of about 50 nm were present that may represent M-Lipidots (Fig. 8h).

Lipidot-PDT affects similar pathways as mTHPC-PDT

To further explore possible differences between mTHPC- and 50 nm M-Lipidot-mediated PDT, we analyzed the expression of 84 known cancer drug target genes by means of qRT-PCR (Fig. 9). Compared to the untreated control, no gross differences in overall expression patterns could be discovered after PDT, since the same 33 genes were upregulated after both regimes. However, the upregulation was generally stronger after mTHPC-PDT. This was e.g. obvious for the expression of PTGS2, TXNRD1, AKT1, NFKB1, EGFR, PIK3C3, NRAS, PLK2, PLK3, RHOB, and HSP90AA1, where a more than two-fold higher upregulation was found after mTHPC-PDT compared to M-Lipidot-PDT. However, it should be noted that the same pathways were affected in the same direction (only upregulation, no downregulation) after both PDT regimes. Among others, we detected signs for abnormal regulation of KRAS and NRAS and an increase of transcription factors ATF2, HIF1A, NFKB1, TP53 despite of the upregulation of histone deacetylases HDAC1, HDAC2 and HDAC4. Genes that were not expressed and/or unaltered after both PDT regimes are summarized in Additional file 1: Table S1.

Discussion

The powerful PS mTHPC is approved in several European countries for palliative PDT of patients with advanced head and neck cancer. However, mTHPC formulations that e.g. improve solubility of this highly hydrophobic drug, reduce its dark toxicity, enhance its intratumoral accumulation and/or increase PDT efficacy would be beneficial for systemic clinical applications [18].

treatment with free mTHPC and irradiation (Fig. 7b). Very low caspase staining occurred after PDT with 120 nm M-Lipidots (Fig. 7d) which was barely more intense than staining of control spheroids (Fig. 7a).

An investigation of CAL-33 spheroids at the ultrastructural level with electron microscopy confirmed different modes of cell death as observed after PDT with mTHPC or 50 nm M-Lipidots (Fig. 8). Untreated

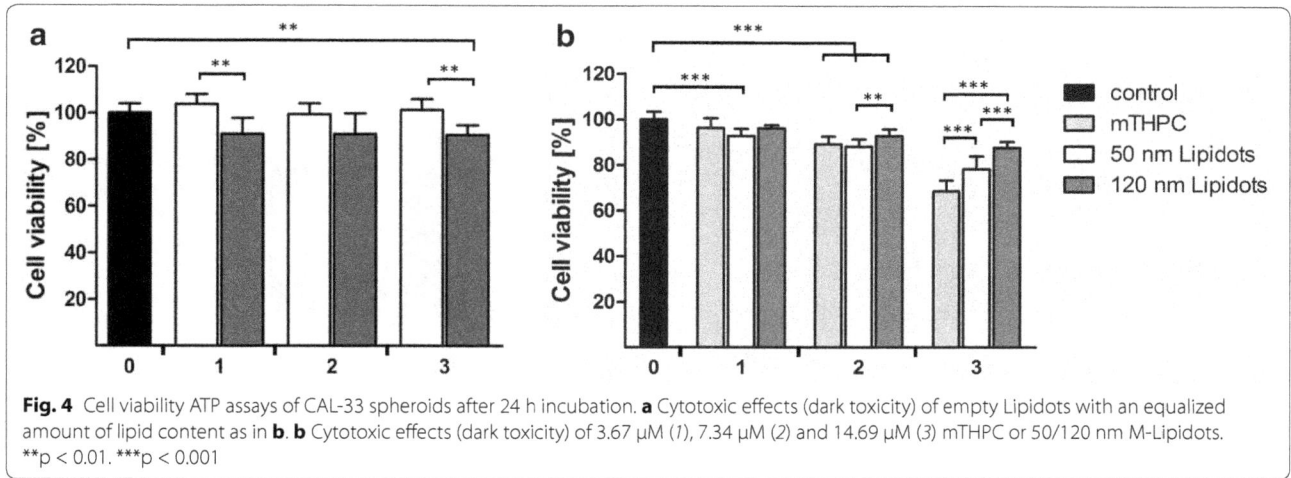

Fig. 4 Cell viability ATP assays of CAL-33 spheroids after 24 h incubation. **a** Cytotoxic effects (dark toxicity) of empty Lipidots with an equalized amount of lipid content as in **b**. **b** Cytotoxic effects (dark toxicity) of 3.67 μM (*1*), 7.34 μM (*2*) and 14.69 μM (*3*) mTHPC or 50/120 nm M-Lipidots. **p < 0.01. ***p < 0.001

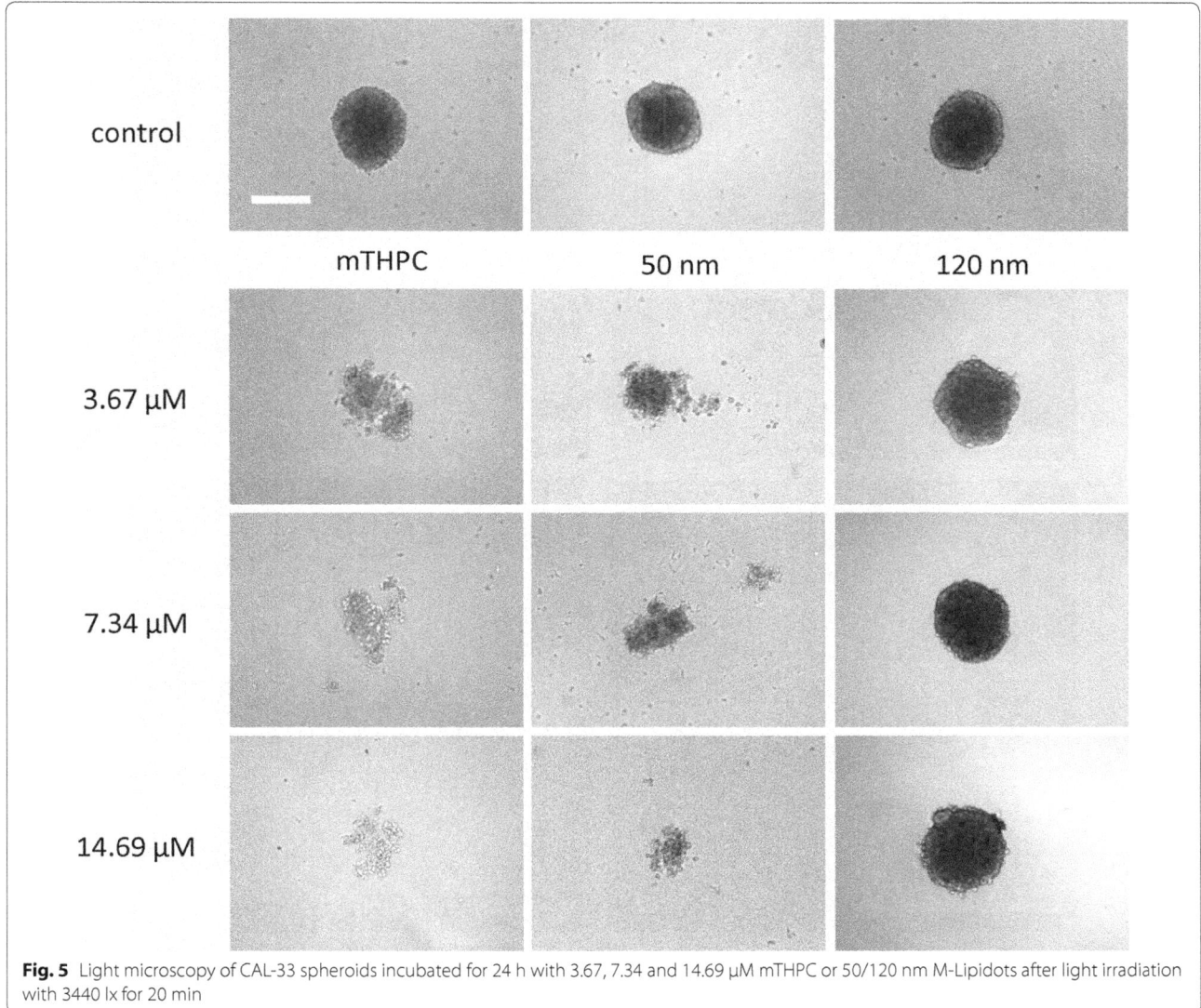

Fig. 5 Light microscopy of CAL-33 spheroids incubated for 24 h with 3.67, 7.34 and 14.69 μM mTHPC or 50/120 nm M-Lipidots after light irradiation with 3440 lx for 20 min

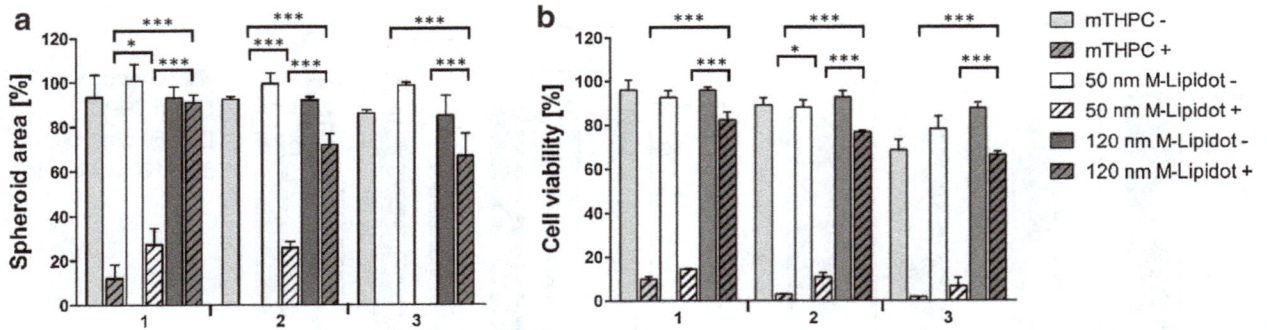

Fig. 6 **a** Light microscopic measurements of spheroid areas of CAL-33 spheroids incubated with 3.67 μM (*1*), 7.34 μM (*2*) and 14.69 μM (*3*) mTHPC or 50/120 nm M-Lipidots with (+) and without (−) light irradiation with 3440 lx for 20 min. **b** Cell viability ATP assays of CAL-33 spheroids incubated at the same conditions as in **a**. *p < 0.05. ***p < 0.001

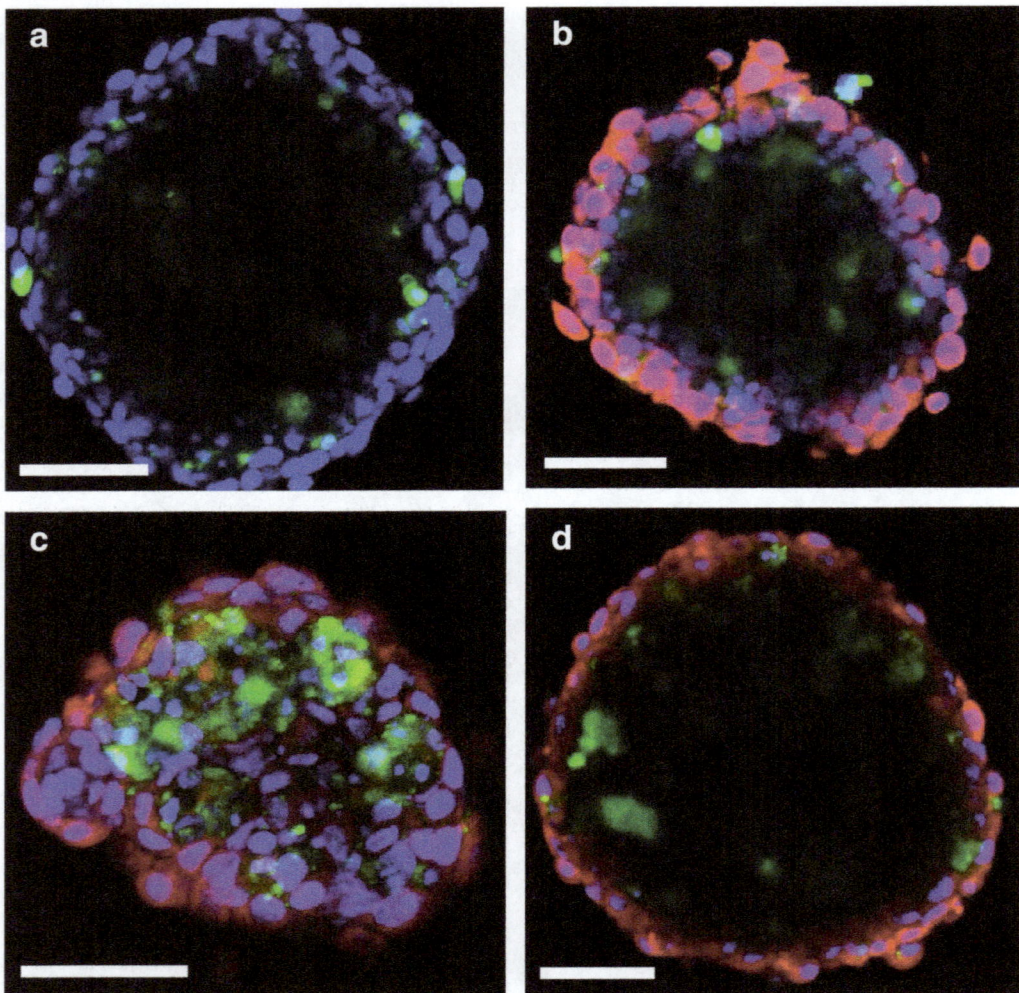

Fig. 7 Confocal laser scanning microscopy images of the fluorescent labeled inhibitor of caspases (FLICA) apoptosis assay after irradiation of CAL-33 spheroids with 3440 lx for 1 min. FLICA: *green*, Hoechst 33342 nuclear stain: *blue*, mTHPC (*red*). Untreated control (**a**) and incubations with mTHPC (**b**), 50 nm M-Lipidots (**c**) or 120 nm M-Lipidots (**d**). Concentration for all treatments: 3.67 μM mTHPC. Incubation time 24 h. *Scale bar* 50 μm

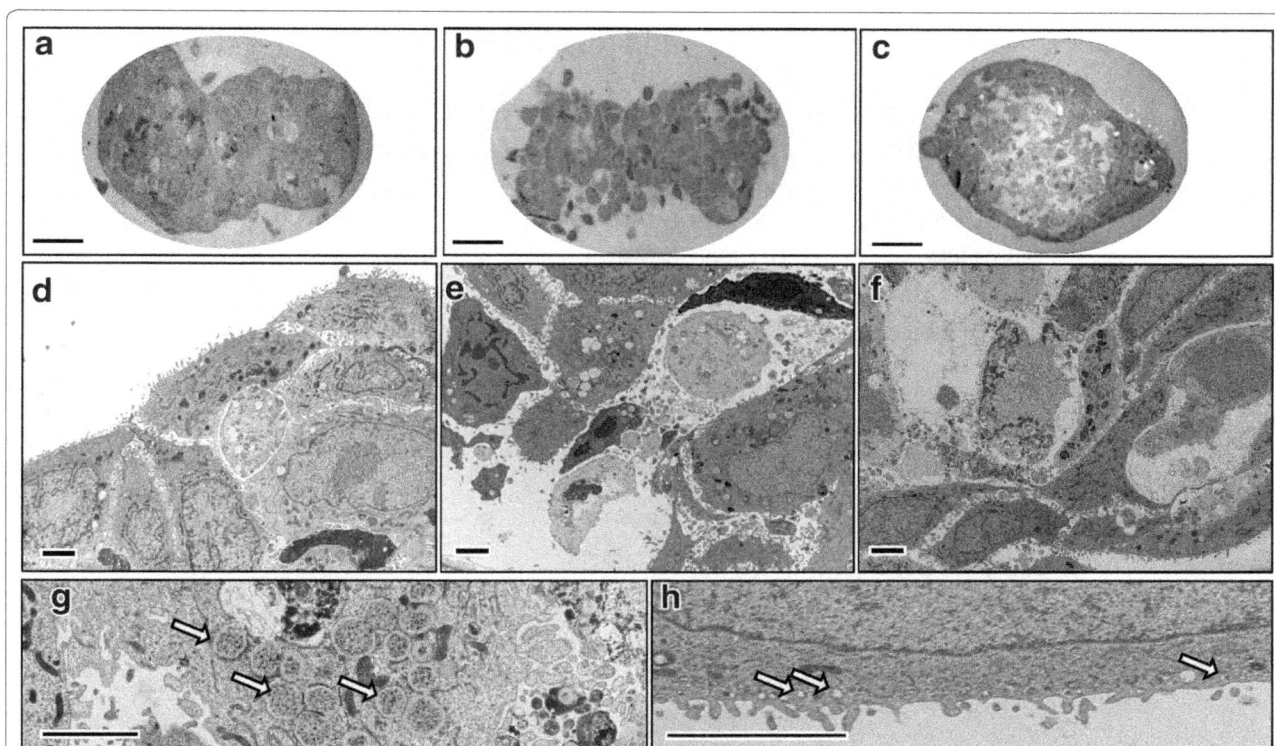

Fig. 8 Transmission electron microscopy images of CAL-33 spheroids. Untreated control (**a**, **d**) and incubations with mTHPC (**b**, **e**, **g**) or 50 nm M-Lipidots (**c**, **f**, **h**). *Arrows* (**g**) vesicles with precipitate. *Arrows* (**h**) engulfed Lipidots. Concentration 3.67 μM mTHPC. Incubation time 24 h. Irradiation 1 min at 3440 lx. *Scale bar* (**a–c**) 50 μm. *Scale bar* (**d–h**) 2.5 μm

Recently, we introduced solid lipid nanoparticles as stable, easy to produce and efficient carriers for mTHPC [15]. However, while physico-chemical and photophysical evaluations indicated their excellent suitability for PDT, only scarce information is available yet with regard to their behavior in biological systems. In the present study we have therefore chosen an advanced in vitro cancer spheroid model to investigate for the first time PDT effects of these particles (called M-Lipidots) at the cellular level and compare it to effects of free mTHPC. Cancer spheroids are multicellular 3D grown minitumors that display features which better mimic the biology of solid tumors than standard monolayer cultures, among others in terms of intercellular contacts, matrix deposition, physiological barriers, cellular inhomogeneity or proliferation properties [19]. Also with regard to ROS diffusion and PS penetration a 3D environment may be advantageous. Spheroids have thus been proposed not only as superior predictive platforms for testing of drugs but also of drug delivery systems [20].

Since diameters of Lipidots can be reliably adjusted between 30 and 120 nm by varying wax, oil and surfactant content [15], we have here included two exemplary sizes of mTHPC-Lipidots, namely 50 and 120 nm.

In both, monolayer cultures (that served as a reference) and spheroids, we found that free mTHPC was taken up in a shorter time frame compared to mTHPC encapsulated into Lipidots. The quicker and higher accumulation of free mTHPC may be explained by the fact that lipophilic PSs can bind to serum proteins and uptake can be mediated by low lipid density protein receptors, which is considered an efficient mechanism [18]. For in vivo applications this slower accumulation of M-Lipidots must of course be considered but may be outweighed by advantages of Lipidot's PEG chains that offer a stealth mechanism to avoid fast recognition by the immune system [21].

Our experiments further indicated favored uptake and superior spheroid penetration properties of the 50 nm M-Lipidots over the 120 nm M-Lipidots. These results are in accordance with most literature reports for other nanocomposites which suggest a size dependency of the uptake behavior and smaller diameters being more readily internalized by cells in monolayers [22]. There are fewer studies investigating size dependent penetration of nanoparticles into spheroids, however, in a work with gold nanocomposites the authors also reported a superior uptake of smaller 50 nm particles over larger 100 nm

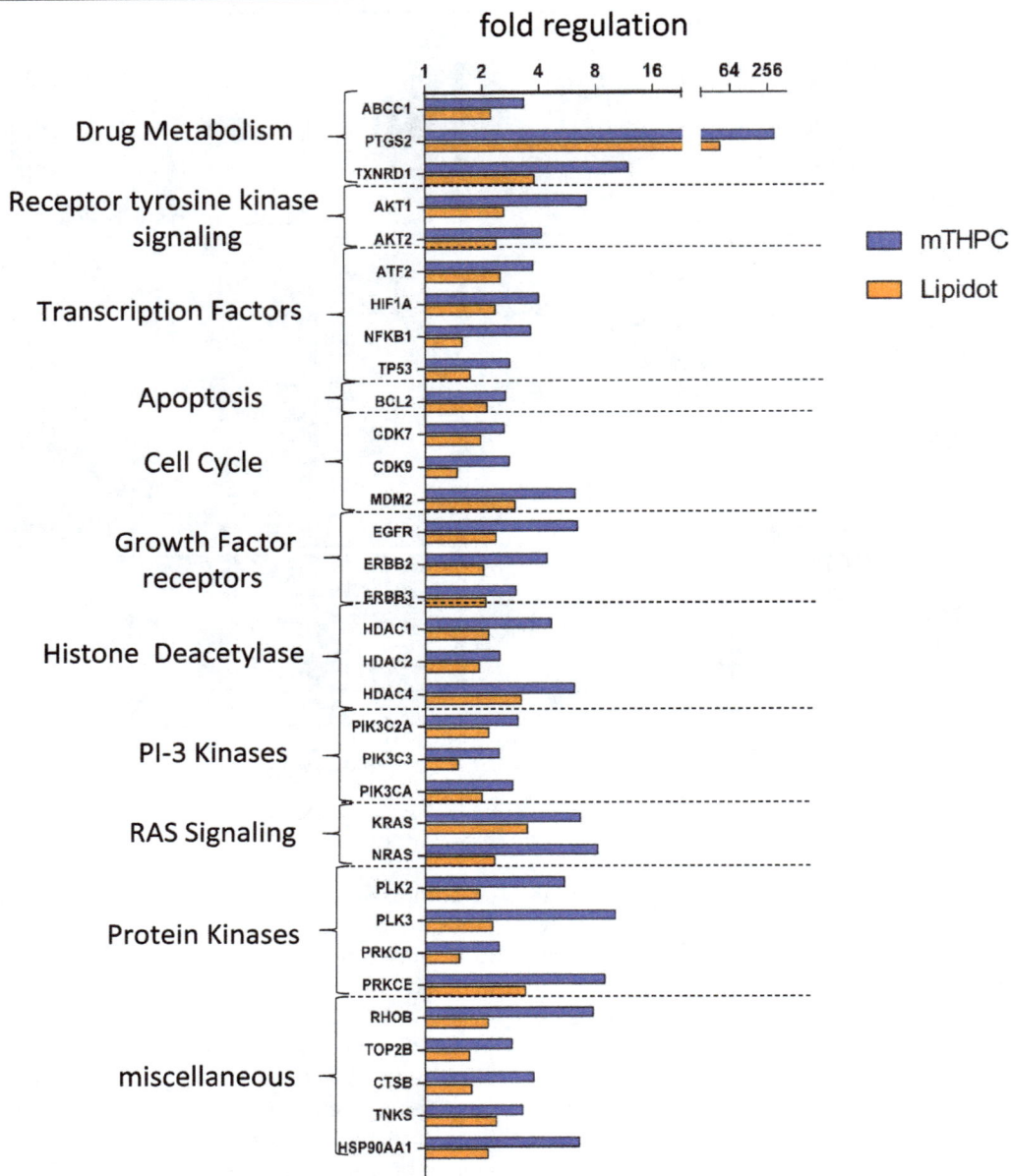

Fig. 9 Fold expression change in spheroids after 24 h incubation with 3.67 µM mTHPC or 50 nm M-Lipidots and light illumination from 2.5 cm above with white light for 1 min at 3440 lx. Gene expression data was normalized against an untreated control and the RPLP0 house keeping gene. *CFTR/MRP* ATP-Binding Cassette, Sub-Family C, *ABCC1* Member 1, *PTGS2* Prostaglandin-Endoperoxide Synthase 2, *TXNRD1* Thioredoxin Reductase 1, *AKT1* V-Akt Murine Thymoma Viral Oncogene Homolog 1, *AKT2* V-Akt Murine Thymoma Viral Oncogene Homolog 2, *ATF2* Activating Transcription Factor 2, *HIF1A* Hypoxia Inducible Factor 1, Alpha Subunit, *NFKB1* Nuclear Factor Of Kappa Light Polypeptide Gene Enhancer In B-Cells 1, *TP53* Tumor Protein P53, *BCL2* B Cell CLL/Lymphoma 2, *CDK7* Cyclin-Dependent Kinase 7, *CDK9* Cyclin-Dependent Kinase 9, *MDM2* MDM2 Proto-Oncogene, E3, *EGFR* Epidermal Growth Factor Receptor, *ERBB2* Erb-B2 Receptor Tyrosine Kinase 2, *ERBB3* Erb-B2 Receptor Tyrosine Kinase 3, *HDAC1* Histone Deacetylase 1, *HDAC2* Histone Deacetylase 2, *HDAC4* Histone Deacetylase 4, *PIK3C2A* Phosphatidylinositol-4-Phosphate 3-Kinase, Catalytic Subunit Type 2 Alpha, *PIK3C3* Phosphatidylinositol 3-Kinase, Catalytic Subunit Type 3, *PIK3CA* Phosphatidylinositol-4,5-Bisphosphate 3-Kinase, Catalytic Subunit Alpha, *KRAS* Kirsten Rat Sarcoma Viral Oncogene Homolog, *V-Ras* Neuroblastoma RAS Viral, *NRAS* Oncogene Homolog, *PLK2* Polo-Like Kinase 2, *PLK3* Polo-Like Kinase 3, *PRKCD* Protein Kinase C, Delta, *PRKCE* Protein Kinase C, Epsilon, *RHOB* Ras Homolog Family Member B, *TOP2B* Topoisomerase (DNA) II Beta 180 kDa, *CTSB* Cathepsin B, *TNKS* Tankyrase, and Heat Shock Protein 90 kDa Alpha (Cytosolic), Class A Member 1 (HSP90AA1)

ones [23]. However, we cannot exclude that the stronger fluorescence signal observed after incubation with 50 nm M-Lipidots may also be due to the fact that five-fold more particles were present in the working solution of 50 nm M-Lipidots compared to 120 nm M-Lipidots. This is related to manufacturing processes and the aim to reach equivalent mTHPC concentrations with both M-Lipidot sizes. Furthermore, fluorescence with the nanoparticles was markedly weaker when compared to free mTHPC which is why we cannot exclude that quenching effects occur in the presence of cells.

For the following PDT experiments, we used a white light source rather than a laser to activate the PS. In a previous study we have shown that this is perfectly feasible and may be advantageous to detect (subtle) differences between effects of treatment regimes [24]. With the aim to preserve some morphology and avoid complete RNA degradation [25, 26] for our microscopic and RNA studies, we furthermore had to reduce the illumination time from 20 to 1 min. We observed a strong and comparable light-induced destruction of spheroids exposed to free mTHPC or 50 nm M-Lipidots. This similar PDT efficiency was despite our observation of a slightly different microscopic fluorescence distribution pattern within the spheroid of free mTHPC and M-Lipidots, respectively. The observed PDT effects complemented our previous study, where we have shown in a cell-free environment that 30, 50 and 100 nm mTHPC-Lipidots are capable of producing high quantum yields after illumination and that singlet oxygen may diffuse through the Lipidot shell to the surrounding [15]. As predicted because of their observed delayed and weaker cellular uptake, 120 nm M-Lipidots caused almost no PDT effects under the applied mild activation conditions. While effects may be improved with stronger illumination regimes, we have shown previously that ROS diffusion from larger Lipidots is anyway worse than from smaller ones [15].

From EM studies and the apoptosis assay, it was evident that spheroid centers were more damaged after PDT with the 50 nm Lipidots, although fluorescence accumulation was highest at the spheroid periphery. We propose that a decreasing nutrient gradient towards the spheroid center may render those cells more susceptible to PDT, and therefore also low PS doses will be sufficient to kill them.

Although PDT with both the free PS as well as the 50 nm M-Lipidots efficiently destroyed spheroids, underlying mechanisms turned out to feature differences under our experimental conditions, i.e. necrosis and apoptosis with mTHPC-PDT, and apoptosis with 50 nm M-Lipidot-PDT. The reasons for that are not clear yet. While it is well known that the subcellular localization of a PS governs PDT cell death pathways [27], we found similar cytoplasmic fluorescence patterns of mTHPC with both formulations. However, necrotic mechanisms have been reported to occur with stronger cellular photodamage [27]. Since light doses were the same, it may therefore be speculated that under the same conditions treatment with M-Lipidots initiated slightly milder PDT effects than free mTHPC. Whether this is a consequence of quantitative PS uptake, exact intracellular distribution or the nanocarrier has to be investigated.

Stronger photodamage after PDT with free mTHPC may also be concluded from our RNA expression studies where we detected always a more pronounced gene regulation. For several genes a more than twofold higher upregulation was found after mTHPC-PDT compared to M-Lipidot-PDT. As the same pathways were affected in the same direction (only upregulation, no downregulation) after both PDT regimes, it indicates common mechanisms of free and Lipidot-encapsulated mTHPC. The changed expression patterns reflect the cell's complex acute responses to (oxidative) stress due to our PDT regimes. Many of the upregulated genes may have dual roles for apoptosis or anti-apoptosis and it is not clear yet whether we observe the cell's efforts to initiate rescue mechanisms or the beginning of cell death. Apparently, many different pathways are dysregulated in parallel. Among others, we detected signs for abnormal regulation of the RAS signalling pathway, chromatin remodeling or an increase of transcription factor RNA despite of the upregulation of histone deacetylases.

In accordance with our previous studies with 30, 50 and 100 nm particles in MCF-7 monolayer cultures [14] the biocompatibility of empty 50 and 120 nm Lipidots could be here confirmed for CAL-33 spheroids. The observed slightly higher cytotoxicity of 120 nm Lipidots may be caused by their increased lipid concentration compared to 50 nm Lipidots, as observed before [14]. However, this difference may not be biologically relevant, leaving more than 90 % of cells vital.

Dark toxicity of PSs is an important issue for clinical PDT applications that may cause detrimental effects on healthy cells. This also applies to the strong PS mTHPC for which cytotoxic effects without light activation are well known. In our spheroid model we could demonstrate that the encapsulation of mTHPC into Lipidots significantly reduced unwanted dark toxicity of this PS at high concentrations. However, we cannot exclude that the lower toxicity is at least partially due to a lower uptake of mTHPC into the cells. Still, considering the outstanding biocompatibility of our carrier it may allow for systemic applications of higher doses of mTHPC for improved PDT without the risk for light-independent effects in patients.

Various different approaches have been proposed in the past, including the development of e.g. liposomal mTHPC formulations [28, 29] or the encapsulation of mTHPC into nanoparticles composed of poly(lactic-co-glycolic acid) [30], poly(lactic-co-glycolic acid)-b-poly(ethylene glycol) [31], poly(ethylene glycol) methacrylate-co-2-(diisopropylamino)ethyl methacrylate copolymers [32], human serum albumin [33], organic-modified silica [34] or calcium phosphate. [35] These studies describe promising carriers for mTHPC by improving solubility and reducing dark toxicity however it is not possible to directly compare them as very different model systems were used in each case. Furthermore, nanotoxicology will be very different depending on the materials used in the formulation and can differ greatly between in vitro and in vivo studies.

The 50 nm Lipidots display several favorable characteristics with regard to in vivo applications. Concerning size Tang et al. [36] e.g. could show in vivo that tumor permeation and retention of 50 nm silica particles (the EPR effect) was superior to smaller 20 nm ones and larger 200 nm ones. Furthermore, in two of our former in vivo studies with Lipidots as carrier for indocyanine green we could report on high chemical stability of the particles of over 6 months and a prolonged tumor labelling of over 1 day [7, 37]. Additionally, Lipidots displayed good long-term plasma stability and tolerability with low hemolytic activity [7, 37].

Conclusions

In conclusion, in an advanced 3D cell culture model, 50 nm Lipidots have presented themselves as nontoxic nanocarriers for hydrophobic photosensitizers such as mTHPC that preserve its functionality in PDT. Lipidots are not only fully biocompatible and easy to produce, but may solve two important problems of mTHPC that currently prevent a more widespread clinical use of this efficient PS by rendering it water soluble and reducing its dark toxicity. The slightly milder PDT effects with M-Lipidots may be beneficial in certain clinical settings, e.g. where an apoptotic cell death (without inflammation) is clinically preferred, such as for tumor ablation.

Abbreviations

PDT: photodynamic therapy; PS: photosensitizer; FDA: (US) food and drug administration; MTHPC: m-tetrahydroxyphenylchlorin; ATP: adenosine triphosphate; EMA: European Medicines Agency; PEG: polyethylene glycol; ROS: reactive oxygen species; DiD: 1,1'-dioctadecyl-3,3,3',3'-tetramethylindodicarbocyanine perchlorate; PBS: phosphate buffered saline; FA: formaldehyde; ROI: region of interest; CDNA: complementary DNA; FLICA: fluorochrome-labeled inhibitor of caspases; M-Lipidots: mTHPC loaded Lipidots; D-Lipidots: dye loaded Lipidots; QRT-PCR: quantitative reverse transcription polymerase chain reaction.

Authors' contributions

DH carried out the culturing of cells, spheroid cell culture, cell toxicity studies, confocal microscopy, uptake studies, phototoxicity studies, apoptosis assay, PCR, preparation of samples for electron microscopy, all data analysis and preparation of the manuscript. FN cultured cells and did flow cytometry analysis, he also helped to improve the manuscript. JST and FM prepared the nanoparticles for the studies and did also the physicochemical characterization under supervision of ACC. ACC also helped to improve the manuscript. AK analyzed the samples for electron microscopy and took images, edited images and helped to improve the sections in the manuscript on microscopy. CM supervised DH and significantly contributed to the writing and adaption of the manuscript. All authors read and approved the final manuscript.

Author details

[1] Institute of Anatomy, University of Zurich, Winterthurerstrasse 190, 8057 Zurich, Switzerland. [2] Technologies for Biology and Healthcare Division, CEA, LETI, MINATEC Campus, Commissariat à l'Énergie Atomique et aux Énergies Alternatives (CEA), 38054 Grenoble, France. [3] Université Grenoble Alpes, 38000 Grenoble, France. [4] Center for Microscopy and Image Analysis, University of Zurich, Winterthurerstrasse 190, 8057 Zurich, Switzerland.

Acknowledgements

We are thankful to the Center for Microscopy and Image Analysis of the University of Zurich (especially to Gery Barmettler, Bruno Guhl and Ursula Lüthi) for their help with electron microscopy. We also want to thank Biolitec Research, Jena, Germany for kindly providing mTHPC for our study. The study was supported by the FP7 ERA-net EuroNanoMed project TARGET-PDT (31NM30-131004/1).

Competing interests

The authors declare that they have no competing interests. Patent: Goutayer M, Navarro F, Robert V, Texier I, Encapsulation of lipophilic or amphiphilic therapeutic agents into nano-emulsions, Granted on 14/08/2008, WO2010018222.

References

1. Samad A, Sultana Y, Aqil M. Liposomal drug delivery systems: an update review. Curr Drug Deliv. 2007;4:297–305.
2. Masood F. Polymeric nanoparticles for targeted drug delivery system for cancer therapy. Mater Sci Eng C Mater Biol Appl. 2016;60:569–78.
3. Sharma H, Mishra PK, Talegaonkar S, Vaidya B. Metal nanoparticles: a theranostic nanotool against cancer. Drug Discov Today. 2015;20:1143–51.
4. Naseri N, Valizadeh H, Zakeri-Milani P. Solid lipid nanoparticles and nanostructured lipid carriers: structure. Preparation and application. Adv Pharm Bull. 2015;5:305–13.
5. Eifler AC, Thaxton CS. Nanoparticle therapeutics: FDA approval, clinical trials, regulatory pathways, and case study. Methods Mol Biol. 2011;726:325–38.
6. Delmas T, Piraux H, Couffin AC, Texier I, Vinet F, Poulin P, et al. How to prepare and stabilize very small nanoemulsions. Langmuir. 2011;27:1683–92.
7. Navarro FP, Mittler F, Berger M, Josserand V, Gravier J, Vinet F, et al. Cell tolerability and biodistribution in mice of indocyanine green-loaded lipid nanoparticles. J Biomed Nanotechnol. 2012;8:594–604.
8. Texier I, Goutayer M, Da Silva A, Guyon L, Djaker N, Josserand V, et al. Cyanine-loaded lipid nanoparticles for improved in vivo fluorescence imaging. J Biomed Opt. 2009;14:54005–11.
9. Delmas T, Couffin AC, Bayle PA, De Crécy F, Neumann E, Vinet F, et al. Preparation and characterization of highly stable lipid nanoparticles with amorphous core of tuneable viscosity. J Colloid Interface Sci. 2011;360:471–81.
10. Benov L. Photodynamic therapy: current status and future directions. Med Princ Pract. 2015;24(Suppl 1):14–28.

11. Triesscheijn M, Baas P, Schellens JHM, Stewart FA. Photodynamic therapy in oncology. Oncologist. 2006;11:1034–44.

12. Dolmans DEJGJ, Fukumura D, Jain RK. Photodynamic therapy for cancer. Nat Rev Cancer. 2003;3:375–80.

13. Bahmani B, Bacon D, Anvari B. Erythrocyte-derived photo-theranostic agents: hybrid nano-vesicles containing indocyanine green for near infrared imaging and therapeutic applications. Sci Rep. 2013;3:2180.

14. Debele TA, Peng S, Tsai H-C. Drug carrier for photodynamic cancer therapy. Int J Mol Sci. 2015;16:22094–136.

15. Navarro FP, Creusat G, Frochot C, Moussaron A, Verhille M, Vanderesse R, et al. Preparation and characterization of mTHPC-loaded solid lipid nanoparticles for photodynamic therapy. J Photochem Photobiol B Biol. 2014;130:161–9.

16. Gravier J, Navarro FP, Delmas T, Mittler F, Couffin A-C, Vinet F, et al. Lipidots: competitive organic alternative to quantum dots for in vivo fluorescence imaging. J Biomed Opt. 2011;16:096013.

17. Besic Gyenge E, Darphin X, Wirth A, Pieles U, Walt H, Bredell M, et al. Uptake and fate of surface modified silica nanoparticles in head and neck squamous cell carcinoma. J Nanobiotechnol. 2011;9:32.

18. Senge MO, Brandt JC. Temoporfin (Foscan®, 5,10,15,20-Tetra(m-hydroxy-phenyl)chlorin)—a second-generation photosensitizer. Photochem Photobiol. 2011;87:1240–96.

19. Leong DT, Ng KW. Probing the relevance of 3D cancer models in nano-medicine research. Adv Drug Deliv Rev. 2014;79–80:95–106.

20. Sambale F, Lavrentieva A, Stahl F, Blume C, Stiesch M, Kasper C, et al. Three dimensional spheroid cell culture for nanoparticle safety testing. J Biotechnol. 2015;205:120–9.

21. Gref R, Lück M, Quellec P, Marchand M, Dellacherie E, Harnisch S, et al. 'Stealth' corona-core nanoparticles surface modified by polyethylene gly-col (PEG): influences of the corona (PEG chain length and surface density) and of the core composition on phagocytic uptake and plasma protein adsorption. Colloids Surf B Biointerfaces. 2000;18:301–13.

22. Shang L, Nienhaus K, Nienhaus GU. Engineered nanoparticles interacting with cells: size matters. J Nanobiotechnol. 2014;12:5.

23. Huo S, Ma B, Huang K, Liu J, Wei T, Jin S, et al. Superior penetration and retention behavior of 50 nm gold nanoparticles in tumors. Cancer Res. 2013;73:319–30.

24. Gyenge EB, Luscher D, Forny P, Antoniol M, Geisberger G, Walt H, et al. Photodynamic mechanisms induced by a combination of hypericin and a chlorin based-photosensitizer in head and neck squamous cell carcinoma cells. Photochem Photobiol. 2013;89:150–62.

25. Buytaert E, Matroule JY, Durinck S, Close P, Kocanova S, Vandenheede JR, et al. Molecular effectors and modulators of hypericin-mediated cell death in bladder cancer cells. Oncogene. 2008;27:1916–29.

26. Song J, Wei Y, Chen Q, Xing D. Cyclooxygenase 2-mediated apoptotic and inflammatory responses in photodynamic therapy treated breast adenocarcinoma cells and xenografts. J Photochem Photobiol B. 2014;134:27–36.

27. Mroz P, Yaroslavsky A, Kharkwal GB, Hamblin MR. Cell death pathways in photodynamic therapy of cancer. Cancers (Basel). 2011;3:2516–39.

28. Buchholz J, Kaser-Hotz B, Khan T, Bley CR, Melzer K, Schwendener R, et al. Optimizing photodynamic therapy: in vivo pharmacokinetics of liposomal meta-(tetrahydroxyphenyl) chlorin in feline squamous cell carcinoma. Clin Cancer Res. 2005;11:7538–44.

29. Molinari A, Colone M, Calcabrini A, Stringaro A, Toccacieli L, Arancia G, et al. Cationic liposomes, loaded with m-THPC, in photodynamic therapy for malignant glioma. Toxicol In Vitro. 2007;21:230–4.

30. Low K, Knobloch T, Wagner S, Wiehe A, Engel A, Langer K, et al. Compari-son of intracellular accumulation and cytotoxicity of free mTHPC and mTHPC-loaded PLGA nanoparticles in human colon carcinoma cells. Nanotechnology. 2011;22:245102.

31. Villa Nova M, Janas C, Schmidt M, Ulshoefer T, Grafe S, Schiffmann S, et al. Nanocarriers for photodynamic therapy-rational formulation design and medium-scale manufacture. Int J Pharm. 2015;491:250–60.

32. Peng C-L, Yang L-Y, Luo T-Y, Lai P-S, Yang S-J, Lin W-J, et al. Development of pH sensitive 2-(diisopropylamino)ethyl methacrylate based nanoparticles for photodynamic therapy. Nanotechnology. 2010;21:155103.

33. Preuss A, Chen K, Hackbarth S, Wacker M, Langer K, Roder B. Photosen-sitizer loaded HSA nanoparticles II: in vitro investigations. Int J Pharm. 2011;404:308–16.

34. Compagnin C, Baù L, Mognato M, Celotti L, Miotto G, Arduini M, et al. The cellular uptake of meta-tetra(hydroxyphenyl)chlorin entrapped in organically modified silica nanoparticles is mediated by serum proteins. Nanotechnology. 2009;20:345101.

35. Haedicke K, Kozlova D, Grafe S, Teichgraber U, Epple M, Hilger I. Multi-functional calcium phosphate nanoparticles for combining near-infrared fluorescence imaging and photodynamic therapy. Acta Biomater. 2015;14:197–207.

36. Tang L, Yang X, Yin Q, Cai K, Wang H, Chaudhury I, et al. Investigating the optimal size of anticancer nanomedicine. Proc Natl Acad Sci USA. 2014;111:15344–9.

37. Navarro FP, Berger M, Guillermet S, Josserand V, Guyon L, Neumann E, et al. Lipid nanoparticle vectorization of indocyanine green improves fluorescence imaging for tumor diagnosis and lymph node resection. J Biomed Nanotechnol. 2012;8:730–41.

15

Cell-based cytotoxicity assays for engineered nanomaterials safety screening: exposure of adipose derived stromal cells to titanium dioxide nanoparticles

Yan Xu[1], M. Hadjiargyrou[2], Miriam Rafailovich[1] and Tatsiana Mironava[1*]

Abstract

Background: Increasing production of nanomaterials requires fast and proper assessment of its potential toxicity. Therefore, there is a need to develop new assays that can be performed in vitro, be cost effective, and allow faster screening of engineered nanomaterials (ENMs).

Results: Herein, we report that titanium dioxide (TiO_2) nanoparticles (NPs) can induce damage to adipose derived stromal cells (ADSCs) at concentrations which are rated as safe by standard assays such as measuring proliferation, reactive oxygen species (ROS), and lactate dehydrogenase (LDH) levels. Specifically, we demonstrated that low concentrations of TiO_2 NPs, at which cellular LDH, ROS, or proliferation profiles were not affected, induced changes in the ADSCs secretory function and differentiation capability. These two functions are essential for ADSCs in wound healing, energy expenditure, and metabolism with serious health implications in vivo.

Conclusions: We demonstrated that cytotoxicity assays based on specialized cell functions exhibit greater sensitivity and reveal damage induced by ENMs that was not otherwise detected by traditional ROS, LDH, and proliferation assays. For proper toxicological assessment of ENMs standard ROS, LDH, and proliferation assays should be combined with assays that investigate cellular functions relevant to the specific cell type.

Keywords: Adipose derived stromal cells, Titanium dioxide, Cytotoxicity assays, Nanomaterials safety screening

Background

The growing annual production of engineered nanomaterials (ENMs) has led to a proportional increase in the chance of occupational and consumer exposure and has raised serious concerns about their environmental, health and safety impact. To be able to properly screen and predict the potential toxicity of ENMs, sensitive and reliable in vitro assays need to be developed as soon as possible [1–3]. Ideally, such assays should be relevant to a real life exposure scenario, be cost effective to allow for massive ENMs screening, and measure common modes of cellular responses [4]. An approach like this will provide researchers with consistent and accurate tools for comparing cellular responses to various ENMs, identify properties of materials causing the response and provide insights into the mechanism of toxicity [2, 5].

However, before developing such methods we need to have a clear strategy as to how to address ENMs risk in relevant cellular models. There are several different cell types that are targeted by ENMs through common routes of exposure: lung epithelium, skin fibroblasts and adipocytes, gastrointestinal tract epithelia, and cells belonging to the reticuloendothelial system such as macrophages [4, 6]. Since all of these cells have unique functions in tissues and organs, there is no universal strategy for ENMs hazard assessment. As such, we should strive to develop appropriate assays for each cell type based on their specific functions. Another important aspect of ENMs safety

*Correspondence: taniamironova@gmail.com
[1] Department of Materials Science and Engineering, Stony Brook University, Stony Brook, NY, USA
Full list of author information is available at the end of the article

evaluation lies in the understanding of how to distinguish between hazardous and safe concentrations and how to ensure that the proposed assays reveal a complete view of cellular changes induced by ENMs.

The current paradigm of ENMs hazard assessment is based on analysis of big data collected using high throughput screening (HTS) methods. However, the set of standard assays used for HTS assessment, only targets the symptoms of the cellular impairment such as decreased proliferation, changes in mitochondrial activity, or excessive reactive oxygen species (ROS) and lactate dehydrogenase (LDH). This approach allows for collection of secondary cellular responses to ENMs and limits our ability to detect earlier cell damage. The novelty of the proposed hazard assessment method is the detection of fundamental changes in early cell function. This approach is extremely sensitive and allows for detection of changes at earlier time prior to any damage detectable by the aforementioned standard assays. This approach also addresses proliferation and spreading of cells with impaired functions that might potentially cause the cascade of long term health problems. We believe that incorporating cell function based assays in the hazard assessment protocols will significantly reduce the number of long-term studies making the assessment process both, cost and time effective.

In this paper we use human ADSCs as a model to demonstrate that at low concentrations of ENMs, assays such as ROS, LDH, and cell proliferation, typically used for ENMs hazard assessment, reveal no damage. In contrast, other assays detect impairment of important cell functions. For this study we used two concentrations of TiO_2 NPs, one that does not show an effect and another that significantly increases both ROS and LDH.

We chose TiO_2 NPs for this study because of their abundance and availability. Previously, it was estimated that by 2015 more than 200,000 metric tons of TiO_2 NPs will be manufactured annually [7]. We investigated two most abundant forms of TiO_2, rutile and anatase, that have a tetragonal crystal structure but different atomic arrangement [8]. Considering, that these NPs are currently being used in products such as pharmaceuticals, personal care, cosmetics, toothpaste, sunscreens and food additives [9–11] it makes the possibility of human exposure guaranteed. Recently, the International Agency for Research on Cancer has classified the TiO_2 particles as *"possibly carcinogenic to humans"* [12]. Hence, there is an urgent need for a proper environmental and health hazard assessment of these NPs with respect to concentration and route of exposure.

Human skin is in constant contact with the external environment and is one of the most important routes of exposure to TiO_2. It is also the predominant organ exposed to high concentrations of TiO_2 NPs used in sunscreens and personal care products [9, 11]. Even though, the latest FDA regulations of over-the-counter sunscreen products for human use allow up to 25% of TiO_2 as an active ingredient, there are no regulations with regard to the size of the TiO_2 or labeling of sunscreens for the inclusion of NPs [13]. Not surprising, recent tests of different sunscreens revealed TiO_2 nanoparticles in each of them [14]. Similarly, tests performed by the Australian government showed that 70% of the sunscreens formulated with TiO_2 in nanoparticulate form [15]. Studies from other countries also indicated that NPs are being used in sunscreen regardless of labeling [14, 16].

Adipose tissue is an abundant, accessible and rich source of adult stem cells suitable for tissue engineering and regenerative medicine. The ADSCs are adult mesenchymal stem cells that can be isolated from subcutaneous adipose tissue (bottom layer of skin) and have a preadipocyte characteristics [17]. They can also be induced to differentiate into adipocytes, bone marrow, neurons and other cell types [18–22]. The main function of these cells in their non-differentiated state is healing of cutaneous injuries while in their differentiated state, to contribute to fat depot maintenance [23].

Even though several research groups showed that nanosized TiO_2 can penetrate through the skin in vivo [24, 25], the majority of reports showed that after application on skin, TiO_2 NPs mostly reside in the stratum corneum and do not reach living skin [26, 27]. The 2013 Scientific Committee on Consumer Safety stated that the use of TiO_2 NPs as a UV-filter in sunscreens, pose no adverse effects in humans when applied on healthy, intact or sunburnt skin [28]. Therefore, this manuscript aims to begin filling the knowledge gap that exists regarding the effects of TiO_2 NPs on compromised skin (due to various diseases or trauma) which enables NPs to penetrate down to the subcutaneous adipose tissue layer. This presents a realistic scenario since exposure to sunlight is known to aggravate skin diseases such as psoriasis, dermatitis, eczema, and acne with daily application of sunscreen recommended [29, 30]. Moreover, TiO_2 is a common ingredient in topical medications (ex. Sorion and Novasone cream) for the aforementioned conditions [31, 32] with a recommended daily application. Here, in addition to the standard ROS, LDH, and proliferation assays, we focused on the effects of TiO_2 exposure on distinct functions of ADSCs: wound healing ability and intracellular lipid accumulation. Altogether, the current study indicates that specialized assays exhibit greater sensitivity in detecting damage induced by TiO_2 NPs in ADSCs as compared to the current set of standard assays. Hence, we propose that for proper ENMs assessment, the standard set of assays needs to be expanded and include tests examining the impairment of characteristic cell functions. Moreover, such approach may establish appropriate guidelines and identify safe concentrations of ENMs.

Methods

Anatase and Rutile TiO$_2$ NPs were purchased from US Cosmetics. Trypsin–EDTA (0.05%) (Cat#: 25300-054) and Dulbecco's Phosphate-Buffered Saline (Cat#: 14190-250) were purchased from Life Technologies. Alexa Fluor 488-Phalloidin (Cat#: A12379), LipidTOX™ red (Cat#: H34476) were purchased from ThermoFisher Scientific.

Cell culture

Primary human ADSCs were obtained from Living Skin Bank (Stony Brook University) and were cultured in basal medium comprising Dulbecco's Modified Eagle's Medium (DMEM) supplemented with 10% fetal bovine serum (FBS; HyClone, Logan, UT) and 1% of penicillin–streptomycin (PS; Sigma, St. Louis, MO). For differentiation, ADSCs were cultured in basal medium supplemented with 250 µM 3-isobutyl-1-methylxanthine (IBMX), 1 µM insulin, 200 µM indomethacin, 33 µM biotin, 17 µM pantothenic acid and 1 µM dexamethasone (adipose induction medium). Medium containing TiO$_2$ NPs (with concentrations of 0.1 and 0.4 mg/mL) was added to each culture plate 24 h after initial cell seeding. The samples were incubated with NPs up to 6 days and then counted or fixed, stained and imaged. Culture medium was replaced every 2–3 days and grown at 37 °C with 5% CO$_2$. NP-free cultures served as controls.

Cell proliferation

To determine cell proliferation, cells were plated at an initial density of 7.5×10^3 cells per well in DMEM supplemented with 10% FBS and 1% of PS in a 12-well tissue culture plate and counted using a hemocytometer at days 1, 2, 3, and 6. Each condition was completed in triplicates (n = 3) and all experiments were conducted three times (n = 3). Medium was changed every 2 days.

Zeta potential and dynamic light scattering

To prepare the samples, 2 µg of TiO$_2$ NPs were placed in 10 mL of deionized water or culture medium and sonicated for 5 min to separate agglomerates. The samples were then diluted ten times in deionized water, briefly sonicated and analyzed. Zeta potentials were measured using Brookhaven Instruments Zeta Plus Zeta Potential Analyzer and particle size measurements were performed using BIC 90Plus dynamic light scattering (DLS) instrument (Brookhaven Instruments, Zeta Plus Zeta Potential Analyzer). The average of 3 measurements of 50 cycles was used as a numerical value of zeta potential.

Transmission electron microscopy (TEM)

TEM analysis was used to assess the size and distribution, as well as intracellular localization of TiO$_2$ NPs. For the particle analyses, one drop of TiO$_2$ NPs suspension was placed on a 300 mesh Formvar coated copper grid and air dried at room temperature. A histogram of the size distribution from approximately 170–200 particles was plotted and fit to a Gaussian distribution from which the mean diameters were obtained.

For intracellular particles distribution, 1×10^5 cells per well were plated in six-well plate, exposed to TiO$_2$ NPs for 3 days and then fixed in a solution of 2.5% paraformaldehyde and 2.5% glutaraldehyde in 0.1 M Phosphate Buffered Saline (PBS). The samples were then dehydrated with ethanol and embedded in propylene oxide. The specimens were then cut into ultrathin (90 nm) sections using a Reichart Ultracut Ultramicrotome, lifted onto uncoated TEM grids and stained with uranyl acetate and lead citrate. The samples were imaged using a FEI Tecnai12 BioTwinG2 transmission electron microscope. Digital images were acquired with an AMT XR-60 CCD Digital Camera System.

Delivered and cellular TiO$_2$ NPs doses

The theoretical estimation of delivered doses (mass of TiO$_2$ NPs deposited per area) was performed using in vitro sedimentation, diffusion and dosimetry (ISDD) model generously provided by Dr. Teeguarden [33]. This computational model of particokinetics (sedimentation, diffusion) estimates the amount of particles reaching cells residing at the bottom of a cell culture dish during a defined exposure period. The model also calculates fraction of particles, surface area, mass and number of particles reaching cells and allows comparison of particle doses among particle types within a system, and among systems with different characteristics (media height, viscosity, orientation).

Effective density by volumetric centrifugation method (VCM)

Effective density of the TiO$_2$ NPs agglomerates was estimated using VCM adopted from Deloid et al. [34]. Briefly, a sample of TiO$_2$ NPs suspension in basal medium was centrifuged in a packed cell volume (PCV) tube (Sigma Aldrich, Cat#: Z760986) at 3000 rpm for 1 h to produce a pellet consisting of packed NPs agglomerates and the media trapped between them. All VCM experiments were performed in triplicates (n = 3) and the average was used to calculate the effective agglomerate density.

The effective density of the TiO$_2$ NPs agglomerates, ρ_{EV}, was calculated using following equation [34]:

$$\rho_{EV} = \rho_{media} + \left[\left(\frac{M_{TiO_2}}{V_{pellet} \cdot SF} \right) \cdot \left(1 - \frac{\rho_{media}}{\rho_{TiO_2}} \right) \right] \quad (1)$$

Where ρ_{TiO2} is TiO$_2$ NPs density, M_{TiO2} is mass of TiO$_2$ NPs, V_{pellet} is the volume of the pellet collected by centrifugation, ρ_{media} is the media density, and SF is stacking factor that depends on the efficiency of agglomerate stacking. In this paper, we used theoretical SF values of 0.634 for anatase and 0.7 for rutile as it was previously recommended as a reasonable approximation by DeLoid et al. [34].

Cell staining for confocal microscopy

Cell area and overall morphology as a function of NP uptake was monitored using a Leica confocal microscope. For these experiments, cells were exposed to TiO$_2$ for 3 weeks of differentiation and then fixed with 3.7% formaldehyde for 15 min. Alexa Fluor 488-Phalloidin was used for actin staining and lipid droplets were visualized using LipidTOX™ red according to the manufacturer's instructions.

Lactate dehydrogenase activity (LDH) measurements

Pierce LDH Cytotoxicity Assay Kit (Cat#: 88953, Life Technology) was used for LDH measurements. Cells were plated with starting density of 8×10^4 per well in six-well plate. After 3 days of incubation with nanoparticles, 50 µL supernatant from each sample were transferred to a 96-well plate in triplicate wells and 50 µL of reaction mixture (lyophilizate mixture) were added. After incubation at room temperature for 30 min, the reaction was stopped by adding 50 µL Stop Solution. Released LDH activity absorbance was measured at 490 and 630 nm respectively.

Reactive oxygen species (ROS) measurement

ROS Detection Reagents (Cat#: C6827, Invitrogen) was used to detect ROS level of ADSCs cells. For this experiment a working solution of 5 µg/mL of 5-(and-6)-chloromethyl-2′,7′-dichlorodihydrofluorescein diacetate, acetyl ester (CM-H$_2$DCFDA) was prepared. Cultures were seeded with starting density of 8×10^4 per well in six-well plate and exposed to TiO$_2$ for 3 days. Cells were then harvested and washed three times with PBS to remove TiO$_2$ NPs from pellets, counted and 5×10^4 cells per well were placed to 96-well dish (each condition had triplicates). Then 100 µL of working solution was added to each well and incubated for 20 min. 100 µL of 20 mM NaN$_3$ were then added to each well and incubated for 2 h. Fluorescence was read at 490 nm excitation and 520 nm emission.

Migration

Cell migration of cultures seeded at 8×10^4 cells per well in six-well plate and treated with TiO$_2$ NPs for 3 days was evaluated using the agarose droplet assay. The agarose gel was prepared by melting a 2% (w/v) agarose stock solution, and diluting it with DMEM to 0.2% (w/v). The 0.2% (w/v) agarose was then used to re-suspend cells to a concentration of 1.5×10^7 cells/mL. After that 1.25 µL drops were placed into each well of a 24-well dish, and allowed to gel at 4 °C for 20 min prior to the addition of 400 µL of DMEM into each well. Following a 24 h incubation at 37 °C, the cells were visualized under phase contrast microscopy. Cell migration from the outer edge of the agarose was quantified using imageJ software.

Collagen gel contraction

Cells seeded at initial density of 8×10^4 per well in six-well plate were exposed to 0.1 and 0.4 mg/mL TiO$_2$ NPs for 3 days. After that cultures were harvested and resuspended in DMEM containing 1.8 mg/mL collagen and 2% BSA at 3.5×10^5 cells/mL. Cell/collagen gel suspensions (0.7 mL) were loaded into each well of 24-well dish pre-coated with 2% BSA in PBS coated (overnight) and incubated at 37 °C to induce gelation. After 2 h the gel was detached by tapping lightly on the wall of the wells and 500 µL DMEM with 2% BSA was added. Detachment was done in order to begin the contraction process. The gels were then incubated for 5 h and imaged by scanning the 24 well plate.

Lipid quantification and visualization

To determine differences in lipid accumulation, cells were differentiated for 1, 2, and 3 weeks in adipose induction media were fixed with 3.7% formaldehyde for 15 min at room temperature and incubated with Oil red O for 2 h. Oil red O was then extracted using isopropanol and the amount of lipids was measured as a function of Oil red O absorbance (510 nm). Lipid amounts were calculated on a per cell basis, where a typical sample contained 1.5–2.0×10^5 million cells per well. The cellular distribution of lipid droplets was visualized using confocal microscopy as described below.

Adiponectin expression

Cell culture media was collected on days 7, 14 and 21 after switching to adipose induction media. The cells were stored at −20 °C in the presence of a protease inhibitor cocktail (Cat#: P8340, Sigma-Aldrich) till assay time. Adiponectin was directly measured using the human adiponectin ELISA kit (Novex®, Cat#: KHP0041). Samples were prepared according to manufacturer's instructions and the absorbance was read at 450 nm using a microplate reader (BioTek EL800).

Collagen and fibronectin expression

Collagen and fibronectin in the cell culture media of ADSCs cultured for 3 days was measured using the

Procollagen Type I C-Peptide EIA Kit (Takara) and human Fibronectin EIA Kit (Takara) as described in instructions provided by the manufacturer. Samples with high concentrations of collagen and fibronectin were diluted with Sample Diluent (Takara) prior to assay. Absorbance was read at 450 nm using a microplate reader (BioTek EL800).

Flow cytometry

Cell were plated, allowed to adhere for 24 h, and exposed to 0.1 and 0.4 mg/mL TiO_2 nanoparticles for another 3 days. The cells were carefully rinsed with PBS three times to remove all the floating particles and were detached by gentle scrapping. The cells were then rinsed twice with BSA (0.2%) in PBS and re-suspended in PBS at a concentration of 10^6 cells/mL. All samples were then analyzed with a Becton–Dickinson FACSCAN analyzers flow cytometer.

Statistical analysis

All experiments were performed in triplicates and repeated at least three times. The results were represented as mean ± SD. A p value <0.05 was considered statistically significant (t test).

Results

Characterization of TiO_2 NPs

Anatase particles have a spherical shape, while rutile particles are rod shape with the aspect ratio of 4 (Fig. 1a, b). From the TEM images, the calculated average diameter of anatase is 136 ± 47 nm and the average length of rutile is 46 ± 28 nm (Fig. 1c, d). X-ray diffraction spectra of both particles are shown on Fig. 1e, f confirming anatase and rutile crystal structures. The surface charges of the particles were measured using zeta potentiometry (Table 1), and were found for particles suspended in deionized water to be −34.75 ± 1.63 and −30.29 ± 0.6 mV for anatase and rutile respectively. After NPs incubation in DMEM for at least 24 h their zeta potential increased to −7.39 ± 0.90 and −14.29 ± 1.73 mV for anatase and rutile respectively. Surface charge of TiO_2 NPs suspended in adipose induction media was −13.85 ± 0.9 and −12.45 ± 2.3 mV for rutile and anatase, respectively.

DLS was performed to determine the hydrodynamic sizes of the TiO_2-NPs in suspension assuming that particles are spherical or can be represented as spheres (average number distributions are reported, Table 1). The dispersions of rutile and anatase TiO_2-NPs were homogeneous and only revealed the presence of

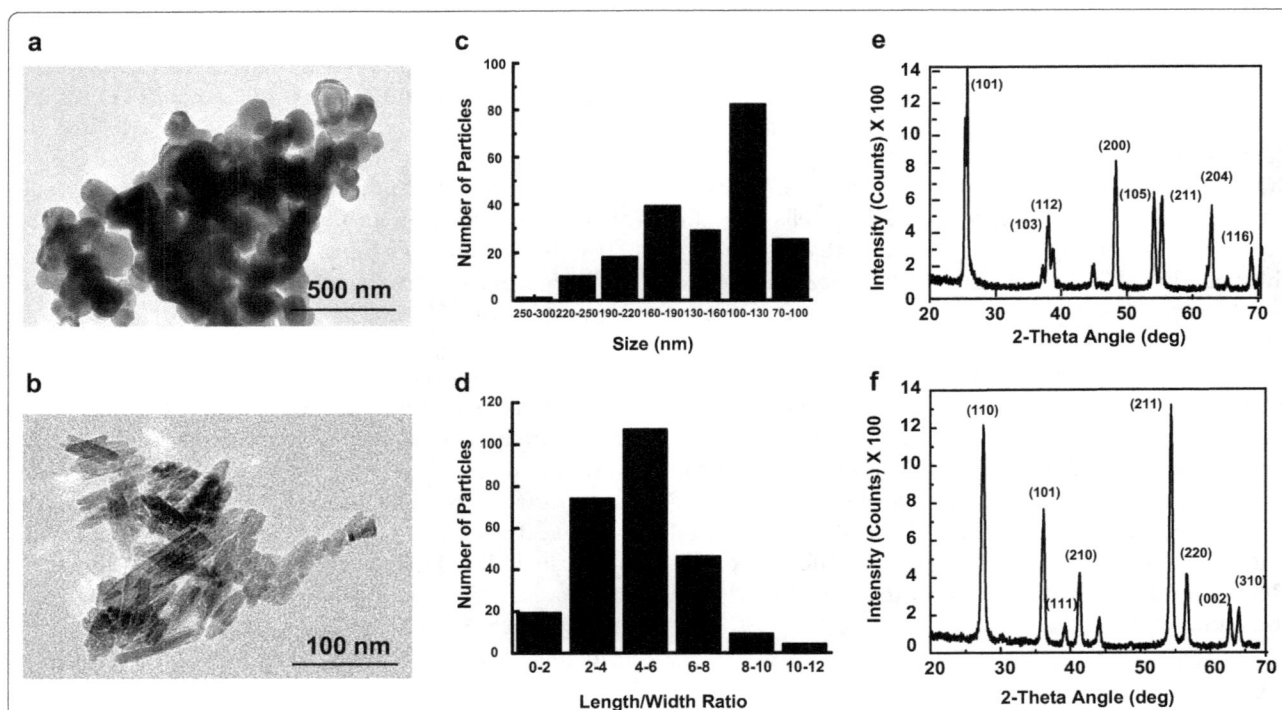

Fig. 1 TiO_2 nanoparticles imaged by TEM, its size distribution histograms and X-ray diffraction spectra. TEM picture of anatase **a** nanoparticles and rutile, **b** TiO_2 nanorods; size distribution histograms of anatase (**c**) and rutile (**d**); X-ray diffraction spectra of anatase (**e**) and rutile (**f**)

Table 1 Properties of TiO$_2$ NPs

	Anatase	Rutile
Zeta potential (mV)		
In DI water	−34.75 ± 1.63	−30.29 ± 0.6
In DMEM	−7.39 ± 0.90	−14.29 ± 1.73
In induction media	−13.85 ± 0.9	−12.45 ± 2.3
DLS (nm)		
In DI water	383 ± 19	640 ± 44
In DMEM	355 ± 37	291 ± 37
In induction media	368 ± 37	408 ± 38
Density of agglomerates (g/mL)		
In DMEM	2.428	3.383
Volume of pellets (µL)		
In DMEM	0.486 ± 0.076	0.766 ± 0.052

secondary aggregates. Anatase TiO$_2$ NPs have aggregates of 383 ± 19 nm in water, 355 ± 37 nm in basal medium and 368 ± 37 nm in adipose induction media. Similarly, aggregates of rutile in water, basal and induction media were 640 ± 44, 291 ± 37, and 408 ± 38 nm, respectively.

Delivered doses of TiO$_2$ NPs

For proper assessment of TiO$_2$ NPs the actual delivered doses that come in contact with cells were estimated using the ISDD model (Table 1). The densities of TiO$_2$ agglomerates were calculated using Eq. 1 (see "Methods") and were 2.428 and 3.383 g/mL for rutile and anatase, respectively. Density of basal media was 1.007 g/mL [33] and volumes of the anatase and rutile pellets, measured by volumetric centrifugation, were 0.486 ± 0.076 and 0.766 ± 0.052 µL, respectively. Finally, delivered fractions of TiO$_2$ NPs after 72 h of incubation were 1 for both anatase and rutile, making the corresponding delivered doses equal to 10 and 40 µg/cm^2 for initial 0.1 and 0.4 mg/mL treatments, respectively. In fact, according to these calculations, 100% of particles were deposited on cells after the first 16 h of incubation.

Proliferation of ADSCs

In order to directly observe the impact of TiO$_2$ NPs exposure on the ADSCs, we measured cell proliferation after incubation with TiO$_2$ NPs for up to 6 days. The data obtained from counting the cells is shown in Fig. 2. It is apparent that exposure to low concentration (0.1 mg/mL) of both rutile and anatase NPs had no effect on cell proliferation (Fig. 2a). In contrast, exposure to 0.4 mg/mL resulted in a moderate decrease in cell number after prolonged exposure. Specifically, on day 4 we observed approximately 11 ± 1% decrease in after exposure to anatase that extended to 16 ± 2% decrease on

day 6. In case of ADSCs exposed to rutile, the decrease was approximately 9 ± 1 and 14 ± 1% on days 4 and 6, respectively. The decrease in cell numbers observed after 6 days of treatment with 0.4 mg/mL TiO$_2$ was statistically significant.

Internalization of TiO$_2$ NPs

Flow cytometry was used to measure cell granularity and confirm internalization of TiO$_2$ NPs by ADSCs. From Fig. 2b, c we can see an increase in side scatter (SSC) and the slight decrease in forward scatter (FSC) in all cultures treated with TiO$_2$ NPs. Also, SSC increased in a NPs dose dependent manner, the overall observed increase was higher for anatase than rutile using identical concentrations.

For accurate characterization of NP-cell interaction it is important to know the localization of NPs within cells. Figure 3 shows TEM images of ADSCs that had been cultured with rutile or anatase NPs for 3 days in basal media or for 3 weeks in adipose induction media. Figure 3a–f shows that after incubation in basal media, both types of TiO$_2$ NPs accumulated in vacuoles. However, the size of the vacuoles in which rutile and anatase TiO$_2$ NPs are stored is significantly different. Rutile-containing vacuoles were 3.2 ± 1.4 µm which is approximately ninefold larger that those filled with anatase (0.29 ± 0.2 µm) (Fig. 3a–c). In addition, the shape of the rutile containing vacuoles is mostly spherical with well-defined edges as oppose to the anatase containing vacuoles that display irregular shape without distinct edges (Fig. 3d–f). Also, inside the vacuole, rutile particles seem to be tightly packed whereas with anatase they are more loose. Cellular compartments such as nuclei, mitochondria, rough endoplasmic reticulum (ER), and Golgi apparatus did not contain TiO$_2$ NPs.

Images of ADSCs after differentiation for 3 weeks in adipose induction media containing TiO$_2$ NPs are shown on Fig. 3g–l. All the TEM images reveal adipose conversion (lipid droplet formation) in cultures treated with TiO$_2$ NPs as well as control. However, significantly smaller amounts of TiO$_2$ NPs can be found inside the cells (Fig. 3j–l). Once again, nuclei, mitochondria, rough ER, and Golgi apparatus were devoid of TiO$_2$ NPs. However, we did observe that both rutile and anatase NPs penetrate lipid droplets as can be seen on Fig. 3k, l as indicated by the arrowheads. It is interesting to note, that TiO$_2$ NPs can percolate the phospholipid monolayer that surrounds lipid droplets, however they are unable to penetrate through the phospholipid bilayer of mitochondria, nuclei, endoplasmic reticulum, and golgi apparatus. Additionally, such structural disturbance of lipid droplets may potentially cause leakage of lipids interfering with fat storage and metabolism.

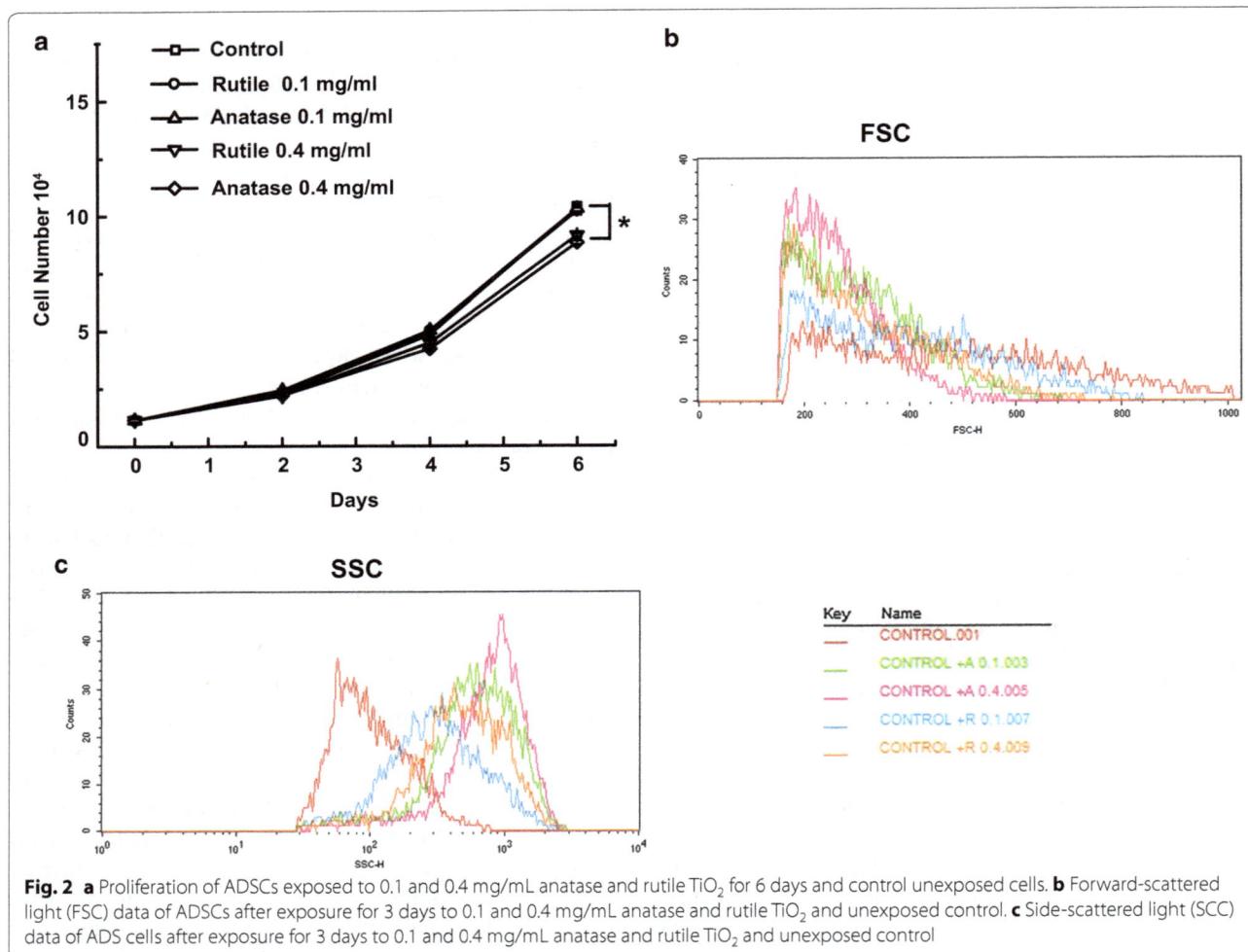

Fig. 2 a Proliferation of ADSCs exposed to 0.1 and 0.4 mg/mL anatase and rutile TiO$_2$ for 6 days and control unexposed cells. **b** Forward-scattered light (FSC) data of ADSCs after exposure for 3 days to 0.1 and 0.4 mg/mL anatase and rutile TiO$_2$ and unexposed control. **c** Side-scattered light (SCC) data of ADS cells after exposure for 3 days to 0.1 and 0.4 mg/mL anatase and rutile TiO$_2$ and unexposed control

TiO$_2$ NPs cytotoxicity

Even though no major changes in cell proliferation were observed with both types and concentrations of TiO$_2$ NPs for short (3 days) exposure, secondary processes triggered by the exposure such as secretory or morphological changes may induce long-term toxicity. To test this, we investigated two of the most common indicators of ENMs cytotoxicity: cellular ROS generation and LDH release. In Fig. 4a we show that no increase in ROS was observed in the cultures treated with 0.1 mg/mL rutile and anatase NPs. On the other hand, cells exposed to 0.4 mg/mL TiO$_2$ NPs for 3 days exhibited 27 ± 5 and 33 ± 4% increase in ROS levels after treatment with rutile and anatase, respectively.

Release of LDH is associated with the loss of cell-membrane integrity and is another important indicator of cellular toxicity induced by ENMs. A recent study reported that LDH binds to TiO$_2$ NPs decreasing the assay readout [35], however concentrations of rutile and anatase used in our experiments were below the minimum concentration

at which differences in LDH readout could be detected. From the Fig. 4b we can see that there are no changes in the extracellular levels of LDH in case of exposure to 0.1 mg/mL rutile and anatase as compared to control culture. In contrast, cultures exposed to 0.4 mg/mL dose of rutile TiO$_2$ NPs exhibit a ~33% increase in extracellular LDH while cultures treated with anatase show an LDH increase of ~43%.

Cell migration and collagen gel contraction

One of the crucial functions of ADSCs is their ability to migrate into a wound and contract newly deposited collagen fibers to close the wound. Hence, we examined this function after ADSCs exposure to TiO$_2$ NPs for 3 days. As shown in Fig. 5a, b, exposure to 0.1 mg/mL TiO$_2$ NPs had no effect on the cell migration speed as well as their ability to contract collagen. However, exposure to 0.4 mg/mL NPs significantly altered both functions. Specifically, it yielded a 15 ± 2 and 27 ± 3% decrease in collagen contraction in ADSCs treated with rutile and anatase,

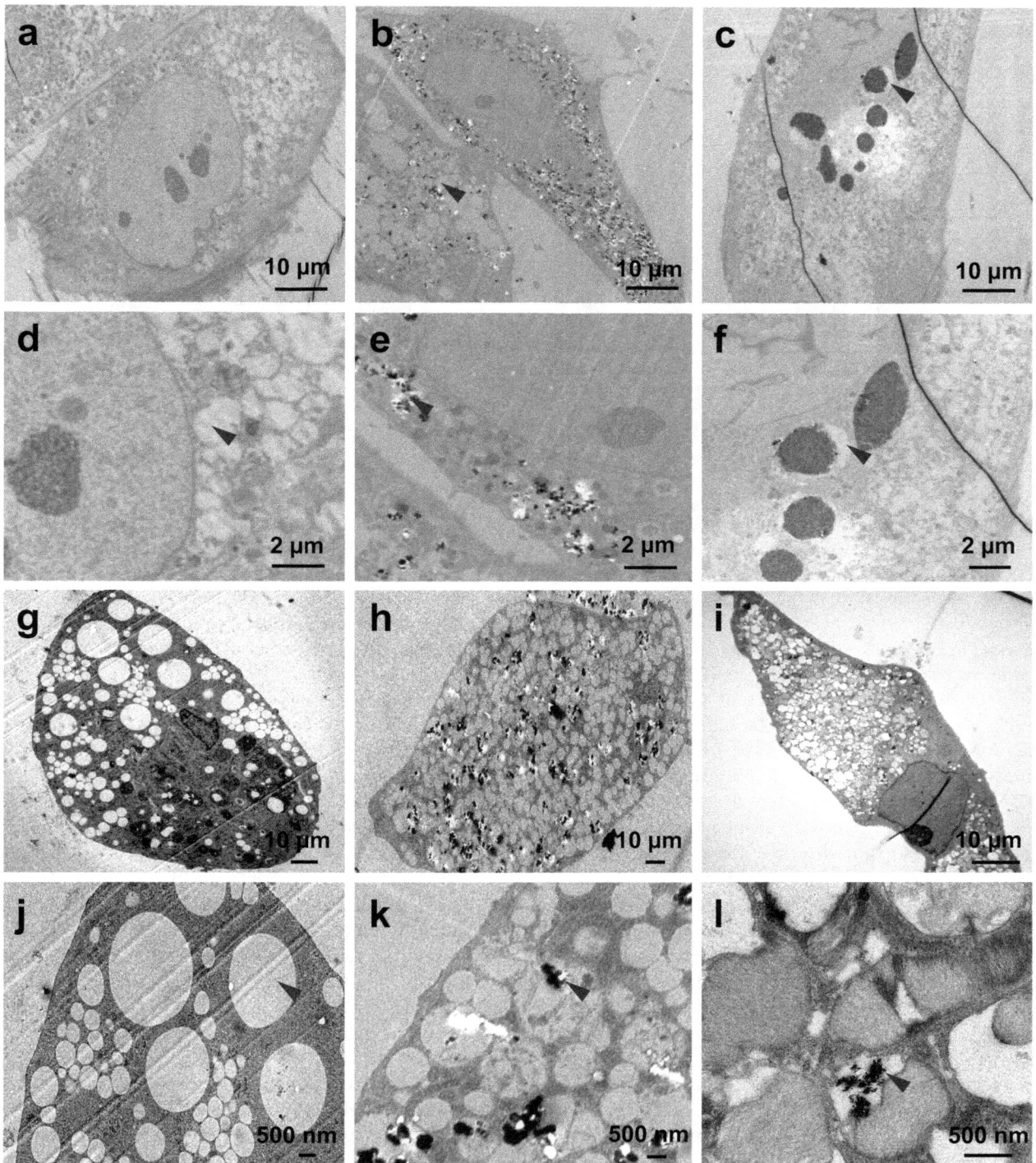

Fig. 3 TEM cross section of ADSCs exposed to TiO₂ NPs in different conditions: in basal media (**a–f**), after 3 weeks in adipose induction media (**g–l**). ADSC control cells in basal media (**a**), ADSCs exposed to 0.4 mg/mL anatase (**b**) and 0.4 mg/mL rutile (**c**) for 3 days in basal media. ADSC control in basal media high magnification (**d**), ADSCs exposed to 0.4 mg/mL anatase (**e**) and 0.4 mg/mL rutile (**f**) for 3 days in basal media high magnification. ADSC control cells in induction media (**g**) for 3 weeks, ADSCs exposed to 0.4 mg/mL anatase (**h**) and 0.4 mg/mL rutile (**i**) for 3 weeks. ADSC control cells in induction media high magnification (**j**), ADSCs exposed to 0.4 mg/mL anatase (**k**) and 0.4 mg/mL rutile (**l**) in induction medium for 3 weeks high magnification. *Black arrows* indicate the vacuoles (**a–f**) and lipid droplets (**g–l**)

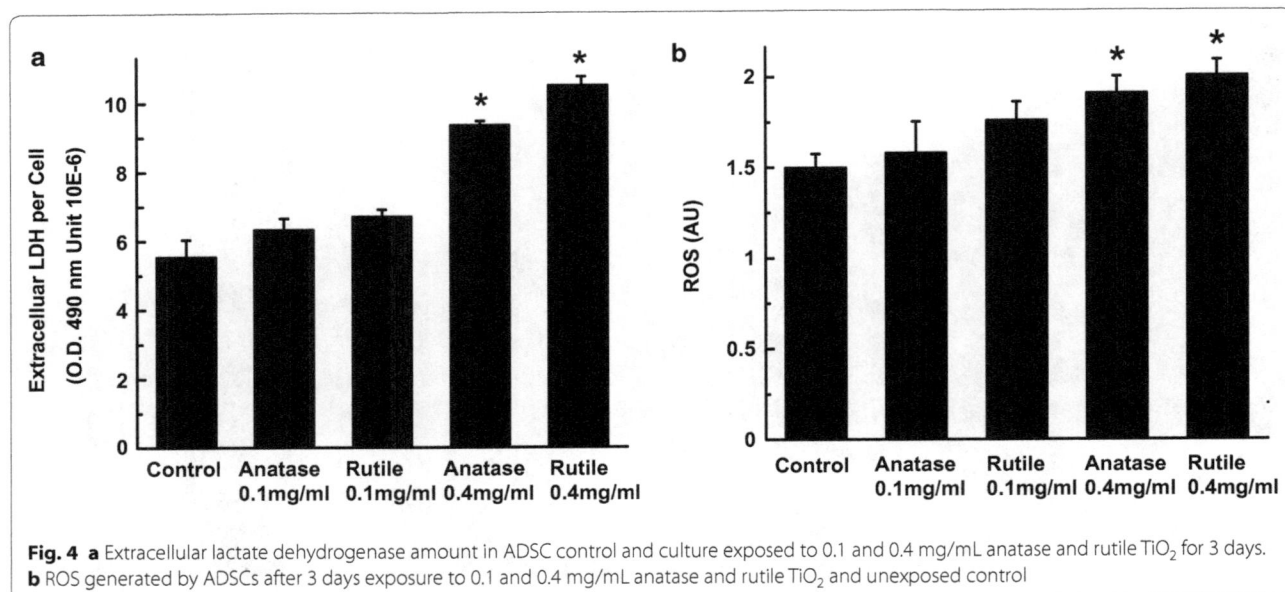

Fig. 4 a Extracellular lactate dehydrogenase amount in ADSC control and culture exposed to 0.1 and 0.4 mg/mL anatase and rutile TiO$_2$ for 3 days. **b** ROS generated by ADSCs after 3 days exposure to 0.1 and 0.4 mg/mL anatase and rutile TiO$_2$ and unexposed control

respectively (Fig. 5a, b). A similar trend was observed in cell migration speed; decrease in speed by 48 ± 13 and 52 ± 15% in cultures exposed to 0.4 mg/mL rutile and anatase, respectively.

Changes in ECM

It is known that extracellular matrix (ECM) proteins play a crucial role in cell behavior and regulate important cellular functions such as proliferation, apoptosis, and migration [36]. Thus, we decided to examine whether exposure of ADSCs to TiO$_2$ NPs affects two major ECM components, collagen type I and fibronectin. Figure 5c shows that there is no change in collagen expression in cultures exposed to low concentrations of rutile and anatase as compared to control. On the other hand, exposure to 0.4 mg/mL of rutile and anatase TiO$_2$ leads to a reduction of collagen by 35 ± 5 and 37 ± 4%, respectively. Interestingly, we did not observe a similar pattern in fibronectin production. Figure 5d indicates that exposure of ADSCs to increasing doses of TiO$_2$ results in small, but steady increase in fibronectin as compared to control. Specifically, our data indicates that the fibronectin concentration increased by 17 ± 2 and 19 ± 2% in cultures treated with 0.1 mg/mL TiO$_2$ and by 26 ± 5 and 32 ± 1% in cultures treated with 0.4 mg/mL for anatase and rutile, respectively.

Adipocyte-specific differentiation

Another important function of ADSCs is their ability to differentiate into adipocytes. Given the appropriate signals ADSCs will differentiate into adipocytes and accumulate lipid droplets as well as express adipocyte specific proteins. Therefore, we investigated the effect

of TiO$_2$ NPs on lipid accumulation and the expression of an adipocyte differentiation marker adiponectin that is involved in several metabolic pathways including glucose and fatty acid catabolism [37]. Experiments were conducted using cultures exposed to rutile and anatase TiO$_2$ NPs for 3 weeks during adipocyte induction.

Lipid accumulation

Lipid accumulation was examined in ADSCs grown in NP-containing induction medium where the NP-free cultures served as control. From Fig. 6a we can see that cultures exposed to 0.1 mg/mL NPs had the same quantity of lipids as control after 1 and 2 weeks of exposure. A small decrease in lipids accumulation of 10 ± 3 and 12 ± 3% was detected after 3 weeks in cultures treated with rutile and anatase (0.1 mg/mL), respectively. Higher reduction in lipids was observed in ADSCs exposed to 0.4 mg/mL TiO$_2$ NPs; exposure to 0.4 mg/mL rutile resulted in 11 ± 1% decrease after 2 weeks of exposure and extended to 15 ± 1% by the end of week 3 (Fig. 6b). Reduction in lipids in cultures treated with 0.4 mg/mL anatase was 22 ± 3% after 3 weeks (Fig. 6b). These data correlate with our qualitative observation via confocal microscopy, where reduction in lipid droplet size was seen in both cultures grown with TiO$_2$ NPs for 3 weeks (Fig. 7g–i); and only cultures exposed to rutile showed decrease in lipids after 2 weeks. No changes were seen at an earlier time point (Fig. 7a–f).

Adipocyte-specific adiponectin secretion

To determine whether TiO$_2$ NPs affect the expression of adipocyte specific cytokines, we measured adiponectin concentration in cultures grown in induction medium

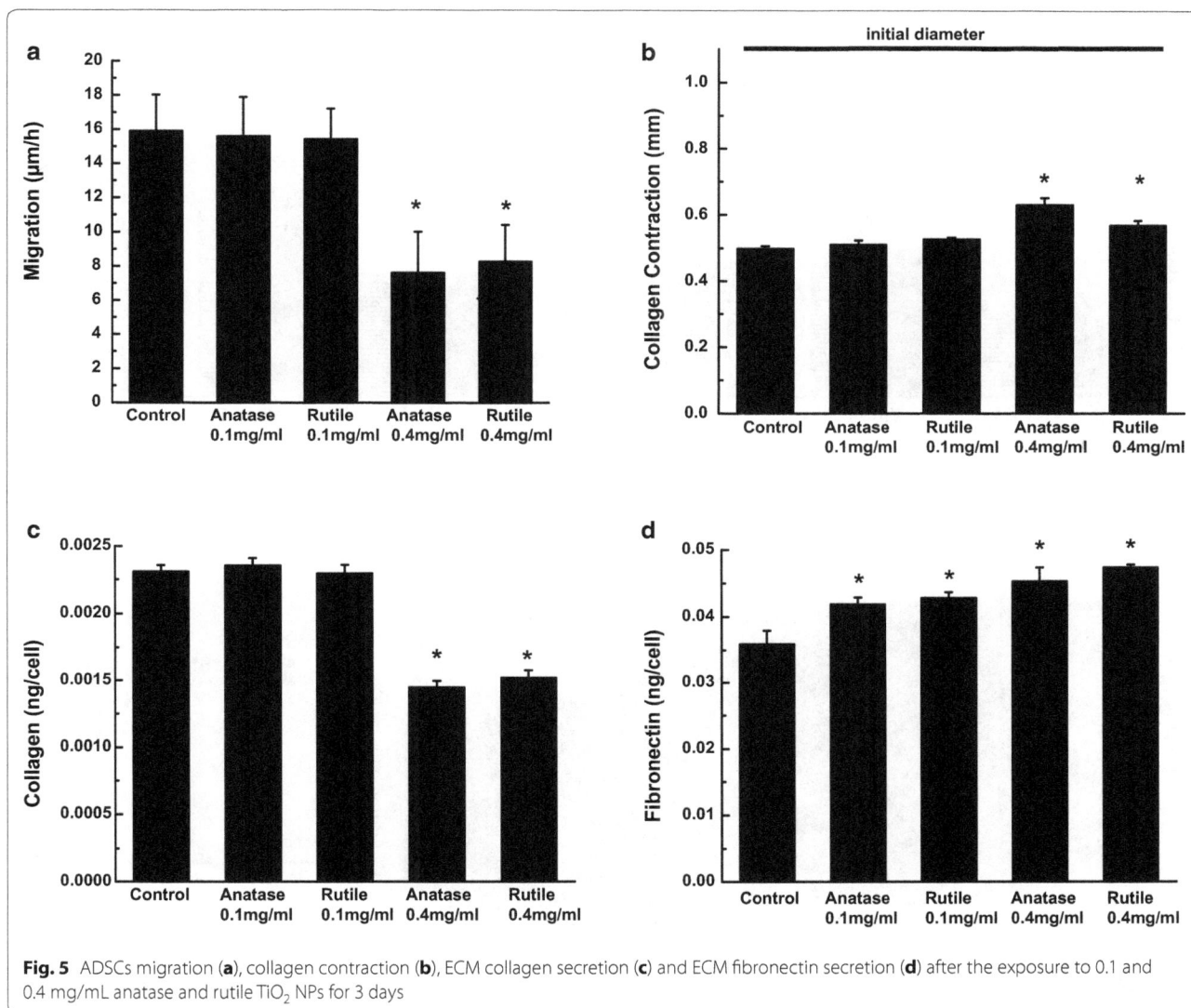

Fig. 5 ADSCs migration (**a**), collagen contraction (**b**), ECM collagen secretion (**c**) and ECM fibronectin secretion (**d**) after the exposure to 0.1 and 0.4 mg/mL anatase and rutile TiO$_2$ NPs for 3 days

containing 0.4 mg/mL TiO$_2$. As expected, after 1 week in induction medium no adiponectin was detected in either control or treated cultures (Fig. 7j). By week two of ADSC differentiation, control cultures showed an increase in adiponectin secretion to 0.61 pg/cell and by week three, secretion reached 1.83 pg/cell. Smaller increases at week two were seen in the cultures incubated with TiO$_2$ NPs, where adiponectin levels were 40 ± 3 and 59 ± 4% less than control in the cultures exposed to anatase and rutile, respectively. At week three, adiponectin concentration further increased in all cultures but resulted in levels that were 30 ± 4 and 42 ± 3% less than control, for those treated with anatase and rutile TiO$_2$ NPs, respectively.

Changes in ECM during differentiation

In addition to proliferation and apoptosis regulation, ECM proteins are also involved in cell differentiation and lipid formation [38, 39]. Data in Fig. 8a reveals that cultures exposed to 0.1 mg/mL TiO$_2$ NPs for 1 week while differentiating show no changes in fibronectin production as compared to control. Exposure of ADSCs to 0.4 mg/mL TiO$_2$ increased fibronectin by 30 ± 10 and 50 ± 12% for anatase and rutile, respectively. Similarly, after 2 and 3 weeks of differentiation, no changes in fibronectin expression were observed for cultures treated with 0.1 mg/mL anatase. In contrast, exposure to 0.1 mg/mL rutile resulted in 40 ± 3% increase in fibronectin after 2 and 3 weeks of differentiation. In all cases, cultures exposed to higher doses of TiO$_2$ exhibited substantial increase in fibronectin expression as compared to control cultures. Specifically, treatment with 0.4 mg/mL TiO$_2$ NPs increased fibronectin production by 30 ± 10 and 50 ± 12% after 1 week, by 25 ± 14 and 79 ± 9% after 2 weeks, and by

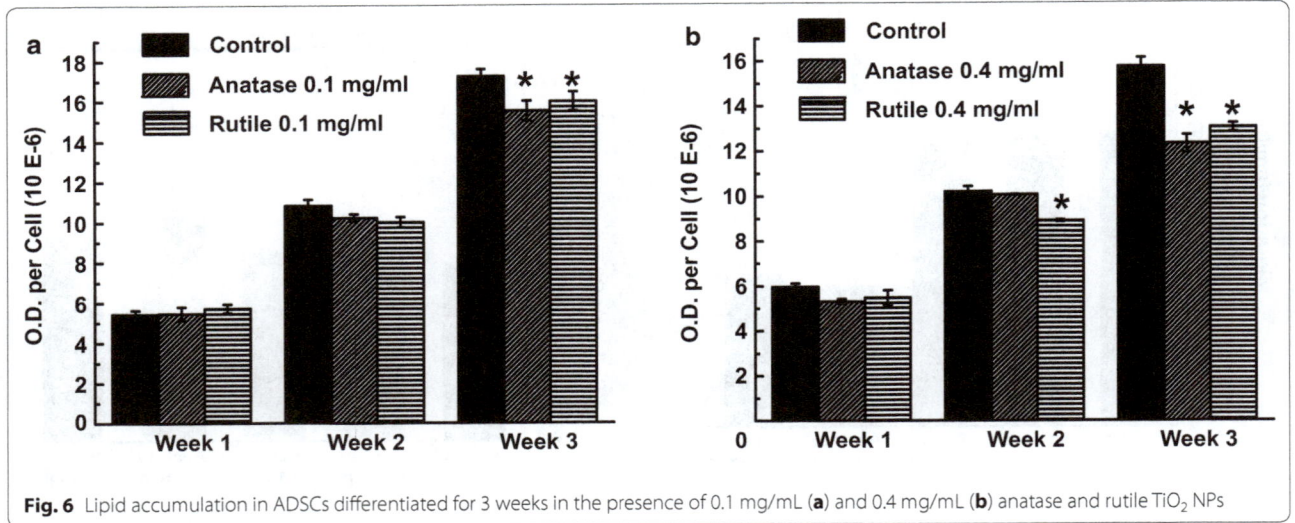

Fig. 6 Lipid accumulation in ADSCs differentiated for 3 weeks in the presence of 0.1 mg/mL (**a**) and 0.4 mg/mL (**b**) anatase and rutile TiO$_2$ NPs

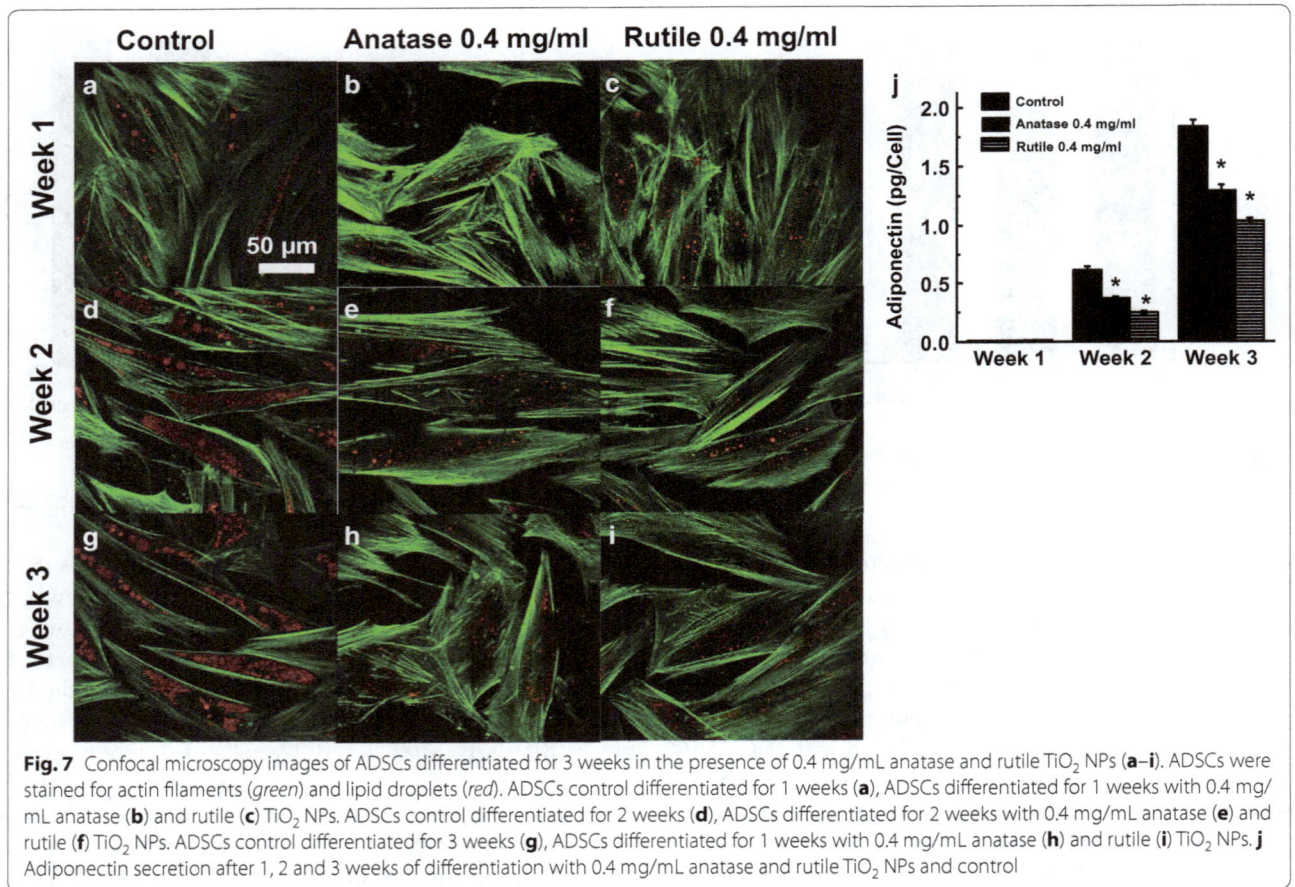

Fig. 7 Confocal microscopy images of ADSCs differentiated for 3 weeks in the presence of 0.4 mg/mL anatase and rutile TiO$_2$ NPs (**a–i**). ADSCs were stained for actin filaments (*green*) and lipid droplets (*red*). ADSCs control differentiated for 1 weeks (**a**), ADSCs differentiated for 1 weeks with 0.4 mg/mL anatase (**b**) and rutile (**c**) TiO$_2$ NPs. ADSCs control differentiated for 2 weeks (**d**), ADSCs differentiated for 2 weeks with 0.4 mg/mL anatase (**e**) and rutile (**f**) TiO$_2$ NPs. ADSCs control differentiated for 3 weeks (**g**), ADSCs differentiated for 1 weeks with 0.4 mg/mL anatase (**h**) and rutile (**i**) TiO$_2$ NPs. **j** Adiponectin secretion after 1, 2 and 3 weeks of differentiation with 0.4 mg/mL anatase and rutile TiO$_2$ NPs and control

50 ± 6 and 150 ± 7% after 3 weeks for anatase and rutile, respectively.

The effect of TiO$_2$ NPs on the expression of collagen is shown on Fig. 8b. Data shows that after 2 week treatment, all cultures exposed to rutile and anatase, at both concentrations, have elevated collagen production. Cultures exposed to 0.1 mg/mL anatase and rutile exhibited an increase in collagen by 32 ± 8 and 44 ± 13%, respectively,

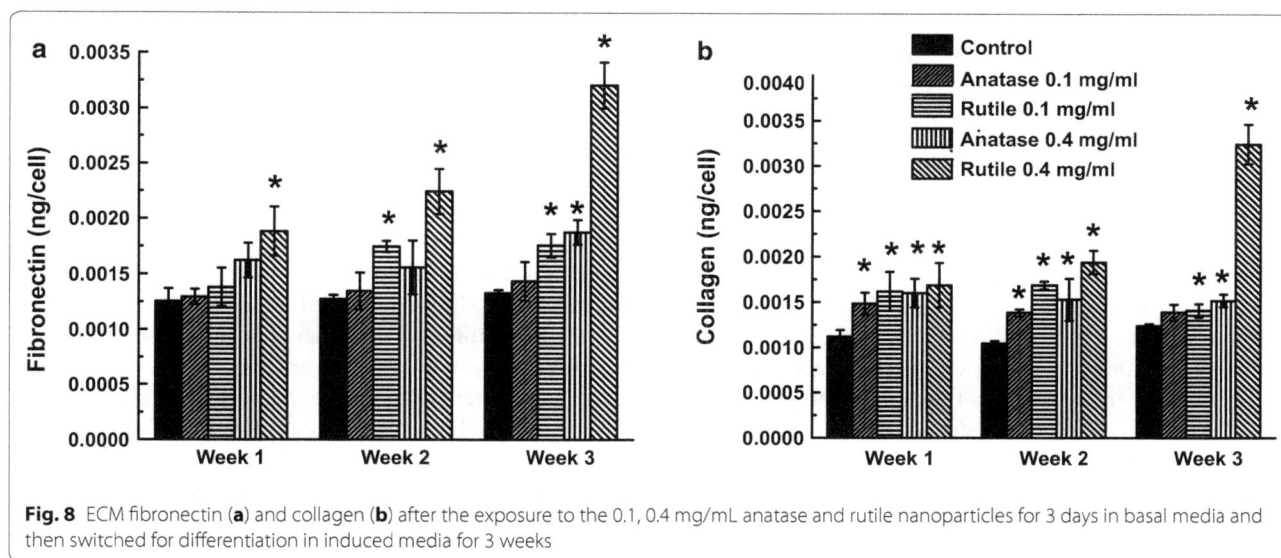

Fig. 8 ECM fibronectin (**a**) and collagen (**b**) after the exposure to the 0.1, 0.4 mg/mL anatase and rutile nanoparticles for 3 days in basal media and then switched for differentiation in induced media for 3 weeks

while exposure to 0.4 mg/mL led to increases by 42 ± 10 and $50 \pm 15\%$, respectively. After 2 weeks of adipose induction cultures exposed to TiO_2 showed a further increase in collagen, ranging from 34 ± 3 to $61 \pm 3\%$ in cultures treated with low concentrations and by 48 and 87% in cultures treated with high concentrations of anatase and rutile, respectively. After 3 weeks of adipogenic induction the increase in collagen in cultures treated with 0.1 mg/mL rutile and anatase was approximately 12%. Cultures treated with 0.4 mg/mL anatase had an increase in collagen by $23 \pm 5\%$ as compared to control TiO_2 untreated cultures. Interestingly, the increase in collagen in culture treated with 0.4 mg/mL rutile TiO_2 after 3 weeks of differentiation was $162 \pm 7\%$ relative to control untreated cultures (Fig. 8b).

Discussion

Increased production of TiO_2 NPs and their application in a variety of consumer products require careful dose and hazard assessment if we are to establish a concentration range that ensures safety for human health and the environment. The first step in hazard assessment should be proper characterization of ENMs in "as synthesized" state (e.g. in water) and, more importantly, in relevant environment or "as administrated" (e.g. in cell media for in vitro experiments). This will help to obtain meaningful and reproducible results and correlate ENMs properties with their potential toxicity in vitro [2, 5].

In this paper, we explored the range of TiO_2 NPs concentrations by testing one concentration that has no effect on cellular behavior based on a standard set of toxicity assays (ROS, LDH, and proliferation) and with another concentration that demonstrates their hazardous potential. In addition to standard assays, we performed

tests addressing the TiO_2 NPs effects on cell function that are characteristic for the chosen model cell line. We successfully demonstrated that for sufficient safety assessment of NPs in ADSCs cultures these assays are necessary as they exhibit greater sensitivity and may be used for identifying the guideline for safe concentrations.

The TiO_2 NPs chosen for this study have chemically unmodified negatively charged surface. It is known that the adsorption of proteins on the surface of the NPs changes the overall particle charge. Both basal and adipose induction medium used in this study were supplemented with 10% FBS that contains significant amount of albumin protein which has high affinity toward TiO_2 [40, 41]. As a result, we observed that protein corona adsorbed on the surface significantly increased the surface charge of both rutile and anatase. These findings are in agreement with previous reports by several research groups who showed that decrease in the surface charge of TiO_2 NPs results from adsorption of albumin and apolipoproteins from media supplemented with FBS [40, 42, 43]. Such a protein corona plays a critical role in NP stability and in the absence of albumin TiO_2 NPs sediment rapidly, as opposed to stable suspension that forms in the presence of albumin [43, 44]. Similarly, in our experiments all media-based TiO_2 NPs suspensions exhibited prolonged stability as compared to water-based suspensions.

Since TiO_2 NPs are suspended in basal or adipose induction medium prior to ADSCs treatment, it is important to determine the size of their aggregates. Therefore, to identify these parameters for in vitro cytotoxicity, we performed TiO_2 NP aggregation assessment by DLS. This is a standard method used to measure the Brownian size of NPs in colloidal suspensions. The average hydrodynamic diameter (secondary TiO_2 NPs size) of anatase aggregates was similar in all conditions revealing small aggregates of 2–3 particles each. On the other hand, aggregates of rutile in both, basal and induction media were determined to be smaller as compared to suspensions in Milli-Q water, indicating the stabilizing effect of the media due to the protein adsorption on the particle surface. This is in agreement with previously reported findings by multiple research groups [45–48].

Once suspended in liquid, TiO_2 NPs form fractal agglomerates [48, 49] decreasing the total number of free particles, and as a result, the total surface area available for bio-interactions [34]. Our measurement of TiO_2 NPs "effective density" confirmed that agglomerates are porous, and contain media trapped inside [33, 49] as their density was significantly smaller than the density of primary particles. Such findings are in alignment with previously published assessment of NPs [34] where authors demonstrated a similar trend; effective density of various NPs agglomerates in culture media is decreased as compared to the density of the material.

Since cells only respond to the NPs they come in contact with, we calculated the actual dose of NPs causing the effect—"delivered dose of NPs". It is known that rates of NP sedimentation, diffusion and agglomeration depend not only on their size, density, but also on surface physicochemistry and media characteristics [50]. All these properties determine the transport of NPs and significantly affect the delivered dose. Recently, a computational model of particokinetics and dosimetry for non-interacting spherical particles has been developed [33]. This ISDD model accounts for Stokes sedimentation and Stokes–Einstein diffusion and was demonstrated to have accurate predictions for particles of different sizes and densities [33]. Our calculations indicated that after 19 h of exposure 100% of administrated TiO_2 NPs mass comes in contact with cells. Therefore, we speculate that the observed toxicological response of the ADSCs treated with 0.1 and 0.4 mg/mL TiO_2 is equivalent to the response to 10 and 40 $\mu g/cm^2$ of TiO_2 NPs applied for 56 h in an in vivo study. It is important to note that according to FDA recommendations the amount of sunscreen that needs to be applied on the skin to achieve labeled SPF rating is 2 mg/cm^2 [51]. Since the FDA approves 25% by weight of TiO_2 in sunscreens [13], the amount of TiO_2 NPs in contact with skin can be as high as 0.5 mg/cm^2. Hence, the TiO_2 doses tested in this study are more than tenfold smaller than what is approved by FDA and thus reflects a real life scenario. Moreover, the FDA recommends re-application of the sunscreen every 2 h, in this case during 8 h of exposure (outdoor workers or sunbathers) the amount of TiO_2 NPs in contact with skin can reaches 4 mg/cm^2. In addition, since sun

exposure as well as medical treatment of skin diseases is likely to be a continuous process requiring at least daily application of sunscreen or medicine for a prolonged time, the exposure time used in our study is certainly representative.

Various research groups reported that TiO_2 NPs aggregates enter cells mainly via endocytosis and reside in the vacuoles and cell cytoplasm [52, 53] Interestingly, in our study, the mechanism of intracellular sequestration appears to be different for rutile and anatase TiO_2 NPs. Close inspection of TEM images reveals that rutile NPs are stored in a few large vacuoles ($\sim$2.7 μm) within the cytoplasm, whereas anatase NPs are sequestered in great number of much smaller vacuoles ($\sim$290 nm) that distributed uniformly across the cytoplasm. Also, similar to previous findings in other cell types [53, 54], we found that TiO_2 NPs did not enter the nucleus or any other organelles. However, TiO_2 NPs penetrated lipid droplets in the differentiated ADSCs, raising the question whether such disruption can cause lipid leakage and contribute to the overall decrease in intracellular lipid accumulation. Similar case of TiO_2 NPs penetrating lipids was reported in the in vivo study where authors observed that TiO_2 NP penetrated oil storage droplets in water flea *Daphnia magna* [55]. Similar to our case, these storage cells mainly contain lipids such as triacylglycerol that are used by *D. Magna* in periods of low food resources. In another report analogous changes were observed in the terrestrial isopod *Porcellio scaber* exposed to TiO_2 NPs [56]. Lastly, it was recently reported that TiO_2 NPs have an affinity to triglycerides and easily absorb them on their surface [57].

The cellular uptake of TiO_2 NPs was confirmed by cell granularity measurements using flow cytometry. Previously, increases in the side scatter intensity (SSC) and decreases in forward scatter intensity (FSC) were shown to correlate with changes in the refractive index of cells containing TiO_2 NPs [58]. Our results reveal shifts in SSC and FSC proving TiO_2 NPs uptake by ADSCs. Specifically, we found that cultures treated with higher concentrations of TiO_2 NPs have a larger increase in SSC as compared to ADSCs exposed to lower amounts of NPs. This trend is expected and related to the larger number of NPs inside the cell that can scatter more of the laser beam. Similarly, the larger increase in SSC intensity for anatase containing cells as compared to those with rutile can be explained by the fact that, unlike rutile NPs that are stored in the few vacuoles, anatase is stored in large number of small vacuoles that are much more uniformly distributed in the cytoplasm and thus increases the scattering. It has been shown that SSC provides information on internal structures and organelles [59], therefore, increase in SCC indicates TiO_2 NPs uptake rather than adhesion to the cell surface. As FSC measures the amount of the laser beam that passes around the cell it can be correlated with relative size of the cell, therefore, a slight decrease in FSC intensity that we observed indicates that cells are roughly the same in size and hence are not apoptotic [58].

Our assessment of TiO_2 NPs cytotoxicity through the ROS, LDH, and proliferation assays revealed the dose-dependent toxic effects on ADSCs. No cytotoxic effects were observed in cultures treated with 0.1 mg/mL TiO_2 NPs regardless of crystal structure. In contrast, higher concentrations (0.4 mg/mL) of both, rutile and anatase TiO_2 NPs, were sufficient to generate oxidative stress and cause LDH release—two hallmarks of NP cytotoxicity. These observations are in agreement with previously reported findings. For example, Shukla et al. reported increased ROS generation in human epidermal cells exposed to TiO_2 NPs for 6 h [60], similarly other research groups reported concentration dependent increase in ROS in human bronchial epithelial cells [61], brain neurons [62], and human amnion epithelial cells [46]. It is interesting to note, that on day 3 of culture treatment with 0.4 mg/mL TiO_2 NPs the proliferation profile of ADSCs was not affected even though we detected significant changes in ROS and LDH. Our data revealed a decrease in the cell numbers only after 4 days of exposure to high TiO_2 NPs concentration confirming greater sensitivity of ROS and LDH assays as compared to a proliferation test.

Similarly, we observed TiO_2 NP dose-dependent impairment of ADSC wound healing ability. This important cell function was studied using a migration assay and a three-dimensional collagen gel contraction model. Culturing cells in collagen gels is a standard procedure used to evaluate multiple aspects of wound repair and tissue remodeling. This model is used to reproduce behavior of mesenchymal cells and fibroblasts in a "tissue-like" environment [63, 64]. In the gel, cells attach to the matrix by integrin-mediated mechanism, exert mechanical tension and contract the matrix mimicking the wound repair process [63]. Our findings of decreased migration speed and collagen gel contraction in cultures exposed to high TiO_2 doses can be correlated and interpreted as a reduction of wound healing rates since these cells are required to move into the wound, deposit collagen, and eventually close the wound by contracting the collagen fibers [65, 66]. Such impairment could contribute to ineffective tissue repair and in turn increase chances of bacterial infection. These findings are in agreement with a previously reported decrease in migration and collagen gel contraction in dermal fibroblasts exposed to TiO_2 [67]. Other nanoparticulate materials, such as SiO_2, gold, and carbon nanotubes were also reported to have inhibitory effects on collagen gel contraction and migration of cells [68–71].

The importance of interaction between cells and the ECM for regulating proliferation, survival, migration, and differentiation is well established [72, 73]. It is also known that alteration of ECM composition can have a profound effect on cell behavior, including cell migration. For example, previous studies showed that migration of different cell types is promoted by expression and secretion of collagen type I [74–76]. Therefore, our findings of decreased collagen I production in cultures treated with 0.4 mg/mL TiO_2 NPs may explain the observed reduction in cell migration speed. Collagens are the most abundant structural components of the ECM that support a vast array of cell and tissue functions, including adhesion, migration, differentiation, morphogenesis, and wound healing [77]. It is interesting to note that moderate increases in fibronectin expression alone (in cultures exposed to 0.1 mg/mL TiO_2) is not enough to alter cell migration. Alternatively, disturbance in the collagen to fibronectin ratio leads to ECM fibrils enriched with fibronectin that makes it softer. Such change in the ECM's mechanical properties might be partially responsible for suppression of cell migration and collagen contraction. It is well documented, that cells have difficulty to exert proper adhesion and traction forces on softer ECM fibrils [78, 79]. Similarly, alteration of collagen/fibronectin ratio was previously found in human dermal fibroblasts exposed to gold nanoparticles [6]. Further, the role of ADSCs in wound healing was also studied by Kim et al. [80] where he identified secretion of various growth factors by ADSCs as being the essential event that promotes wound closure and re-epithelialization through a paracrine mechanism. For example, different cues secreted by ADSCs activate dermal fibroblasts and keratinocytes which accelerate wound healing in vivo by stimulating collagen expression and migration of dermal fibroblasts, and also protect dermal fibroblasts from oxidative stress [81, 82]. Therefore, we would like to suggest that changes in ADSCs observed in our study may have more profound effects on the wound healing process in vivo due to the complexity of cell–cell interactions and increased secretory load of ADSCs.

Another essential physiological function of ADSCs is their ability to differentiate into adipose tissue and store energy via accumulation of lipids. Adipocyte differentiation is a complex process that occurs via a chain of transcriptional and post-transcriptional events that coordinate changes in cell morphology, hormone sensitivity and gene expression [83, 84]. In this study, we found a delayed lipid accumulation and adipokine secretion in ADSCs that demonstrate the inhibitory effects of TiO_2 NPs on cell differentiation. Our findings are in agreement with previous reports (including our own previous data) demonstrating similar inhibitory effects of various nanomaterials on adipogenic conversion of ADSCs and mesenchymal stem cells [71, 85, 86]. A recent study evaluated the cytotoxic effects of TiO_2 nanorods in mesenchymal stem cells [87], however, no changes in adipogenic differentiation were observed. These results may be explained by the low concentration of TiO_2 nanorods chosen for the study (10 μg/mL).

Since adiponectin is an important mediator of many physiologically relevant processes that help regulate whole body energy expenditure [88], its reduction in various fat depots could potentially generate profound local and systemic effects. For example, it can modify energy metabolism, induce insulin resistance, or promote cardiovascular disease [23, 89–91]. In addition, adipogenesis of dermal adipocytes occurs following injury where during the proliferative stage of the healing process adiponectin-expressing adipocytes repopulate skin wounds [92]. Therefore, reduced adipogenesis of ADSCs exposed to TiO_2 NPs may also adversely affect skin wound healing in addition to previously observed alterations in ECM expression, migration, and collagen contraction.

The transition of ADSCs from fibroblast-like state to an adipocyte phenotype is a complex event guided by remodeling of the ECM [93] and as such, we studied two main structural proteins of the ECM—fibronectin and collagen during the adipogenic differentiation process. A hallmark of this transition is the degradation of ECM and moderate decrease in fibronectin and collagen amounts. Even though we only observed small fluctuations in fibronectin and collagen content in the control samples, we observed sufficient increase in the amount of both proteins in cultures differentiated in the presence of TiO_2 NPs. The overproduction of ECM may explain the delayed differentiation observed in the cultures treated with rutile and anatase. On the other hand, such a change in the collagen/fibronectin ratio during differentiation results in stiffer collagen enriched ECM which has been previously reported to limit adipocyte growth [94]. It is interesting that abnormal collagen deposition is a hallmark of fibrosis development in adipose tissue and is tightly linked to tissue inflammation due to infiltration of immune cells [95]. Moreover, increased ECM deposition combined with its decreased flexibility were recently shown to cause adipocyte metabolic dysfunction and obesogenic adipose tissue remodeling [95, 96].

Conclusions

The discrepancy between in vitro and in vivo cytotoxicity tests results is a bottleneck in developing efficient screening methods to address the safety of a constantly growing number of ENMs. Thus, evaluation of ENMs cytotoxicity based on delivered doses should help to eliminate this difference and enable efficient and reliable in vitro screening methods. However, before developing a new generation of assays the relevance and sensitivity of these tools needs to be carefully assessed.

Here, we have demonstrated greater sensitivity of assays based on specialized cell functions for assessing damage induced by TiO_2 NPs exposure in ADSCs as compared to the standard assays (ROS, LDH, and proliferation). We observed significant changes in ECM protein secretion and reduction in angiogenesis of cultures treated with low concentrations of TiO_2 NPs in contrast to no damage detected by standard assays. Such changes potentially can cause impairment of two most important ADSCs functions and lead to unintentional harm to human health. In addition, we demonstrated that TiO_2 NPs induce cytotoxicity in ADSCs in a concentration-dependent manner by increasing ROS, extracellular LDH, altering ECM protein secretion, and decreasing wound healing ability and differentiation.

Our approach of addressing ENMs toxicity is based on changes in crucial cellular functions which is more sensitive than the set of standardized assays and depicts functional alterations that may have serious health implications. As different cell types will react differently to ENMs exposure, we envision that the proposed approach will help in hazardous ranking of ENMs and also become an essential tool for the development of novel and safer-by-design ENMs.

Abbreviations

ADSC: adipose derived stromal cells; TiO_2: titanium dioxide; NPs: nanoparticles; ENMs: engineered nanomaterials; ENM: engineered nanomaterial; PBS: phosphate buffered saline; FBS: fetal bovine serum; DMEM: Dulbecco's modified eagle's medium; IBMX: 3-isobutyl-1-methylxanthine; PS: penicillin-streptomycin; ROS: reactive oxygen species; LDH: lactate dehydrogenase; ECM: extracellular matrix; DLS: dynamic light scattering; TEM: transmission electron microscopy; ISDD: in vitro sedimentation, diffusion and dosimetry; VCM: volumetric centrifugation method; PCV: packed cell volume; SSC: side scatter; FSC: forward scatter.

Authors' contributions

YX participated in the design of the studies, performed cell experiments, conducted biological assays, collected and processed particle size distribution data, characterized particle charge and stability, and analyzed final data. MH participated in the design of the study, helped with interpretation of results, and edited final manuscript. MR helped to design the study and assisted with data interpretation. TM initiated the study, participated in the study design and data analysis, performed ELISA assays, lipids staining experiments, oversaw cell exposure studies, other biological assays, particle characterization, and drafted the manuscript. All authors read and approved the final manuscript.

Author details

[1] Department of Materials Science and Engineering, Stony Brook University, Stony Brook, NY, USA. [2] Department of Life Sciences, New York Institute of Technology, Old Westbury, NY, USA.

Acknowledgements

Not applicable.

Competing interests
The authors declare that they have no competing interests.

Funding
This study was partially supported by the NSF-Inspire program Grant DMR-1344267.

References

1. Holsapple MP, Farland WH, Landry TD, Monteiro-Riviere NA, Carter JM, Walker NJ, Thomas KV. Research strategies for safety evaluation of nano-materials, part II: toxicological and safety evaluation of nanomaterials, current challenges and data needs. Toxicol Sci. 2005;88:12–7.
2. Oberdorster G, Oberdorster E, Oberdorster J. Nanotoxicology: an emerging discipline evolving from studies of ultrafine particles. Environ Health Perspect. 2005;113:823–39.
3. Nel AE. Implementation of alternative test strategies for the safety assessment of engineered nanomaterials. J Intern Med. 2013;274:561–77.
4. Nel A, Xia T, Madler L, Li N. Toxic potential of materials at the nanolevel. Science. 2006;311:622–7.
5. Powers KW, Brown SC, Krishna VB, Wasdo SC, Moudgil BM, Roberts SM. Research strategies for safety evaluation of nanomaterials. Part VI. Characterization of nanoscale particles for toxicological evaluation. Toxicol Sci. 2006;90:296–303.
6. Mironava T, Hadjiargyrou M, Simon M, Jurukovski V, Rafailovich MH. Gold nanoparticles cellular toxicity and recovery: effect of size, concentration and exposure time. Nanotoxicology. 2010;4:120–37.
7. Robichaud CO, Uyar AE, Darby MR, Zucker LG, Wiesner MR. Estimates of upper bounds and trends in nano-TiO$_2$ production as a basis for exposure assessment. Environ Sci Technol. 2009;43:4227–33.
8. Occupational exposure to titanium dioxide. vol. 63. Current Intelligence Bulletin: Department of Health and Human Services; 2011.
9. Weir A, Westerhoff P, Fabricius L, Hristovski K, von Goetz N. Titanium dioxide nanoparticles in food and personal care products. Environ Sci Technol. 2012;46:2242–50.
10. Odonoghue MN. Sunscreen—the ultimate cosmetic. Dermatol Clin. 1991;9:99–104.
11. Wang JJ, Sanderson BJS, Wang H. Cyto- and genotoxicity of ultrafine TiO$_2$ particles in cultured human lymphoblastoid cells. Mutat Res. 2007;628:99–106.
12. Kuempel ED, Ruder R. Titanium dioxide (TiO$_2$). 2010. https://monographs.iarc.fr/ENG/Publications/techrep42/TR42-4.pdf. Accessed 6 July 2017.
13. Regulations C-CoF. Sunscreen drug products for over-the-counter human use. 2015.
14. Earth Fot. 2012.
15. Newman MD, Stotland M, Ellis JI. The safety of nanosized particles in titanium dioxide- and zinc oxide-based sunscreens. J Am Acad Dermatol. 2009;61:685–92.
16. Barker PJ, Branch A. The interaction of modern sunscreen formulations with surface coatings. Prog Org Coat. 2008;62:313–20.
17. Tholpady SS, Llull R, Ogle RC, Rubin JP, Futrell JW, Katz AJ. Adipose tissue: stem cells and beyond. Clin Plast Surg. 2006;33:55.
18. Pittenger MF, Mackay AM, Beck SC, Jaiswal RK, Douglas R, Mosca JD, Moorman MA, Simonetti DW, Craig S, Marshak DR. Multilineage potential of adult human mesenchymal stem cells. Science. 1999;284:143–7.
19. Zuk PA, Zhu M, Mizuno H, Huang J, Futrell JW, Katz AJ, Benhaim P, Lorenz HP, Hedrick MH. Multilineage cells from human adipose tissue: implications for cell-based therapies. Tissue Eng. 2001;7:211–28.
20. Lin YF, Luo E, Chen XZ, Liu L, Qiao J, Yan ZB, Li ZY, Tang W, Zheng XH, Tian WD. Molecular and cellular characterization during chondrogenic differentiation of adipose tissue-derived stromal cells in vitro and cartilage formation in vivo. J Cell Mol Med. 2005;9:929–39.
21. De Ugarte DA, Morizono K, Elbarbary A, Alfonso Z, Zuk PA, Zhu M, Dragoo JL, Ashjian P, Thomas B, Benhaim P, et al. Comparison of multi-lineage cells from human adipose tissue and bone marrow. Cells Tissues Organs. 2003;174:101–9.
22. Bunnell BA, Flaat M, Gagliardi C, Patel B, Ripoll C. Adipose-derived stem cells: isolation, expansion and differentiation. Methods. 2008;45:115–20.
23. Trujillo ME, Scherer PE. Adipose tissue-derived factors: impact on health and disease. Endocr Rev. 2006;27:762–78.
24. Wu JH, Liu W, Xue CB, Zhou SC, Lan FL, Bi L, Xu HB, Yang XL, Zeng FD. Toxicity and penetration of TiO$_2$ nanoparticles in hairless mice and porcine skin after subchronic dermal exposure. Toxicol Lett. 2009;191:1–8.
25. Tan MH, Commens CA, Burnett L, Snitch PJ. A pilot study on the percutaneous absorption of microfine titanium dioxide from sunscreens. Australas J Dermatol. 1996;37:185–7.
26. Popov AP, Kirillin MY, Priezzhev AV, Lademann J, Hast J, Myllyla R. Optical sensing of titanium dioxide nanoparticles within horny layer of human skin and their protecting effect against solar UV radiation. In: Priezzhev AV, Cote GL, editors. Optical diagnostics and sensing, vol. 5702. Proceedings of the Society of Photo-Optical Instrumentation Engineers (Spie). 2005. p. 113–22.
27. Durand L, Habran N, Henschel V, Amighi K. In vitro evaluation of the cutaneous penetration of sprayable sunscreen emulsions with high concentrations of UV filters. Int J Cosmet Sci. 2009;31:279–92.
28. SCCS. Opinion on titanium dioxide (nano form). 2013.
29. Association NE. Sunscreens.
30. Foundation NP. Treating psoriasis with sunlight.
31. GmbH RH. Product description, Sorion Cream. 2016.
32. Information CM. Novasone (Mometasone furoate). 2011.
33. Hinderliter PM, Minard KR, Orr G, Chrisler WB, Thrall BD, Pounds JG, Teeguarden JG. ISDD. A computational model of particle sedimentation, diffusion and target cell dosimetry for in vitro toxicity studies. Part Fibre Toxicol. 2010;7:36.
34. DeLoid G, Cohen JM, Darrah T, Derk R, Rojanasakul L, Pyrgiotakis G, Wohlleben W, Demokritou P. Estimating the effective density of engineered nanomaterials for in vitro dosimetry. Nat Commun. 2014;5:3514.
35. Han X, Gelein R, Corson N, Wade-Mercer P, Jiang J, Biswas P, Finkelstein JN, Elder A, Oberdoerster G. Validation of an LDH assay for assessing nanoparticle toxicity. Toxicology. 2011;287:99–104.
36. Geiger B, Yamada KM. Molecular architecture and function of matrix adhesions. Cold Spring Harb Perspect Biol. 2011;3:a005033.
37. Ahima RS. Metabolic actions of adipocyte hormones: focus on adiponectin. Obesity. 2006;14:9S–15S.
38. Sampaolesi M, Torrente Y, Innocenzi A, Tonlorenzi R, D'Antona G, Pellegrino MA, Barresi R, Bresolin N, De Angelis MGC, Campbell KP, et al. Cell therapy of α-sarcoglycan null dystrophic mice through intra-arterial delivery of mesoangioblasts. Science. 2003;301:487–92.
39. Jiang YH, Vaessena B, Lenvik T, Blackstad M, Reyes M, Verfaillie CM. Multipotent progenitor cells can be isolated from postnatal murine bone marrow, muscle, and brain. Exp Hematol. 2002;30:896–904.
40. Oliva FY, Avalle LB, Camara OR, De Pauli CP. Adsorption of human serum albumin (HSA) onto colloidal TiO$_2$ particles, Part I. J Colloid Interface Sci. 2003;261:299–311.
41. Klinger A, Steinberg D, Kohavi D, Sela MN. Mechanism of adsorption of human albumin to titanium in vitro. J Biomed Mater Res. 1997;36:387–92.
42. Deng ZJ, Mortimer G, Schiller T, Musumeci A, Martin D, Minchin RF. Differential plasma protein binding to metal oxide nanoparticles. Nanotechnology. 2009;20:455101.
43. Allouni ZE, Cimpan MR, Hol PJ, Skodvin T, Gjerdet NR. Agglomeration and sedimentation of TiO$_2$ nanoparticles in cell culture medium. Colloids Surf B Biointerfaces. 2009;68:83–7.
44. Vamanu CI, Hol PJ, Allouni ZE, Elsayed S, Gjerdet NR. Formation of potential titanium antigens based on protein binding to titanium dioxide nanoparticles. Int J Nanomed. 2008;3:69–74.
45. Singh S, Shi TM, Duffin R, Albrecht C, van Berlo D, Hoehr D, Fubini B, Martra G, Fenoglio I, Borm PJA, Schins RPF. Endocytosis, oxidative stress and IL-8 expression in human lung epithelial cells upon treatment with fine and ultrafine TiO$_2$: role of the specific surface area and of surface methylation of the particles. Toxicol Appl Pharmacol. 2007;222:141–51.
46. Saquib Q, Al-Khedhairy AA, Siddiqui MA, Abou-Tarboush FM, Azam A, Musarrat J. Titanium dioxide nanoparticles induced cytotoxicity, oxidative

stress and DNA damage in human amnion epithelial (WISH) cells. Toxicol In Vitro. 2012;26:351–61.

47. Xia T, Kovochich M, Brant J, Hotze M, Sempf J, Oberley T, Sioutas C, Yeh JI, Wiesner MR, Nel AE. Comparison of the abilities of ambient and manufactured nanoparticles to induce cellular toxicity according to an oxidative stress paradigm. Nano Lett. 2006;6:1794–807.

48. Verma A, Stellacci F. Effect of surface properties on nanoparticle–cell interactions. Small. 2010;6:12–21.

49. Cohen J, DeLoid G, Pyrgiotakis G, Demokritou P. Interactions of engineered nanomaterials in physiological media and implications for in vitro dosimetry. Nanotoxicology. 2013;7:417–31.

50. Teeguarden JG, Hinderliter PM, Orr G, Thrall BD, Pounds JG. Particokinetics in vitro: dosimetry considerations for in vitro nanoparticle toxicity assessments. Toxicol Sci. 2007;95:300–12.

51. (CDER) USDoHaHSFaDACfDEaR. Guidance for industry labeling and effectiveness testing: sunscreen drug products for over the-counter human use—small entity compliance guide. 2012.

52. Hussain S, Boland S, Baeza-Squiban A, Hamel R, Thomassen LCJ, Martens JA, Billon-Galland MA, Fleury-Feith J, Moisan F, Pairon JC, Marano F. Oxidative stress and proinflammatory effects of carbon black and titanium dioxide nanoparticles: role of particle surface area and internalized amount. Toxicology. 2009;260:142–9.

53. Xu Y, Wei MT, Ou-Yang HD, Walker SG, Wang HZ, Gordon CR, Guterman S, Zawacki E, Applebaum E, Brink PR, et al. Exposure to TiO$_2$ nanoparticles increases Staphylococcus aureus infection of HeLa cells. J Nanobiotechnol. 2016;14:34.

54. Tucci P, Porta G, Agostini M, Dinsdale D, Iavicoli I, Cain K, Finazzi-Agro A, Melino G, Willis A. Metabolic effects of TiO$_2$ nanoparticles, a common component of sunscreens and cosmetics, on human keratinocytes. Cell Death Dis. 2013;4:e549.

55. Rosenkranz P, Chaudhry Q, Stone V, Fernandes TF. A comparison of nanoparticle and fine particle uptake by Daphnia magna. Environ Toxicol Chem. 2009;28:2142–9.

56. Srpcic AM, Drobne D, Novak S. Altered physiological conditions of the terrestrial isopod Porcellio scaber as a measure of subchronic TiO$_2$ effects. Protoplasma. 2015;252:415–22.

57. Jebali A, Kazemi B. Triglyceride-coated nanoparticles: skin toxicity and effect of UV/IR irradiation on them. Toxicol In Vitro. 2013;27:1847–54.

58. Zucker RM, Massaro EJ, Sanders KM, Degn LL, Boyes WK. Detection of TiO$_2$ nanoparticles in cells by flow cytometry. Cytom Part A. 2010;77A:677–85.

59. Steen HB. Flow cytometer for measurement of the light scattering of viral and other submicroscopic particles. Cytom Part A. 2004;57A:94–9.

60. Shukla RK, Sharma V, Pandey AK, Singh S, Sultana S, Dhawan A. ROS-mediated genotoxicity induced by titanium dioxide nanoparticles in human epidermal cells. Toxicol In Vitro. 2011;25:231–41.

61. Park EJ, Yi J, Chung YH, Ryu DY, Choi J, Park K. Oxidative stress and apoptosis induced by titanium dioxide nanoparticles in cultured BEAS-2B cells. Toxicol Lett. 2008;180:222–9.

62. Liu SC, Xu LJ, Zhang T, Ren GG, Yang Z. Oxidative stress and apoptosis induced by nanosized titanium dioxide in PC12 cells. Toxicology. 2010;267:172–7.

63. Grinnell F. Fibroblast biology in three-dimensional collagen matrices. Trends Cell Biol. 2003;13:264–9.

64. Dallon JC, Ehrlich HP. A review of fibroblast-populated collagen lattices. Wound Repair Regen. 2008;16:472–9.

65. Cherubino M, Rubin JP, Miljkovic N, Kelmendi-Doko A, Marra KG. Adipose-derived stem cells for wound healing applications. Ann Plast Surg. 2011;66:210–5.

66. Dujmovic TB, Clark S. Assessment of dermal fibroblast and myofibroblast migration during wound healing. Wound Repair Regen. 2009;17:A38.

67. Pan Z, Lee W, Slutsky L, Clark RAF, Pernodet N, Rafailovich MH. Adverse effects of titanium dioxide nanoparticles on human dermal fibroblasts and how to protect cells. Small. 2009;5:511–20.

68. Kim HJ, Liu XD, Kobayashi T, Kohyama T, Wen FQ, Romberger DJ, Conner H, Gilmour PS, Donaldson K, MacNee W, Rennard SI. Ultrafine carbon black particles inhibit human lung fibroblast-mediated collagen gel contraction. Am J Respir Cell Mol Biol. 2003;28:111–21.

69. Wilson CG, Sisco PN, Goldsmith EC, Murphy CJ. Glycosaminoglycan-functionalized gold nanorods: interactions with cardiac cells and type I collagen. J Mater Chem. 2009;19:6332–40.

70. Zhang YY, Hu L, Gao CY. Effect of cellular uptake of SiO$_2$ particles on adhesion and migration of HepG2 cells. Acta Polym Sin. 2009;8:815–22.

71. Mironava T, Hadjiargyrou M, Simon M, Rafailovich MH. Gold nanoparticles cellular toxicity and recovery: adipose derived stromal cells. Nanotoxicology. 2014;8:189–201.

72. Khalili AA, Ahmad MR. A review of cell adhesion studies for biomedical and biological applications. Int J Mol Sci. 2015;16:18149–84.

73. Hynes RO. The extracellular matrix: not just pretty fibrils. Science. 2009;326:1216–9.

74. Rocnik EF, Chan BMC, Pickering JG. Evidence for a role of collagen synthesis in arterial smooth muscle cell migration. J Clin Investig. 1998;101:1889–98.

75. Basson MD, Turowski G, Emenaker NJ. Regulation of human (Caco-2) intestinal epithelial cell differentiation by extracellular matrix proteins. Exp Cell Res. 1996;225:301–5.

76. Fu XL, Xu M, Liu J, Qi YM, Li SH, Wang HJ. Regulation of migratory activity of human keratinocytes by topography of multiscale collagen-containing nanofibrous matrices. Biomaterials. 2014;35:1496–506.

77. Aumailley M, Gayraud B. Structure and biological activity of the extracellular matrix. J Mol Med. 1998;76:253–65.

78. Daley WP, Peters SB, Larsen M. Extracellular matrix dynamics in development and regenerative medicine. J Cell Sci. 2008;121:255–64.

79. Bischofs IB, Schwarz US. Cell organization in soft media due to active mechanosensing. Proc Natl Acad Sci USA. 2003;100:9274–9.

80. Kim WS, Park BS, Sung JH, Yang JM, Park SB, Kwak SJ, Park JS. Wound healing effect of adipose-derived stem cells: a critical role of secretory factors on human dermal fibroblasts. J Dermatol Sci. 2007;48:15–24.

81. Kim WS, Park BS, Kim HK, Park JS, Kim KJ, Choi JS, Chung SJ, Kim DD, Sung JH. Evidence supporting antioxidant action of adipose-derived stem cells: protection of human dermal fibroblasts from oxidative stress. J Dermatol Sci. 2008;49:133–42.

82. Kim WS, Park BS, Park SH, Kim HK, Sung JH. Antiwrinkle effect of adipose-derived stem cell: activation of dermal fibroblast by secretory factors. J Dermatol Sci. 2009;53:96–102.

83. Rosen ED, Spiegelman BM. Molecular regulation of adipogenesis. Annu Rev Cell Dev Biol. 2000;16:145–71.

84. Rangwala SM, Lazar MA. Transcriptional control of adipogenesis. Annu Rev Nutr. 2000;20:535–59.

85. Son MJ, Kim WK, Kwak M, Oh KJ, Chang WS, Min JK, Lee SC, Song NW, Bae KH. Silica nanoparticles inhibit brown adipocyte differentiation via regulation of p38 phosphorylation. Nanotechnology. 2015;26:435101.

86. Rocca A, Mattoli V, Mazzolai B, Ciofani G. Cerium oxide nanoparticles inhibit adipogenesis in rat mesenchymal stem cells: potential therapeutic implications. Pharm Res. 2014;31:2952–62.

87. Shrestha S, Mao Z, Fedutikb Y, Gao C. Influence of titanium dioxide nanorods with different surface chemistry on the differentiation of rat bone marrow mesenchymal stem cells. J Mater Chem B. 2016;4:6955–66.

88. Lafontan M, Viguerie N. Role of adipokines in the control of energy metabolism: focus on adiponectin. Curr Opin Pharmacol. 2006;6:580–5.

89. Lin ZF, Tian HS, Lam KSL, Lin SQ, Hoo RCL, Konishi M, Itoh N, Wang Y, Bornstein SR, Xu AM, Li XK. Adiponectin mediates the metabolic effects of FGF21 on glucose homeostasis and insulin sensitivity in mice. Cell Metab. 2013;17:779–89.

90. Bastard JP, Maachi M, Lagathu C, Kim MJ, Caron M, Vidal H, Capeau J, Feve B. Recent advances in the relationship between obesity, inflammation, and insulin resistance. Eur Cytokine Netw. 2006;17:4–12.

91. Schondorf T, Maiworm A, Emmison N, Forst T, Pfutzner A. Biological background and role of adiponectin as marker for insulin resistance and cardiovascular risk. Clin Lab. 2005;51:489–94.

92. Schmidt BA, Horsley V. Intradermal adipocytes mediate fibroblast recruitment during skin wound healing. Development. 2013;140:1517–27.

93. Lee J, Wu YY, Fried SK. Adipose tissue remodeling in pathophysiology of obesity. Curr Opin Clin Nutr Metab Care. 2010;13:371–6.

94. Khan T, Muise ES, Iyengar P, Wang ZV, Chandalia M, Abate N, Zhang BB, Bonaldo P, Chua S, Scherer PE. Metabolic dysregulation and adipose tissue fibrosis: role of collagen VI. Mol Cell Biol. 2009;29:1575–91.

95. Sun K, Kusminski CM, Scherer PE. Adipose tissue remodeling and obesity. J Clin Investig. 2011;121:2094–101.

96. Sun K, Tordjman J, Clement K, Scherer PE. Fibrosis and adipose tissue dysfunction. Cell Metab. 2013;18:470–7.

16

Understanding cellular internalization pathways of silicon nanowires

Kelly McNear[1†], Yimin Huang[1†] and Chen Yang[1,2*]

Abstract

Background: Understanding how cells interact with nanomaterials is important for rational design of nanomaterials for nanomedicine and transforming them for clinical applications. Particularly, the mechanism for one-dimensional (1D) nanomaterials with high aspect ratios still remains unclear.

Results: In this work, we present amine-functionalized silicon nanowires (SiNW-NH$_2$) entering CHO-β cells via a physical membrane wrapping mechanism. By utilizing optical microscopy, transmission electron microscopy, and confocal fluorescence microscopy, we successfully visualized the key steps of internalization of SiNW-NH$_2$ into cells.

Conclusion: Our results provide insight into the interaction between 1D nanomaterials and confirm that these materials can be used for understanding membrane mechanics through physical stress exerted on the membrane.

Keywords: Cellular interaction, Silicon nanowires, Membrane wrapping

Background

As nanotechnology advances as an innovative option in clinical settings, researchers continue to explore a wide array of nanomaterials for applications as imaging and anti-cancer agents, for drug delivery purposes, and for therapeutics. While this progress has been exciting for the future of medicine, these materials have not overcome the barrier of translating from benchtop to clinic. In order for nanomaterials to advance as viable options for biological applications, further understanding of the basic interactions between mammalian cells and nanomaterials must be achieved.

In the past few decades, to understand how the cellular membrane can respond to the entry of external nanomaterials research has been mainly focused on finding the endocytosis pathways of zero dimensional (0D) nanomaterials [1–5]. Limited efforts have also been made to understand the uptake of various 1D nanomaterials into cells. For example, gold nanorods and magnetic nanowires have been heavily studied for imaging and tracking

purposes [3, 6–8], but their specific uptake pathways were not well-studied. Single-walled carbon nanotubes (SWCNTs) have also been of interest due to their high aspect ratio and uniqueness as a material. Yaron and coworkers showed that the uptake of SWCNTs was energy-dependent, suggesting that the pathway is endocytosis and not membrane penetration, but the surface and size is so dissimilar to that of other longer 1D materials that these uptake pathways cannot be translated [9]. Additionally, Kostarelos and co-authors investigated previously investigated functionalized carbon nanotubes (*f*-CNTs) with a variety of functional groups, including an ammonium functionalization. The authors observed that the ammonium functionalized wires enter the cells at both 37 and 4 °C, ruling out a receptor-mediated pathway [10]. Notably, the dimensions of the *f*-CNTs that were studied were on the scale of 1 nm in diameter and 1000 nm in length, making them unique from other 1D nanomaterials.

Theoretical studies on both 0D and 1D nanostructures suggest that understanding the membrane mechanics during endocytosis is a critical aspect to explain internalization pathways of these nanostructures. Huang and co-authors designed a nanorod model using coarse-grained molecular dynamics to demonstrate that for endocytosis such model system needs to initially bind in an upright docking

*Correspondence: yang@purdue.edu

†Kelly McNear and Yimin Huang contributed equally to the work

¹ Department of Chemistry, Purdue University, 560 Oval Drive, West Lafayette, IN 47907, USA

Full list of author information is available at the end of the article

position, on the membrane plane, and then be wrapped by the membrane in order to proceed through a laying-down-then-standing-up sequence to enter cells [11]. Based on this work, it is reasonable to believe that membrane wrapping occurs as one of the first steps in endocytosis. Shi and co-authors used multi-walled carbon nanotubes to experimentally and theoretically illustrate that the cell entry of 1D nanomaterials can occur by tip recognition and rotation, but the authors do not delve into the details of internalization [12]. Yi and co-workers theoretically proved that cell membrane internalizes 1D nanomaterials following a near-perpendicular entry mode at small membrane tension but switches to a near-parallel interaction mode at large membrane tension [13]. These theoretical models illustrate the necessity to experimentally understand the membrane interactions of high aspect ratio 1D nanomaterials.

Due to the advantages of anisotropy and higher surface area to volume ratios than 0D nanomaterials, 1D nanomaterials can produce a stronger interaction with cells during the entering process. These features indicate that 1D nanomaterials can be considered as a better system to explore the possible membrane wrapping mechanism in the uptake pathways of nanomaterials [14, 15]. To this end, in this work we use multiple microscopy methods to visualize key steps during the cellular internalization process of silicon nanowires. Since silicon nanowires are fabricated with a complete control of dimensions [16, 17], have flexible chemistry for surface modification [18], have unique optical properties for in vitro and in vivo imaging [19–21], they are excellent candidates for the studies of cellular uptake pathways. Silicon nanowires modified with amine groups are the focus of the study, as compared to as-grown SiNWs with hydroxyl groups and SiNWs with specific targeting groups conjugated, SiNW-NH$_2$ are able to be internalized in cells without targeting receptor mediated processes [22]. We demonstrate the uptake pathway of 5 μm SiNW-NH$_2$ to be a physical membrane wrapping mechanism using CHO-β and HeLa cell lines. Studies at two different incubation temperatures, 37 and 4 °C, were carried out in order to evaluate temperature dependence of the membrane mechanics as well as to elucidate that the process is physically driven rather than receptor-mediated. We chose 4 °C because it is well understood that many endocytic pathways are temperature dependent and that these pathways are limited to high temperature due to the large activation barrier, so uptake at a lower temperature would indicate that the mechanism is physically driven [23].

Methods
Synthesis and functionalization of SiNWs
Silicon nanowires were synthesized by chemical vapor deposition (CVD) with 40 nm gold nanoparticles as growth catalysts and silane as a precursor. SiNWs in length of 5 μm were chosen in this works because of the appropriate aspect ratio, low cytotoxicity and considerable high endocytosis rate to study the cellular internalization [22]. The CVD growth was carried out in with a growth pressure of 100 Torr, silane flow rate of 5 sccm, and growth time of 5 min [22]. A representative TEM micrograph (Additional file 1: Figure S1) shows that the dimensions of the wires are consistent with the growth conditions.

The as-grown SiNWs were first treated with thermal oxidation in atmosphere at 900 °C for 2 min in order to clean and oxidize the surface for further modification. The thermally oxidized SiNWs were then submerged in a solution of 1% (3-aminopropyl) trimethoxysilane (APTMS) (Sigma-Aldrich, St. Louis, MO, USA) in pure ethanol overnight at room temperature. The substrates were then rinsed and submerged in pure ethanol at 80 °C in order to stabilize the functional groups. After 2 h, the substrates were removed and dried under a stream of nitrogen. SiNWs-NH$_2$ were then removed from the substrate into pure ethanol via sonication for several seconds. The SiNW-NH$_2$ suspensions were centrifuged at 14,000 rpm three times to wash away impurities. The wires were then resuspended in a complete Roswell Park Memorial Institute (RPMI) 1640 cell culture medium with 10% FBS, 1% L-glutamine, 1% penicillin/streptococcus for future use.

In order to confirm that the amine modification was successful, unmodified SiNWs and SiNWs-NH$_2$ were sonicated in ultrapure water and placed in disposable folded capillary cells for zeta potential measurements. The concentration of SiNWs and SiNW-NH$_2$ in each capillary cell was estimated to be 10 μg/mL. Measurements were carried out on a Malvern Zetasizer Nano Series.

To further confirm and quantify amine functionalization on the modified silicon nanowires, the authors performed Fourier transform infrared spectroscopy (FTIR) (Thermo Nicolet Nexus FTIR) as well as X-ray photoelectron spectroscopy (XPS) analysis. As grown nanowires on growth substrates were modified with amine and used as the modified samples for measurements here. Comparison between the FTIR spectrum for the unmodified silicon nanowires (blue curve in Additional file 1: Figure S2) and the FTIR for the amine-modified wires (red curve in Additional file 1: Figure S2) shows that a peak near 1600 cm^{-1} in the spectrum for the amine-modified wires is visible and corresponds to the NH$_2$ bending mode, confirming the presence of amine groups. The XPS spectrum measured from the modified SiNW-NH$_2$ sample (Additional file 1: Figure S3) shows atomic percentages of N and Si of 1.2 and 20.3%, respectively. Based on this result, the amine coverage was estimated to be 0.61 mol/

nm^2, which is close to the coverage previously modeled based on the total covalently bonded APTES coverage on silica [24].

Cell culture

We cultured immortalized Chinese hamster ovary cells transfected with folate receptor beta (CHO-β) for our cellular interaction studies. Our previous study investigated the uptake of SiNW-NH$_2$ using both CHO and CHO-β cells. The findings showed that SiNW-NH$_2$ were successfully internalized by both cell types. Therefore, CHO-β was chosen as the cell line of choice due to their success in internalizing SiNW-NH$_2$ [22]. Typically, CHO-β cells were cultured in the complete RPMI 1640 mentioned above at 37 °C in a humidified atmosphere with 5% CO$_2$. Cell viability was maintained and confirmed by cell morphology under optical microscopy during the testing periods discussed in the work [25, 26]. For optical studies, the CHO-β cells were cultured on sterile glass cover slides in 35 × 10 mm tissue culture dishes with one million cells per milliliter. Once they came to confluency, the cells were ready for microscopy studies.

Optical microscopy

To prepare samples for the optical images, the medium was removed from the culture dish and the dish was rinsed with 1 mL of phosphate buffered saline (PBS). 1 mL of the prepared nanowire solution was added to the dish and the cells were incubated with the wires for various time points.

Before imaging, the SiNW-NH$_2$ solution was removed and the dishes were washed with 1 mL PBS. To achieve better focus of live cells, we prepare samples as follows. A microscope slide was prepared by adhering double sided tape along the long edges of the slide. The glass cover slide taken from the cell dish was then placed on top of the prepared microscope slide. One side of the cover slide was sealed with nail varnish. Fresh medium was added between the cover glass and the slide followed by sealing the other side. Bright field, fluorescent, and dark field images were collected using an Olympus BX-51 optical microscope.

Transmission electron microscopy

Once the cells were incubated with the wires for a given amount of time, the nanowire solution was removed via pipette and rinsed with 1 mL of PBS. The PBS was replaced with 2 mL of 2.5% glutaraldehyde and 0.1 M sodium cacodylate buffer solution (fix). This solution was allowed to sit for several minutes before being poured off. Another 2 mL of the fix was added and the cells were scraped from the bottom of the dish and transferred to a small conical centrifuge tube. The cells were spun down,

the old fix was removed and 1.5 mL of fresh fix was added.

The remainder of the processing was done at the Purdue Life Sciences Microscopy Facility. Cells were embedded in 2% agarose, and post-fixed in buffered 1% osmium tetroxide containing 0.8% potassium ferricyanide. Cells were then *en bloc* stained in 4% uranyl acetate, dehydrated with a graded series of ethanol, and embedded in LX-112 resin. Sections with a 90 nm thickness for the 37 °C and 180 nm for the 4 °C samples were cut using a Reichert-Jung Ultracut E ultramicrotome and stained with 2% uranyl acetate and lead citrate. Images were acquired on a FEI Tecnai G220 electron microscope equipped with a LaB6 source and operated at 100 kV.

Confocal microscopy

After incubation, the SiNW-NH$_2$ solution was removed and the dishes were washed twice with 1 mL PBS. PBS was removed and 1 mL of 5 μM 1,1′-dioctadecyl-3,3,3′,3′-tetramethylindocarbocyanine perchlorate (DiI) was added and allowed to incubate at 37 °C for 20 min. The DiI was then removed and the dish was rinsed twice with 1 mL of PBS. PBS was replaced with 2 mL of fresh culture medium and the cells were incubated at 37 °C for another 10 min to rinse off excess DiI. The old medium was removed before cells were imaged.

The glass cover slide was placed on a microscope slide using the aforementioned preparation. Images were taken using an Olympus FluoView 300 Confocal Laser Scanning Microscope with a 543 nm excitation. For each area, approximately 100 images were taken at 10 μm steps through the Z-axis and the images obtained towards the center of the top and bottom of the cells are presented here.

Results and discussion

Zeta potential measurement confirms successful surface functionalization

The as-grown SiNWs were surface modified with amine groups to introduce a positive surface charge. The zeta potential measurements were carried out in triplicate to confirm the success of functionalization. The zeta potential of the unmodified SiNWs was measured to be −29.7 ± 7.85 mV, due to the oxide layer on the SiNW surface. The zeta potential for the SiNW-NH$_2$ was measured to be +28.1 ± 5.11 mV, consistent with that the amine groups are protonated under neutral conditions. Such change in zeta potential after the modification confirms a successful functionalization. Since the cell membrane has slightly negative charge [27], the positive charge of the SiNW-NH$_2$ will ensure charge–charge interaction between the wires and cell membrane.

The interaction between SiNW-NH$_2$ and cells is insensitive to temperature

As we previously reported [22], no obvious binding or internalization was observed in the CHO-β cells treated with the same concentration of unmodified SiNWs even after 11 h incubation (see Additional file 1: Figure S4). We attributed the observation of no interaction to the lack of charge–charge interaction between the unmodified SiNWs and the cell membrane [22]. This is the also the reason why we chose to focus on amine-functionalized SiNWs in this paper.

Bright field and dark field transmission optical imaging was utilized not only to confirm the interaction of the SiNW-NH$_2$, but also to investigate if temperature played a role in membrane interaction. The bright field images in Fig. 1 were used to visualize the cells, while the dark field images were taken to confirm the presence of the SiNW-NH$_2$. SiNW-NH$_2$ were co-cultured with cells at both 37 and 4 °C for 3, 5 and 10 h, respectively. PBS was used to wash the cells prior to imaging, so any wires in the medium or weakly interacting wires were removed. Notably, due to the limitation of the optical transmission imaging, bound wires and internalized wires cannot be differentiated, therefore both scenarios were considered as the wires interacting with the cells. To directly observe the internalized wires, we also used the TEM and confocal microscopy imaging and results are presented later in this manuscript. The control groups were cells incubated at 37 and 4 °C without the presence of wires for 3 h. As shown in Fig. 1a, after incubation for 3 h at 37 °C, SiNW-NH$_2$ were observed binding to the membrane of the cells. As the incubation time increased, more wires could be seen interacting with the cells. Figure 1b depicts the wire interactions at 4 °C and we noticed similar trends of interaction, albeit with less wires observed than the 37 °C group. Binding was also observed after the 3-h incubation period in both the 37 and 4 °C optical images, a slight increase in the number of wires interacting with the cells can be seen in the 5 h panels of both the 37 and 4 °C optical images, and many wires were still visible in the image after 10 h.

Figure 2 plotted the quantitative analysis of wires interacting with cells as a function of incubation time at 37 and 4 °C. Wires were counted from the overlay of the bright and dark field images. Five areas measuring 70 μm × 50 μm were examined for each sample. Numbers of wires interacting per cell measured and standard deviation of the data presented as error bars are plotted

in Fig. 2. Figure 2 showed similar trends for both temperatures as functions of the incubation period with more wires interacting with cells at 37 °C. The average rate of interaction over the first 5 h of the incubation period was estimated to be 0.5 wires per cell per hour at 37 °C (green curve), while 0.4 wires per cell per hour at 4 °C, likely indicating a slightly slower internalization for 4 °C. Such small difference can be attributed to the fact that the cells incubated at 4 °C were not as active as they were at 37 °C. It is well-known that receptor mediated endocytosis processes do not happen at low temperatures [28–30]. A small decrease in the number of wires interacting with cells after 5 h at 37 °C was observed. We attributed the decrease after 5 h in the number of wires interacting with cells to exocytosis, as it is consistent with the reported values between 6 and 8 h at 37 °C for the typical time scale for exocytosis of nanoparticles and nanorods [29]. Since it is a competing process between endocytosis and exocytosis once the SiNWs enter the cells, we attributed the decrease in the number of wires interacting with cells after 5 h at 37 °C to exocytosis. The interaction between SiNW-NH$_2$ and cells was found to be insensitive to temperature in our studies, indicating that the charge–charge attraction induced interaction between SiNW-NH$_2$ and cells is not through the receptor mediated pathway but a physically driven process.

Cross-section TEM images indicate membrane wrapping

We utilized cross-sectional TEM studies to visualize the relative position of bound and internalized nanowires to cell membrane. Specifically, cross-sectional TEM images can confirm that the SiNW-NH$_2$ are internalized by the cells and can be used to examine how the membrane of the cells interacts with the wires, offering insight into the uptake pathway. Figure 3 shows cross-sectional TEM images of CHO-β cells cultured with SiNW-NH$_2$ and fixed at 2, 3, 5, and 10 h. It represents snapshots of the internalization process at 37 °C and highlights how the membrane of the cells interacts with the wires. In the first 2 h, the wires were just starting to interact with the cell membrane via a non-specific charge–charge interaction process. In Fig. 3a, the wire is observed to be tangent with the membrane, but not yet entering, showing the binding stage of the internalization process. After 3 h of incubation, the membrane is shown changing its shape in Fig. 3b in order to accommodate the wires. After 5 and 10 h, depicted in Fig. 3c, d, respectively, the membrane has closely wrapped around the wires and pockets

---

(See figure on next page.)
Fig. 1 Optical microscopy images of SiNW-NH$_2$ after **a** 37 °C incubation and **b** 4 °C incubation (*bottom*) with CHO-β cells. The images of control group, after 3 h of incubation group, after 5 h of incubation group, and after 10 h of incubation group are displayed, respectively. *Scale bars* 20 μm

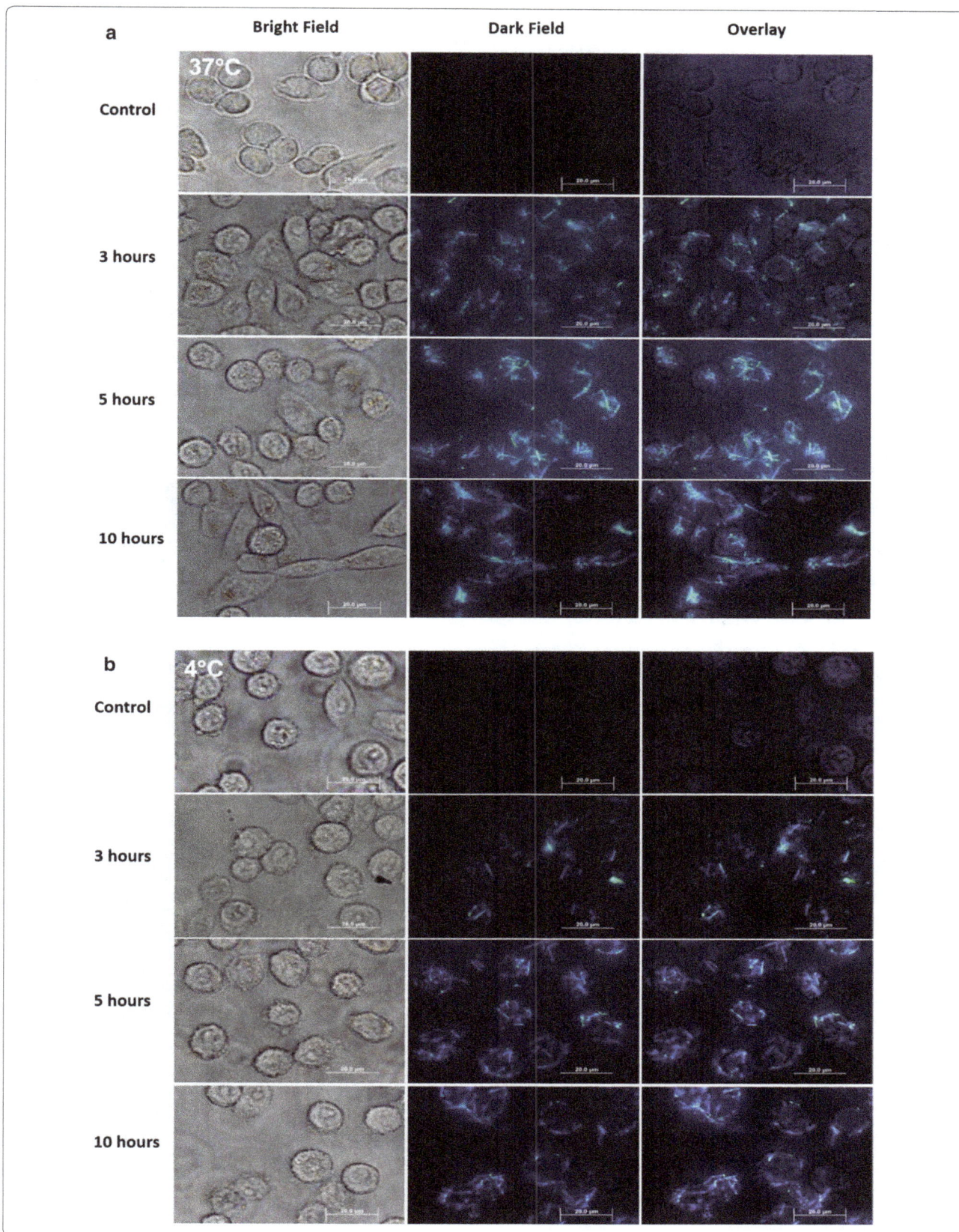

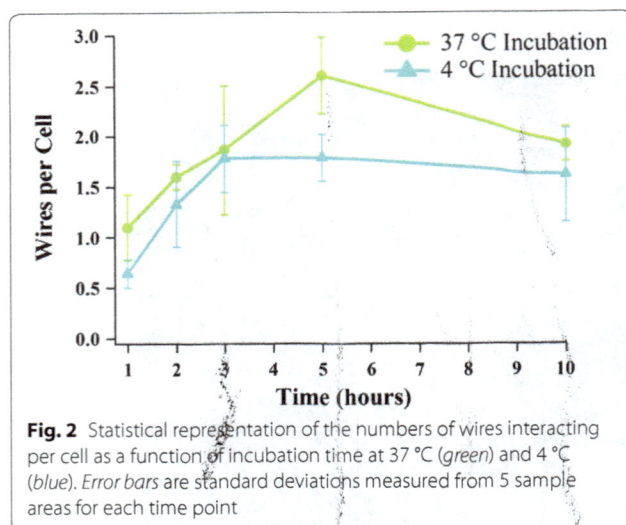

Fig. 2 Statistical representation of the numbers of wires interacting per cell as a function of incubation time at 37 °C (*green*) and 4 °C (*blue*). *Error bars* are standard deviations measured from 5 sample areas for each time point

have formed, with an average gap of 18 nm between the wires and the membrane, as a means to "swallow" the wires. The formation of the pockets and the fact that the wires were not to target a specific receptor confirms that the uptake pathway of the SiNW-NH$_2$ at 37 °C is via the membrane wrapping mechanism.

To confirm the internalization of SiNW-NH$_2$ is also occurring at a lower temperature than 37 °C, we performed the cross-sectional TEM experiment at 4 °C. The SiNW-NH$_2$ were co-cultured with cells in 4 °C and fixed at different incubation periods up to 24 h. The uptake and internalization of the SiNW-NH$_2$ at 4 °C is depicted in Fig. 4. The key steps during this process, including the initial interaction induced wire-membrane binding (Fig. 4a), the membrane accommodating the wires (Fig. 4b), and the internalization with the formed pocket of membrane closely surrounding the wires (Fig. 4c) are

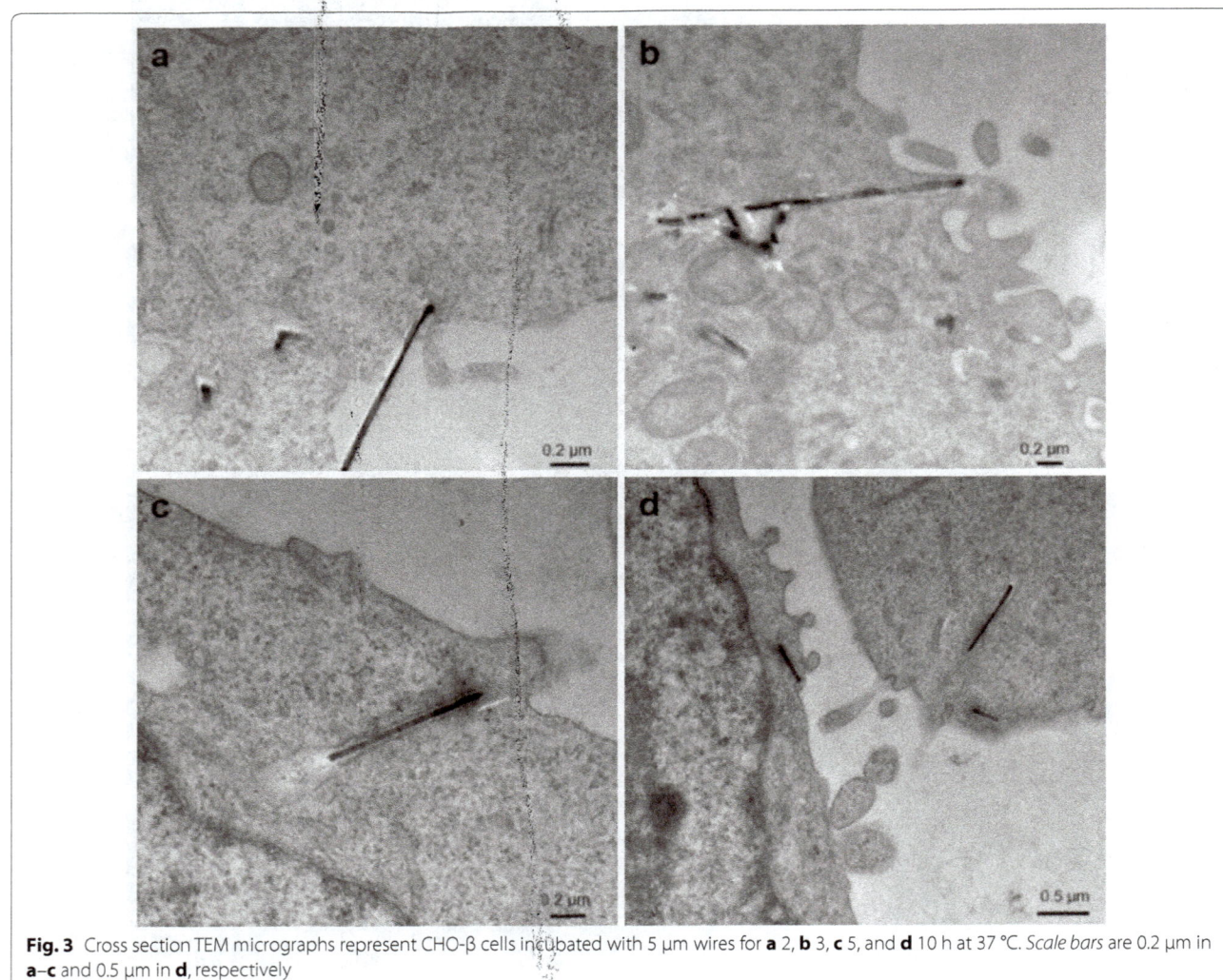

Fig. 3 Cross section TEM micrographs represent CHO-β cells incubated with 5 μm wires for **a** 2, **b** 3, **c** 5, and **d** 10 h at 37 °C. *Scale bars* are 0.2 μm in **a–c** and 0.5 μm in **d**, respectively

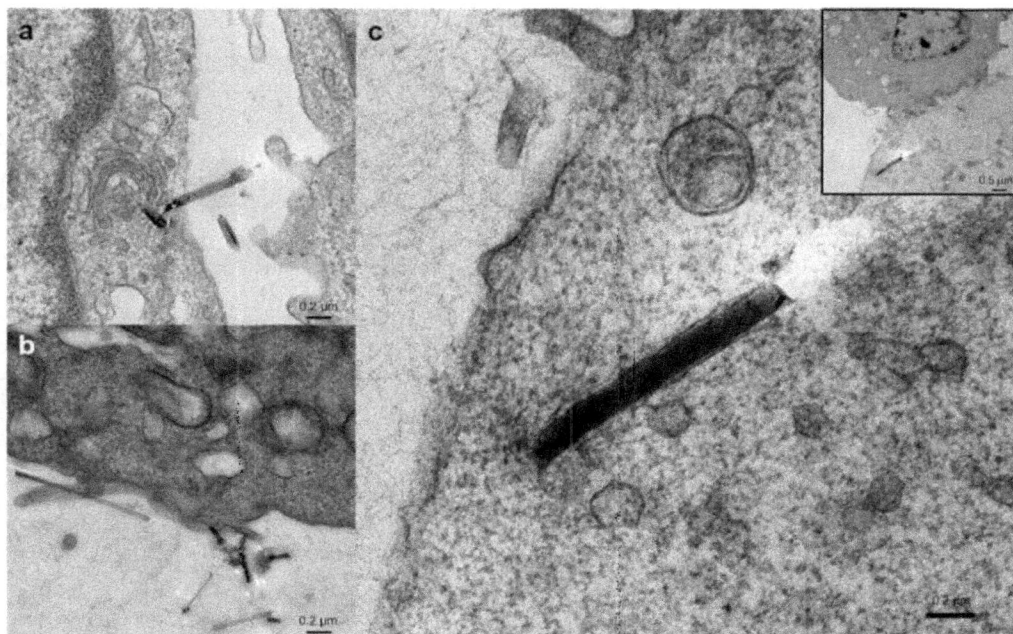

Fig. 4 Cross section TEM micrographs represent CHO-β cells incubated with SiNW-NH$_2$ for **a** 2, **b** 12, **c** 24 h at 4 °C. *Inset* in **c** represents the lower magnification image of the wire in **c**. *Scale bars* are 0.2 μm in **a–c** and 0.5 μm in the inset, respectively

observed at 4 °C. Representative images obtained indicate that the binding is consistent with the optical microscopy data in Fig. 2. The internalization was confirmed at 24 h incubated samples at 4 °C, later than the observation from optical and fluorescence images, which could be attributed to limitation of the TEM cross section sample preparation and small sampling acquired through TEM. In summary, the temperature independent observations of the internalization and the formation of the membrane pockets are strong evidence to support the membrane wrapping mechanism and that this mechanism is, indeed, due to a physical force exerted on the membrane by the wires.

Confocal fluorescence confirms the physical membrane wrapping

We performed confocal fluorescence microscopy on the cell samples cultured with SiNW-NH$_2$, in which 3D information about nanowires and cells can be obtained through scanning along the z-axis and used to confirm the internalization of the wires. DiI (red) was used to stain the cell membrane. First, Fig. 5a shows a control study in which SiNWs were incubated with DiI in the cell medium without the presence of the cells. After one hour of incubation, the dark field image (Fig. 5a left) shows the presence of wires and the fluorescence image confirms that the wires do not fluoresce, suggesting the DiI does not label the surface of nanowires.

Figure 5b, c show bright field and fluorescence images obtained from the control group without wires (Fig. 5b) and the CHO-β cells co-cultured with SiNW-NH$_2$ for 10 h at 37 °C (Fig. 5c). Bright field optical images on the left column of Fig. 5 show the presence of wires and the fluorescence images on the right column were used to locate the cell membrane, therefore indicating the interaction between wires and cellular membrane. Specifically, since the images plotted here were obtained towards the center of the top and bottom of the cells along the z scanning under the confocal microscope, the wires shown in focus in the bright field are in the focal plane therefore considered to be inside the cell. Clearly different from Fig. 5b, c demonstrated internalization of SiNW-NH$_2$ by CHO-β cells. Importantly, Fig. 5c also shows that wire-shaped fluorescent signals present in the cells, indicating the internalized SiNW-NH$_2$ became fluorescent. Together with observation that SiNWs were not directly labeled by DiI from Fig. 5a, these results also suggest that internalized SiNW-NH$_2$ is through membrane wrapping. More fluorescence images obtained from 2- and 5-h incubation periods along with a plot showing average

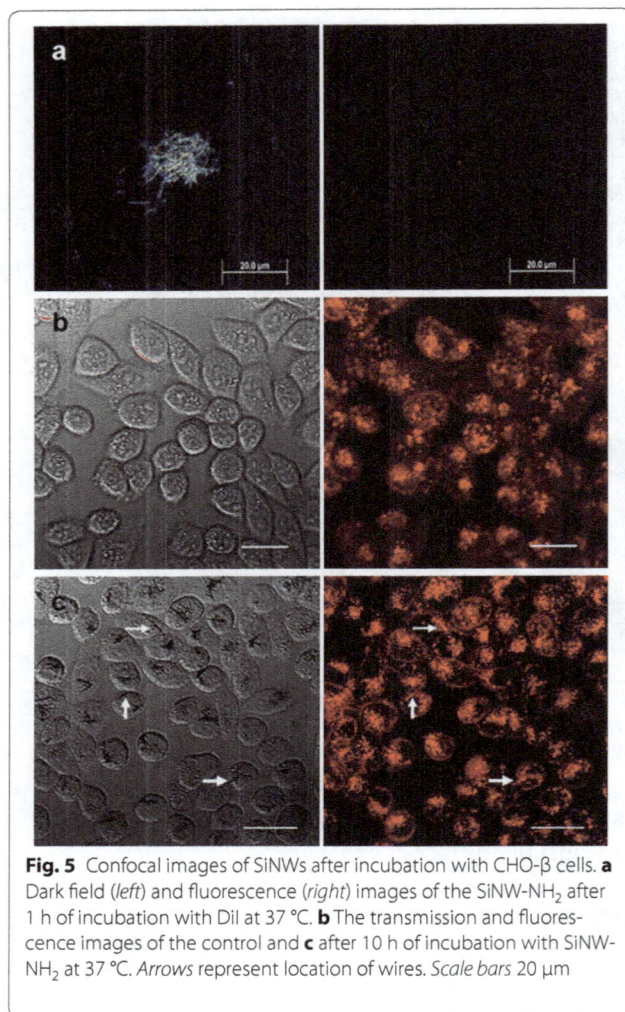

Fig. 5 Confocal images of SiNWs after incubation with CHO-β cells. **a** Dark field (*left*) and fluorescence (*right*) images of the SiNW-NH₂ after 1 h of incubation with DiI at 37 °C. **b** The transmission and fluorescence images of the control and **c** after 10 h of incubation with SiNW-NH₂ at 37 °C. *Arrows* represent location of wires. *Scale bars* 20 μm

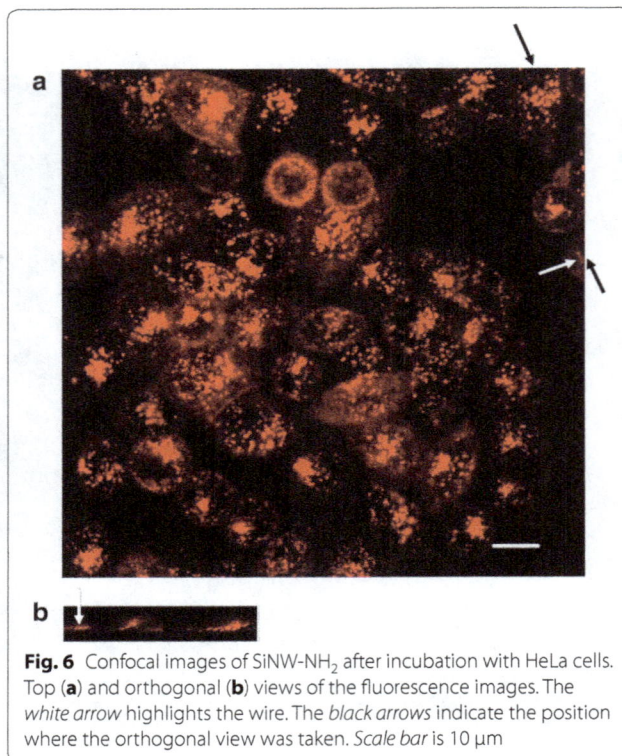

Fig. 6 Confocal images of SiNW-NH₂ after incubation with HeLa cells. Top (**a**) and orthogonal (**b**) views of the fluorescence images. The *white arrow* highlights the wire. The *black arrows* indicate the position where the orthogonal view was taken. *Scale bar* is 10 μm

number of wires per cell internalized as a function of incubation time period are included in Additional file 1 (Figures S5, S6, respectively).

The confocal experiment was also performed using the immortalized HeLa cell line using both unmodified SiNWs and SiNW-NH₂. The results showed that, much like we saw with the CHO-β cells previously [22], the unmodified silicon nanowires did not interact with the cells, even after 5 h at both 37 and 4 °C (Additional file 1: Figures S7, S8). For amine-modifed SiNWs, Fig. 6 shows top and orthogonal views, demonstrating the co-localization of amine-modified SiNWs with cell membranes after 2 h of incubation at 37 °C, which is consistent with our findings with CHO-β cells. These results demonstrate that the SiNW-NH₂ are internalized via a physically driven, membrane wrapping pathway, rather than an energy-dependent, receptor-mediated process.

Conclusion

We have successfully visualized the key steps of internalization of SiNW-NH₂ into cells as illustrated in Scheme 1, using optical, confocal fluorescence and transmission electron microscopies. Specifically, we confirmed that SiNW-NH₂ are internalized by CHO-β and HeLa cells via a membrane wrapping mechanism for the first time and that this process is most likely due to physical force exerted on the membrane. Optical microscopy images at 37 °C incubation as well as 4 °C incubation show that the charge–charge induced binding and interactions occur regardless of temperature, indicating that the mechanism is physical rather than receptor mediated. Further, TEM micrographs captured the key steps of interactions between the cell membrane and the wires showing that the internalized wires are surrounded by pocket of the membrane. The confocal images independently demonstrate that the wires are successfully internalized. Fluorescence of the wires internalized confirms that the mechanism is membrane wrapping. This work provides better understanding of cellular uptake pathways for 1D nanomaterials without specific targeting ligands. Since the interaction presented here is physical, this work opens up the potential for using 1D materials to understand membrane mechanics through physical stress exerted on the membrane.

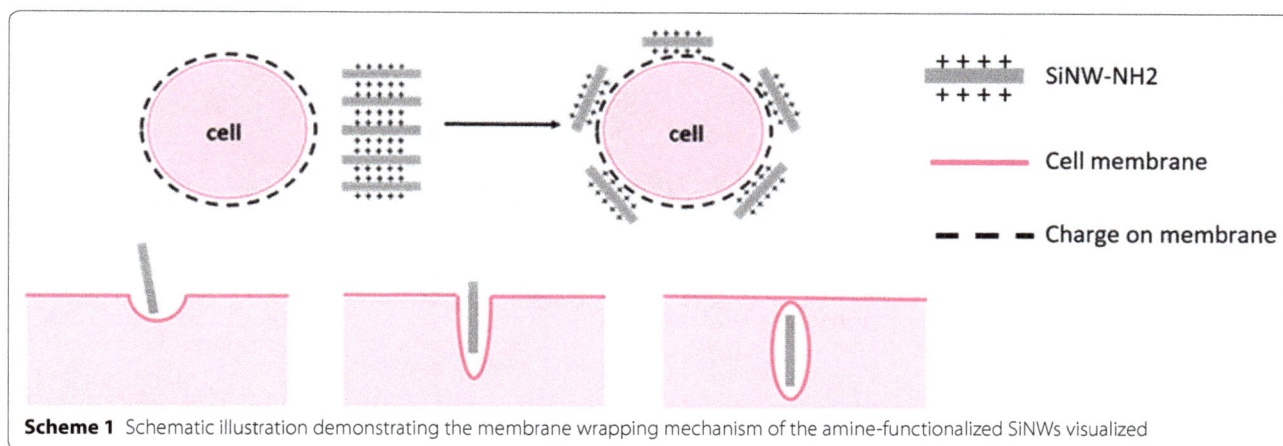

Scheme 1 Schematic illustration demonstrating the membrane wrapping mechanism of the amine-functionalized SiNWs visualized

Additional file

Additional file 1. Wire characterization including representative TEM of unmodified SiNWs and SiNW-NH$_2$, FTIR analysis of unmodified SiNWs and SiNW-NH$_2$, and wide XPS spectrum for an amine-modified silicon nanowires. Confocal images of SiNW-NH$_2$ after incubation with CHO-β cells at 2 and 5-hour incubation periods and a graph plotting the average number of wires internalized per cell as a function of incubation time at 37 °C. Confocal images of HeLa cells at 2 and 5-h incubation time at 37 and 4 °C with unmodified SiNWs.

Abbreviations

1D: one-dimensional; SiNWs: silicon nanowires; SiNW-NH$_2$: amine-functionalized silicon nanowires; TEM: transmission electron microscopy; 0D: zero dimensional; SWCNTs: single-walled carbon nanotubes; CVD: chemical vapor deposition; CHO-β: Chinese hamster ovary cells transfected with folate receptor beta; APTMS: (3-aminopropyl) trimethoxysilane; RPMI: Roswell Park Memorial Institute; PBS: phosphate buffered saline; Dil: 1,1'-dioctadecyl-3,3,3',3'-tetramethylindocarbocyanine perchlorate; FTIR: Fourier transform infrared spectroscopy; XPS: X-ray photoelectron spectroscopy.

Authors' contributions

KM and YH carried out experiments, analyzed data and wrote the manuscript. CY supervised the entire project, was involved in the design of all experiments and revised the manuscript. All authors read and approved the final manuscript.

Author details

[1] Department of Chemistry, Purdue University, 560 Oval Drive, West Lafayette, IN 47907, USA. [2] Department of Physics and Astronomy, Purdue University, West Lafayette, IN 47907, USA.

Acknowledgements

Authors gratefully acknowledge the Purdue University Life Sciences Microscopy Facility, Purdue University Amy Facility, Dr. Dmitry Zemlyanov at the Surface Analysis Facility of the Birck Nanotechnology Center, Purdue University for XPS analysis, and Dr. J. X. Cheng for access to measurement facilities. The authors would also like to acknowledge Justine Arrington for helpful discussions.

Competing interests

The authors declare that they have no competing interests.

Funding

This work was funded by Purdue University.

References

1. Shukla R, Bansal V, Chaudhary M, Basu A, Bhonde RR, Sastry M. Biocompatibility of gold nanoparticles and their endocytotic fate inside the cellular compartment: a microscopic overview. Langmuir. 2005;21:10644–54.
2. Nativo P, Prior IA, Brust M. Uptake and intracellular fate of surface-modified gold nanoparticles. ACS Nano. 2008;2:1639–44.
3. Chithrani DB. Intracellular uptake, transport, and processing of gold nanostructures. Mol Membr Biol. 2010;27:299–311.
4. Xu S, Olenyuk BZ, Okamoto CT, Hamm-Alvarez SF. Targeting receptor-mediated endocytic pathways with nanoparticles: rationale and advances. Adv Drug Deliv Rev. 2013;65:121–38.
5. Liu X, Huang N, Li H, Jin Q, Ji J. Surface and size effects on cell interaction of gold nanoparticles with both phagocytic and nonphagocytic cells. Langmuir. 2013;29:9138.
6. Bartneck M, Keul HA, Singh S, Czaja K, Bornemann J, Bockstaller M, Moeller M, Zwadlo-Klarwasser G, Groll J. Rapid uptake of gold nanorods by primary human blood phagocytes and immunomodulatory effects of surface chemistry. ACS Nano. 2010;4:3073–86.
7. Safi M, Yan M, Guedeau-Boudeville MA, Conjeaud H, Garnier-Thibaud V, Boggetto N, Baeza-Squiban A, Niedergang F, Averbeck D, Berret JF. Interactions between magnetic nanowires and living cells: uptake, toxicity, and degradation. ACS Nano. 2011;5:5354–64.
8. Prina-Mello A, Diao Z, Coey JMD. Internalization of ferromagnetic nanowires by different living cells. J Nanobiotechnol. 2006;4:9.
9. Yaron PN, Holt BD, Short PA, Lösche M, Islam MF, Dahl K. Single wall carbon nanotubes enter cells by endocytosis and not membrane penetration. J Nanobiotechnol. 2011;9:45.
10. Bianco A, Kostarelos K, Partidos CD, Prato M. Biomedical applications of functionalised carbon nanotubes. Chem Commun (Camb). 2005;(5):571-577. doi:10.1039/b410943k
11. Huang C, Zhang Y, Yuan H, Gao H, Zhang S. Role of nanoparticle geometry in endocytosis: laying down to stand up. Nano Lett. 2013;13:4546–50.
12. Shi X, von dem Bussche A, Hurt RH, Kane AB, Gao H. Cell entry of one-dimensional nanomaterials occurs by tip recognition and rotation. Nat Nanotechnol. 2011;6:714–9.
13. Yi X, Shi X, Gao H. A universal law for cell uptake of one-dimensional nanomaterials. Nano Lett. 2014;14:1049–55.
14. Gao J, Xu B. Applications of nanomaterials inside cells. Nano Today. 2009;4:37–51.

15. Hernandez-Velez M. Nanowires and 1D arrays fabrication: an overview. Thin Solid Films. 2006;495:51–63.

16. Cui Y, Lauhon LJ, Gudiksen MS, Wang J, Lieber CM. Diameter-controlled synthesis of single-crystal silicon nanowires. Appl Phys Lett. 2001;78:2214–6.

17. Zhong Z, Yang C, Lieber CM, Vijay K. Nanosilicon, vol. 5. Oxford: Elsevier; 2008. p. 176–216

18. Boehm H-P. The chemistry of silica. Solubility, polymerization, colloid and surface properties, and biochemistry. Von RK Iler. John Wiley and Sons, Chichester 1979. XXIV, 886 S. Angew Chemie. 1980;92:328.

19. Park H, Crozier KB. Multispectral imaging with vertical silicon nanowires. Sci Rep. 2013;3:2460.

20. Dovrat M, Arad N, Zhang XH, Lee ST, Sa'ar A. Optical properties of silicon nanowires from cathodoluminescence imaging and time-resolved photoluminescence spectroscopy. Phys Rev B Condens Matter Mater Phys. 2007;75:1–5.

21. Jung Y, Tong L, Tanaudommongkon A, Cheng J-X, Yang C. In vitro and in vivo nonlinear optical imaging of silicon nanowires. Nano Lett. 2009;9:2440–4.

22. Zhang W, Tong L, Yang C. Cellular binding and internalization of functionalized silicon nanowires. Nano Lett. 2012;12:1002–6.

23. Silverstein SC, Cohn ZA, Steinman RM. Endocytosis. Ann Rev Biochem. 1977;46:669–722.

24. Vrancken KC, Possemiers K, Van Der Voort P, Vansant EF. Surface modification of silica gels with aminoorganosilanes. Colloids Surfaces A Physicochem Eng Asp. 1995;98:235–41.

25. Stephens DJ, Allan VJ. Light microscopy techniques for live cell imaging. Science. 2008;300:82.

26. Thomas C, Daly A, Suresh S, Burg K, Harrison GM, Smith DW. Amido-modified polylactide for potential tissue engineering applications. J Biomater Sci Polymer Edition. 2004;15:595–606.

27. Danon D, Goldstein L, Marikovsky Y, Skutelsky E. Use of cationized ferritin as a label of negative charges on cell surfaces. J Ultrastruct Res. 1972;38:500–10.

28. Yan L, Zhang J, Lee C-S, Chen X. Micro- and nanotechnologies for intracellular delivery. Small. 2014;10:4487–504.

29. Chithrani BD, Chan WCW. Elucidating the mechanism of cellular uptake and removal of protein-coated gold nanoparticles of different sizes and shapes. Nano Lett. 2007;7:1542–50.

30. Kam NWS, Liu Z, Dai H. Carbon nanotubes as intracellular transporters for proteins and DNA: an investigation of the uptake mechanism and pathway. Angew Chemie Int Ed. 2006;45:577–81.

17

Alternative moth-eye nanostructures: antireflective properties and composition of dimpled corneal nanocoatings in silk-moth ancestors

Mikhail Kryuchkov[1], Jannis Lehmann[2], Jakob Schaab[2], Vsevolod Cherepanov[3], Artem Blagodatski[1,3], Manfred Fiebig[2] and Vladimir L. Katanaev[1,3]*

Abstract

Moth-eye nanostructures are a well-known example of biological antireflective surfaces formed by pseudoregular arrays of nipples and are often used as a template for biomimetic materials. Here, we provide morphological characterization of corneal nanostructures of moths from the *Bombycidae* family, including strains of domesticated *Bombyx mori* silk-moth, its wild ancestor *Bombyx mandarina*, and a more distantly related *Apatelodes torrefacta*. We find high diversification of the nanostructures and strong antireflective properties they provide. Curiously, the nano-dimple pattern of *B. mandarina* is found to reduce reflectance as efficiently as the nanopillars of *A. torrefacta*. Access to genome sequence of *Bombyx* further permitted us to pinpoint corneal proteins, likely contributing to formation of the antireflective nanocoatings. These findings open the door to bioengineering of nanostructures with novel properties, as well as invite industry to expand traditional moth-eye nanocoatings with the alternative ones described here.

Keywords: Moth-eye structures, Antireflective nanocoatings, Biomimetic materials, Silkmoth

Introduction

Moth-eye nanostructures—pseudo-regular arrays of nanopillars first described on corneal surfaces of the nocturnal moth *Spodoptera eridania* [1]—served as a paradigm for bio-inspired antireflective coating with broad technological applications, from solar cells to art paintings [2–5]. Gradually matching the refractive index of air to that of the lens material, these nanostructures minimize light reflectance and maximize perception. While physics of the anti-reflectance by nanopillar arrays is well-understood, permitting reliable simulations [6, 7], nature also designed other types of corneal nanocoatings [8], of which some were shown to play the antireflective function too [9, 10]. Numerous studies on artificial nanocoatings showed that the shape, dimensions, as well as the packing order of the nanostructures contribute to the anti-reflectivity [11–14].

Model insect organisms, permitting genetic manipulations and/or providing complete information on their genome sequences, have been instrumental in advancing the research on insect molecular, developmental, and cell biology [15]. Following our research on the corneal nanocoatings in *Drosophila melanogaster* [16, 17], we are now turning to another famous model insect—the silkmoth *Bombyx mori*.

More than 1000 different *B. mori* silkworm strains exist worldwide, having different phenotypic and genomic features [18, 19]. The silkworm was domesticated in East Asia from wild *B. mandarina* moths some 5000 years ago, losing several features essential in the wild habitat on the expense of maximizing silk production [20]. *Bombyx* moths' genomes have been fully sequenced [18, 19], increasing their importance as genetic model organisms. While many aspects of the *Bombyx* biology have been analyzed, these insects have not been previously studied

*Correspondence: vladimir.katanaev@unil.ch
[1] Department of Pharmacology and Toxicology, University of Lausanne, Rue du Bugnon 27, 1011 Lausanne, Switzerland
Full list of author information is available at the end of the article

in terms of their corneal morphology and properties. We hypothesized that distinct corneal nanocoatings may be found in the wild vs. domesticated silkmoths.

Results and discussion

To explore differences between corneal nanocoatings in wild and domestic silkmoths, an atomic-force microscopy (AFM) analysis has been performed, comparing the corneal surfaces of *B. mandarina*, to the samples from two different *B. mori* strains obtained from Japan (Jp) and Vietnam (Vn, see "Materials and methods"; Fig. 1a–c). Unusually for Lepidopterans, different species of which so far have displayed different varieties of nanopillars, in some species fused into mazes or parallel strands [8], corneal surfaces of *B. mandarina* reveal a clear nano-dimpled pattern (Fig. 1d), previously described in insects

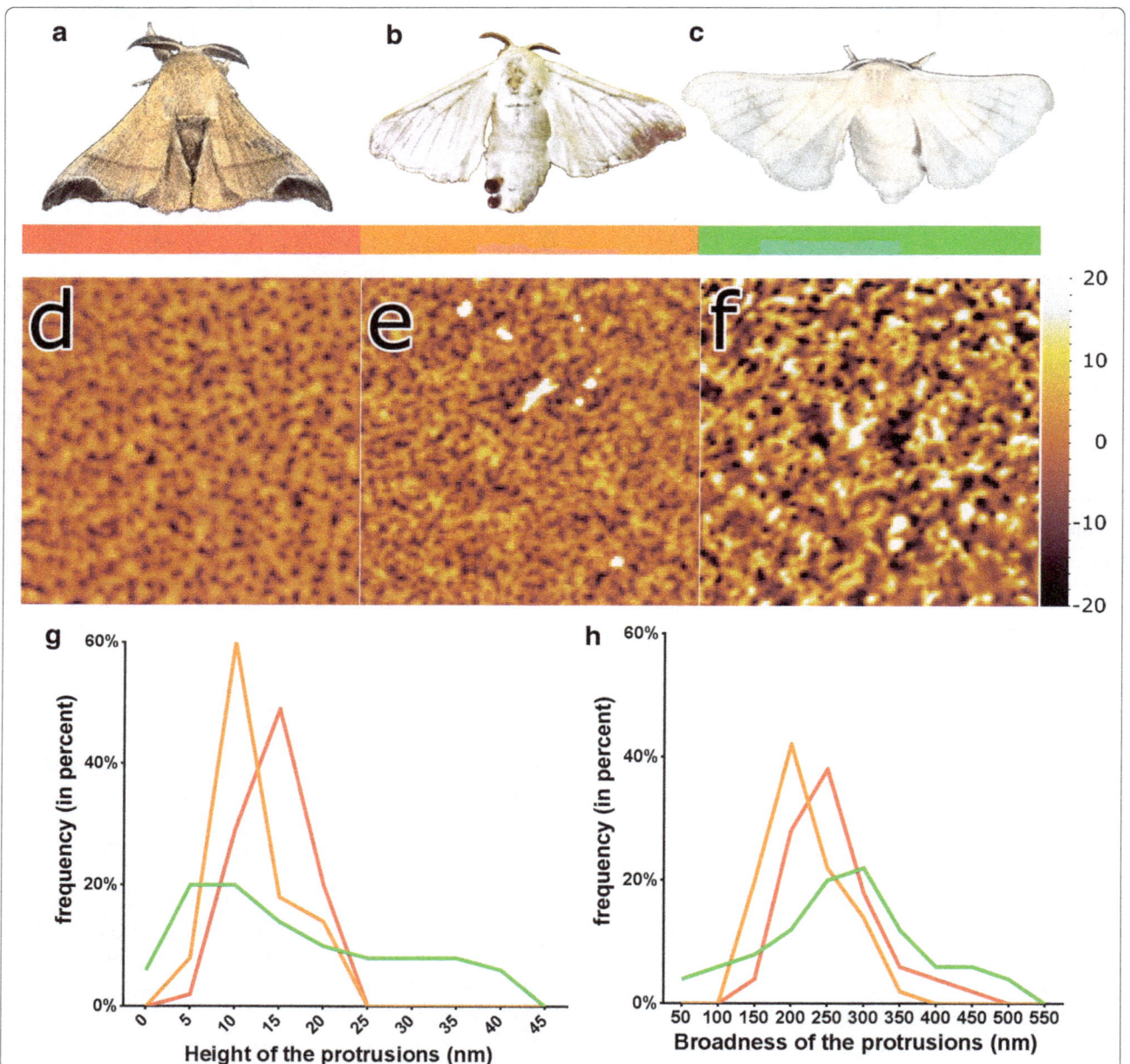

Fig. 1 Corneal nanocoatings in *Bombyx* moths. **a–c** Photographs of *B. mandarina* (**a**), *B. mori* from Vietnam (**b**) and *B. mori* from Japan (**c**). **d–f** Representative AFM scans (5 × 5 μm) of corneal surfaces of the *Bombyx* species presented in **a–c**. The height dimension of the surface (in nm) is indicated by the *color scale* next to (**f**) with the mean set to zero. **g, h** Calculation of the height of protrusions (from the lowest point up to the next highest point (**g**) and their broadness (**h**) of *B. mandarina* (in *red*), *B. mori* [Vn] (in *orange*), and *B. mori* [Jp] (in *green*); n = 50

of other orders, such as earwigs or *Carabidae* beetles [8]. Curiously, corneae of the two *B. mori* strains reveal different degrees of corruption of this pattern, with the Vn strain depicting a dimple-to-maze transition previously seen in other insects (e.g. bug from *Pyrrhocoridae* family [8]), and the Jp strain—a complete degeneration into sporadic irregularities (Fig. 1e–h). Fourier analysis [16, 21, 22] confirms this visual inspection, indicating that the *B. mandarina* and *B. mori* [Vn] corneal structures are close to quasi-random, while the *B. mori* [Jp]—to completely random structures (Additional file 1: Figure S1). Fully random structures are known to show less anti-reflectance than the quasi-random structures [11, 12], such as those we see in *B. mandarina* and *B. mori* [Vn] (Additional file 1: Figure S1). As *B. mori* [Jp] corneae further possess randomization in the broadness and depth of the nanostructures (Fig. 1), we may expect even stronger loss of the anti-reflectivity. Using finite-difference time domain-based approximations of the Maxwell function, simulations of the reflectance pattern of structured surfaces were found to match the results received in the process of the optical experiments [6, 7, 23, 24]. However, such simulations in general do not predict antireflective properties of nanostructures, whose dimensions are below 30–50 nm [7, 25], such as those we see in the *Bombyx* species.

Given these uncertainties, we decided to directly measure reflectivity from corneal surfaces of the *Bombyx* moths. This analysis reveals strong reduction of the reflected light in the broad visible spectrum from the dimpled corneal surfaces of *B. mandarina* and the dimple-to-maze surfaces of *B. mori* [Vn], as compared to the irregularly rough surface of *B. mori* [Jp] (Fig. 2). Remarkably, the antireflective properties provided by the *Bombyx* nano-dimpled coating even exceeded those provided by the nano-pillar arrays of another *Bombycidae* moth, *Apatelodes torrefacta* (Fig. 2d, e). While the nano-dimpled *B. mandarina* arrays (as the nano-pillar *A. torrefacta* arrays) appear to provide uniform broadband anti-reflectivity, the dimple-to-maze *B. mori* [Vn] structures are efficient at low wavelengths and start to decrease anti-reflectivity at >650 nm (Fig. 2e). With the theoretical assessment of these findings currently missing, we are left to speculate that either the lack of uniformity in the overall morphology of the dimple-to-maze nanostructures (Figs. 1e, 2b), or their on average smaller dimensions (Fig. 1g, h) could be the cause of this difference of *B. mori* [Vn] from its wild ancestor. Our findings provide the first demonstration of the antireflective capacity of nano-dimpled surfaces less than 100 nm in depth, indicating that nature has found a relatively low-cost and unpredicted solution to the anti-reflectance.

Corneal nanostructures are built from proteins and lipids placed on the chitin background [26, 27]. The

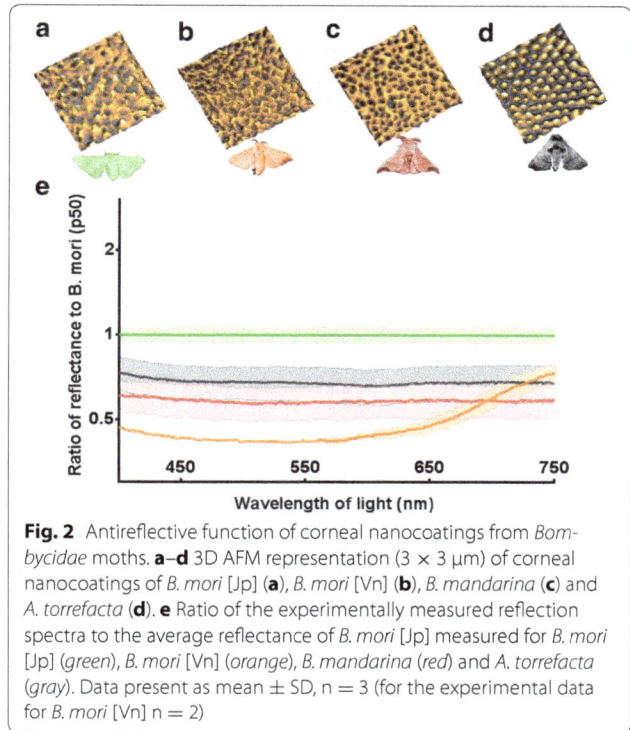

Fig. 2 Antireflective function of corneal nanocoatings from *Bombycidae* moths. **a–d** 3D AFM representation (3 × 3 μm) of corneal nanocoatings of *B. mori* [Jp] (**a**), *B. mori* [Vn] (**b**), *B. mandarina* (**c**) and *A. torrefacta* (**d**). **e** Ratio of the experimentally measured reflection spectra to the average reflectance of *B. mori* [Jp] measured for *B. mori* [Jp] (green), *B. mori* [Vn] (orange), *B. mandarina* (red) and *A. torrefacta* (gray). Data present as mean ± SD, n = 3 (for the experimental data for *B. mori* [Vn] n = 2)

Turing reaction–diffusion model was proposed to govern the formation and diversity of the insect corneal nanostructures [8], which predicts corneal protein(s) to serve as the key slow-diffusing component of this reaction and the building blocks of the resulting structures. To prove directly that corneal proteins are important for formation of the antireflective nanocoatings, we treated corneae of *B. mandarina* with a detergent (see "Materials and methods"), and analyzed the resulting samples for their morphology and anti-reflectance. Remarkably, we find that removal of proteins purges away the nano-dimpled coatings (Additional file 2: Figure S2), correlating with a strong loss of the anti-reflective potential of the corneal surface (Additional file 2: Figure S2b). These features, together with the detailed inspection of the remnant nanostructures' height and broadness (Additional file 2: Figure S2c, d), indicate that protein extraction brings the surface of *B. mandarina* close to that of *B. mori* [Jp]. These findings prompted us to pinpoint corneal proteins, which may show differential abundance in the wild vs. the domesticated *Bombyx* moths.

The choice of *B. mori* and *B. mandarina* insects for our analysis was to a large extent dictated by the fact that genomes of these insects have been fully sequenced [18, 19], opening the possibility to identify corneal proteins involved in formation of the nanocoatings—the task not achievable for many other insects (such as *A. torrefacta*) whose genomes have not been sequenced. Thus, we analyzed the protein samples obtained by the detergent extraction (Additional file 2: Figure S2a) of *B. mori* and *B. mandarina*, as well as samples from their underlying retina, by SDS-PAGE followed by mass-spectrometry identification of the most prominent corneal-specific bands (Fig. 3a). This proteomic analysis identified a number of cuticular proteins (CPs, Additional file 3: Table S1), such as CPRs (classical CPs with the chitin-binding domain), and CPHs (hypothetical CPs) [28]. Various chitin-binding domains, RR motifs (where RR2 mediates association with the hard cuticle such as head capsule, and RR1—with soft intermediate membranes), and 18 amino acid repeats can be distinguished in CPs [28].

The major proteins, comprising each >10% and together ca. 50% of the total corneal load in *B. mandarina* corneae, are CPR83, CPR150, CPR19 and CPH30. Remarkably, we find substantial reduction of these proteins in the domesticated moth *B. mori* [Jp] (Fig. 3a, b). CPR83, CPR150, and CPR19 are the regular CPRs containing the chitin-binding domain 4, whereas CPH30 does not contain a chitin-binding domain but has the 18-residue repeats [29]. CPR83 contains an RR2, while CPR150 and CPR19—the RR1 motifs, and thus may mediate interactions with different types of cuticle [28]. Analysis of the corneal proteome of *B. mori* [Vn] revealed similar to *B. mandarina* levels of CPR83 and CPH30, but significantly lower levels of the RR1-family proteins CPR150 and CPR19 (Fig. 3b).

We hypothesize that these changes in the corneal CPs, seen among different *Bombyx* specimen, underlie the differences in the nanostructures that decorate corneae of these strains. Specifically, we propose that the hard-cuticle binding CPR83, possessing the RR2 motif, and the 18-residue repeat-containing CPH30 are the first proteins to be "gained" in *B. mori* [Vn] as compared to *B. mori* [Jp], providing structuring on the corneal surface and strong anti-reflectance (Fig. 3c, d). Next, the soft material-interacting RR1 motif is "recruited" in the form of the CPR19 and CPR150 in corneae of *B. mandarina*, mediating formation of the well-formed nano-dimpled coatings (Fig. 3e). Genetic manipulations (loss-of-function and overexpression) of these proteins in *B. mori* or another insect model is required for the unambiguous

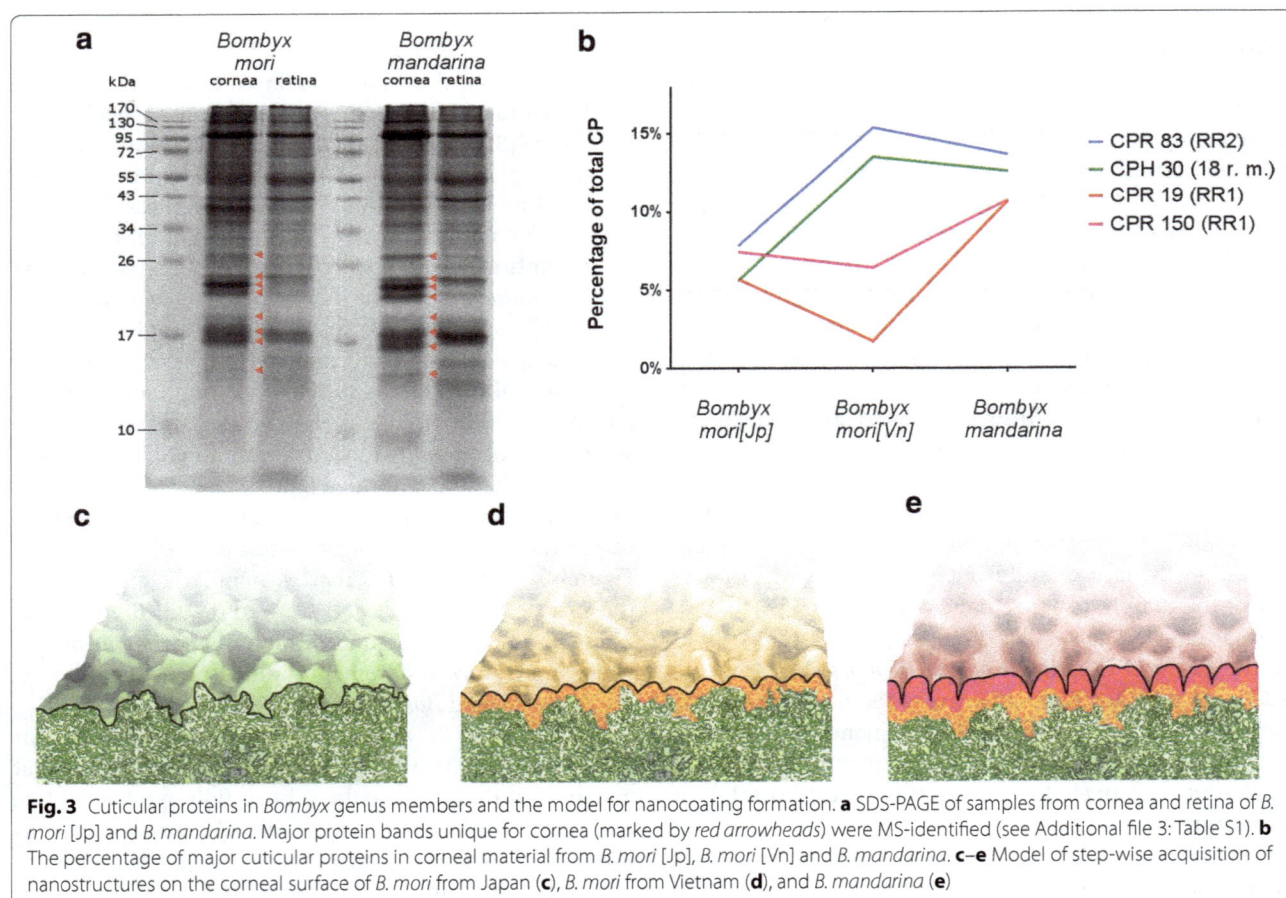

Fig. 3 Cuticular proteins in *Bombyx* genus members and the model for nanocoating formation. **a** SDS-PAGE of samples from cornea and retina of *B. mori* [Jp] and *B. mandarina*. Major protein bands unique for cornea (marked by *red arrowheads*) were MS-identified (see Additional file 3: Table S1). **b** The percentage of major cuticular proteins in corneal material from *B. mori* [Jp], *B. mori* [Vn] and *B. mandarina*. **c–e** Model of step-wise acquisition of nanostructures on the corneal surface of *B. mori* from Japan (**c**), *B. mori* from Vietnam (**d**), and *B. mandarina* (**e**)

testing of our hypothesis, and will be subject of future investigations.

Identification of the candidate proteins governing formation of the *Bombyx* corneal nanocoatings permits creation of novel nanocoatings. Indeed, cloning of the genes encoding for these proteins and their targeted mis- and over-expression in other hosts, such as the genetic model insect *Drosophila melanogaster* [16, 30] is expected to produce novel corneal nanocoatings—potentially with improved physical properties such as anti-reflectance, self-cleaning, anti-bacterial, etc. Together with the similar analysis of the protein composition of corneal nanostructures in other arthropods, this approach opens the door to a high-throughput bioengineering of nanocoatings, which may eventually lead to generation of nanostructures with features interesting for industrial applications.

In regard to the interest for industry, we wish to stress that the nano-dimpled moth-eye nanostructures we have described here display unexpectedly good anti-reflective properties. Given the high cost of industrial generation of the nano-pillar coatings and their sensitivity to physical damage, these nano-dimpled coatings may represent an attractive industrial alternative, given the low cost fabrication of nanohole surfaces [31].

Materials and methods

The samples of *Bombyx mori* (p50 strain) and *Bombyx mandarina* (Oki line) were provided by the National Bio-Resource Project (NBRP) of the Ministry of Education, Science, Sports and Culture of Japan. Additionally, we studied Vietnamese *B. mori* obtained from the Cuong Hoan Silk Factory, Trung Vuong, Lam Dong Province, Vietnam. The samples of *Apatelodes torrefacta* (mature adults from Louisiana, USA) were obtained from the online shop http://www.thebugmaniac.com.

Preparation and analysis of corneal and retinal samples: Corneal and retinal samples were prepared by cutting off the eyes with a scalpel from the heads of mature adult guillotined *Bombycidae* moths. The retinal material was removed from the immobilized samples into a drop of water by washing and very gentle scrupulous scratching. Upon the separation, the corneal material was further washed 3 times in water. The corneal and retinal samples were extracted from the material of 10 eyes. The samples were boiled for 60 min in the Sample Buffer (62.5 mM Tris–HCl pH 6.8; 10% glycerol; 2% SDS; 1% β-mercaptoethanol; trace of bromophenol blue) prior to separation by 15% SDS-PAGE (Fig. 3) or AFM and reflectance measurement (Additional file 2: Figure S2). Ratio of bands to total protein were counted by the ImageJ software.

Mass-spectrometry (MS): in-gel trypsin digestion and MS were performed by the Protein Analysis Facility of University of Lausanne (Switzerland). The Scaffold viewer software was used for the data analysis with the following parameters: protein threshold 90%, minimum of number of peptides 1, peptide threshold 90%, minimum of probability 90%. The amount of cuticular proteins was counted by using the following formula:

$$\frac{\sum \left(\frac{a}{b} \times 100 \times c\right)_n}{\sum \left(\sum \left(\frac{a}{b} \times 100 \times c\right)_n\right)_m} \times 100,$$

where a is the percentage of total spectra from protein of interest (particular CP) in one band, b is the percentage of total spectra of all proteins in one band, c is the ratio of band to total protein in the gel, n is the number of bands, and m is the number of CPs.

The corneal samples of *B. mori* from Vietnam, due to scarceness of material, were directly sent to MS and the results were analyzed as the ratio of the percentage of the total spectra from protein of interest (particular CP) to the percentage of the total spectra of all CPs.

For AFM, corneal samples prepared as described above were attached to a coverslip by a double-sided bonding tape. Microscopy was performed by the NTegra-Prima microscopes (NT-MDT, Zelenograd, Russia) using the contact procedure with the long NSG 11 cantilever (NT-MDT) and the BioScope Resolve, Bruker, with cantilever SCANASYST-AIR. The Gwyddion software [32] was used for visualization and for Fourier analysis.

The same samples were also used for the reflectance measurements, using the JASCO MSV-370 micro-spectrophotometer in the reflection geometry. Using a non-dispersive Schwarzschild-objective and an aperture, the region of interest was set to an area of 300 × 300 μm. The spectral region from beginning of visible spectrum (400 nm) to near infrared (750 nm). The data is used to visualize the spectral ratio $\left(R_{(of\ interest)}/R_{(B.\ mori\ [Jp])}\right)$ between the two species.

Additional files

Additional file 1: Figure S1. Fourier analysis of the *Bombyx* corneal nanostructures. **a–f**, Example of structures with increasing degree of disordering (artificial sets for panels **a** to **c**: panel size 1.3 μm, height 30 nm) and their Fourier analysis (**d–f**). **g–l**, AFM scans (**g–i**) of corneal surfaces of different *Bombyx* samples (**g**, *B. mandarina*, **h**, *B. mori* [Vn], **i**, *B. mori*[Jp]) and the corresponding Fourier spectra (**j–l**). This analysis reveals quasi-random vs. random structures in different samples. Quasi-random structures retain a stable period between any two neighboring peaks resulting in a ring of reflexes in their Fourier analysis (**a**, **d**). Alternatively, quasi-random structures can lack a stable period but fill the space uniformly, producing a large lattice covered by reflexes on the Fourier spectrum (**b**, **e**). Instead, random structures fill the space lopsidedly and despite the fact that the ratio of large and small objects and distances between them may exactly match those of the quasi-random structures, the Fourier spectrum shows a significant decrease of the lattice size (**c**, **f**). Fourier analysis of the *B. mori*

[Jp] corneal nanocoatings shows that the size of lattice (**i, l**) of reflexes is just 12 μm^{-1}, significantly smaller than in the case of those *B. mori* [Vn] and *B. mandarina* (17 μm^{-1}, **g, h, j, k**). In the absence of clearly defined reflections, these two last-mentioned structures could be recognized as the quasi-random dimpled patterns, while the lattice size of Fourier spectrum of *B. mori* [Jp] indicates absence of any ordering.

Additional file 2: Figure S2. Detergent treatment removes nanostructures and anti-reflectivity in *B. mandarina* corneae. **a,** Detergent treatment purges away the nano-dimpled pattern of *B. mandarina* corneae. Images pre- and post-treatment are 3.5 × 3.5 µm. The height dimension of the surface (in nm) is indicated by the color scale at the right panel with the mean set to zero. **b,** Ratio of the experimentally measured reflection spectra to the average reflectance of *B. mori*[Jp] measured for *B. mori*[Jp] (green), *B. mandarina* (red) and detergent-treated *B. mandarina* (blue). Data present as mean ± SD, n = 3. **c, d,** Calculation of the height of protrusions (from the lowest point up to the next highest point, **c**) and their broadness (**d**) of *B. mori* [Jp] (in green), *B. mandarina* (in red), and detergent-treated *B. mandarina* (in blue); n = 50.

Additional file 3: Table S1. Ratio of individual CPs to total amount of CPs.

Abbreviations

AFM: atomic-force microscopy; CPs: cuticular proteins; MS: mass-spectrometry.

Authors' contributions

MK, JL, JS, VC, AB and MF performed the experiments, MK and VLK wrote the manuscript, VLK supervised the research. All authors read and approved the final manuscript.

Author details

[1] Department of Pharmacology and Toxicology, University of Lausanne, Rue du Bugnon 27, 1011 Lausanne, Switzerland. [2] Department of Materials, ETH Zurich, Vladimir-Prelog-Weg 1-5/10, 8093 Zurich, Switzerland. [3] School of Biomedicine, Far Eastern Federal University, Sukhanova Street 8, Vladivostok 690922, Russian Federation.

Acknowledgements

We thank Prof. Yutaka Banno for providing the *B. mandarina* and *B. mori* (p50 strain) samples.

Competing interests

The authors declare that they have no competing interests.

References

1. Bernhard CG, Miller WH. A corneal nipple pattern in insect compound eyes. Acta Physiol Scand. 1962;56:385–6.
2. Cai JG, Qi LM. Recent advances in antireflective surfaces based on nanostructure arrays. Mater Horiz. 2015;2:37–53.
3. Raut HK, Ganesh VA, Nair AS, Ramakrishna S. Anti-reflective coatings: a critical, in-depth review. Energy Environ Sci. 2011;4:3779–804.
4. Han ZW, Wang Z, Feng XM, Li B, Mu ZZ, Zhang JQ, Niu SC, Ren LQ. Antireflective surface inspired from biology: a review. Biosurf Biotribol. 2016;2:137–50.
5. Palasantzas G, De Hosson JTM, Michielsen KFL, Stavenga DG: Optical properties and wettability of nanostructured biomaterials: moth eyes, lotus leaves, and insect wings. In: Nalwa HS, editor. Handbook of nanostructured biomaterials and their applications in nanobiotechnology. Volume 1. London: American Scientific Publishers; 2005. p. 273–301.
6. Deinega A, Valuev I, Potapkin B, Lozovik Y. Minimizing light reflection from dielectric textured surfaces. J Opt Soc Am Opt Image Sci Vis. 2011;28:770–7.
7. Dewan R, Fischer S, Meyer-Rochow VB, Ozdemir Y, Hamraz S, Knipp D. Studying nanostructured nipple arrays of moth eye facets helps to design better thin film solar cells. Bioinspir Biomim. 2012;7:016003.
8. Blagodatski A, Sergeev A, Kryuchkov M, Lopatina Y, Katanaev VL. Diverse set of turing nanopatterns coat corneae across insect lineages. Proc Natl Acad Sci USA. 2015;112:10750–5.
9. Blagodatski A, Kryuchkov M, Sergeev A, Klimov AA, Shcherbakov MR, Enin GA, Katanaev VL. Under- and over-water halves of Gyrinidae beetle eyes harbor different corneal nanocoatings providing adaptation to the water and air environments. Sci Rep. 2014;4:6004.
10. Kryuchkov M, Lehmann J, Schaab J, Fiebig M, Katanaev VL. Antireflective nanocoatings for UV-sensation: the case of predatory owlfly insects. J Nanobiotechnol. 2017;15:52.
11. Oskooi A, Favuzzi PA, Tanaka Y, Shigeta H, Kawakami Y, Noda S. Partially disordered photonic-crystal thin films for enhanced and robust photovoltaics. App Phys Lett. 2012;100:18110.
12. Pratesi F, Burresi M, Riboli F, Vynck K, Wiersma DS. Disordered photonic structures for light harvesting in solar cells. Opt Express. 2013;21:A460–8.
13. Xin Y, Jin H, Feng G, Hongjie L, Laixi S, Lianghong Y, Xiaodong J, Weidong W, Wanguo Z. High power laser antireflection subwavelength grating on fused silica by colloidal lithography. J Phys D Appl Phys. 2016;49:265104.
14. Daglar B, Khudiyev T, Demirel GB, Buyukserin F, Bayindir M. Soft biomimetic tapered nanostructures for large-area antireflective surfaces and SERS sensing. J Mater Chem C. 2013;1:7842–8.
15. Fraser MJ Jr. Insect transgenesis: current applications and future prospects. Annu Rev Entomol. 2012;57:267–89.
16. Kryuchkov M, Katanaev VL, Enin GA, Sergeev A, Timchenko AA, Serdyuk IN. Analysis of micro- and nano-structures of the corneal surface of Drosophila and its mutants by atomic force microscopy and optical diffraction. PLoS ONE. 2011;6:e22237.
17. Sergeev A, Timchenko AA, Kryuchkov M, Blagodatski A, Enin GA, Katanaev VL. Origin of order in bionanostructures. Rsc Adv. 2015;5:63521–7.
18. Xia Q, Guo Y, Zhang Z, Li D, Xuan Z, Li Z, Dai F, Li Y, Cheng D, Li R, et al. Complete resequencing of 40 genomes reveals domestication events and genes in silkworm (Bombyx). Science. 2009;326:433–6.
19. Xia Q, Li S, Feng Q. Advances in silkworm studies accelerated by the genome sequencing of Bombyx mori. Annu Rev Entomol. 2014;59:513–36.
20. Banno Y, Shimada T, Kajiura Z, Sezutsu H. The silkworm-an attractive BioResource supplied by Japan. Exp Anim. 2010;59:139–46.
21. Martins ER, Li J, Liu Y, Depauw V, Chen Z, Zhou J, Krauss TF. Deterministic quasi-random nanostructures for photon control. Nature communications. 2013;4:2665.
22. van Lare MC, Polman A. Optimized scattering power spectral density of photovoltaic light-trapping patterns. ACS Photonics. 2015;2:822–31.
23. Yu YF, Zhu AY, Paniagua-Dominguez R, Fu YH, Luk'yanchuk B, Kuznetsov AI. High-transmission dielectric metasurface with 2 phase control at visible wavelengths. Laser Photonics Rev. 2015;9:412–8.
24. Aghaeipour M, Anttu N, Nylund G, Samuelson L, Lehmann S, Pistol M-E. Tunable absorption resonances in the ultraviolet for InP nanowire arrays. Opt Express. 2014;22:29204–12.
25. Ji S, Park J, Lim H. Improved antireflection properties of moth eye mimicking nanopillars on transparent glass: flat antireflection and color tuning. Nanoscale. 2012;4:4603–10.
26. Anderson MS, Gaimari SD. Raman-atomic force microscopy of the ommatidial surfaces of Dipteran compound eyes. J Struct Biol. 2003;142:364–8.
27. Nickerl J, Tsurkan M, Hensel R, Neinhuis C, Werner C. The multi-layered protective cuticle of Collembola: a chemical analysis. J R Soc Interface. 2014;11:20140619.
28. Willis JH. Structural cuticular proteins from arthropods: annotation, nomenclature, and sequence characteristics in the genomics era. Insect Biochem Mol Biol. 2010;40:189–204.
29. Guo Y, Shen YH, Sun W, Kishino H, Xiang ZH, Zhang Z. Nucleotide diversity and selection signature in the domesticated silkworm, *Bombyx mori*, and wild silkworm, *Bombyx mandarina*. J Insect Sci. 2011;11:155.
30. Katanaev VL, Kryuchkov MV. The eye of Drosophila as a model system for studying intracellular signaling in ontogenesis and pathogenesis. Biochemistry (Mosc). 2011;76:1556–81.
31. Son J, Verma LK, Danner AJ, Bhatia CS, Yang H. Enhancement of optical transmission with random nanohole structures. Opt Express. 2011;19:A35–40.
32. Necas D, Klapetek P. Gwyddion: an open-source software for SPM data analysis. Cent Eur J Phys. 2012;10:181–8.

Permissions

List of Contributors

Nadezda Shershakova, Elena Baraboshkina, Sergey Andreev, Daria Purgina, Irina Struchkova, Oleg Kamyshnikov, Alexandra Nikonova and Musa Khaitov
NRC Institute of Immunology FMBA of Russia, Moscow, Russia

Ariane Thérien, Mikaël Bédard, Damien Carignan, Gervais Rioux, Louis Gauthier-Landry, Marie-Ève Laliberté-Gagné, Marilène Bolduc and Denis Leclerc
Department of Microbiology, Infectiology and Immunology, Infectious Disease Research Center, Laval University, 2705 Boul. Laurier, Quebec City, PQ G1V 4G2, Canada

Pierre Savard
Neurosciences, Laval University, 2705 Boul. Laurier, Québec City, PQ G1V 4G2, Canada

Fang Yang and René Riedel
Department of Chemistry, University of Marburg, Marburg, Germany

Pablo del Pino, Beatriz Pelaz, Alaa Hassan Said, Mahmoud Soliman and Neus Feliu
Department of Physics, University of Marburg, Marburg, Germany.

Shashank R. Pinnapireddy and Udo Bakowsky
Department of Pharmacy, University of Marburg, Marburg, Germany

Wolfgang J. Parak
Department of Physics, University of Marburg, Marburg, Germany
CIC bioma-GUNE, San Sebastián, Spain

Norbert Hampp
Department of Chemistry, University of Marburg, Marburg, Germany
Material Science Center, University of Marburg, Marburg, Germany

Biswaranjan Pradhan, Dipanjan Guha, Krushna Chandra Murmu, Abhinav Sur, Pratikshya Ray, Debashmita Das and Palok Aich
School of Biological Sciences, National Institute of Science Education and Research (NISER), HBNI, P.O. Bhimpur-Padanpur, Khurdha, Jatni, Odisha 752050, India

Augusto Márquez, Krisztina Kocsis, Gregor Zickler, Gilles R. Bourret, Andrea Feinle, Nicola Hüsing, Thomas Berger and Oliver Diwald
Department of Chemistry and Physics of Materials, Paris Lodron University of Salzburg, Jakob-Haringer-Strasse 2a, 5020 Salzburg, Austria

Martin Himly and Albert Duschl
Department of Molecular Biology, Paris Lodron University of Salzburg, Hellbrunnerstrasse 34/III, 5020 Salzburg, Austria

Atul A. Chaudhari, D'andrea Ashmore, Vida Dennis, Shree R. Singh and Shreekumar R. Pillai
Center for Nanobiotechnology Research, Alabama State University, Montgomery, AL, USA

Subrata deb Nath and Kunal Kate
Department of Mechanical Engineering, University of Louisville, Louisville, KY, USA

Don R. Owen and Chris Palazzo
Therapeutic Peptides Inc., 7053 Revenue Drive, Baton Rouge, LA 70809, USA

Robert D. Arnold
Department of Drug Discovery and Development, Auburn University, Auburn, AL, USA

Michael E. Miller
Research Instrumentation Facility, Auburn University, Auburn, AL, USA

Erika Bruni
Department of Biology and Biotechnology C. Darwin, Sapienza University of Rome, Piazzale Aldo Moro 5, Rome, Italy

Elena Zanni and Daniela Uccelletti
Department of Biology and Biotechnology C. Darwin, Sapienza University of Rome, Piazzale Aldo Moro 5, Rome, Italy
Research Center on Nanotechnology Applied to Engineering of Sapienza (CNIS), SNNLab, Sapienza University of Rome, Piazzale Aldo Moro 5, Rome, Italy

Chandrakanth Reddy Chandraiahgari, Maria Sabrina Sarto, Agnese Bregnocchi and Giovanni De Bellis
Research Center on Nanotechnology Applied to Engineering of Sapienza (CNIS), SNNLab, Sapienza University of Rome, Piazzale Aldo Moro 5, Rome, Italy
Department of Astronautical, Electrical and Energy Engineering, Sapienza University of Rome, Via Eudossiana 18, Rome, Italy

Maria Grazia Santangelo and Maurizio Leone
Department of Physics and Chemistry, University of Palermo, Palermo, Italy

Patrizia Mancini
Department of Experimental Medicine, Sapienza University of Rome, Viale Regina Elena 324, Rome, Italy

Uwe Himmelreich and Stefaan J. Soenen
Biomedical NMR Unit/MoSAIC, KU Leuven Campus Gasthuisberg, Herestraat 49, 3000 Louvain, Belgium

Gareth J. S. Jenkins and Shareen H. Doak
Institute of Life Science, Swansea University Medical School, Singleton Park, Swansea SA2 8PP, UK

Stefaan C. De Smedt and Jo Demeester
Faculty of Pharmaceutical Sciences, Ghent University, Harelbekestraat 72, 9000 Ghent, Belgium.

Bella B. Manshian
Biomedical NMR Unit/MoSAIC, KU Leuven Campus Gasthuisberg, Herestraat 49, 3000 Louvain, Belgium
Institute of Life Science, Swansea University Medical School, Singleton Park, Swansea SA2 8PP, UK

Thomas F. Martens and Kevin Braeckmans
Faculty of Pharmaceutical Sciences, Ghent University, Harelbekestraat 72, 9000 Ghent, Belgium.
Center of Nano- and Biophotonics, Ghent University, Harelbekestraat 72, 9000 Ghent, Belgium

Karsten Kantner and Beatriz Pelaz
Philipps University of Marburg, Renthof 7, 35032 Marburg, Germany

Wolfgang J. Parak
Philipps University of Marburg, Renthof 7, 35032 Marburg, Germany
CICBiomagune, San Sebastian, Spain

Jean-Pierre Brog, Joël Seydoux, Nam Hee Kwon, Katharina M. Fromm, Benoît Baichette and Sivarajakumar Maharajan
Department of Chemistry, University of Fribourg, Chemin du Musée 9, 1700 Fribourg, Switzerland

Aurélien Crochet
Fribourg Center for Nanomaterials FriMat, University of Fribourg, Chemin du Musée 9, 1700 Fribourg, Switzerland

Martin J. D. Clift, Barbara Rothen-Rutishauser and Hana Barosova
Adolphe Merkle Institute, University of Fribourg, 1700 Fribourg, Switzerland

Pierre Brodard
College of Engineering and Architecture of Fribourg, University of Applied Sciences of Western Switzerland, Boulevard de Pérolles 80, 1705 Fribourg, Switzerland

Mariana Spodaryk and Andreas Züttel
Laboratory of Materials for Renewable Energy (LMER), ISIC-SB, École Polytechnique Fédérale de Lausanne (EPFL), Valais/Wallis Energypolis, Rue de l'Industrie 17, 1951 Sion, Switzerland

Xiugong Gao, Vanessa D. Topping, Zachary Keltner, Robert L. Sprando and Jeffrey J. Yourick
Division of Applied Regulatory Toxicology, Office of Applied Research and Safety Assessment, Center for Food Safety and Applied Nutrition, U.S. Food and Drug Administration, 8301 Muirkirk Road, Laurel, MD 20708, USA

Sarit Cohen, Chen Gelber, Enav Corem-Salkmon and Shlomo Margel
Department of Chemistry, The Institute of Nanotechnology and Advanced Materials, Bar-Ilan University, 52900 Ramat Gan, Israel

Michal Natan and Ehud Banin
The Mina and Everard Goodman Faculty of Life Sciences, The Institute for Advanced Materials and Nanotechnology, Bar-Ilan University, 52900 Ramat Gan, Israel

Longyan Chen, Longyi Chen and Jin Zhang
Department of Chemical and Biochemical Engineering, University of Western Ontario, 1151 Richmond St., London, ON N6A 5B9, Canada

Michelle Dotzert and C. W. James Melling
School of Kinesiology, Faculty of Health Sciences, University of Western Ontario, London, ON N6A 5B9, Canada

Anaïs Vaissière, Prisca Boisguerin, Sébastien Deshayes, Karidia Konate, Mattias F. Lindberg, Carole Jourdan, Anthony Telmar and Quentin Seisel
Centre de Recherche de Biologie cellulaire de Montpellier, UMR 5237 CNRS, 1919 Route de Mende, 34293 Montpellier, France

Gudrun Aldrian
Sys2Diag, UMR 9005-CNRS/ALCEDIAG, 1682 Rue de la Valsiere, 34184 Montpellier, France

Frédéric Fernandez and Véronique Viguier
Microscopie Électronique et Analytique, Université de Montpellier, Place Eugène Bataillon, 34095 Montpellier, France

Coralie Genevois and Franck Couillaud
EA 7435 IMOTION (Imagerie moléculaire et thérapies innovantes en oncologie), Université de Bordeaux, 146 rue Leo Saignat, 33076 Bordeaux, France

Doris Hinger and Caroline Maake
Institute of Anatomy, University of Zurich, Winterthurerstrasse 190, 8057 Zurich, Switzerland

Fabrice Navarro, Jean-Sébastien Thomann, Frédérique Mittler and Anne-Claude Couffin
Technologies for Biology and Healthcare Division, CEA, LETI, MINATEC Campus, Commissariat à l'Énergie Atomique et aux Énergies Alternatives (CEA), 38054 Grenoble, France
Université Grenoble Alpes, 38000 Grenoble, France

Andres Käch
Center for Microscopy and Image Analysis, University of Zurich, Winterthurerstrasse 190, 8057 Zurich, Switzerland

Yan Xu, Miriam Rafailovich and Tatsiana Mironava
Department of Materials Science and Engineering, Stony Brook University, Stony Brook, NY, USA

M. Hadjiargyrou
Department of Life Sciences, New York Institute of Technology, Old Westbury, NY, USA

Kelly McNear and Yimin Huang
Department of Chemistry, Purdue University, 560 Oval Drive, West Lafayette, IN 47907, USA

Chen Yang
Department of Chemistry, Purdue University, 560 Oval Drive, West Lafayette, IN 47907, USA
Department of Physics and Astronomy, Purdue University, West Lafayette, IN 47907, USA

Mikhail Kryuchkov
Department of Pharmacology and Toxicology, University of Lausanne, Rue du Bugnon 27, 1011 Lausanne, Switzerland

Jannis Lehmann, Manfred Fiebig and Jakob Schaab
Department of Materials, ETH Zurich, Vladimir-Prelog-Weg 1-5/10, 8093 Zurich, Switzerland

Vsevolod Cherepanov
School of Biomedicine, Far Eastern Federal University, Sukhanova Street 8, Vladivostok 690922, Russian Federation

Artem Blagodatski and Vladimir L. Katanaev
Department of Pharmacology and Toxicology, University of Lausanne, Rue du Bugnon 27, 1011 Lausanne, Switzerland
School of Biomedicine, Far Eastern Federal University, Sukhanova Street 8, Vladivostok 690922, Russian Federation

Index